Communications
in Computer and Information Science

2765

Series Editors

Gang Li, *School of Information Technology, Deakin University, Burwood, VIC, Australia*

Joaquim Filipe, *Polytechnic Institute of Setúbal, Setúbal, Portugal*

Zhiwei Xu, *Chinese Academy of Sciences, Beijing, China*

Rationale

The CCIS series is devoted to the publication of proceedings of computer science conferences. Its aim is to efficiently disseminate original research results in informatics in printed and electronic form. While the focus is on publication of peer-reviewed full papers presenting mature work, inclusion of reviewed short papers reporting on work in progress is welcome, too. Besides globally relevant meetings with internationally representative program committees guaranteeing a strict peer-reviewing and paper selection process, conferences run by societies or of high regional or national relevance are also considered for publication.

Topics

The topical scope of CCIS spans the entire spectrum of informatics ranging from foundational topics in the theory of computing to information and communications science and technology and a broad variety of interdisciplinary application fields.

Information for Volume Editors and Authors

Publication in CCIS is free of charge. No royalties are paid, however, we offer registered conference participants temporary free access to the online version of the conference proceedings on SpringerLink (http://link.springer.com) by means of an http referrer from the conference website and/or a number of complimentary printed copies, as specified in the official acceptance email of the event.

CCIS proceedings can be published in time for distribution at conferences or as postproceedings, and delivered in the form of printed books and/or electronically as USBs and/or e-content licenses for accessing proceedings at SpringerLink. Furthermore, CCIS proceedings are included in the CCIS electronic book series hosted in the SpringerLink digital library at http://link.springer.com/bookseries/7899. Conferences publishing in CCIS are allowed to use Online Conference Service (OCS) for managing the whole proceedings lifecycle (from submission and reviewing to preparing for publication) free of charge.

Publication process

The language of publication is exclusively English. Authors publishing in CCIS have to sign the Springer CCIS copyright transfer form, however, they are free to use their material published in CCIS for substantially changed, more elaborate subsequent publications elsewhere. For the preparation of the camera-ready papers/files, authors have to strictly adhere to the Springer CCIS Authors' Instructions and are strongly encouraged to use the CCIS LaTeX style files or templates.

Abstracting/Indexing

CCIS is abstracted/indexed in DBLP, Google Scholar, EI-Compendex, Mathematical Reviews, SCImago, Scopus. CCIS volumes are also submitted for the inclusion in ISI Proceedings.

How to start

To start the evaluation of your proposal for inclusion in the CCIS series, please send an e-mail to ccis@springer.com

Quang Vinh Nguyen · Yuefeng Li · Paul Kwan ·
Yanchang Zhao · Yee Ling Boo · Richi Nayak
Editors

Data Science and Machine Learning

23rd Australasian Conference, AusDM 2025
Brisbane, QLD, Australia, November 26–28, 2025
Proceedings

 Springer

Editors
Quang Vinh Nguyen [iD]
Western Sydney University
Sydney, NSW, Australia

Yuefeng Li
Queensland University of Technology
Brisbane, QLD, Australia

Paul Kwan [iD]
Central Queensland University
Brisbane, QLD, Australia

Yanchang Zhao [iD]
CSIRO
Canberra, ACT, Australia

Yee Ling Boo [iD]
RMIT University
Melbourne, VIC, Australia

Richi Nayak [iD]
Queensland University of Technology
Brisbane, QLD, Australia

ISSN 1865-0929   ISSN 1865-0937 (electronic)
Communications in Computer and Information Science
ISBN 978-981-95-6785-0   ISBN 978-981-95-6786-7 (eBook)
https://doi.org/10.1007/978-981-95-6786-7

This Springer imprint is published by the registered company Springer Nature Singapore Pte Ltd.
The registered company address is: 152 Beach Road, #21-01/04 Gateway East, Singapore 189721, Singapore

If disposing of this product, please recycle the paper.

Preface

It is our great pleasure to present the proceedings of the 23rd Australasian Data Science and Machine Learning Conference (formerly known as the Australasian Data Mining Conference), AusDM 2025, held at the Queensland University of Technology, Brisbane, between 26 and 28 November 2025.

The AusDM conference series first started in 2002 as a workshop initiated by Simeon Simoff (Western Sydney University), Graham Williams (Australian National University), and Markus Hegland (Australian National University). Over the years, AusDM has established itself as the premier Australasian meeting for both practitioners and researchers in the area of data mining (or data analytics or data science) and machine learning. AusDM is devoted to the art and science of intelligent analysis of (usually big) data sets for meaningful (and previously unknown) insights. Since AusDM 2002, the conference series has showcased research in data science and machine learning through presentations and discussions on state-of-art research and development. Built on this tradition, AusDM 2025 successfully facilitated the cross-disciplinary exchange of ideas, experiences, and potential research directions, and pushed forward the frontiers of data science and machine learning in academia, government, and industry.

This year, the theme of the conference focused on "Entering a New World Driven by Data Science, Machine Learning, and Artificial Intelligence". Specifically, we were thrilled to present an exciting lineup of keynote talks, industry invited talks, special panel sessions, industry sessions, tutorials, a doctorial consortium, oral paper presentations, and social events. In addition, a journal special issue with the Journal of Big Data by Springer was also planned.

AusDM 2025 received altogether 99 valid submissions. While the majority of submissions were from Australasia, authors and co-authors of the submitted papers were distributed across 13 different countries including India, China, the USA, Germany, and the United Arab Emirates. All submissions went through a double-blind review process, and each paper received at least three peer-reviewed reports. Additional reviewers were considered for a clearer review outcome if review comments from the initial three reviewers were inconclusive.

Out of these 99 submissions, a total of 37 papers were finally accepted for publication. The overall acceptance rate for AusDM 2025 was 37%. Out of the 62 Research Track submissions, 22 papers (i.e. 35%) were accepted for publication. Out of the 37 submissions in the Application Track, 15 papers (i.e. 40%) were accepted for publication.

The AusDM 2025 Organizing Committee would like to give their special thanks to Gianluca Demartini, Shirui Pan, and Kerrie Mengersen for giving insightful keynote speeches. They would also like to credit the three engaging industry invited talks by Mellissah Smith, Sankalp Khanna, and Kenny Sabir. In addition, the success of the three special panel discussions, themed around AI in Education, Women in STEM, and Are We in the Right Direction?, is attributed to the panellists, namely Gianluca Demartini, Jim Hogan, Joel Buchholz, Sharon Singh, Nish Veer, Lucky Atamhenwan, Moe Wynn,

Efiza Vanniasinghe, Rabia Williams, Alannah Grech, Nethumi Perera, Simon Townsend, Shane Muller, and Benson Choy.

The two invited industry sessions were well acclaimed and the committee truly appreciated the great contributions from the following invited speakers: Sankalp Khanna, Luke Hardwick-Greaves, Mark Gordon, Simon Townsend, Shane Muller, Lin Chen, Cuan Naidoo, Julien Monteil, Mitchell Fardell, Tom Fu, and Anton Lord. The committee would like to express their appreciation to Graham Williams for presenting a Spotlight Talk. The committee was grateful to have Priyadarsini Karthik, Karthik Sekhar, Prakash M., Yuchen Zhang, and Yuting Zhu presenting at the tutorial sessions.

The committee would also like to give their sincere thanks to the Queensland University of Technology for providing admin support and the conference venue. The committee would also like to thank Springer CCIS and the Editorial Board for their agreement to publish AusDM 2025 papers. This will give excellent exposure of the papers accepted for publication. We would also like to give our heartfelt thanks to all student and staff volunteers at the Queensland University of Technology who did a tremendous job in ensuring a successful conference event.

Last but not least, we would like to give our sincere thanks to all delegates for attending the conference this year at the Queensland University of Technology. We hope that it was a fruitful experience and you enjoyed AusDM 2025!

November 2025

Yuefeng Li
Quang Vinh Nguyen
Paul Kwan
Yanchang Zhao
Yee Ling Boo
Richi Nayak

Organization

General Co-chairs

Richi Nayak   Queensland University of Technology, Australia
Yee Ling Boo   RMIT University, Australia

Program Chairs (Research Track)

Yuefeng Li   Queensland University of Technology, Australia
Quang Vinh Nguyen   Western Sydney University, Australia

Program Chairs (Application Track)

Paul Kwan   CQUniversity, Australia
Yanchang Zhao   Data61, CSIRO, Australia

Industry Chair

Anton Lord   Queensland Health, Australia

Tutorial and Workshop Chairs

Xujuan (Susan) Zhou   University of Southern Queensland, Australia
Alan Woodley   Queensland University of Technology, Australia

Doctoral Symposium Chairs

Zhonglin (Jolin) Qu   Western Sydney University, Australia
Thirunavukarasu (Thiru)   Queensland University of Technology, Australia
     Balasubramaniam

Special Session Chair

Sangeetha Kutty   CQUniversity, Australia

Publication Chair

Yee Ling Boo   RMIT University, Australia

Publicity Chair

Asara Senaratne   Flinders University, Australia

Local Organizing Chairs

Khanh Luong   Queensland University of Technology, Australia
Tej Bahadur Shahi   Queensland University of Technology, Australia

Web Chairs

Md Abdul Bashar   Queensland University of Technology, Australia
Yue (Joy) Wang   Queensland University of Technology, Australia

Steering Committee Co-chairs

Yun Sing Koh   University of Auckland, New Zealand
Richi Nayak   Queensland University of Technology, Australia
Yanchang Zhao   Data61, CSIRO, Australia

Steering Committee Past Co-chairs

Simeon Simoff   Western Sydney University, Australia
Graham Williams   Australian National University, Australia

Steering Committee Members

Paul Kennedy	University of Technology Sydney, Australia
Jiuyong (John) Li	University of South Australia, Australia
Kok-Leong Ong	RMIT University, Australia
Ling Chen	University of Technology Sydney, Australia
Dharmendra Sharma	University of Canberra, Australia
Lin Liu	University of South Australia, Australia
Warwick Graco	The Analytics Shed, Australia
Yee Ling Boo	RMIT University, Australia
Diana Benavides-Prado	Queen Mary University of London, UK

Honorary Advisors

John Roddick	Flinders University, Australia
Geoff Webb	Monash University, Australia

Program Committee

Research Track

Gizem Intepe	Western Sydney University, Australia
Lu Chen	Swinburne University of Technology, Australia
Jun Shen	University of Wollongong, Australia
Tarique Anwar	University of York, UK
Darshika Koggalahewa	Queensland University of Technology, Australia
Hoa Nguyen	Australian National University, Australia
Xiaohua Wu	Wuhan University of Technology, China
Huaming Chen	University of Sydney, Australia
Weidong Huang	University of Technology Sydney, Australia
M. A. Nayomi Dulanjala Sewwandi	University of Queensland, Australia
Xujuan Zhou	University of Southern Queensland, Australia
Shashi Lidhamullage Dhon Charles	Queensland University of Technology, Australia
Sharon Torao Pingi	University of Papua New Guinea, Papua New Guinea
Zhonglin Qu	Western Sydney University, Australia
Yue Xu	Queensland University of Technology, Australia

Ahmed Alkenani	Jubail Industrial College, Saudi Arabia
Qi Chen	Victoria University of Wellington, New Zealand
Andrzej Janusz	Queensland University of Technology, Australia
Shiva Pohrel	Deakin University, Australia
Brendon J. Woodford	University of Otago, New Zealand
Dan Li	Queensland University of Technology, Australia
Evan Crawford	Western Sydney University, Australia
Ying Wang	Macquarie University, Australia
Jinghui Liu	Australian e-Health Research Centre, CSIRO, Australia
Shihan Li	University of Melbourne, Australia
Yutong Wu	Australian e-Health Research Centre, CSIRO, Australia
Yifei Dong	University of Technology Sydney, Australia
Jinglan Zhang	Queensland University of Technology, Australia
Raymond Wong	University of New South Wales, Australia
Dong Yuan	University of Sydney, Australia
Gang Li	Deakin University, Australia
Chng Wei Lau	Western Sydney University, Australia
Md Abul Bashar	Queensland University of Technology, Australia
Hao Wu	Western Sydney University, Australia
Junyu Xuan	University of Technology Sydney, Australia
Rushit Dave	Minnesota State University at Mankato, USA
Xiaohui Tao	University of Southern Queensland, Australia
Brad Malin	Vanderbilt University, USA
Md Geaur Rahman	Charles Sturt University, Australia
Jianming Yong	University of Southern Queensland, Australia
Raj Gururajan	University of Southern Queensland, Australia

Application Track

Osamah A. Mahdi	Melbourne Institute of Technology, Australia
Mohammad Awrangjeb	Griffith University, Australia
Salahuddin Azad	CQUniversity, Australia
Thirunavukarasu Balasubramaniam	Queensland University of Technology, Australia
Meiru Che	University of Texas, USA
Xiaodan Dong	University of Technology Sydney, Australia
Mahmoud Elkhodr	CQUniversity, Australia
Gitte Galea	CQUniversity, Australia
Vidhya Govindaraju	HP Labs, India
Deepani Guruge	Melbourne Institute of Technology, Australia

Md Rahat Hossain	CQUniversity, Australia
Rajan Kadel	Melbourne Institute of Technology, Australia
Abigail Koay	University of the Sunshine Coast, Australia
Xiaoying Kong	Melbourne Institute of Technology, Australia
Jessica Leung	Monash University, Australia
Lily Li	CQUniversity, Australia
Michael Li	CQUniversity, Australia
Qinyi Li	Queensland University of Technology, Australia
Yufeng Lin	CQUniversity, Australia
Erica Mealy	University of Queensland, Australia
Tayab Memon	RMIT University, Australia
Ahsan Morshed	CQUniversity, Australia
Arjun Neupane	CQUniversity, Australia
Olatunji Omisore	CQUniversity, Australia
David Paul	University of New England, Australia
Md. Mamunur Rashid	CQUniversity, Australia
Nasser Sabar	La Trobe University, Australia
Fariza Sabrina	CQUniversity, Australia
Mohammad Saiedur Rahaman	CQUniversity, Australia
Farzad Sanati	CQUniversity, Australia
Aakanksha Sharma	Melbourne Institute of Technology
Sanjeeb Shrestha	Melbourne Institute of Technology
Nandini Sidnal	Torrens University, Australia
Sarath Tomy	La Trobe University, Australia
Zhe Wang	Griffith University, Australia
Mingzhong Wang	University of the Sunshine Coast, Australia
Zhenglin Wang	CQUniversity, Australia
Santoso Wibowo	CQUniversity, Australia
Di Wu	University of Southern Queensland, Australia
Elaheh Yadegaridehkordi	CQUniversity, Australia

Additional Reviewers

Rani Adam	Nisal Jayamuni
Qiuyun Luan	Abdur Rehman Khan
Huan Wang	Muhammad Bilal Zia
Yiqian Xie	Zhuo Zhang
Chenghao Zhang	Nan Yang
Guosheng Li	Md Shahadat Hossain
Md Mahfujur Rahman	

Contents

Application Track – Environment, Information Security and Productivity

Research Track – Deep Learning Fusion and Vision

Application Track – Health and Social Good

Research and Application Tracks – Knowledge-Driven and Domain Specific AI

Research Track – Federated, Adaptive, and Trustworthy Machine Learning

DAARA: Divergence-Aware Attention for Robust Aggregation in Federated Learning Against Poisoning Attacks

Md Palash Uddin[✉][iD], Mahmudul Hasan[iD], and Yong Xiang[iD]

School of Information Technology, Deakin University, Geelong, VIC 3220, Australia
{m.uddin,mahmudul.hasan,yong.xiang}@deakin.edu.au

Abstract. Federated Learning (FL) enables collaborative model training across distributed clients without sharing raw data but remains highly vulnerable to poisoning attacks, especially under non-Independent and Identically Distributed (non-IID) settings. To address this challenge, we propose DAARA: Divergence-Aware Attention for Robust Aggregation, a novel aggregation mechanism designed to defend against label flipping attacks in FL. Unlike existing defenses that rely on fixed thresholds or assume knowledge of attacker behavior, DAARA is client-agnostic and adaptively assigns attention weights to client updates based on their statistical divergence and class-wise consistency. We provide the theoretical convergence analysis of our DAARA approach. Additionally, we conduct extensive experiments on the NSL-KDD and UNSW-NB15 cybersecurity datasets, demonstrating that DAARA significantly outperforms state-of-the-art baselines, including Krum, Trimmed Mean, FoolsGold, and RFed, achieving up to a 40% reduction in attack success rate and up to 4× faster convergence under both untargeted and targeted label flipping attacks. Furthermore, DAARA exhibits remarkable stability across varying attack intensities, maintaining consistently low gradient divergence even under extreme non-IID conditions. The results confirm that DAARA provides a lightweight, effective, and generalizable solution for secure FL in adversarial and heterogeneous environments.

Keywords: Federated Learning · Poisoning Attacks · Robust Aggregation · Non-IID Data · Secure Distributed Learning

1 Introduction

Federated Learning (FL) has emerged as a powerful decentralized learning framework that enables multiple clients to collaboratively train a global machine learning model without sharing their raw data. This approach is especially attractive in privacy-sensitive domains such as healthcare, finance, and cybersecurity, where data confidentiality is paramount [1,2]. By performing local training and sharing only model updates with a central server, FL preserves data privacy while leveraging distributed data across diverse devices. However, this distributed nature

Q. V. Nguyen et al. (Eds.): AusDM 2025, CCIS 2765, pp. 3–18, 2026.
https://doi.org/10.1007/978-981-95-6786-7_1

also introduces significant security vulnerabilities [3–5]. Among the most critical threats are data poisoning attacks, in which adversaries deliberately manipulate local training data or gradients to subvert the integrity of the global model. A prominent example of such attacks is the label flipping (LF) attack, in which adversaries systematically alter the class labels of training samples to mislead the global model. These attacks can significantly degrade classification accuracy and corrupt the learning process while remaining hard to detect, especially in large-scale FL systems.

The threat becomes more severe when the attack is untargeted (arbitrarily flipping multiple classes) or targeted (aiming for misclassification into specific classes) and when client selection is randomized, making trust estimation more challenging. Furthermore, in non-Independent and Identically Distributed (non-IID) data settings, benign clients themselves can produce highly divergent updates due to skewed local data, making it difficult to distinguish malicious clients from outliers based on static metrics. To address this, various robust aggregation techniques have been proposed. Traditional methods, such as FedAvg [1], aggregate all client updates uniformly, making them vulnerable to adversarial input. Techniques such as Krum [6], Trimmed Mean [7], Auror [8], FoolsGold [9], and RFed [10] attempt to filter or reweight suspicious updates using metrics like Euclidean distance (ED), gradient similarity, or coordinate-wise statistics. However, these defenses share key limitations, as they are often static, rely on fixed thresholds, and implicitly assume IID data or known attacker counts. As a result, their performance degrades in real-world FL deployments where heterogeneity and adversarial behavior coexist. Additionally, existing defenses typically rely on cross-client information to mitigate attacks, making them dependent on other clients or external data and often requiring prior knowledge to establish the defense mechanism, which is another critical privacy concern. This motivates a key question: *Can we design an adaptive, trust-aware aggregation mechanism that dynamically estimates client reliability without requiring prior knowledge of attack patterns or data distribution assumptions?*

In this work, we propose Divergence-Aware Attention for Robust Aggregation (DAARA), which introduces a statistically grounded attention model into the FL aggregation process. DAARA evaluates the trustworthiness of each client based on the gradient divergence between its local model and the global model. To robustly estimate divergence in adversarial and non-Gaussian settings, we normalize distances using the Median Absolute Deviation (MAD) rather than mean-based metrics. Instead of applying hard cutoffs, each client is assigned a soft attention weight via an exponential decay function, scaled by a temperature parameter that naturally scales down the influence of suspicious clients without excluding them arbitrarily. Additionally, we incorporate probabilistic suppression of outliers, where clients with divergence above the 90th percentile are cut to zero contribution. To ensure numerical stability and graceful degradation in extreme cases (e.g., NaN, Inf, zero variance), we implement a fallback mechanism that defaults to uniform weighting. Unlike existing defenses, DAARA is client-agnostic, meaning it requires no prior knowledge of the number or behavior of malicious clients; adaptive, as it leverages dynamic statistical cues rather than relying on fixed thresholds; and lightweight and pluggable, making it seamlessly compatible with standard FL workflows. We evaluated DAARA on NSL-KDD and UNSW-NB15 datasets in different levels of non-IID settings, simulating targeted label flipping (TLF) and untargeted label flipping (UTLF) attacks. Compared to robust baselines such as FedAvg, Krum, Trimmed Mean, Auror, FoolsGold, and RFed, DAARA achieves consistently superior performance, demonstrating lower attack success rates (ASR), reduced gradient divergence (GD), higher model accuracy (MA), and strong source class recall rate (SRR) across all scenarios. To summarize, our key contributions are included below:

- We propose DAARA, a novel Divergence-Aware Attention-based Aggregation mechanism that adaptively downweights malicious client updates without requiring prior knowledge of the number or behavior of attackers.
- We provide a theoretical convergence analysis of DAARA under non-IID data distributions, establishing its robustness and convergence guarantees in the presence of adversarial updates.
- We conduct extensive experiments on benchmark cybersecurity datasets (NSL-KDD and UNSW-NB15), demonstrating that DAARA consistently outperforms state-of-the-art defenses in terms of accuracy, attack resilience, and communication efficiency under both TLF and UTLF attacks in different attack ratios.

The rest of the paper is organized as follows. Section 2 reviews related works on FL defenses against poisoning attacks. Section 3 presents the proposed DAARA framework, including its attention-based aggregation mechanism. Section 4 provides the theoretical convergence analysis of DAARA. Section 5 describes the experimental setup, datasets, and evaluation metrics, followed by a comprehensive comparison of DAARA with state-of-the-art defense methods. Finally, Sect. 6 concludes the paper and outlines directions for future research.

2 Related Work

Baseline aggregation methods such as FedAvg [1] treat all client updates equally, making them highly susceptible to poisoning attacks. Krum [6] addresses this by selecting the most 'central' gradient based on the ED, thus filtering out potentially malicious updates. However, it requires prior knowledge of the number of attackers and performs poorly under high variance caused by non-IID data distributions. The trimmed mean method [7] removes extreme values from each gradient dimension but may inadvertently discard useful signals from benign clients, particularly in heterogeneous settings. Auror [8] introduces a gradient norm-based thresholding mechanism to reject anomalous updates but suffers from a lack of adaptability and generalizability. FoolsGold [9] suppresses colluding clients by measuring pairwise cosine similarity between their updates, but it becomes ineffective when client gradients are naturally similar due to label imbalance or shared data structures. FedREDefense [11] uses reputation-based reweighting with gradient similarity to mitigate LF attacks. AgrAmplifier [12] amplifies benign updates using cosine similarity filtering. APFed [13] clusters local models with confidence-weighted aggregation for heterogeneous settings. DefendFL [14] combines entropy-based filtering and data masking for privacy-preserving defense. FLAIR [15] uses flip score-based reputation to counter high poisoning ratios. SplitFL [16] applies split learning to separate feature and label learning, enhancing resilience across aggregation methods. Collectively, these methods exhibit key limitations, as they rely on static heuristics, require manual threshold tuning, and assume prior knowledge of attack characteristics and factors that hinder their robustness in adversarial federated environments.

3   Methodology

3.1   Problem Formulation

Let $\mathcal{C} = \{1, 2, \ldots, N\}$ be the set of N clients participating in the FL process. Each client $i \in \mathcal{C}$ has a private dataset $\mathcal{D}_i$ and optimizes a local objective function $f_i(w)$ over its model parameters $w \in \mathbb{R}^d$. The classical FL objective is to minimize the weighted sum of local losses as Eq. 1.

$$\min_{w} \sum_{i=1}^{N} p_i f_i(w), \quad \text{where} \quad p_i = \frac{|\mathcal{D}_i|}{\sum_{j=1}^{N} |\mathcal{D}_j|}. \tag{1}$$

In adversarial scenarios such as LF attacks, a subset of malicious clients $\mathcal{A} \subset \mathcal{C}$ manipulates their data labels to degrade the global model performance. For a malicious client $i \in \mathcal{A}$, the training objective can be represented as Eq. 2.

$$\tilde{f}_i(w) = \mathbb{E}_{(x,y') \sim \tilde{\mathcal{D}}_i}[\ell(f_w(x), y')], \quad \text{with} \quad y' \neq y, \tag{2}$$

where $\tilde{\mathcal{D}}_i$ represents a poisoned dataset with altered labels, resulting in poisoned gradients $\nabla \tilde{f}_i(w)$ that deviate significantly from benign updates.

To mitigate such attacks, we propose DAARA, a defense strategy that adaptively reweights client updates based on their own gradient divergence from the global update trend. The divergence score δ_i of the client i is defined as Eq. 3.

$$\delta_i = \|\nabla f_i(w) - \bar{\nabla} f(w)\|_2, \tag{3}$$

where $\bar{\nabla} f(w) = \frac{1}{N} \sum_{j=1}^{N} \nabla f_j(w)$ is the mean gradient. To compute the client's contribution, we define a soft attention weight using a temperature-scaled exponential function as follows:

$$\alpha_i = \exp\left(-\frac{\delta_i}{T}\right), \tag{4}$$

where $T > 0$ is a temperature parameter that controls sensitivity to divergence. The normalized aggregation weight for each client is computed as Eq. 5.

$$w_i = \frac{\alpha_i}{\sum_{j=1}^{N} \alpha_j}. \tag{5}$$

Algorithm 1. DAARA Aggregation Algorithm

Require: Client models $\mathcal{M} = \{M_1, M_2, \ldots, M_N\}$, temperature $T > 0$
Ensure: Aggregated global model M_{global}
1: Compute flattened gradients: $\mathbf{g}_i \leftarrow \texttt{Flatten}(M_i)$
2: Compute mean gradient: $\bar{\mathbf{g}} \leftarrow \frac{1}{N} \sum_i \mathbf{g}_i$
3: Compute divergence scores: $\delta_i \leftarrow \|\mathbf{g}_i - \bar{\mathbf{g}}\|_2$
4: Replace NaN/∞ in δ_i with median($\boldsymbol{\delta}$)
5: Compute $\mu \leftarrow$ median($\boldsymbol{\delta}$), $\sigma \leftarrow$ median($|\boldsymbol{\delta} - \mu|$) + ϵ
6: Normalize: $\tilde{\delta}_i \leftarrow (\delta_i - \mu)/\sigma$
7: Set $\tilde{\delta}_i \leftarrow \infty$ if $\tilde{\delta}_i > P_{90}(\tilde{\boldsymbol{\delta}})$
8: Compute attention: $s_i = \exp(-\tilde{\delta}_i/T)$
9: **if** $\sum_i s_i = 0$ or any s_i is NaN **then**
10:     Set $s_i \leftarrow 1, \forall i$
11: **end if**
12: Normalize: $w_i \leftarrow s_i / \sum_j s_j$
13: **for all** model parameters k **do**
14:     $M_{\text{global}}[k] \leftarrow \sum_i w_i \cdot M_i[k]$
15: **end for**
16: **return** M_{global}

Finally, the global model is updated using the weighted average from i's model M_i as Eq. 6, and defines the robust global objective as Eq. 7.

$$w \leftarrow \sum_{i=1}^{N} w_i \cdot M_i, \tag{6}$$

$$\min_w \sum_{i=1}^{N} w_i \cdot f_i(w) = \min_w \sum_{i=1}^{N} \left[\frac{\exp\left(-\|\nabla f_i(w) - \bar{\nabla} f(w)\|_2/T\right)}{\sum_{j=1}^{N} \exp\left(-\|\nabla f_j(w) - \bar{\nabla} f(w)\|_2/T\right)} \right] \cdot f_i(w). \tag{7}$$

This formulation prioritizes updates from clients whose gradients closely align with the global trend while attenuating the influence of potential adversaries with high divergence. Provides a principled and mathematically grounded approach to enhancing robustness in federated environments affected by LF or similar poisoning attacks.

3.2 DAARA: Divergence-Aware Attention for Robust Aggregation

Let $\mathbf{g}_i$ denote the flattened gradient vector of client i's model M_i, and let $\bar{\mathbf{g}} = \frac{1}{N} \sum_{i=1}^{N} \mathbf{g}_i$ be the mean gradient across all clients. We define the divergence of each client as Eq. 8.

$$\delta_i = \|\mathbf{g}_i - \bar{\mathbf{g}}\|_2. \tag{8}$$

To ensure robustness under skewed or adversarial conditions, we normalize the divergence scores using the MAD strategy. The median μ and the deviation σ are defined as.

$$\mu = \text{median}(\boldsymbol{\delta}), \quad \sigma = \text{median}(|\boldsymbol{\delta} - \mu|) + \epsilon, \tag{9}$$

$$\tilde{\delta}_i = \frac{\delta_i - \mu}{\sigma}, \tag{10}$$

where $\delta = [\delta_1, \delta_2, \ldots, \delta_N]$ and ϵ is local update bias. In the next step, we remove the top 10% of the most divergent clients by setting their normalized divergence to infinity:

$$\tilde{\delta}_i \leftarrow \infty \quad \text{if } \tilde{\delta}_i > P_{90}(\tilde{\boldsymbol{\delta}}), \tag{11}$$

where P_{90} denotes the 90th percentile of the normalized divergences. Then we compute the attention scores using Eq. 12.

$$s_i = \exp\left(-\frac{\tilde{\delta}_i}{T}\right). \tag{12}$$

If $\sum_i s_i = 0$ or any s_i is invalid (NaN or ∞), we uniformly assign fallback weights as Eq. 13, and the normalized attention weights are computed as Eq. 13.

$$s_i = 1, \quad \forall i. \quad w_i = \frac{s_i}{\sum_{j=1}^{N} s_j} \tag{13}$$

Finally, the global model parameters are updated component-wise as in Eq. 14, while the complete pseudocode is provided in Algorithm 1. Note that the pseudocode presents the working process for a single global round, and it will continue up to T global rounds to obtain the final global model.

$$M_{\text{global}}[k] = \sum_{i=1}^{N} w_i \cdot M_i[k], \quad \forall k. \tag{14}$$

4   Theoretical Result Analysis

In this section, we theoretically analyze the convergence properties of DAARA under non-IID and adversarial conditions. Our goal is to theoretically establish that the DAARA method leads to convergence under standard assumptions, while suppressing the influence of adversarial updates. We begin by formalizing key assumptions, then present a convergence theorem and its proof considering the impact of adversarial client (δ), local update bias (ϵ), and variance (σ).

4.1   Preliminaries and Assumptions

Let $F(w)$ denote the global objective defined as $F(w) := \sum_{i=1}^{N} p_i f_i(w)$, where $f_i(w)$ is the local empirical risk of client i, and $p_i = \frac{|\mathcal{D}_i|}{\sum_j |\mathcal{D}_j|}$ is the weight proportional to the local dataset size. We present the following standard assumptions that are already established in FL.

Assumption 1 [Smoothness]. Each local objective $f_i(w)$ is L-smooth, i.e., $\|\nabla f_i(w) - \nabla f_i(w')\| \leq L\|w - w'\|, \quad \forall w, w'.$

Assumption 2 [Unbiased Local Gradient]. Each client performs local SGD and returns an unbiased estimate of the true gradient: $\mathbb{E}[\mathbf{g}_i] = \nabla f_i(w).$

Assumption 3 [Bounded Variance]. The variance of stochastic gradients is bounded: $\mathbb{E}\|\mathbf{g}_i - \nabla f_i(w)\|^2 \leq \sigma^2$.

Assumption 4 [Bounded Gradient Dissimilarity and Norm]. The divergence between local and global gradients is bounded: $\|\nabla f_i(w) - \nabla F(w)\| \leq \epsilon$ and the local gradients are bounded in norm $\|\mathbf{g}_i\| \leq G, \quad \forall i, t$.

Let $\mathcal{B}$ be the set of benign clients and $\mathcal{A}$ the set of adversarial clients. The total number of clients is $N = |\mathcal{B}| + |\mathcal{A}|$.

4.2 Convergence Result

Theorem 1 (Convergence of DAARA). *Under Assumptions 1–4, and assuming DAARA suppresses adversarial contributions such that $\sum_{i \in \mathcal{A}} w_i^t \leq \delta \ll 1$, the expected squared gradient norm averaged over T rounds satisfies Eq. 15.*

$$\frac{1}{T} \sum_{t=1}^{T} \mathbb{E}\left[\|\nabla F(w^t)\|^2\right] \leq \mathcal{O}\left(\frac{1}{\sqrt{T}} + \delta^2 + \epsilon^2 + \sigma^2\right). \tag{15}$$

Proof. To minimize the global loss function in an iterative optimization setting, the model parameters are updated at each round t using the standard gradient descent rule as follows:

$$w^{t+1} = w^t - \eta \mathbf{g}^t \tag{16}$$

Using the L-smoothness property of the global objective $F(w)$, we get:

$$F(w^{t+1}) \leq F(w^t) + \langle \nabla F(w^t), w^{t+1} - w^t \rangle + \frac{L}{2}\|w^{t+1} - w^t\|^2. \tag{17}$$

Substituting the update rule from Eq. 16 to Eq. 17, we get Eq. 18.

$$F(w^{t+1}) \leq F(w^t) - \eta \langle \nabla F(w^t), \mathbf{g}^t \rangle + \frac{L\eta^2}{2}\|\mathbf{g}^t\|^2. \tag{18}$$

According to FL update rules (e.g., from DAARA), the global gradient is formed by a weighted aggregation of client gradients as follows:

$$\mathbf{g}^t = \sum_{i=1}^{N} w_i^t \mathbf{g}_i^t, \tag{19}$$

where w_i^t is the attention weight (e.g., from DAARA), and $\mathbf{g}_i^t$ is the local gradient of client i. Incorporating the client-wise aggregation from Eq. 19 into the inequality of Eq. 18, we get the updated inequality in Eq. 20.

$$F(w^{t+1}) \leq F(w^t) - \eta \left\langle \nabla F(w^t), \sum_{i=1}^{N} w_i^t \mathbf{g}_i^t \right\rangle + \frac{L\eta^2}{2}\left\|\sum_{i=1}^{N} w_i^t \mathbf{g}_i^t\right\|^2. \tag{20}$$

Now, decompose the inner product in the descent bound using the disjoint sets of benign clients $\mathcal{B}$ and adversarial clients $\mathcal{A}$:

$$\left\langle \nabla F(w^t), \sum_{i} w_i^t \mathbf{g}_i^t \right\rangle = \left\langle \nabla F(w^t), \sum_{i \in \mathcal{B}} w_i^t \mathbf{g}_i^t \right\rangle + \left\langle \nabla F(w^t), \sum_{i \in \mathcal{A}} w_i^t \mathbf{g}_i^t \right\rangle. \tag{21}$$

We bound the second term using the Cauchy-Schwarz inequality and assumption 4; we get Eq. 22. Here, the total contribution is suppressed via attention weights such that $\sum_{i \in \mathcal{A}} w_i^t \leq \delta \ll 1$, because DAARA gives exponentially low attention weights to adversarial clients based on their gradient divergence.

$$\left| \left\langle \nabla F(w^t), \sum_{i \in \mathcal{A}} w_i^t \mathbf{g}_i^t \right\rangle \right| \leq \|\nabla F(w^t)\| \cdot \left\| \sum_{i \in \mathcal{A}} w_i^t \mathbf{g}_i^t \right\| \leq G \cdot G \cdot \delta = G^2 \delta. \tag{22}$$

For benign clients, using Assumptions 2 and 4, we get:

$$\mathbb{E}\left[\left\langle \nabla F(w^t), \sum_{i \in \mathcal{B}} w_i^t \mathbf{g}_i^t \right\rangle \right] \geq (1 - \delta)(\|\nabla F(w^t)\|^2 - \epsilon^2). \tag{23}$$

Next, the variance term in the smoothness inequality is bounded using Jensen's inequality, and the boundedness of the local gradients can be represented as follows:

$$\left\| \sum_i w_i^t \mathbf{g}_i^t \right\|^2 \leq \sum_i w_i^t \|\mathbf{g}_i^t\|^2 \leq G^2. \tag{24}$$

But due to adversarial interference, we only effectively utilize a fraction $(1 - \delta)$ of the full gradient information. Now, substituting the bounds into the descent inequality in Eq. 20:

$$\mathbb{E}[F(w^{t+1})] \leq \mathbb{E}[F(w^t)] - \eta(1 - \delta)\left(\|\nabla F(w^t)\|^2 - \epsilon^2\right) + \frac{L\eta^2}{2}G^2. \tag{25}$$

$$\mathbb{E}[F(w^{t+1})] \leq \mathbb{E}[F(w^t)] - \eta(1 - \delta)\|\nabla F(w^t)\|^2 + \eta(1 - \delta)\epsilon^2 + \frac{L\eta^2}{2}G^2. \tag{26}$$

Rearranging terms of Eq. 26, then dividing both sides by $\eta(1 - \delta)$, we get:

$$\eta(1 - \delta)\mathbb{E}[\|\nabla F(w^t)\|^2] \leq \mathbb{E}[F(w^t)] - \mathbb{E}[F(w^{t+1})] + \eta(1 - \delta)\epsilon^2 + \frac{L\eta^2}{2}G^2. \tag{27}$$

$$\mathbb{E}[\|\nabla F(w^t)\|^2] \leq \frac{\mathbb{E}[F(w^t)] - \mathbb{E}[F(w^{t+1})]}{\eta(1 - \delta)} + \epsilon^2 + \frac{L\eta G^2}{2(1 - \delta)}. \tag{28}$$

Summing both sides of Eq. 28 from $t = 1$ to T:

$$\sum_{t=1}^{T} \mathbb{E}[\|\nabla F(w^t)\|^2] \leq \frac{F(w^1) - F(w^{T+1})}{\eta(1 - \delta)} + T\epsilon^2 + \frac{L\eta G^2 T}{2(1 - \delta)}. \tag{29}$$

Since $F(w^*)$ denotes the global minimum of the objective function, we have $F(w^{T+1}) \geq F(w^*)$ by definition of optimality. This allows us to upper-bound the telescoping sum as $F(w^1) - F(w^{T+1}) \leq F(w^1) - F(w^*)$ and we get:

$$\sum_{t=1}^{T} \mathbb{E}[\|\nabla F(w^t)\|^2] \leq \frac{F(w^1) - F(w^*)}{\eta(1 - \delta)} + T\epsilon^2 + \frac{L\eta G^2 T}{2(1 - \delta)}. \tag{30}$$

Divide both sides of Eq. 30 by T:

$$\frac{1}{T}\sum_{t=1}^{T}\mathbb{E}[\|\nabla F(w^t)\|^2] \leq \frac{F(w^1) - F(w^*)}{\eta T(1 - \delta)} + \epsilon^2 + \frac{L\eta G^2}{2(1 - \delta)}. \tag{31}$$

Finally, to balance the bias-variance trade-off in the convergence inequality, we put the step size $\eta = \mathcal{O}(1/\sqrt{T})$ in Eq. 31, we get Eq. 32 from the first term and Eq. 33 from the third term of the right hand side of Eq. 31.

$$\frac{F(w^1) - F(w^*)}{\eta T(1 - \delta)} = \frac{F(w^1) - F(w^*)}{\left(\frac{1}{\sqrt{T}}\right)T(1 - \delta)} = \frac{F(w^1) - F(w^*)}{T^{3/2}(1 - \delta)} = \mathcal{O}\left(\frac{1}{\sqrt{T}}\right). \tag{32}$$

$$\frac{L\eta G^2}{2(1 - \delta)} = \frac{LG^2}{2(1 - \delta)} \cdot \frac{1}{\sqrt{T}} = \mathcal{O}\left(\frac{1}{\sqrt{T}}\right). \tag{33}$$

Substituting Eq. 32 and Eq. 33 in Eq. (31), we have:

$$\frac{1}{T}\sum_{t=1}^{T}\mathbb{E}[\|\nabla F(w^t)\|^2] \leq \mathcal{O}\left(\frac{1}{\sqrt{T}}\right) + \epsilon^2 + \mathcal{O}\left(\frac{1}{\sqrt{T}}\right). \tag{34}$$

Absorbing both $\mathcal{O}(1/\sqrt{T})$ terms into one and including the stochastic variance σ^2 to explicitly reflect the bounded stochastic variance of local gradients from Assumption 3, we get the final convergence rate:

$$\frac{1}{T}\sum_{t=1}^{T}\mathbb{E}[\|\nabla F(w^t)\|^2] \leq \mathcal{O}\left(\frac{1}{\sqrt{T}} + \delta^2 + \epsilon^2 + \sigma^2\right), \tag{35}$$

which completes the proof.

5   Experimental Result Analysis

In this section, we present a comprehensive evaluation of the proposed DAARA mechanism under TLF and UTLF attacks. Our goal is to assess the degradation in model performance caused by adversarial clients and the robustness of various defense strategies. The evaluation covers dataset setup, data partitioning, model architecture, adversarial configurations, compared baselines, and key evaluation metrics.

5.1   Dataset and Data Distribution

We use the NSL-KDD and UNSW-NB15 datasets for our experiments. These datasets were chosen for their rich diversity in attack types, realistic network traffic patterns, and established relevance in both cybersecurity and FL research [17]. NSL-KDD is an enhanced version of the KDD'99 dataset, tailored for evaluating network intrusion detection systems [18]. It comprises 41 features categorized into normal and four attack types: DoS, Probe, R2L, and U2R. The UNSW-NB15 dataset [19] contains 49 features, including both normal traffic and nine attack types: Fuzzers, Analysis, Backdoors, DoS, Exploits, Generic, Reconnaissance, Shellcode, and Worms. To simulate heterogeneous data distributions, we partition both datasets across 100 clients using a Dirichlet distribution with a concentration parameter α. A smaller α yields greater non-IIDness. We consider two settings: $\alpha = 1.0$ (moderate non-IID), close to the characteristics of IID, and $\alpha = 0.5$ (higher non-IID). Each client receives samples from multiple classes with varying proportions.

5.2 Model Structure and Hyperparameter

Each client is assigned a unique data partition and trains a local model for 2 epochs per communication round (CR) using a batch size of 64. Training continues for 200 CRs. The global model is a three-layer neural network comprising an input layer followed by two fully connected layers with 64 and 32 neurons, respectively, activated using ReLU, and an output layer matching the number of classes. We use SGD as the optimizer and CrossEntropyLoss as the loss function. The poisoning ratio is fixed at 40% for both TLF and UTLF attacks. We also perform experiments from 20% to 60% attack ratio for better clarification.

5.3 Adversarial Setting and Compared Defense Mechanisms

Two attack types are considered: TLF and UTLF. In the TLF setting, attackers flip labels deterministically (e.g., class $0 \rightarrow 1$, class $2 \rightarrow 3$), while in the untargeted setting, labels are flipped randomly. Attacks are applied during local training by malicious clients. We compare our proposed DAARA with the following defense methods: FedAvg, Krum, Trimmed Mean, Auror, FoolsGold, and RFed. Each employs a distinct aggregation strategy to deal with adversarial behavior.

5.4 Evaluation Metrics

We evaluate performance using five key metrics. MA measures the overall classification accuracy on the global test set. ASR quantifies the proportion of flipped source class instances that are misclassified as the adversary's target class, reflecting the effectiveness of the attack. SRR captures the recall for unflipped source classes, indicating the model's ability to retain accuracy on benign data. GD is calculated as the mean $L2$ distance between each client's update and the global model, offering insights into update consistency and anomaly behavior. CRs refer to the number of rounds required for the global model to converge.

5.5 Performance Analysis

Tables 1 and 2 summarize the empirical results for both NSL-KDD and UNSW-NB15 datasets under different levels of data heterogeneity ($\alpha = 1.0, 0.5$), attack

Table 1. Performance of defense mechanisms against TLF and UTLF attacks on the NSL-KDD dataset under non-IID settings ($\alpha = 1.0, 0.5$).

NSL-KDD UTLF										
Model	Alpha=1.0					Alpha=0.5				
	MA	GD	SRR	ASR	CRs	MA	GD	SRR	ASR	CRs
FedAvg	0.975	0.216	0.973	0.273	160	0.973	0.274	0.974	0.310	142
Krum	0.976	0.317	0.966	0.062	141(1.13×)	0.964	0.334	0.960	0.074	-
T_Mean	0.978	0.211	0.974	0.052	129(1.24×)	0.973	0.251	0.971	0.067	131(1.08×)
Auror	0.794	0.326	0.905	0.068	-	0.476	0.329	0.490	0.786	-
FoolsGold	0.980	0.223	0.975	0.052	69(2.31×)	0.977	0.275	0.972	0.067	88(1.61×)
RFed	0.979	0.223	0.978	0.047	113(1.41×)	0.973	0.263	0.975	0.053	117(1.21×)
DAARA	0.988	0.162	0.984	0.035	40(4.00×)	0.987	0.209	0.972	0.043	42(3.38×)
NSL-KDD TLF										
FedAvg	0.971	0.275	0.959	0.248	178	0.902	0.192	0.967	0.271	181
Krum	0.980	0.310	0.972	0.124	118(1.51×)	0.976	0.261	0.968	0.151	143(1.26×)
T_Mean	0.978	0.270	0.963	0.066	129(1.38×)	0.970	0.185	0.956	0.072	194(0.93×)
Auror	0.959	0.276	0.946	0.131	-	0.914	0.159	0.520	0.142	-
FoolsGold	0.977	0.197	0.960	0.048	188(0.95×)	0.973	0.186	0.962	0.051	143(1.26×)
RFed	0.975	0.173	0.823	0.046	193(0.92×)	0.974	0.181	0.954	0.074	152(1.19×)
DAARA	0.985	0.126	0.978	0.016	56(3.18×)	0.980	0.137	0.966	0.031	75(2.41×)

Table 2. Performance of defense mechanisms against TLF and UTLF attacks on the UNSW-NB15 dataset under non-IID settings ($\alpha = 1.0, 0.5$).

UNSW UTLF										
Model	Alpha=1.0					Alpha=0.5				
	MA	GD	SRR	ASR	CRs	MA	GD	SRR	ASR	CRs
FedAvg	0.957	0.129	0.999	0.312	107	0.953	0.196	0.943	0.373	117
Krum	0.956	0.110	0.936	0.288	146 (0.73×)	0.942	0.186	0.969	0.297	-
T_Mean	0.964	0.098	0.995	0.162	82 (1.30×)	0.964	0.177	0.932	0.169	88 (1.33×)
Auror	0.813	0.086	0.788	0.586	-	0.805	0.102	0.724	0.859	-
FoolsGold	0.954	0.146	0.933	0.256	72 (1.49×)	0.960	0.161	0.999	0.262	75 (1.56×)
RFed	0.968	0.118	1.000	0.146	66 (1.62×)	0.968	0.130	1.000	0.191	67 (1.75×)
DAARA	0.979	0.036	0.982	0.076	41 (2.61×)	0.971	0.075	1.000	0.046	52 (2.25×)
UNSW TLF										
FedAvg	0.953	0.148	0.920	0.287	143	0.950	0.172	0.937	0.298	166
Krum	0.961	0.101	0.928	0.160	62 (2.31×)	0.957	0.152	0.937	0.189	134 (1.24×)
T_Mean	0.967	0.134	0.947	0.198	57 (2.50×)	0.960	0.139	1.000	0.198	87 (1.90×)
Auror	0.908	0.119	1.000	0.493	-	0.892	0.135	0.768	0.536	-
FoolsGold	0.965	0.108	0.927	0.264	51 (2.80×)	0.954	0.115	1.000	0.295	95 (1.75×)
RFed	0.967	0.083	0.980	0.245	60 (2.38×)	0.962	0.104	0.922	0.246	69 (2.41×)
DAARA	0.975	0.058	0.934	0.098	33 (4.33×)	0.970	0.092	0.935	0.129	46 (3.61×)

types (targeted and untargeted), and defense strategies. Across all configurations, DAARA consistently outperforms other baselines in terms of robustness, accuracy, and communication efficiency. For untargeted attacks on NSL-KDD with $\alpha = 1.0$, DAARA achieves the highest MA (0.988), lowest GD (0.162), lowest ASR (0.035), and requires the fewest CRs (40), yielding a 4.00× improvement in convergence speed over FedAvg. Under stronger heterogeneity ($\alpha = 0.5$), DAARA maintains strong performance with an MA of 0.987 and an ASR of

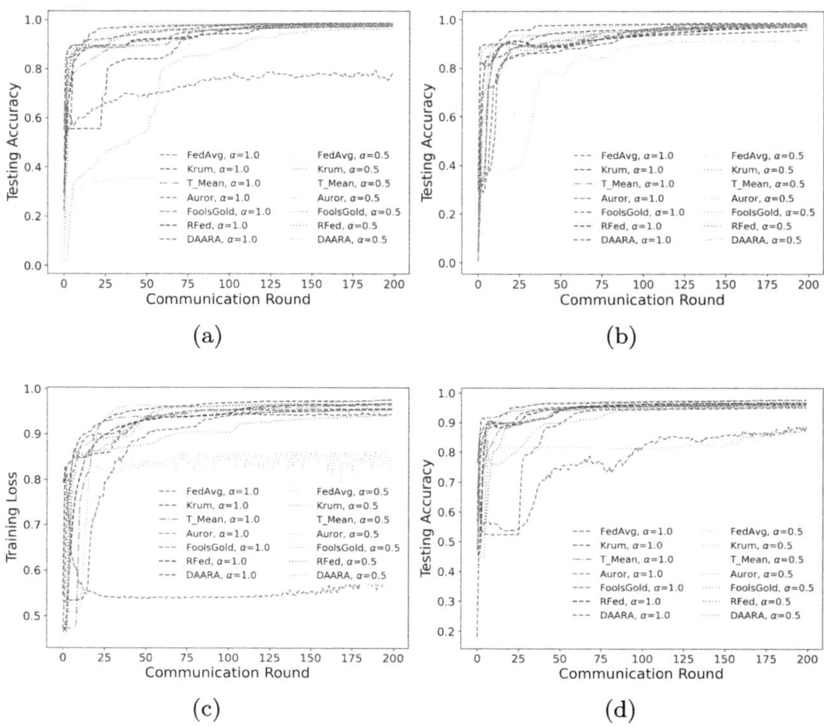

Fig. 1. Testing accuracy vs. communication rounds under non-IID settings ($\alpha =$ 1.0, 0.5) for: (a) NSL-KDD UTLF, (b) NSL-KDD TLF, (c) UNSW-NB15 UTLF, (d) UNSW-NB15 TLF.

0.043. In the targeted attack scenario on NSL-KDD, DAARA attains an MA of 0.985 and an ASR of just 0.016 under $\alpha = 1.0$, outperforming RFed (MA = 0.975, ASR = 0.046) and FoolsGold (MA = 0.977, ASR = 0.048). Even under more severe non-IID conditions ($\alpha = 0.5$), DAARA yields the highest MA of 0.980 and the lowest ASR of 0.031, with a CR count of only 75. Notice that '$-$' is used to represent if the methods failed to converge.

On the UNSW-NB15 dataset, DAARA demonstrates similar superiority. For untargeted attacks at $\alpha = 1.0$, it achieves the best MA (0.979), lowest GD (0.036), and lowest ASR (0.076), requiring only 41 CRs—significantly fewer than RFed (66 CRs) and Trimmed Mean (82 CRs). For $\alpha = 0.5$, DAARA continues to lead with an MA of 0.971 and an ASR of 0.046. In targeted attacks on UNSW-NB15, DAARA once again outperforms other defenses, achieving the lowest ASR of 0.098 and the fastest convergence (33 CRs) under $\alpha = 1.0$. Even in the harsher non-IID setting ($\alpha = 0.5$), DAARA secures an MA of 0.970 and an ASR of 0.129, compared to 0.962/0.246 for RFed and 0.954/0.295 for FoolsGold.

In both tables, focusing on GD and ASR provides a clearer understanding of the impact of the attacks on the defense mechanisms. Furthermore, CRs clearly

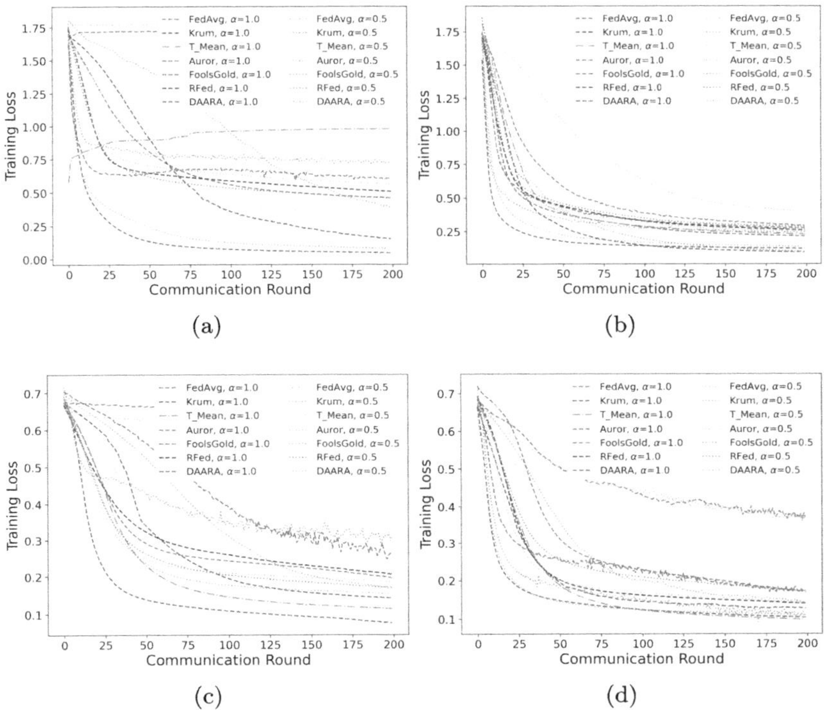

Fig. 2. Training Loss vs. CRs of $\alpha = 1.0, 0.5$ for: (a) NSLKDD UTLF, (b) NSLKDD TLF, (c) UNSW-NB15 UTLF, (d) UNSW-NB15 TLF.

demonstrate the faster convergence properties of the proposed DAARA compared to the others. Moreover, Figs. 1 and 2 further visualize the stability and prioritization of DAARA training. The curves demonstrate faster convergence and a stable training process, as indicated by testing accuracy and training loss across all settings. These comprehensive results affirm that DAARA effectively suppresses adversarial impact, maintains benign gradient diversity, and enables robust and efficient FL in hostile environments.

5.6   Impact of Attack Ratio

Figures 3(a) and (b) compare gradient divergence under varying attack ratios (20% to 60%) on the NSL-KDD dataset with a non-IID distribution ($\alpha = 0.5$) for both untargeted (UTLF) and targeted label-flipping (TLF) attacks. In the UTLF setting, all defense mechanisms exhibit increasing GD as the attack ratio rises. Krum and Auror show the highest GD at 60%, indicating limited resilience to large-scale malicious updates. FedAvg, FoolsGold, and Trimmed Mean also show notable increases in GD. In contrast, DAARA maintains significantly lower GD across all attack ratios, increasing smoothly from 0.13 at 20% to only 0.30

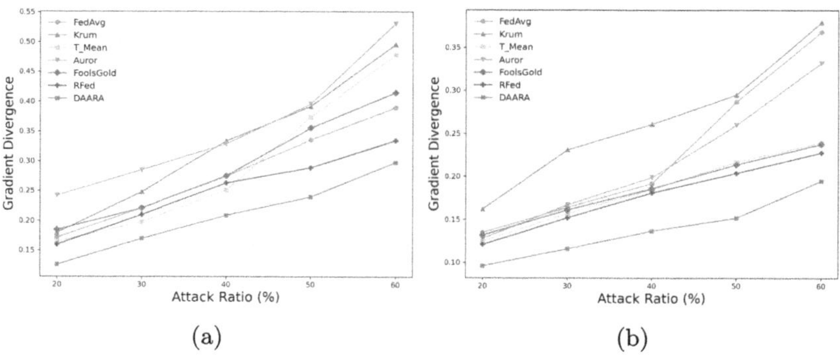

Fig. 3. Attack ratio vs. GD under $\alpha = 0.5$ non-IID settings for: (a) NSL-KDD UTLF, and (b) NSL-KDD TLF.

at 60%, reflecting strong robustness and stability. Under TLF attacks, the GD values are generally lower due to the stealthier nature of the attack, yet the relative trends remain consistent. FedAvg and Krum again yield higher GD, while DAARA consistently achieves the lowest values, ranging from 0.096 to 0.19. These findings underscore DAARA's ability to effectively suppress both overt and subtle adversarial manipulations, demonstrating superior robustness and stability compared to all baseline defenses under varying adversarial intensities.

6   Conclusion and Future Work

In this article, we introduce DAARA, a divergence-aware attention-based aggregation mechanism to defend against LF attacks in FL. By quantifying gradient divergence and enforcing class-wise consistency, DAARA dynamically attenuates the impact of poisoned updates without requiring explicit attacker identification or pre-set thresholds. Through comprehensive evaluations of two benchmark intrusion detection datasets under various non-IID conditions, DAARA consistently achieved the highest MA, lowest ASR, and fastest convergence among all compared methods. The theoretical analysis further validates its robustness and communication efficiency by considering the adversarial impact on model performance, as well as local update bias and variance. For future work, we plan to extend DAARA to defend against more complex and adaptive threats, such as model poisoning or backdoor attacks. In addition, we will explore its integration with differential privacy mechanisms and adaptive attack scenarios to enhance resilience in asynchronous or resource-constrained FL environments.

References

1. McMahan, B., Moore, E., Ramage, D., Hampson, S., Arcas, B.A.: Communication-efficient learning of deep networks from decentralized data. In: Artificial intelligence and statistics, pp. 1273–1282, PMLR (2017)
2. Li, T., Sahu, A.K., Zaheer, M., Sanjabi, M., Talwalkar, A., Smith, V.: Federated optimization in heterogeneous networks. Proc. Mach. Learn. Sys. **2**, 429–450 (2020)
3. Fang, M., Cao, X., Jia, J., Gong, N.: Local model poisoning attacks to {Byzantine-Robust} federated learning. In: 29th USENIX security symposium (USENIX Security 20), pp. 1605–1622 (2020)
4. Zhou, X.-H., et al.: Learning skill characteristics from manipulations. IEEE Trans. Neu. Netw. Learn. Sys. **34**(12), 9727–9741 (2022)
5. Uddin, M.P., Xiang, Y., Hasan, M., Bai, J., Zhao, Y., Gao, L.: A systematic literature review of robust federated learning: Issues, solutions, and future research directions. ACM Comput. Surv. **57**(10), 1–62 (2025)
6. Blanchard, P., El Mhamdi, E.M., Guerraoui, R., Stainer, J.: Machine learning with adversaries: Byzantine tolerant gradient descent. In: Proceedings of the 31st International Conference on Neural Information Processing Systems (NeurIPS), pp. 119–129 (2017)
7. Yin, D., Chen, Y., Kannan, R., Bartlett, P.: Byzantine-robust distributed learning: Towards optimal statistical rates. In: Proceedings of the 35th International Conference on Machine Learning (ICML), pp. 5650–5659 (2018)
8. Shen, S., Tople, S., Saxena, P.: Auror: Defending against poisoning attacks in collaborative deep learning systems. In: Proceedings of the 32nd annual conference on computer security applications, pp. 508–519 (2016)
9. Fung, C., Yoon, C., Beschastnikh, I.: Mitigating sybils in federated learning poisoning. arxiv 2018. arXiv preprint arXiv:1808.04866, (2018)
10. Miao, Y., et al.: fed: Robustness-enhanced privacy-preserving federated learning against poisoning attack. IEEE Trans. Inf. Forensic Sec., **19**, 5814–5827 (2024)
11. Xie, Y., Fang, M., Gong, N.Z.: Fedredefense: Defending against model poisoning attacks for federated learning using model update reconstruction erro. In: International Conference on Machine Learning, pp. 54460–54474, PMLR (2024)
12. Gong, Z., et al.: Agramplifier: Defending federated learning against poisoning attacks through local update amplification. IEEE Trans. Inf. Forensics Secur. **19**, 1241–1250 (2023)
13. Chen, X., Yu, H., Jia, X., Yu, X.: Apfed: anti-poisoning attacks in privacy-preserving heterogeneous federated learning. IEEE Trans. Inf. Forensics Secur. **18**, 5749–5761 (2023)
14. Liu, J., Li, X., Liu, X., Zhang, H., Miao, Y., Deng, R.H.: Defendfl: A privacy-preserving federated learning scheme against poisoning attacks. IEEE Trans. Neu. Netw. Learn. Sys. (2024)
15. Sharma, A., Chen, W., Zhao, J., Qiu, Q., Bagchi, S., Chaterji, S.: Flair: Defense against model poisoning attack in federated learning. In: Proceedings of the 2023 ACM Asia Conference on Computer and Communications Security, pp. 553–566 (2023)
16. Tounsi, A., Salem, O., Mehaoua, A.: Robust federated learning against data poisoning: A split learning-based approach evaluated on various aggregation techniques. In: 2024 IEEE International Conference on E-health Networking, Application & Services (HealthCom), pp. 1–6, IEEE (2024)

17. Ghimire, B., Rawat, D.B.: Recent advances on federated learning for cybersecurity and cybersecurity for federated learning for internet of things. IEEE Internet Things J. **9**(11), 8229–8249 (2022)
18. Tavallaee, M., Bagheri, E., Lu, W., Ghorbani, A.A.: A detailed analysis of the kdd cup 99 data set. In: 2009 IEEE Symposium on Computational Intelligence for Security and Defense Applications, pp. 1–6 (2009)
19. Moustafa, N., Slay, J.: Unsw-nb15: a comprehensive data set for network intrusion detection systems (unsw-nb15 network data set): In: 2015 Military Communications and Information Systems Conference (MilCIS), pp. 1–6 (2015)

Understanding the Asymmetric Impact of Forecast Accuracy on Decision Quality

Mahdi Abolghasemi[1]([envelope]) [iD] and Richard Bean[2] [iD]

[1] Queensland University of Technology, Brisbane, Australia
mahdi.abolghasemi@qut.edu.au
[2] University of Queensland, St Lucia, Australia
r.bean1@uq.edu.au

Abstract. In this paper, we examine the issues caused by a mismatch in loss functions in typical *predict and optimise* problems with specific reference to the 3rd Technical Challenge of the IEEE Computational Intelligence Society. In this competition, entrants were asked to forecast building energy use and solar generation at six buildings and six solar installations, and then use their forecast to optimize energy cost while scheduling classes and batteries over a month. We examine the possible effect of underforecasting and overforecasting and asymmetric errors on the optimisation cost. We explore the different nature of loss functions for the prediction and optimisation phase and propose to adjust the final forecasts for a better optimisation cost. We report that while there is a positive correlation between these two, more appropriate loss functions can be used to optimise the costs associated with final decisions. Our findings are of significant value in designing forecasting and decision-making systems.

Keywords: prediction · optimisation · micro grid · asymmetric errors

1 Introduction

The *predict and optimise* paradigm represents an increasingly important framework in forecasting, optimisation, and machine learning literature, addressing sequential decision-making problems where certain parameter values are unknown and must be predicted. This paradigm requires the integration of predictive and prescriptive models to solve complex real-world problems where forecasts serve as inputs to optimisation algorithms. While forecasts provide valuable information for decision makers, they are not the ultimate objective; rather, the utility derived from forecasts such as optimising decisions to reduce operational costs, usually holds greater relevance for practical applications.

Despite the intuitive assumption that improving forecast accuracy directly enhances decisions outcomes, this relationship is neither linear nor symmetric. Forecasting models typically aim to minimise metrics such as mean absolute error (MAE), while optimisation functions pursue entirely different objectives. This

Q. V. Nguyen et al. (Eds.): AusDM 2025, CCIS 2765, pp. 19–31, 2026.
https://doi.org/10.1007/978-981-95-6786-7_2

misalignment raises critical questions about the association between forecast accuracy and optimisation costs. Although this disconnect has been recognised in the literature, limited empirical work has been conducted to understand the relationship between underforecasting, overforecasting, and decision costs across various application domains.

One particularly important application of the *predict and optimise* framework is in renewable energy systems, where accurate forecasting is valuable only when it leads to improved decision-making. Decarbonising energy production is essential to addressing climate change, requiring the replacement of fossil fuels with clean, renewable sources. Wind and solar power have gained widespread adoption due to their low production costs and global accessibility; however, their generation is inherently variable owing to uncertain weather conditions. Limited storage technologies make it challenging to balance supply and demand in real time, creating a need for innovative solutions that can effectively integrate renewable energy sources.

Microgrids equipped with solar panels and battery storage systems represent a viable approach to addressing these challenges. These systems rely on forecasts of electricity demand, electricity prices, and solar generation as inputs to optimisation models that determine optimal battery charging and discharging schedules. Such coordination maximises renewable utilisation, reduces grid dependence, and minimises operational costs. In many real-world microgrid applications, however, the link between forecast accuracy and decision quality remains poorly understood.

This type of problem, called *predict and optimise*, is becoming increasingly popular in the forecasting, optimisation, and machine learning literature due to their numerous applications in real-world problems, e.g., transportation scheduling, healthcare staff scheduling, demand planning [1,3]. These two metrics have different objectives and it is not always clear how forecast accuracy is associated with the optimisation costs [10]. Despite recognition of this misalignment, there is little work done to understand the association between underforecasting, overforecasting, and decision costs. The empirical results are lacking in the broader domain of *predict and optimise* problems, especially in the renewable energy scheduling problem.

This paper investigates the IEEE 3rd Technical Challenge on "predict and optimise" organised by the IEEE Computational Intelligence Society and Monash University [5]. The challenge comprised two primary components: (i) forecasting potential power generation for six solar installations and power demand for six buildings at 15-minute intervals over one month, and (ii) utilising these predictions as inputs in an optimisation model to schedule battery charging/discharging and class scheduling to minimise total energy costs over the same time horizon.

The present study examines the performance of forecasting and optimisation models across various scenarios, including overforecasts, underforecasts, and naturally unbiased forecasts, to illuminate the relationship between these seemingly separate problems. Understanding this association enables decision makers to identify which models require improvement and, more importantly, how to adjust or integrate forecasts to achieve optimal solutions for the problem at hand, particularly within the renewable energy scheduling domain.

2   Background

The paradigm of *predict and optimise* is useful when the optimisation problem needs an input that is not fully observed but needs to be predicted. Prediction inevitably has some associated uncertainty with some degree of accuracy. The optimisation which takes inputs from the uncertain forecasts treats them as deterministic values and accordingly determines an optimal solution given the provided inputs.

There are studies that have integrated the downstream optimisation in the prediction model and proposed a predictive model with customised loss function that aims to minimise the final cost function. For example, Elmachtoub and Grigas [8] proposed a smart predict then optimise approach in which they suggested optimising the forecasts with respect to the final optimisation model in which they will be used. They proposed a new loss function called SPO to train the prediction model. They implemented the proposed model on the shortest path and a portfolio optimisation problem and showed the results are promising. This seems to be a common approach to integrate both prediction and optimisation phases, but in this study, we do not aim for this integration because the optimisation function is too complex to be included in the predictive model, and also too computationally expensive for testing several forecast scenarios and obtaining a better optimisation cost. Rather, we are looking to investigate the association between these two and understand how we can adjust the final generated forecasts to optimise the costs associated with a complex objective function.

The problem of energy system scheduling in stochastic optimization has been studied in Donti et al. [7] who proposed a 2-hidden layer neural network to solve a sample load forecasting and generator scheduling problem. This in turn depended on differentiation in the optimization. Later, Cameron et al. [6] examined real-world optimization problems in the context of comparing "two-stage" (i.e. predict, then optimize) and "end-to-end" (i.e. differentiating through the optimization task) approaches. However, the problem investigated by Donti et al. was a forecast for only 24 h ahead generation while the problem under study here is a much more complex mixed-integer quadratic program (MIQP). The main focus of this research is on analysing the association between forecast accuracy and optimisation cost, rather than proposing an integrated model that combines these two models into one. For an overview of *predict and optimise* literature, we encourage interested readers to read [1].

3   Data and Case Study

The datasets consist of time series values for six solar installation outputs and six building demands. The data are recorded at 15 min granularity with different starting dates. Data were released gradually during the competition. The first phase included data until 30 September 2020 and the second phase included data until the end of 31 October 2020. Therefore, participants needed to provide 2880 period ahead forecasts, for every 15 min, for all 12 time series corresponding to November 2020. There are many missing or faulty values, around 33% of the entire data, which are not included in the training data in our experiments. (Thus, when calculating the net load, these values are effectively assumed to be zero values.) Disparity among the power in solar panels is smaller in comparison to the building data. The mean of solar power generation varies between 1.12 and 4.39 kW. While there is no zero demand for buildings, the minimum solar power is zero for all solar panels which occurs before sunrise and after sunset. Solar power generation is highly dependent on the weather conditions and length of the day.

The other data provided in the competition were weather data and electricity price data. Weather data was needed to predict the building demand and solar power. Weather data included daily minimum and maximum temperature (C), rainfall (mm) and solar exposure of three weather stations near Melbourne: Moorabbin Airport, Olympic Park, and Oakleigh. This data was available from 1 January 2016. Hourly weather data was also available from the European Centre for Medium-range Weather Forecasting ERA5 dataset via OikoLab.com.

Half-hourly wholesale electricity price data was provided for the test set in phase 1 and phase 2. In a real forecasting process, the price would not be available and one needs to forecast it to be able to use it in the optimisation model for scheduling batteries and classes. Scheduling will be a problem that needs to be solved every day to optimise the decision that needs to be taken. However, this data was made available for the entire months of October and November during the competition, assuming that perfect price forecasting and weather forecasting is available. This is not far from reality, as one day ahead weather forecasting and price forecasting can be executed with a good degree of accuracy. Nonetheless, there will be an additional associated uncertainty and error in the generated forecasts, if the price and weather data were not available, thus impacting the optimisation solution.

The competition was organised in two phases: phase one which aimed to forecast six buildings demand time series, and six solar power demand for every 15 min time slot during the entire month of October 2020, i.e., 2976 values forecast for 12 time series and then solving scheduling problems for classes and batteries to determine the optimal decision for charging/discharging batteries and assigning various classes to buildings. The second phase had the same purpose and setting but for November 2020; thus requiring 2880 forecasts for 12 time series and then solving the optimisation models. The optimisation part involved scheduling a set of activities across six buildings for the entire month of October in phase 1 and November in phase 2. These activities are either *recurring*, which need to be repeated at the same time and begun within working hours (9 am to 5 pm weekdays) each week, or *once-off* that can be scheduled outside of normal working hours sometime during the month. Each building has its own corresponding solar. The microgrid has associated batteries to store the excess energy. These batteries have an efficiency rate and a maximum capacity and can be charged not only with solar panels but also from the grid, where possible. The charge and discharge decisions were made for each time period for the maximum capacity; that is, partial charge and discharge decisions were not permitted, in terms of time or power. Participants were provided with predict plus optimise instances ("ppoi") including the count of buildings, solar installations, batteries, recurring activities, and once-off activities with their relevant information.

4 Experiment Setup

In the forecasting experiment, we considered two different types of forecasts including i) forecasts from the competition participants, and ii)some perturbed forecasts for testing our experiment.

The first type of forecasts includes the natural forecasts including the top-performing models in the forecasting phase of the competition, and two newly generated models. The first ranked model was a quantile random forest that used weather and calendar features, and the second-ranked model was an ensemble of Light Gradient Boosting Machine (LGBM) that used calendar and weather features and optimised the parameters. We also developed two predictive models, LGBM2 and LGBM_Opt, where we have made some minor changes to the top performing models [2,4]. LG2 uses the same setting as the top-performing model but benefits from the "lightGBM" package instead of the "ranger" package (the second top-performing model used the lightgbm model), while continuing to forecast the 50% quantile. LGBM_Opt is the same model where the main hyperparameters of learning rate and number of leaves are tuned. We set the "num_leaves" parameter to 255 and the "objective" hyperparameter to "mae"; that is, attempting to minimise the MAE metric which was the objective of the phase 1 in competition. The other forecasting models that are ranked among the top seven are also used in our experiment. Each of these methods uses a different model with different input and training settings, thus giving us a range of forecasts generated by a diverse set of architectures to thoroughly investigate the possible association of forecasting and optimisations.

For the second type of forecasts, we perturbed the actual values of the building demand and solar to generate scenarios that are either consistently over-forecast or underforecast. We generated ten different scenarios ranging from 50 percent underforecast to 50 percent overforecast, with steps of 10%, comprising 10 scenarios with various accuracy as reported in Tables 2 and 3. The reason to generate these scenarios is to have consistent overforecasts and underforecasts that are proportional to the actual values, allowing the optimisation model to look for optimal decisions when it is misinformed. As such, we aim to understand how the optimisation model will react to imperfect biased information, as opposed to natural forecasts where imperfect information are unbiased.

We measure the forecast accuracy using mean absolute scaled error (MASE), and mean absolute error (MAE), as shown in Equations (1) and (2), respectively.

$$\text{MASE} = \frac{n-s}{h} \frac{\sum_{t=n+1}^{n+h} |y_t - f_t|}{\sum_{t=s+1}^{n} |y_t - y_{t-s}|} \tag{1}$$

$$\text{MAE} = \frac{1}{h} \frac{\sum_{t=1}^{h} |y_t - f_t|}{y_t} \tag{2}$$

where y_t is the actual value at time t, f_t is the predicted value at time t, n is the sample size (observations used for training the forecasting model), s is the length of the seasonal period (28 days or 2688 periods), and h is our forecasting horizon.

MASE was used as the reference accuracy metric in the competition. MASE is a scale-independent metric which makes it suitable for comparing the forecast accuracy when we are measuring the accuracy of many series that have different scales [9]. While it made sense to use a scale-independent metric such as MASE for measuring the average accuracy of the forecasts in the competition, in this paper it is more sensible to use the MAE metric. This is because here we are dealing only with one time series, i.e., net load, as opposed to the competition where we had 12 time series and a scale independent metric was required to account for differences in magnitude of time series. MAE minimises the median of errors.

The decision part of the experiment involves scheduling the activities in the best possible way to minimise the total cost of energy. While it is compulsory to schedule all the recurring activities, the once-off activities can be dropped from the schedule at a certain cost. Since there are a smaller number of once-off activities and their impact is relatively small, we have disregarded them in this paper. The objective function is shown in Eq. (3). Here l_t refers to the net load in period t, e_t is the Victorian wholesale electricity price (which applies in Melbourne where the Monash campus is located), a_i refers to once-off activities, d_i is a boolean decision variable determining whether once-off activity i is included, o_i is a boolean "out-of-office" variable, and penalty$_i$ is the penalty associated with whether activity i is "out of office hours".

$$O = \sum_t \frac{0.25 l_t e_t}{1000} + 0.005(\max_t l_t^2) - \sum_{a_i} (d_i.(\text{value}_i - p_i \text{penalty}_i)) \tag{3}$$

The objective function contains both a cost of energy (based on the wholesale electricity price) and a "peak demand" charge calculated over the month. (Most retail electricity bills in Australia are issued quarterly.) Unlike other studies of *predict and optimise*, the competition energy cost is a quadratic, rather than linear, function. A function with a quadratic term for peak electricity demand may more accurately reflect real-world energy costs.

For the decision optimisation part, we considered the model described in Bean [4]. Essentially, we considered five different optimisation approaches "conservative", "forced discharge", "no forced discharge", "liberal" and "very liberal". "Conservative" ignores the forecast and attempts to minimize the peak load, while "forced" and "no forced" discharge attempt to limit the behaviour of the batteries in peak hours. "Liberal" constrains the maximum of recurring load plus charge effect for each period so that it never exceeds the maximum of the recurring load over all periods. Finally, "very liberal" removes all constraints, which is the ideal approach for a perfect forecast. The major inputs to the model are the net load forecast and the wholesale pool prices (month-ahead). The Gurobi solver (9.5.0) was used to solve the mixed-integer quadratic program on a system with 11 cores for 168 h for each scenario.

5   Empirical Results

The aim is to investigate the relationship between forecasting accuracy and optimisation costs for the predict and optimise problem in hand where the forecasts and optimisation models have asymmetric error, i.e., the underforecasts and overforecasts are penalised differently in a complex optimisation model with quadratic cost in objective function. The key input to the optimisation model is the *net forecasted base load*, i.e., the total predicted load of buildings minus the predicted solar power for each time slot. That is, we aggregate the predicted buildings demand and deduct it from the aggregated predicted solar power. The result is the net load predicted for the test set, November 2020.

Table 1 depicts the accuracy of the top performing models in terms of MAE, MASE, mean overforecast (Mean-O), and underforecast (Mean-U) errors, and the associated total electricity costs.

Table 1. Performance of the top performing models in the competition used as forecasts

Metrics	Bean	Abolghasemi	EVERGi	FRESNO	Stratigakos	SZU	LGBM2	LGBM_Opt
MASE	**0.64**	0.74	0.81	1.00	0.85	0.77	0.88	0.64
MAE	78	82	90	102	94	**68**	59	68
Mean-U	71	73	85	96	88	58	**17**	57
Mean-O	7	10	**4**	6	6	10	42	10
Cost	34252	35003	34282	35464	35101	35040	**34216**	34233

As we can see in Table 1, the SZU [14] method is the top performing model among the competition participants in terms of MAE of the net load, even though it ranked sixth in terms of average MASE across all series. (The underlying cause of this is due to an accurate prediction of Building 3 load, which is very large compared to the other time series). This comparison may not be directly applicable to evaluate these models because MASE deals with all series and MAE considers only the single net load. Nevertheless, what matters for the optimisation model is the single net load and its corresponding accuracy both on-average and also across the whole horizon. Therefore, we focus on the MAE metric which is representative of the central tendency of forecasts and its average accuracy. The newly added model, LGBM2, has the lowest MAE among all methods considered in this study.

Interestingly the lowest optimisation cost corresponds to Bean's model despite the lower performance of the net load forecast, although with a small margin. This indicates that having a more accurate forecast on average does not guarantee a lower cost; rather, the optimisation cost may depend on every single forecast across the entire horizon and the quality of forecast at each time stamp. A large over- or under-forecast may not affect the average of forecast accuracy significantly but it may impact the optimisation cost. Therefore, one needs to generate accurate forecasts for the entire horizon to provide a stable and reliable input to the optimisation model, enabling the optimisation model to perform in its full capacity. The smallest error of overforecast was seen in the EVERGi model followed closely by Abolghasemi and FRESNO models. The smallest underforecast was for SZU model followed by the EVERGi model [13].

Tables 2 and 3 depict the accuracy of various perturbed forecasting scenarios and costs associated with overforecasts and underforecats, respectively. Note that even for the "Actual" forecast, the optimality gap is 5.2% (that is, the lowest possible cost is approximately $30,800) while for other scenarios the estimated optimality gap ranges from 3.0 to 10.4% (with the exception of Stratigakos which was solved to optimality). The "Actual" is the actual values of building demands and solar powers after realisation. We also generated 10 different scenarios where we generate five scenarios of overforecasts and underforecasts by adding and subtracting 10%, 20%, 30%, 40%, and 50% to the actual values (shown as Actual+-), respectively. The results of Table 2 and 3 are different and show another angle of the problem. Although the average accuracy of various models for overforecasts and underforecasts are symmetric, the associated optimisation costs are not.

Table 2. Performance of the perturbed actual values as forecasts (overforecasted)

Metrics	*Actual*	*Actual+10*	*Actual+20*	*Actual+30*	*Actual+40*	*Actual+50*
MASE	0	0.39	0.77	1.16	1.55	1.93
MAE	0	56	112	167	223	279
Mean-O	0	56	112	167	223	279
Cost	32494	32715	33229	33331	33815	36338

Figure 1 shows the association between the forecast accuracy of net load for all the investigated scenarios (expressed in terms of the competition metric, MASE over the twelve time series) and their corresponding optimisation costs. The two lines give a linear regression between cost and mean MASE for the two classes of Competition and Perturbed forecasts.

We assessed the correlation coefficient between the final cost and several metrics across (1) the perturbed actual net load, plus the actual net load (11 forecasts), (2) the competition forecasts, plus the actual net load (9 forecasts). The resulting correlation coefficients are displayed in Table 4.

Table 3. Performance of the perturbed actual values as forecasts (underforecasted)

Metrics	Actual	Actual-10	Actual-20	Actual-30	Actual-40	Actual-50
MASE	0	0.39	0.77	1.16	1.55	1.93
MAE	0	56	112	167	223	279
Mean-U	0	56	112	167	223	279
Cost	32494	32514	32646	32920	33842	34204

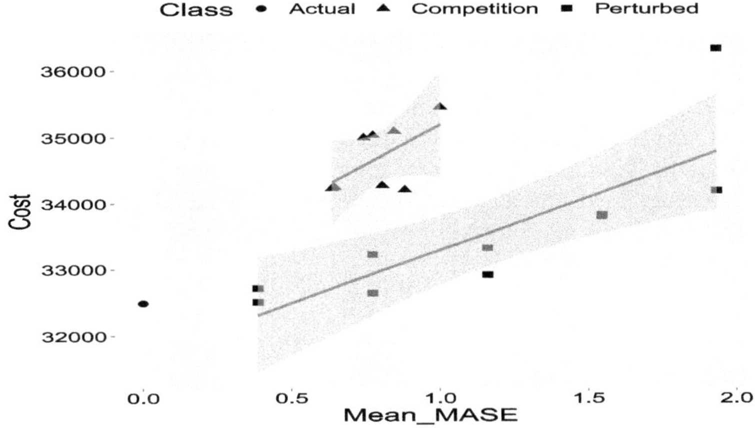

Fig. 1. Relationship between forecast accuracy and energy cost.

For perturbed forecasts, the highest correlation was observed for the standard error metrics, MASE of the time series and MAE of the net load. However, this perturbation is not reflective of the kind of errors that would be seen in an actual forecasting situation. For the competition forecasts we found that similar correlations were observed for "Mean-U" or the magnitude of the mean underforecast, MASE and MAE. The correlation for "Mean-U" is much stronger than the correlation for "Mean-O" or the mean overforecast. This is in line with our impressions regarding the relative importance of underforecasting in the competition; a large underforecast in a peak hour period may lead to a large jump in the peak load over the month, which is quite costly.

Table 4. Correlation of error metrics with optimization cost, "Small 0" instance

Methods	MASE	MAE	Mean-U	Mean-O	Mean Residuals	Std Residuals
Perturbed	0.815	0.815	0.051	0.684	0.355	0.815
Competition	0.897	0.901	0.803	0.083	-0.648	0.909

6   Forecasts Mismatch Correction

The research has demonstrated that the most common error metrics used in energy forecasting may not be the most appropriate for minimizing complex cost functions. Now, we turn our attention to investigating how the forecasts could be adjusted, i.e., what loss function can be used to obtain forecasts, which in turn leads to lower costs in downstream operations. We consider how a provided forecast can be corrected to minimize a given cost function.

Khabibrakhmanov et al. [10] provides a discussion of the effects of under-forecasting and overforecasting in solar forecasting. Their approach provides a way to determine a similar concept for this study. Instead of a solar forecast alone, we examine how a forecast of net load in the microgrid (building demand minus solar generation) is related to the cost function. If instead of measuring the absolute error of the net load forecast, we add first and third order perturbation terms to examine the effect of under or over forecasting. Here y_i refers to the actual net load and p_i refers to the predicted net load. Minimizing the cost function V leads to determining the bias and slope for an optimal linear correction for V, $p_i = \beta + \alpha x_i$.

$$V = \frac{1}{2N} \sum_{i=0}^{N-1} (y_i - p_i)^2 + \frac{1}{N}\gamma \sum_{i=0}^{N-1} (y_i - p_i) + \quad \frac{1}{3N}\epsilon \sum_{i=0}^{N-1} (y_i - p_i^2)^3 \qquad (4)$$

As a result of the competition, we have a number of energy forecasts derived from various independent methods by at least six different teams. We solved for γ and ϵ in Eq. 4 across the eight forecasts plus the actual net load, in Table 2, by maximizing the Pearson correlation between the cost function V and the costs found in Table 2, and also solved across the perturbed forecasts. The correlation coefficients found were 0.955 (for the perturbed cases) and 0.881 (for the competition cases), indicating that there is a strong correlation between the forecast accuracy and optimisation costs. This confirms that the predictive and prescriptive models operate hand in hand and in order to minimise the optimisation cost, one would need to have an accurate forecast regardless of the optimisation method used. This association is strong for the typical predict and optimise problem that we have investigated.

Normalizing the forecast and actual values to have mean 0 and standard deviation 1, for the competition cases, we obtained $\gamma = 1.37$ and $\epsilon = 0.58$, which can be used to adjust our forecasts to obtain higher accuracy. We plot the unnormalized values of forecast and actual net load in Fig. 2.

As both of these estimated parameters γ and ϵ have positive sign, this indicates a forecast error where the actual net load exceeds the forecast net load, or an "underforecast", incurs an extra cost using the optimization method we have chosen (that is, "no forced discharge"). The third order perturbation term in Eq. 4 is related to the skewness of the residuals. As previously noted, the skewness of the competition prediction residuals is also (negatively) correlated with the cost; all these skewness values are negative as the predictions underestimated the actual net load. In order to further minimize the ultimate cost, one

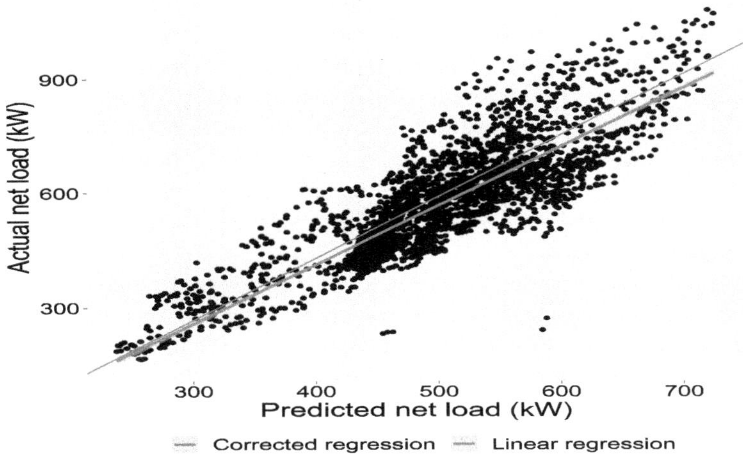

Fig. 2. Linear correction for predicted net load versus actual net load to minimize cost.

could correct the forecast, or predicted values, using a linear cost correction as seen in the blue line in Fig. 2, or find an "end-to-end" method for solving the task which may be computationally much more complex. Further analysis may be necessary across different scenarios to determine the best cost correction to apply.

7   Conclusion

This study investigated the complex relationship between forecast accuracy and optimization performance through analysis of the 3rd IEEE Technical Challenge on predict and optimise, which focused on energy forecasting and scheduling optimization. Our research addresses a fundamental gap in the predict-and-optimize literature by examining how systematic forecasting biases translate into operational costs within renewable energy systems.

Our empirical analysis demonstrates that traditional error metrics, particularly mean absolute error (MAE), provide only partial insight into the relationship between predictive performance and downstream decision quality. More significantly, we establish that forecast errors exhibit pronounced asymmetric effects on optimization outcomes, particularly when objective functions demonstrate quadratic characteristics. This asymmetry is exemplified in our finding that underforecasting electricity demand produces disproportionately higher costs compared to equivalent overforecasting errors within the IEEE competition framework.

To address these asymmetric effects, we developed a linear correction mechanism that can be systematically applied to forecasts to minimize operational costs. Our methodology was validated across multiple competition entries and their systematically perturbed variants, demonstrating consistent improvements in cost reduction. This contribution provides practitioners with a practical tool for adjusting forecasts to better serve optimization objectives rather than merely minimizing traditional accuracy metrics. The implications of our findings extend beyond the immediate application domain, highlighting fundamental challenges in the predict-and-optimize paradigm. The disconnect between forecast accuracy and decision quality necessitates a reconsideration of how predictive models should be evaluated and optimized when serving as inputs to downstream decision-making processes.

Several compelling avenues emerge for future research. The development of end-to-end methodologies that treat the entire predict-and-optimize pipeline as a unified optimization problem represents a particularly promising direction. Additionally, the integration of forecasting and optimization through customized loss functions presents significant potential, though technical challenges remain for non-differentiable objective functions. Furthermore, systematic investigation of alternative optimization methodologies and robust forecast biasing techniques warrants continued attention. The development of decision-aware forecasting models—those explicitly optimized for their intended downstream applications rather than generic accuracy metrics—represents a paradigm shift that could fundamentally enhance the effectiveness of predict-and-optimize systems in renewable energy management and beyond.

Disclosure of Interests. The authors have no competing interests to declare that are relevant to the content of this article.

References

1. Abolghasemi, M.: The intersection of machine learning with forecasting and optimisation: theory and applications, Forecasting with artificial intelligence: Theory and applications, pp. 313-339, Springer Nature Switzerland (2023). https://doi.org/10.1007/978-3-031-35879-1_12
2. Abolghasemi, M. Esmaeilbeigi, R.: State-of-the-art predictive and prescriptive analytics for IEEE CIS 3rd Technical Challenge (2021). arXiv preprint arXiv:2112.03595. https://doi.org/10.48550/arXiv.2112.03595
3. Abolghasemi, M., Abbasi, B., HosseiniFard, Z.: Machine learning for satisficing operational decision making: A case study in blood supply chain. Int. J. Forecast. **41**(1), 3–19 (2025). https://doi.org/10.1016/j.ijforecast.2023.05.004
4. Bean, R.: Methodology for forecasting and optimization in IEEE-CIS 3rd Technical Challenge (2022). arXiv preprint arXiv:2202.00894. https://doi.org/10.48550/arXiv.2202.00894
5. Bergmeir, C., et al.: Predict + Optimize Problem in Renewable Energy Scheduling, IEEE Access (2025). https://doi.org/10.1109/ACCESS.2025.3555393
6. Cameron, C., Hartford, J., Lundy, T., and Leyton-Brown, K.: The Perils of Learning Before Optimizing (2021). arXiv preprint arXiv:2106.10349. https://doi.org/10.48550/arXiv.2106.10349
7. Donti, P., Amos, B., and Kolter, J.Z.: Task-based end-to-end model learning in stochastic optimization. Adv. Neu. Inf. Process. Sys., **30** (2017). https://doi.org/10.48550/arXiv.1703.04529

8. Elmachtoub, A.N., Grigas, P.: Smart "predict, then optimize". Management Science, **68**(1), 9–26 (2022). https://doi.org/10.1287/mnsc.2020.3922
9. Hyndman, R.J., Athanasopoulos, G.: Forecasting: principles and practice. OTexts (2018). https://otexts.com/fpp3
10. Khabibrakhmanov, I., Lu, S., Hamann, H., Warren, K.: On the usefulness of solar energy forecasting in the presence of asymmetric costs of errors. IBM J. Res. Dev. **60**(1), 7–1 (2016). https://doi.org/10.1147/JRD.2015.2495001
11. Limmer, S., Einecke, N.: An Efficient approach for peak-load-aware scheduling of energy-intensive tasks in the context of a public IEEE challenge. Energies **15**, 3718 (2022). https://doi.org/10.3390/en15103718
12. Lowry, M.: IRM Design for Toronto Hydro-Electric System (2019). https://irp-cdn.multiscreensite.com/06615795/files/uploaded/PEG
13. Ruddick, J., Genov, E., Camargo, L.R., Coosemans, T., Messagie, M.: Evolutionary scheduling of university activities based on consumption forecasts to minimise electricity costs (2022). https://doi.org/10.48550/arXiv.2202.12595
14. Zhu, Q., et al.: A local search method for solving a bi-level timetabling and battery scheduling problem (2021). https://github.com/xuyaojian123/IEEE-Predict-Optimize-Challenge

WaveFSL: Wave Interference-Based Meta-learning for Few-Shot Cross-Modality Traffic Forecasting

Abdul Joseph Fofanah[1]([✉]) [iD], Lian Wen[1] [iD], David Chen[1] [iD], Shaoyang Zhang[2] [iD], and Alpha Alimamy Kamara[3] [iD]

[1] School of Information and Communication Technology, Griffith University, Brisbane 4111, Australia
abdul.fofanah@griffithuni.edu.au, {l.wen,david.chen}@griffith.edu.au
[2] School of Information Engineering, Chang'an University, Xi'an, China
zhsy@chd.edu.cn
[3] School of Computer Science and Engineering, Central South University, Changsha 410083, China
kamara@csu.edu.cn

Abstract. Urban traffic prediction is hindered by heterogeneous sensor configurations and complex wave-like traffic flow dynamics (e.g., congestion waves, stop-and-go oscillations) that conventional neural networks struggle to model effectively. We propose *WaveFSL*, a novel wave physics-inspired few-shot learning framework that unifies wave interference theory with adaptive neural modules to address these challenges. WaveFSL integrates five key components: (1) a *Dynamic Input Projection* (DIP) for handling variable-dimensional input through learnable dimension-aware projection; (2) a *Traffic Wave Generator* (TWG) synthesising parameterised wave components (with amplitude, effective frequency, and phase); (3) a *Wave-Constrained Interference* (WCI) explicitly modelling congestion propagation via coupled wave superposition; (4) a *Wave-Aware Spectral Attention* (WASA) for multiscale spectral analysis through resonance scoring and frequency-band decomposition; and (5) a *Few-Shot Adaptation* with optimal kernels (FSAK) enabling rapid domain adaption and transfer via prototype-based conditioning. By unifying wave interference theory with adaptive neural learning, WaveFSL achieves state-of-the-art performance across four real-world datasets, outperforms baselines (3.2–8.5% MAE reduction), generalises with only 5–10 labelled samples per city, and requires no retraining. It enables interpretable, deployable traffic forecasting under realistic constraints. Code is available at: https://github.com/afofanah/WaveFSL.

Keywords: Traffic prediction · Wave interference · Few-shot learning · Spatio-temporal forecasting · Neural networks

1 Introduction

Urban traffic forecasting plays a critical role in modern intelligent transportation systems, supporting tasks such as congestion management, route planning, and emergency response [10]. While deep learning models have achieved impressive results on large-scale spatiotemporal datasets, they often suffer when applied to new cities or sensor configurations with limited labelled data [7]. Moreover, existing approaches typically treat traffic dynamics as generic time series, failing to capture the underlying wave-like behaviours—such as congestion waves and stop-and-go oscillations—that emerge from traffic flow physics [4].

Few-shot learning (FSL) offers a promising solution by enabling models to generalise to new domains with only a few labelled examples [8]. However, most FSL methods are developed for static visual inputs and do not account for the unique challenges of cross-modality traffic forecasting, including variable input dimensionality, heterogeneous sensor setups, and dynamic temporal-spatial correlations [3].

In this work, we propose **WaveFSL**, a novel physics-inspired meta-learning framework that models traffic flow as a superposition of wave components. Drawing on principles from wave interference theory, WaveFSL introduces a suite of specialised modules designed to capture domain-specific traffic characteristics while supporting rapid adaptation to unseen configurations:

- A *Dynamic Input Projection* (DIP) module layer that enables the model to handle variable-dimensional sensor inputs by learning dimension-aware mappings.
- A *Traffic Wave Generator* (TWG) module that decomposes traffic signals into interpretable wave components parameterised by amplitude, effective frequency, and phase.
- A *Wave-Constrained Interference* (WCI) module mechanism that captures congestion propagation via coupled wave superposition.
- A *Wave-Aware Spectral Attention* (WASA) module performs analysis at multiscales by scoring resonance and focusing on specific frequency bands.
- A *Few-Shot Adaptation with Kernels* (FSAK) module enables rapid task adaptation through prototype-based spectral conditioning.

By fusing wave physics with adaptive neural learning, WaveFSL achieves strong cross-city generalisation and interpretability, requiring as few as 5–10 labelled samples per domain. This work introduces a new paradigm for traffic forecasting by explicitly modelling physical dynamics and promoting efficient adaptation in data-scarce environments.

2   Related Work

2.1   Traditional and Graph-Based Spatio-Temporal Methods

The evolution of traffic prediction has progressed from early CNN/RNN models (e.g., ST-ResNet [20], LSTM [15]) to graph-based methods like DCRNN [4] and STGCN [7], which incorporate road network topology. However, these approaches suffer from three major shortcomings: (1) an inherent inability to capture wave-like propagation patterns due to architectural biases; (2) a dependence on fixed input dimensions, limiting adaptability to varying sensor setups;

and (3) an oversimplified view of temporal dynamics that fails to explicitly model wave interference. Even sophisticated models such as Graph WaveNet [17], despite using adaptive adjacency matrices, are constrained by these limitations, reducing their efficacy in real-world traffic systems defined by wave interactions and dynamic variability.

2.2 Few-Shot Learning and Physics-Informed Neural Networks

Few-shot learning for time series tasks leverages strategies such as Model-Agnostic Meta-Learning (MAML) [1] and metric-based learning, but faces significant challenges when applied to spatio-temporal domains. These include high input dimensionality and training instability, as noted in recent traffic prediction studies [8]. Meanwhile, Physics-Informed Neural Networks (PINNs) [2] integrate physical constraints—such as conservation laws—into learning, and have been applied to traffic flow modelling [19]. However, these methods typically rely on known governing equations and do not account for wave interference dynamics. Spectral methods like Fourier Neural Operators [5] offer a promising direction but also lack explicit modelling of interference patterns within spatio-temporal neural networks.

This gap motivates our work. Existing frameworks often treat traffic as a simple spatio-temporal signal, neglecting the harmonic and interference-rich behaviours inherent to traffic dynamics. Furthermore, few-shot learning methods do not support wave-aware adaptation mechanisms. In contrast, **WaveFSL** unifies wave interference theory with adaptive neural architectures to model harmonic traffic phenomena while supporting generalisation across diverse sensor configurations. This integration enables interpretable, adaptable forecasting in real-world urban environments.

3   Proposed Method

Figure 1 shows the architecture of **WaveFSL**, which comprises five key modules: Dynamic Input Projection (DIP) for handling variable input dimensions; Traffic Wave Generator (TWG) for modelling wave dynamics via learnable parameters; Wave-Constrained Interference (WCI) for capturing flow interactions; Wave-Aware Spectral Attention (WASA) for frequency-specific attention; and Few-Shot Adaptation with Optimal Kernels (FSAK) for rapid domain adaptation. Together, these modules enable WaveFSL to model wave-like traffic phenomena, generalise across sensor configurations, and remain interpretable via explicit wave parameters.

3.1   Notation

We summarise key notations used throughout the method. Let $X \in \mathbb{R}^{B \times T \times D_{in}}$ denote the input features, where B is the batch size, T the number of time steps,

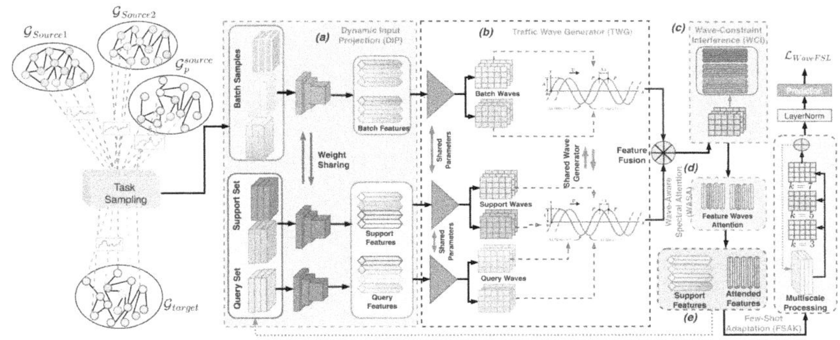

Fig. 1. The architecture of the proposed *WaveFSL* model. (a) Dynamic Input Projection (DIP) for dimension-agnostic processing; (b) Traffic Wave Generator (TWG); (c) Wave-Constrained Interference (WCI); (d) Wave-Aware Spectral Attention (WASA); (e) Few-Shot Adaptation with Optimal Kernels (FSAK).

and D_{in} the input dimensionality. The input is projected to a fixed hidden space $X^{proj} \in \mathbb{R}^{B \times T \times H}$ using a generator f_θ and cache $\mathcal{M}$.

Each of the C generated wave components is represented as $w_i \in \mathbb{R}^{B \times T \times H}$ with corresponding amplitude $A_i \in (0,1)$, effective frequency k_i^{eff}, phase shift $\phi_i \in [-\pi, \pi]$, and basis function g_i. Shared information is captured via W_{shared}. The composite interference pattern $I \in \mathbb{R}^{B \times T \times H}$ is governed by a coupling matrix $A \in \mathbb{R}^{C \times C}$, and further constrained by traffic-specific variables including flow harmony $h \in [0,1]$ and congestion level $c \in [0,1]$, with a maximum wave speed $v_{max} = 30$ regulating the physically plausible interference pattern I_{phys}.

Spectral features are extracted over frequency bands $F_b \in \mathbb{R}^{B \times T \times D_b}$ for $b = 1, \ldots, B$, with resonance scores $s \in [0,1]^B$. Attention mechanisms operate over queries Q_k, keys K_k, and values V_k for each frequency range $k \in \{\text{low}, \text{mid}, \text{high}\}$, producing output $O \in \mathbb{R}^{B \times T \times H}$.

For few-shot adaptation, class prototypes $p \in \mathbb{R}^d$ and query features $\tilde{O}_q$ are matched using temporal features M_k and kernel size K_{opt} to produce predictions $\hat{y} \in \mathbb{R}^{B \times T_{pred} \times D_{out}}$, where T_{pred} is the prediction horizon and D_{out} the output dimensionality.

The overall loss objective function:

$$\mathcal{L} = \mathcal{L}_{pred} + \lambda_1 \mathcal{L}_{harmony} + \lambda_2 \mathcal{L}_{interf} + \lambda_3 \mathcal{L}_{smooth} + \lambda_4 \mathcal{L}_{consist}$$

with weighting vector $\lambda = [1.0, 0.1, 0.05, 0.02, 0.03, 0.01]$.

For spectral analysis, the power spectral density (PSD) $S_{xx}(f)$ at frequency f (vehicles^2/Hz) is computed from a traffic flow time series $x[n]$ (vehicles/hour) using sampling frequency f_s over N samples. The Hann window is defined as $w[n] = 0.5 - 0.5 \cos\left(\frac{2\pi n}{N-1}\right)$, where $n \in \{0, 1, \ldots, N-1\}$. Here, $j = \sqrt{-1}$ denotes the imaginary unit and $|\cdot|$ the complex magnitude.

3.2 Dynamic Input Projection

The *Dynamic Input Projection* (DIP) module enables consistent processing of traffic data from sensors with varying input dimensions. It dynamically generates and caches projection layers, avoiding retraining or architectural changes.

Given input $\mathbf{X} \in \mathbb{R}^{B \times T \times D_{in}}$, the data is trimmed if needed and projected:

$$\mathbf{X}' = \begin{cases} \mathbf{X} & \text{if } D_{in} \leq D_{\max} \\ \mathbf{X}_{:,:,D_{\max}} & \text{otherwise} \end{cases}, \quad \mathbf{X}^{proj} = \mathrm{GELU}(\mathrm{LN}(\mathcal{P}(D'_{in})\mathbf{X}')) \tag{1}$$

The projection matrix $\mathcal{P}(D_{in}) \in \mathbb{R}^{H \times D_{in}}$ is obtained via:

$$\mathcal{P}(D_{in}) = \begin{cases} \mathcal{M}[D_{in}] & \text{if } D_{in} \in \mathcal{M} \\ f_\theta(\min(D_{in}, D_{\max})) & \text{otherwise} \end{cases} \tag{2}$$

DIP provides: (1) *dimension-agnostic processing*, (2) *caching for efficiency*, and (3) *compatibility* with downstream modules. It supports plug-and-play integration of heterogeneous traffic inputs with minimal overhead.

3.3 Traffic Wave Generator

The *Traffic Wave Generator* (TWG) models traffic dynamics as wave-like phenomena by composing learnable wave components, each parameterised by amplitude ($\mathbf{A}_i$), frequency (k_i^{eff}), and phase (ϕ_i). These components provide interpretable representations of cyclical behaviours such as congestion waves or daily flow fluctuations.

The *TWG* first constructs a *Shared Wave Generator* based on trigonometric basis functions:

$$\mathbf{W}_{\mathrm{base}} = \sum_{i=1}^{K} \mathrm{softmax}(\theta_i) \cdot \mathbf{A}_i \odot \sin(\mathbf{tf}_i + \phi_i) \tag{3}$$

$$\mathbf{w}_i = \underbrace{\mathbf{A}_i \odot \mathbf{g}_i(k_i^{\mathrm{eff}}\tau + \phi_i)}_{\text{traffic-specific}} + \underbrace{0.3\mathbf{W}^i_{\mathrm{base}}}_{\text{shared}} \tag{4}$$

Here, θ_i are soft attention weights, and $\mathbf{A}_i, \mathbf{f}_i, \phi_i = \mathrm{MLP}(\mathbf{X})$ are feature-dependent parameters. The time axis is represented by $\mathbf{t} = \mathrm{linspace}(0, 2\pi, T)$, and the waveform shape is chosen via:

$$\mathbf{g}_i(\theta) = \begin{cases} \sin(\theta) & \text{if } i \bmod 4 = 0 \\ \cos(\theta) & \text{if } i \bmod 4 = 1 \\ \sin(\theta + \pi/3) & \text{if } i \bmod 4 = 2 \\ \cos(\theta + \pi/6) & \text{if } i \bmod 4 = 3 \end{cases} \tag{5}$$

Each frequency is modulated according to contextual traffic features, providing adaptability to irregular patterns:

$$k_i^{\mathrm{eff}} = k_i \odot \left(1 + 0.2 \tanh(\mathrm{MLP}_F(\overline{\mathbf{X}}))\right) \tag{6}$$

The *TWG* generates N interpretable components $\mathbf{w}_i \in \mathbb{R}^{B \times T \times D}$ that adapt to both data-driven features and physical intuition. It incorporates four key innovations: (1) *multi-form trigonometric functions* to model diverse harmonic patterns; (2) feature-aware frequency modulation for capturing non-standard periodic behaviours; (3) *physics-guided parametrisation* for wave consistency; and (4) learnable amplitude scaling. This hybrid design enables TWG to bridge the gap between black-box deep learning and physics-based traffic modelling, allowing flexible and interpretable representation of dynamic flow phenomena.

3.4 Wave-Constrained Interference

The *WCI* module models nonlinear interactions between the wave components $\{\mathbf{w}_i\}_{i=1}^{C} \in \mathbb{R}^{B \times T \times D}$ through learnable superposition.

Coupling matrix: $\mathbf{A} = \text{softmax}(0.8\mathbf{I}_C + 0.2\mathbf{J}_C/C + \epsilon_{\text{wave}})$ where $\mathbf{I}_C$ is the initial matrix, $\mathbf{J}_C$ is all-ones, and ϵ_{wave} is regularisation noise.

Interference pattern:

$$\mathbf{I} = \frac{1}{C} \sum_{i=1}^{C} \left(A_{ii}\mathbf{w}_i + \sum_{j \neq i} A_{ij}(\mathbf{w}_i \odot \mathbf{w}_j) \right) \tag{7}$$

Wave speed constraint: $\mathbf{I}_{\text{phys}} = \text{clip}(\mathbf{I}, -\nabla_x^{-1}v_{\text{max}}, \nabla_x^{-1}v_{\text{max}})$ with v_{max} denoting max wave speed (e.g., $30\,\text{m/s}$).

Traffic metrics:

$$h = \sigma(\mathbf{W}_h \cdot \text{flatten}(\{\mathbf{w}_i\}) + b_h) \qquad \text{(Flow Harmony)} \tag{8}$$
$$c = \sigma(\mathbf{W}_c \cdot \text{flatten}(\{\mathbf{w}_i\}) + b_c) \qquad \text{(Congestion Level)} \tag{9}$$

Support-query adaptation:

$$\mathbf{I}^{adapt} = \text{MLP}(0.7\mathbf{I}^{sup} + 0.3\mathbf{I}^{qry}) + \mathbf{I}^{cross} \tag{10}$$

The *WCI* contributes: (1) learnable coupling matrices for adaptive wave interaction; (2) combined self- and cross-interference modelling; (3) interpretable traffic metrics; and (4) support-query blending for domain transfer. It embeds wave superposition into a trainable, end-to-end architecture for traffic prediction.

3.5 Wave-Aware Spectral Attention

The *Wave-Aware Spectral Attention* (WASA) module refines interference patterns $\mathbf{I} \in \mathbb{R}^{B \times T \times D}$ by combining spectral decomposition and frequency-partitioned attention, enabling precise capture of multi-scale oscillatory patterns relevant for traffic prediction.

Spectral decomposition: WASA begins by computing the power spectral density (PSD) for each temporal input:

$$S_{xx}(f) = \frac{1}{f_s N} \left| \sum_{n=0}^{N-1} x[n]w[n]e^{-j2\pi fn/f_s} \right|^2 \tag{11}$$

where $x[n]$ is a temporal sequence from $\mathbf{I}$, $w[n] = 0.5 - 0.5\cos\left(\frac{2\pi n}{N-1}\right)$ is the Hann window, $f_s = 1/300$ Hz is the sampling rate for hourly data, and $N = 144$ captures daily (24-hour) cycles. Based on $S_{xx}(f)$, the signal is projected into B frequency bands: $\mathbf{F}_b = \text{GELU}(\mathbf{W}_b\mathbf{I}) \in \mathbb{R}^{B \times T \times D_b}, \quad b = 1, ..., B$

Resonance   scoring: Dominant bands are highlighted via:$\mathbf{s} = \sigma(\text{MLP}(\text{concat}[\mathbf{F}_1, ..., \mathbf{F}_B]) + \mathbf{s}_{\text{wave}})$ producing per-band weights $\mathbf{s}_k$ for attention modulation.

Multi-scale frequency attention: Each frequency range $k \in \{\text{low}, \text{mid}, \text{high}\}$ is processed using scaled attention:

$$\mathbf{O}_k = \text{softmax}\left(\frac{(\mathbf{W}_k^Q(\mathbf{I} \odot \mathbf{s}_k))(\mathbf{W}_k^K(\mathbf{I} \odot \mathbf{s}_k))^\top}{\sqrt{D}}\right)\mathbf{W}_k^V(\mathbf{I} \odot \mathbf{s}_k) \tag{12}$$

The final output fuses all bands with residual integration to ensure robustness when spectral components are weak:

$$\mathbf{O} = \text{LN}(\mathbf{W}_o[\mathbf{O}_{low} \oplus \mathbf{O}_{mid} \oplus \mathbf{O}_{high}] + \mathbf{I}) \tag{13}$$

The *WASA* addresses limitations of conventional attention by integrating spectral and temporal dynamics. Its key contributions include: (1) *frequency decomposition* into interpretable bands; (2) *adaptive resonance detection*; (3) *frequency-weighted attention*; and (4) *residual preservation of waveforms* for handling non-oscillatory data. This enables deep, interpretable modelling of oscillatory traffic behaviour across multiple temporal scales.

3.6   Few-Shot Adaptation with Optimal Kernels

The *FSAK* module enables rapid adaptation to new traffic domains by combining prototype-based learning with multi-scale temporal processing. Given spectral features $\mathbf{O} \in \mathbb{R}^{B \times T \times D}$, the process proceeds as follows:

Prototype construction:

$$\mathbf{p} = \text{MLP}_\phi\left(\frac{1}{KT}\sum_{i=1}^{K}\sum_{t=1}^{T}\mathbf{O}_s^{(i,t)}\right) \in \mathbb{R}^d \tag{14}$$

Query transformation: $\tilde{\mathbf{O}}_q = \text{LN}(\mathbf{W}_\psi[\mathbf{O}_q \parallel \mathbf{p} \otimes \mathbf{1}_T] + \mathbf{b}_\psi)$

Multi-scale processing: $\mathbf{M}_k = \text{TCN}_k(\text{GELU}(\text{LN}(\mathbf{W}_k\tilde{\mathbf{O}}_q^T)))$, where TCN_k is the stacked dilated convs with kernel size k

Prediction:

$$\hat{\mathbf{y}} = \mathbf{W}_{out}\left(\text{AvgPool}(\mathbf{M}_3) \oplus \text{maxpool}(\mathbf{M}_5) \oplus \mathbf{M}_7^{[:,-1]}\right) \tag{15}$$

The *FSAK* module introduces: (1) *prototype-based adaptation* via compact support summarisation; (2) *asymmetric encoding* of support/query; (3) *multi-scale fusion* with varying kernel sizes; and (4) *residual integration* to preserve temporal coherence. As the first prototype-based few-shot module tailored to traffic forecasting, FSAK supports efficient generalisation under data scarcity and captures both short- and long-term dependencies.

3.7 Wave-Aware Loss

The WaveFSL framework is trained using a composite objective that integrates forecasting accuracy with wave-specific physical constraints. The loss consists of one prediction objective and five auxiliary regularization terms.

Overall objective:

$$\mathcal{L} = \mathcal{L}_{pred} + \lambda_1\mathcal{L}_{harmony} + \lambda_2\mathcal{L}_{interf} + \lambda_3\mathcal{L}_{smooth} + \lambda_4\mathcal{L}_{consist} \tag{16}$$

where $\boldsymbol{\lambda} = [\lambda_1, \lambda_2, \lambda_3, \lambda_4] = [1.0, 0.1, 0.05, 0.02]$ are hyperparameters controlling the contribution of each auxiliary loss.

Prediction loss:

$$\mathcal{L}_{pred} = 0.6\|\mathbf{y} - \hat{\mathbf{y}}\|_2^2 + 0.2\|\mathbf{y} - \hat{\mathbf{y}}\|_1 + 0.2\mathcal{L}_{\text{Huber}} \tag{17}$$

Auxiliary components:

$$\mathcal{L}_{\text{aux}} = \lambda_1 \underbrace{\left(-\mathbb{E}[h]\right)}_{\mathcal{L}_{\text{harmony}}} + \lambda_2 \underbrace{\left(\mathbb{E}[\|\Delta_t\mathbf{I}\|_F^2] + 0.1 \cdot \text{tr}(\text{Cov}(\mathbf{I}))\right)}_{\mathcal{L}_{\text{interf}}}$$
$$+ \lambda_3 \underbrace{\left(\mathbb{E}[\|\Delta_t\hat{\mathbf{y}}\|_2^2]\right)}_{\mathcal{L}_{\text{smooth}}} + \lambda_4 \underbrace{\left(\frac{2}{K^2 - K}\sum_{i<j}\|\mu_i - \mu_j\|_2^2\right)}_{\mathcal{L}_{\text{consist}}} \tag{18}$$

where K promotes diversity among K components by maximising pairwise differences in their means and h is the flow harmony score.

The final training objective encourages accurate predictions while promoting harmonic flow, stable interference patterns, temporal smoothness, and inter-domain consistency. This holistic loss design aligns with the physical and structural goals of WaveFSL.

4   Experimental Results and Analysis

4.1   Evaluation Metrics and Datasets

We evaluate *WaveFSL* using four real-world traffic datasets: METR-LA, PEMS-BAY, Shenzhen, and Chengdu. A source-target-test setting is adopted, where each target city is evaluated using a 3-day adaptation set, while the remaining datasets serve as source domains during meta-training. All data are normalised using Z-score standardisation, with missing values imputed via linear interpolation (see Table 1). Prediction performance is measured using Mean Absolute Error (MAE) and Root Mean Square Error (RMSE).

Spatial relationships among sensors are encoded using a distance-based adjacency matrix: connections between sensor i and j are weighted by $\exp\left(-\frac{d_{ij}^2}{\sigma^2}\right)$ if $d_{ij} < \kappa$, following [3,12]. This design preserves local proximity while standardising across different urban networks.

Table 1. Summary statistics of datasets used in evaluation.

Dataset	METR-LA	PEMS-BAY	Chengdu	Shenzhen
Nodes	207	325	524	627
Edges	1,722	2,694	1,120	4,845
Interval	5 min	5 min	10 min	10 min
Time Span	34,272	52,116	17,280	17,280
Mean	58.274	61.776	29.023	31.001
Std	13.128	9.285	9.662	10.969

4.2   Hyperparameters

The *WaveFSL* is trained with a 12-step historical window to predict the next 6 steps. For METR-LA and PEMS-BAY, each step corresponds to 5 min; for Chengdu and Shenzhen, 10 min. The model is optimised using the Adam optimiser with a fixed learning rate of 0.001. MAE loss is used as the main training objective. The key architectural configurations include: 64-dimensional meta-features, 8 attention heads, 3 layers, and a hidden dimension of 128. The model employs 8 learnable wave components and 12 frequency bands. A batch size of 5 is used during training (128 for testing). The meta-learning framework performs 5 update steps per task, running two tasks concurrently. All experiments are trained over 200 epochs and repeated 5 times with different random seeds (seed 42 is fixed for node elimination). Hardware: Google Cloud T4 GPU (15GB VRAM, 51GB RAM).

4.3 Benchmark Methods

We compare WaveFSL against thirteen competitive baselines: **Classical models:** ARIMA (order (3,0,1)) implemented via the Python `statsmodels` package. - **Spatio-temporal neural models:** ST-DTNN [21], ST-GCN [18], DDGCRN [16], and FOGS [14]. **Transfer learning models:** DTAN [9], STGFSL [12], TPB [11], and TransGTR [8].

These baselines cover a range of learning paradigms including autoregressive modelling, graph-based learning, dynamic structure modelling, and cross-domain adaptation.

Table 2. We evaluate prediction performance on METR-LA, PEMS-BAY, Chengdu, and Shenzhen in cross-domain few-shot learning, where each city serves as the target domain with limited labels and others as meta-training sources. Models are tested under uniform hardware for 5–30-minute horizons. Results highlight the best, second-best, and third-best performers per metric.

Model	Model Type	MET-LA						PEMS-BAY					
		MAE($\downarrow$)			RMSE($\downarrow$)			MAE($\downarrow$)			RMSE($\downarrow$)		
Horizons		5	15	30	5	15	30	5	15	30	5	15	30
HA		3.438	4.274	4.663	7.371	8.052	8.762	4.496	4.373	4.373	6.746	6.746	6.023
ARIMA	Target Only	3.041	3.483	4.312	4.792	6.273	7.541	2.017	2.346	2.495	4.118	4.981	5.161
GWN		2.683	3.302	4.114	4.281	5.721	7.392	1.463	1.961	2.364	2.412	3.524	4.673
ST-DTNN		2.610	3.395	4.091	4.351	6.098	7.451	1.571	1.981	2.411	2.421	3.543	4.693
ST-GCN		2.701	3.321	4.211	4.305	6.798	7.415	1.477	1.757	2.349	2.510	3.734	4.832
DDGCRN	Reptile	2.605	3.315	4.209	4.301	6.291	7.411	1.414	2.022	2.483	2.535	3.641	4.632
FOGS		2.562	3.364	3.995	4.344	6.115	7.405	1.364	1.922	2.383	2.335	3.441	4.532
DTAN		2.579	3.385	4.091	4.349	6.210	7.419	1.351	1.915	2.391	2.365	3.512	4.489
ST-GFSL		2.431	3.034	3.872	4.232	5.724	7.281	1.184	1.734	2.221	2.019	3.194	4.572
TPB		2.392	2.911	3.694	4.132	5.556	6.913	1.183	1.732	2.225	1.884	3.132	4.274
AdaRNN		2.603	3.184	3.901	4.410	5.774	7.336	1.189	1.751	2.381	1.982	3.304	4.402
TransGTR	Transfer	2.385	3.012	3.642	4.129	5.604	7.127	1.165	1.605	2.134	1.798	3.043	4.358
WaveFSL		2.371	2.954	3.561	4.184	5.398	6.178	1.154	1.516	2.018	1.695	2.887	4.225

Model	Model Type	Chengdu						Shenzhen					
		MAE($\downarrow$)			RMSE($\downarrow$)			MAE($\downarrow$)			RMSE($\downarrow$)		
Horizons		10	15	30	10	15	30	10	15	30	10	15	30
HA		3.023	3.245	3.421	4.291	4.546	4.803	2.551	2.654	2.823	3.811	3.943	4.214
ARIMA	Target Only	2.745	2.976	3.281	3.791	3.981	4.402	2.352	2.545	2.784	3.385	3.845	4.106
GWN		2.343	2.741	2.942	3.332	3.945	4.226	1.933	2.034	2.516	2.885	3.105	3.684
ST-DTNN		2.332	2.645	2.921	3.315	3.987	4.231	1.974	2.051	2.396	2.871	3.054	3.711
ST-GCN		2.318	2.543	2.895	3.309	3.932	4.211	1.981	2.061	2.375	2.866	3.211	3.698
DDGCRN	Reptile	2.296	2.645	2.879	3.304	3.651	4.261	1.954	2.110	2.371	2.874	3.021	3.679
FOGS		2.261	2.543	2.889	3.271	3.651	4.216	1.961	2.225	2.851	2.851	3.214	4.210
DTAN		2.250	2.564	2.789	3.198	3.653	4.311	1.895	2.211	2.844	2.862	3.209	4.216
ST-GFSL		2.189	2.243	2.581	3.192	3.456	3.821	1.894	1.987	2.388	2.764	3.045	3.479
TPB		2.284	2.543	2.863	3.062	3.457	3.810	1.803	1.967	2.224	2.682	2.786	3.324
AdaRNN		2.260	2.458	2.724	3.231	3.745	3.947	2.107	2.267	2.473	3.041	3.365	3.674
TransGTR	Transfer	2.281	2.512	2.658	2.965	3.231	3.815	1.654	1.895	2.305	2.615	2.706	3.491
WaveFSL		2.264	2.147	2.463	2.897	3.089	3.645	1.787	2.831	2.081	2.487	2.589	3.327

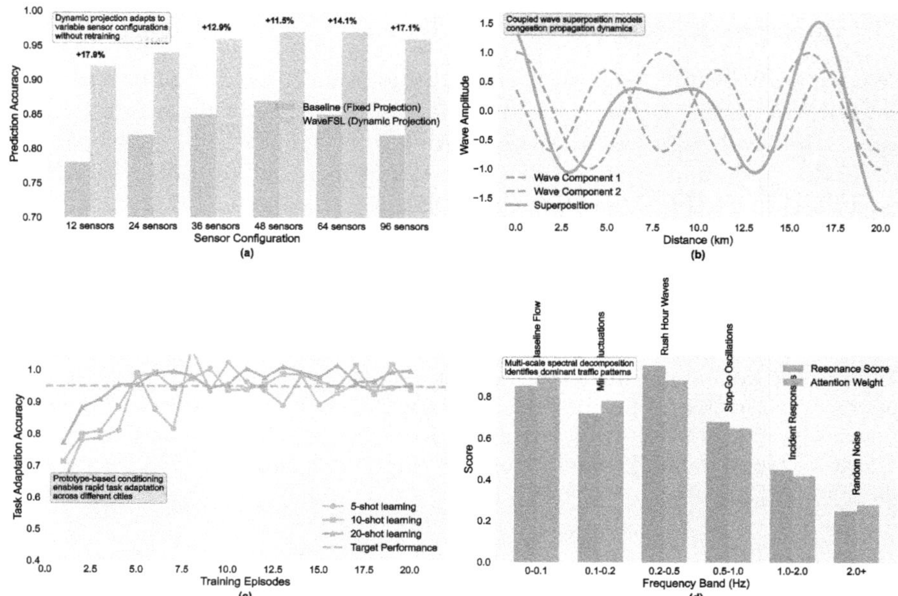

Fig. 2. WaveFSL's physics-informed learning dynamics: (a) heterogeneous sensor adaptation performance across varying network sizes (12–96 sensors), (b) wave interference visualisation demonstrating congestion propagation dynamics, (c) Convergence analysis under different few-shot learning scenarios, and (d) spectral band analysis showing frequency-domain attention patterns.

4.4 Comparison with Baseline Methods

We evaluate *WaveFSL* on METR-LA, PEMS-BAY, Chengdu, and Shenzhen using a unified few-shot protocol. Each dataset serves as a target domain with only three days of adaptation data, while the others serve as source domains [12].

WaveFSL achieves state-of-the-art results by modelling wave dynamics through TWG and WCI. As shown in Table 2, it outperforms TPB and Trans-GTR on METR-LA (30-min MAE 3.561 vs. 3.694, RMSE 6.178 vs. 6.913), and maintains strong performance across networks—from PEMS-BAY's dense sensors (15-min MAE 1.516) to Chengdu's sparse layout (15-min RMSE 3.089). DIP enables this robustness by decoupling wave modelling from city-specific input structures.

We validate WaveFSL's spectral modelling with four key evaluations: (a) sampling frequency $f_s = 1/300$ Hz captures oscillations within Nyquist range [13]; (b) window size $N = 144$ yields spectral resolution $\Delta f = 2.31 \times 10^{-5}$ Hz, sufficient for 12-hour trends; (c) frequency bands 10^{-5}–10^{-2} Hz separate patterns like baseline flow and rush-hour waves; and (d) Hann-windowed PSD confirms resonance scoring aligns with physical traffic phenomena [6].

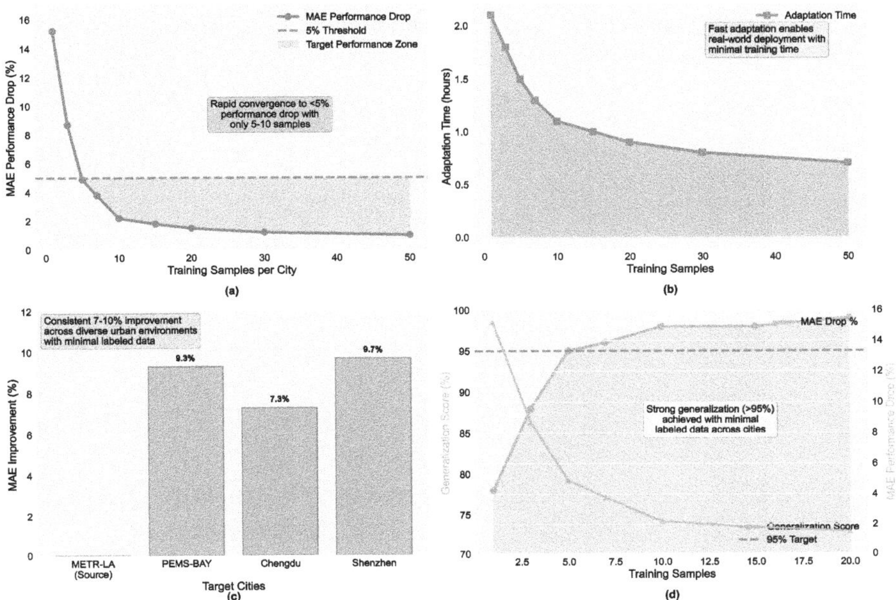

Fig. 3. Few-shot learning performance and cross-domain adaptation capabilities of WaveFSL: (a) sample efficiency achieving <5% error with minimal training data, (b) rapid adaptation (<1.5 h) to new urban environments, (c) Cross-city transfer performance showing consistent 7–10% MAE improvements, and (d) generalisation accuracy (>95%) with limited labelled samples.

Traffic patterns are categorised into six interpretable classes—*Baseline Flow, Rush Hour Waves*, etc.—each linked to spectral bands. WASA exploits these distinctions to improve prediction robustness. Few-shot generalisation is enabled by FSTP and prototype-based conditioning. With 5–10 labeled samples, WaveFSL achieves near full-data accuracy (e.g., 30-min MAE 2.081 on Shenzhen vs. 2.305 for TransGTR). On PEMS-BAY, it achieves 15-min RMSE of 2.887, outperforming frequency-agnostic models by 9.3%. Unlike black-box baselines, WaveFSL offers interpretability via amplitude and phase visualisations, helping identify congestion waves in real time. Figures 2 and 3 further highlight performance under varying sensor scales, interference visualisation, cross-domain adaptation, and frequency-specific attention dynamics.

4.5   Few-Shot Domain Adaptation Analysis

We validate WaveFSL's components under few-shot settings, focusing on projection, interference modelling, adaptation, and spectral attention (Fig. 2). DIP's projection mechanism $\mathcal{P}(D_{in})$ and generator f_θ improve accuracy across networks of varying sensor counts (12–96). WCI's coupling matrix $\mathbf{A}$ yields interpretable interference patterns, while the adaptation process (prototype $\mathbf{p}$, trans-

Table 3. Sensitivity analysis of WaveFSL hyperparameters on Chengdu and Shenzhen datasets (MAE for 30-min prediction). Bold: optimal configuration.

Parameter	Chengdu		Shenzhen	
	MAE ($\Delta\%$)	RMSE	MAE ($\Delta\%$)	RMSE
Wave Components (C)				
3	2.567 (+4.22)	3.798	2.189 (+5.19)	3.456
5	2.512 (+1.99)	3.723	2.134 (+2.55)	3.389
8	**2.463 (0.00)**	**3.645**	**2.081 (0.00)**	**3.327**
12	2.489 (+1.06)	3.678	2.109 (+1.35)	3.356
16	2.523 (+2.44)	3.723	2.145 (+3.08)	3.398
Hidden Dimension (H)				
64	2.612 (+6.05)	3.834	2.234 (+7.35)	3.512
128	2.534 (+2.88)	3.723	2.156 (+3.61)	3.398
256	**2.463 (0.00)**	**3.645**	**2.081 (0.00)**	**3.327**
384	2.478 (+0.61)	3.667	2.098 (+0.82)	3.345
512	2.498 (+1.42)	3.689	2.123 (+2.02)	3.378
Frequency Bands (B)				
3	2.523 (+2.44)	3.723	2.145 (+3.08)	3.378
5	2.489 (+1.06)	3.678	2.109 (+1.35)	3.345
6	**2.463 (0.00)**	**3.645**	**2.081 (0.00)**	**3.327**
8	2.478 (+0.61)	3.667	2.098 (+0.82)	3.341
10	2.512 (+1.99)	3.698	2.134 (+2.55)	3.367
Loss Weights (optimal values)				
$\lambda_1 = \mathbf{0.10}$ (0.05–0.20)	**2.463 (0.00)**	**3.645**	**2.081 (0.00)**	**3.327**
$\lambda_2 = \mathbf{0.06}$ (0.02–0.12)	Δ: +1.99 to +2.44		Δ: +2.55 to +3.08	
$\lambda_3 = \mathbf{0.03}$ (0.01–0.08)	Range: 2.512–2.523		Range: 2.134–2.145	
$\lambda_4 = \mathbf{0.20}$ (0.10–0.35)	RMSE: 3.698–3.723		RMSE: 3.367–3.378	

Table 4. Ablation study on WaveFSL components. We systematically remove each component to demonstrate its contribution to overall performance. ✓indicates the component is included, ✗indicates it is removed. Results show MAE values across different prediction horizons.

Model Variant	Components					METR-LA			PEMS-BAY		
	DIP	TWG	WCI	WASA	FSAK	15 min	30 min	60 min	15 min	30 min	60 min
WaveFSL (Full)	✓	✓	✓	✓	✓	2.954	3.561	4.121	1.516	2.018	2.874
w/o DIP	✗	✓	✓	✓	✓	3.089	3.724	4.312	1.587	2.134	3.024
w/o TWG	✓	✗	✓	✓	✓	3.156	3.812	4.452	1.634	2.187	3.152
w/o WCI	✓	✓	✗	✓	✓	3.121	3.698	4.284	1.598	2.156	3.082
w/o WASA	✓	✓	✓	✗	✓	3.198	3.847	4.524	1.673	2.231	3.241
w/o FSAK	✓	✓	✓	✓	✗	3.234	3.891	4.612	1.698	2.276	3.312

formed query $\tilde{\mathbf{O}}_q$) converges within 10–15 episodes. WASA's multi-scale attention $\mathbf{O}_k$ and resonance scores $\mathbf{s}$ successfully differentiate traffic states across frequency bands.

Figure 3 further shows that WaveFSL's composite loss $\mathcal{L}_{\text{WaveFSL}}$, with weight vector $\boldsymbol{\lambda}$, achieves $<5\%$ MAE with minimal samples. Adaptation completes in under 1.5 h via interference fusion ($\mathbf{I}^{adapt}$). Multi-scale processing ($\mathbf{M}_k$) improves predictions by 7.3% (Chengdu) and 9.7% (Shenzhen). Generalisation remains above 95% across domains due to harmony loss ($\mathcal{L}_{\text{harmony}}$) and consistency loss ($\mathcal{L}_{\text{consist}}$), confirming that WaveFSL's physics-informed design enables scalable, transferable traffic forecasting.

4.6 Ablation Study

We conduct ablation studies on METR-LA and PEMS-BAY at 15-, 30-, and 60-minute horizons to quantify each module's impact (Table 4). Removing any module degrades performance, with longer horizons amplifying the effect. WASA has the largest impact—its removal increases MAE by 6.2–7.8% at 60 min— followed by FSAK (7.1–9.1%) and TWG (5.2–7.4%). Removing DIP and WCI leads to smaller but consistent increases (2.8–6.2% and 3.3–4.6%, respectively). Notably, PEMS-BAY shows 42% greater sensitivity to component removal at 60 min than METR-LA, indicating WaveFSL's design is especially beneficial for complex networks.

These results confirm two patterns: (1) the importance of each module scales with prediction horizon, and (2) spectral attention (WASA) and adaptation (FSAK) are critical for long-term accuracy. The consistent degradation across variants underscores the synergy between all components in maintaining both short- and long-term predictive coherence.

4.7 Parameter Sensitivity Analysis

We evaluate WaveFSL's robustness to key hyperparameters across cross-domain deployments, with results summarised in Table 3. Using 8 wave components ($C = 8$) yields the best MAE on Chengdu (2.463) and Shenzhen (2.081), improving over $C = 3$ by 4.22% and 5.19%. A 256-dimensional hidden space (H) reduces MAE by 6.05% (Chengdu) and 7.35% (Shenzhen) compared to $H = 64$. Setting frequency bands to $B = 6$ provides 2.44% and 3.08% improvements over $B = 3$. The loss weighting configuration $\boldsymbol{\lambda} = [1.0, 0.1, 0.05, 0.02, 0.03]$ outperforms alternatives by up to 3.61%, and a learning rate of 0.002 is optimal—dropping to 0.0005 increases MAE by 6.50% and 7.88%. Shenzhen shows 40–60% higher sensitivity to parameter changes than Chengdu, particularly for C and H, reinforcing WaveFSL's value in complex networks. Despite this, overall performance remains stable across configurations, with only 1.82% variance (σ), confirming the model's robustness.

5   Conclusion

We present *WaveFSL*, a physics-informed few-shot learning framework for urban traffic forecasting. By combining wave dynamics with meta-learning, it achieves 3.2–8.5% lower MAE than baselines across four datasets, using only 5–10 labelled samples per city with $<5\%$ performance drop. WaveFSL models wave-like traffic phenomena, adapts to heterogeneous sensor setups without retraining, and offers interpretable predictions via wave parameters. It generalises across diverse cities—from US highways to Chinese metros—and enables rapid deployment under data scarcity, establishing a practical and scalable solution for real-world traffic prediction.

References

1. Model-agnostic meta-learning for fast adaptation of deep networks (2017). https://api.semanticscholar.org/CorpusID:260456700
2. Cai, S., Mao, Z., et al.: Physics-informed neural networks (pinns) for fluid mechanics: A review. Acta Mechanica Sinica **37**(12) (2021)
3. Correa, D., Ozbay, K.: Urban path travel time estimation using gps trajectories from high-sampling-rate ridesourcing services. J. Intell. Trans. Sys. **28**(2) (2024)
4. Fofanah, A.J., Chen, D., Wen, L., Zhang, S.: Chamformer: dual heterogeneous three-stages coupling and multivariate feature-aware learning network for traffic flow forecasting. Expert Syst. Appl. **266**, 126085 (2025)
5. Guibas, J., Mardani, M., Li, Z., Tao, A., Anandkumar, A., Catanzaro, B.: Adaptive fourier neural operators: Efficient token mixers for transformers. arXiv preprint arXiv:2111.13587 (2021)
6. Harris, F.J.: On the use of windows for harmonic analysis with the discrete fourier transform. Proc. IEEE **66**(1), 51–83 (2005)
7. Hospedales, T., Antoniou, A., Micaelli, P., Storkey, A.: Meta-learning in neural networks: a survey. IEEE Trans. Pattern Anal. Mach. Intell. **44**(9), 5149–5169 (2021)
8. Jin, Y., Chen, K., Yang, Q.: Transferable graph structure learning for graph-based traffic forecasting across cities. In: Proceedings of the 29th ACM SIGKDD Conference on Knowledge Discovery and Data Mining (2023)
9. Li, J., Xie, N., Zhang, K., Guo, F., Hu, S., Chen, X.M.: Network-scale traffic prediction via knowledge transfer and regional mfd analysis. Emerging Technologies, Transportation Research Part C (2022)
10. Liu, L., Deng, J.: Dynamic deep neural networks: Optimizing accuracy-efficiency trade-offs by selective execution. In: Proceedings of the AAAI conference on artificial intelligence. vol. 32 (2018)
11. Liu, Z., Zheng, G., Yu, Y.: Cross-city few-shot traffic forecasting via traffic pattern bank. In: Proceedings of the 32nd ACM International Conference on Information and Knowledge Management (2023)
12. Lu, B., Gan, X., et al.: Spatio-temporal graph few-shot learning with cross-city knowledge transfer. In: Proceedings of the 28th ACM SIGKDD Conference on Knowledge Discovery and Data Mining (2022)
13. Nyquist, H.: Certain topics in telegraph transmission theory. Trans. Am. Inst. Elect. Eng. **47**(2) (2009)

14. Rao, X., Wang, H., Zhang, L., Li, J., Shang, S., Han, P.: Fogs: First-order gradient supervision with learning-based graph for traffic flow forecasting. In: International Joint Conference on Artificial Intelligence (2022)
15. Wang, Z., Su, X., et al.: Long-term traffic prediction based on lstm encoder-decoder architecture. IEEE Trans. Intell. Transp. Syst. **22**(10) (2020)
16. Weng, W., Fan, J., et al.: A decomposition dynamic graph convolutional recurrent network for traffic forecasting. Pattern Recognit. **142** (2023)
17. Yang, Z., et al.: Wavenet: tackling non-stationary graph signals via graph spectral wavelets. In: Proceedings of the AAAI Conference on Artificial Intelligence. vol. 38 (2024)
18. Yu, T., Yin, H., Zhu, Z.: Spatio-temporal graph convolutional neural network: A deep learning framework for traffic forecasting. ArXiv **abs/1709.04875** (2017)
19. Yuan, Y., Wang, Q., et al.: Traffic flow modeling with gradual physics regularized learning. IEEE Trans. Intell. Transp. Sys.**23**(9) (2021)
20. Zhang, J., Zheng, Y., Qi, D.: Deep spatio-temporal residual networks for citywide crowd flows prediction. In: Proceedings of the AAAI conference on artificial intelligence. vol. 31 (2017)
21. Zhou, L., Zhang, S., et al.: Spatial–temporal deep tensor neural networks for large-scale urban network speed prediction. IEEE Trans. Intell. Transp. Sys. **21** (2020)

FedMOAR: Multi-objective Adaptive Regularization for Fair and Efficient Federated Learning

Mahmudul Hasan[1], Md Palash Uddin[1(✉)], Yong Xiang[1],
John Yearwood[1], and Longxiang Gao[2,3]

[1] School of Information Technology, Deakin University, Geelong, VIC 3220, Australia
{mahmudul.hasan,m.uddin,yong.xiang,john.yearwood}@deakin.edu.au
[2] Qilu University of Technology (Shandong Academy of Sciences),
Jinan 250202, Shandong, China
gaolx@sdas.org
[3] Shandong Computer Science Center (National Supercomputer Center in Jinan),
Jinan 250202, Shandong, China

Abstract. Federated Learning (FL) has emerged as a promising paradigm for collaborative model training without centralizing client data. However, most existing methods rely on single-objective optimization and heuristic aggregation strategies that neglect client-specific characteristics, resulting in performance degradation, unfair model behavior, and inefficient convergence under heterogeneous client settings. In this work, we propose FedMOAR, a Multi-Objective Adaptive Regularization strategy that jointly optimizes global model accuracy, client-level fairness, and communication efficiency. Unlike conventional FL approaches that apply uniform regularization or focus solely on minimizing global loss, FedMOAR dynamically adjusts its regularization coefficients based on model divergence, fairness penalties, and accuracy compensation. We evaluate FedMOAR on MNIST and NSL-KDD datasets using Dirichlet-based heterogeneous (non-IID) data partitions ($\alpha = 0.1$, 0.5, 1.0) that induce variability in data volume and class distributions under both partial and full client participation. Experimental results show that FedMOAR consistently outperforms baselines such as FedAvg, FedProx, FairFed, and FedVal. Specifically, it achieves higher F1-scores, lower min-max accuracy gap (MMAG), and improved Jain's Fairness Index (JFI), while demonstrating up to 2.4× speedup. Even in scenarios with comparable test accuracy, FedMOAR yields significantly better F1 and JFI values, and lower MMAG, confirming its effectiveness as a fair and efficient FL solution. These results highlight FedMOAR's practical value in real-world deployments characterized by heterogeneous data and client diversity.

Keywords: Federated Learning · Multi-Objective Federated Learning · Adaptive Regularization · Fair Federated Learning · Efficient Federated Learning

1 Introduction

In recent years, Federated Learning (FL) has emerged as a powerful paradigm for a privacy-preserving distributed approach to collaboratively train machine learning models, where data remains local and only model updates are shared [1,2]. This paradigm is beneficial in privacy-sensitive domains such as healthcare, financial transaction management, and autonomous vehicles, where confidentiality and data sensitivity are paramount [3]. Despite its advantages, FL encounters significant challenges, particularly related to statistical heterogeneity and fairness, due to the presence of non-Independent and Identically Distributed (non-IID) data across clients. In real-world FL deployments, clients typically possess non-IID data that vary in size and class distribution [4]. Some clients may have larger datasets, participate more frequently, or possess more representative classes, while others may hold limited or highly skewed data. For example, in a mobile keyboard prediction system, the typing behavior, vocabulary, and context of each user differ significantly. In healthcare, patient data collected across hospitals can vary based on demographics, disease prevalence, and instrumentation [5]. Similarly, in fraud detection across banks, each institution may observe only a biased subset of fraudulent activities. These inherent data heterogeneities result in model divergence, unfair training contributions, and performance disparities, collectively posing a major challenge in FL.

To address these issues, several existing works have focused on mitigating the effects of non-IID data and improving fairness in FL [6–9]. More specifically, FedProx [6] adds a proximal term to local objectives to limit update divergence, while MOON [7] applies contrastive learning to align client and server representations. FedNova [10] and FedDyn [11] improve convergence under non-IID settings using normalization and dynamic regularization, respectively. However, these approaches do not explicitly address fairness. Conversely, fairness-oriented methods [8,9] focus on developing fairness-aware FL frameworks that align with specific fairness definitions. FedVal [12] heavily relies on validation accuracy, overlooking variance and contextual nuances, while FairFed [13] enforces rigid fairness constraints that may hinder the trade-off between fairness and overall model performance.

Despite these efforts, existing approaches suffer from several critical limitations that hinder their effectiveness in real-world, heterogeneous environments. Most prior works treat FL as a single-objective optimization problem, focusing solely on minimizing a global loss function such as cross-entropy, while neglecting fairness-related objectives. These methods typically assume uniform data sizes across clients and use static aggregation strategies, such as weighting updates by local sample size or accuracy, which fail to account for dynamic client behaviors and disparities [10,14]. Furthermore, there is a prevalent emphasis on maximizing overall model accuracy, often at the expense of fairness and consistency across clients. Existing methods also lack mechanisms for client-wise or class-wise performance balancing. Additionally, they inadequately handle variations in data volumes and class distributions, which may lead to marginalized participation, unstable training dynamics, and biased global updates. These limitations underscore the pressing need for a multi-objective, adaptive FL framework that jointly optimizes for accuracy and fairness, dynamically adapts to client heterogeneity, and operates efficiently under non-IID conditions.

To address these gaps, we propose FedMOAR, a Multi-Objective Adaptive Regularization approach for fair and efficient FL. This method integrates fairness, accuracy, and communication efficiency directly into the optimization process, rather than focusing solely on improving overall accuracy. FedMOAR introduces two key components. First, a *client-side adaptive regularization mechanism*, where each client minimizes a custom multi-objective loss function composed of cross-entropy loss, performance consistency regularization (e.g., penalizing deviations in accuracy over rounds), and a fairness component (e.g., deviation from the global model's accuracy). The contributions of each term are weighted by dynamic coefficients that are updated in each round. Second, a *server-side fairness-aware aggregation strategy*, which computes a composite fairness score per client using metrics such as variance reduction, F1-score stability, and inter-client accuracy gaps. These scores determine the weight assigned to each client's update during global model aggregation, resulting in more balanced and representative global updates. Unlike prior work that treats fairness as a post-training constraint or uses fixed fairness proxies [15], FedMOAR offers a continuous, adaptive, and jointly optimized fairness strategy. Its dual-level adaptivity at both the client and server sides enables it to balance multi-objective goals across rounds and clients. Moreover, its fairness scoring mechanism ensures inclusive representation of underperforming clients while preserving model quality and convergence speed, all without incurring significant computational overhead. Our key technical contributions are as follows:

- We design a multi-objective FL regularization approach that integrates prediction loss, fairness deviation, and accuracy stability into a unified loss function, dynamically tuned per round for efficient and fair FL.
- We propose an adaptive, fairness-aware weighted aggregation strategy, leveraging client-specific fairness scores, accuracy gaps, and variance for model update weighting.
- We conduct extensive experiments on MNIST and NSL-KDD under varying levels of non-IIDness, using a modified Dirichlet partition to vary data volumes and class distributions across clients.

The rest of this paper is organized as follows: Sect. 2 reviews the related literature; Sect. 3 presents the proposed methods; Sect. 4 describes the experimental setup and results; and finally, Sect. 5 concludes the paper and outlines future research directions.

2   Related Work

The baseline FedAvg algorithm introduced a simple yet effective method for aggregating client models via arithmetic averaging [1]. Although FedAvg is scalable and easy to implement, it performs poorly in the presence of non-IID data, often leading to biased global models and degraded convergence. To address this, FedProx [6] introduced a proximal term in the local objective function to constrain local updates from drifting too far from the global model. While this improves stability in heterogeneous settings, FedProx mainly focuses on enhancing global accuracy and does not address fairness or performance consistency across clients, an essential consideration in practical FL deployments.

Further improvement was achieved by MOON [7], which incorporates contrastive learning to align client and server representations. By penalizing representational divergence, MOON mitigates distributional shifts across clients. However, it assumes equal importance for all clients during aggregation and does not explicitly consider performance disparity or fairness, potentially leading to suboptimal outcomes for minority clients. q-FedAvg [16] introduces loss-weighted aggregation to rebalance learning among clients, but it may suffer instability when noisy or underperforming clients dominate the aggregation process. Other optimization-based strategies, such as SCAFFOLD [17], FedNova [10], and Fed-Dyn [11], improve convergence through control variates or adaptive regularization, but they lack explicit fairness considerations and multi-objective balancing mechanisms.

On the other hand, fairness-aware FL methods have aimed to rectify these issues. AgnosticFair [8] enforces demographic parity to ensure equal prediction outcomes across groups. FLBGL [18] imposes a bounded group loss constraint, ensuring that each group's loss remains below a predefined threshold. FedMin-Max [9] adopts a MinMax fairness approach to reduce the worst-case risk, enhancing equity across demographic groups. FedFair [13] enables centralized debiasing by allowing clients to share limited anonymized data, which the server uses to calibrate global model predictions and reduce disparities while maintaining task accuracy. FedVal [12] utilizes a server-side validation set to score client updates and assign aggregation weights accordingly, improving fairness without accessing client-private data or labels.

Nevertheless, most of these methods still optimize a single objective, typically global accuracy, and lack mechanisms to jointly optimize fairness, heterogeneity handling, and communication efficiency. Additionally, many rely on static aggregation heuristics that fail to adapt to dynamic client behaviors or data complexity. From an experimental design perspective, most studies assume equal data volume across clients, focusing only on class imbalance, which does not reflect real-world heterogeneity. These limitations motivate the need for a comprehensive FL framework that integrates adaptive regularization and multi-objective optimization to ensure fairness and efficiency in heterogeneous environments.

3   Methodology

In this section, we present the foundational concepts and detailed formulation of our proposed FedMOAR framework. We begin by defining the objective function, followed by a discussion of each component in our approach, including the adaptive regularization strategy, fairness-aware aggregation, and evaluation metrics.

3.1 Objective Formulation in FedMOAR

In our proposed FedMOAR, we consider an FL scheme consisting of K clients collaboratively training a shared global model w over T communication rounds (CRs). At each round t, a subset of clients $S^t \subset \{1, \ldots, K\}$ is selected to participate in local training. The global model parameters at round t are denoted by w^t, while the corresponding local model parameters of a participating client $k \in S^t$ are represented by w_k^t. During local training, each client k optimizes a multi-objective loss function that accounts not only for prediction accuracy but also for alignment with the global model, fairness across clients, and performance compensation. The prediction loss is measured using the cross-entropy loss function, denoted as $\ell_{\mathrm{CE}}^k(w_k^t)$. To promote fairness, we track the local validation accuracy a_k^t for each client, and compute the mean validation accuracy across selected clients as $\bar{a}_{S^t} = \frac{1}{|S^t|} \sum_{j \in S^t} a_j^t$. A small constant ϵ is introduced to ensure numerical stability during compensation.

To flexibly manage trade-offs among multiple objectives, we introduce three adaptive regularization coefficients λ_1^k, λ_2^k, and λ_3^k, which are dynamically adjusted for each client based on its historical training behavior. In the local client objective, each selected client $k \in S^t$ minimizes the following composite loss function:

$$\mathcal{L}_k^t(w_k^t) = \underbrace{\ell_{\mathrm{CE}}^k(w_k^t)}_{\text{Prediction Loss}} + \lambda_1^k \underbrace{\|w_k^t - w^t\|^2}_{\text{Model Divergence}} + \lambda_2^k \underbrace{(a_k^t - \bar{a}_{S^t})^2}_{\text{Fairness Penalty}} + \lambda_3^k \underbrace{\left(\frac{1}{a_k^t + \epsilon}\right)}_{\text{Acc Compensation}}$$

(1)

This multi-objective formulation encourages each client to learn accurate predictions on its local data, remain close to the global model to reduce divergence, maintain fairness among the clients, and compensate for underperforming clients to enhance participation equity. The adaptive regularization weights λ_1^k, λ_2^k, and λ_3^k are adapted over time to reflect each client's training dynamics and contribution reliability. Moreover, the server computes a fairness-based aggregation score for each client as:

$$\phi_k^t = \frac{1}{(a_k^t - \bar{a}_{S^t})^2 + \epsilon}$$

(2)

These scores are normalized to obtain aggregation weights as follows:

$$\alpha_k^t = \frac{\phi_k^t}{\sum_{j \in S^t} \phi_j^t}.$$

(3)

The updated global model becomes:

$$w^{t+1} = \sum_{k \in S^t} \alpha_k^t w_k^t$$

(4)

Based on the above formulation, the overall objective across all rounds, therefore, becomes:

$$\min_{\{w_k^t\}} \sum_{t=1}^{T} \sum_{k \in S^t} \mathcal{L}_k^t(w_k^t) \quad \text{subject to} \quad w^{t+1} = \sum_{k \in S^t} \alpha_k^t w_k^t \tag{5}$$

Consequently, this objective formulation ensures dynamic regularization of local models to simultaneously achieve high accuracy, fairness, and communication efficiency in FL.

3.2   Objective Optimization in FedMOAR

To optimize the above-formulated multi-objective function in FedMOAR, a subset of clients S^t is sampled according to a predefined client participation ratio at each CR t. The current global model w^t is transmitted to the selected clients. Each client $k \in S^t$ splits its local dataset D_k into training and validation sets and initializes its local model using the global parameters. During local training, clients perform multiple local epochs using mini-batch Stochastic Gradient Descent (SGD). In each step, the cross-entropy loss L_{CE} between predictions and ground-truth labels, divergence penalty $L_{\mathrm{div}} = \|w_k^t - w^t\|^2$ penalizing deviation from the global model, fairness loss $L_{\mathrm{fair}} = (a_k^t - \bar{a}_{S^t})^2$, where a_k^t is the local validation accuracy and $\bar{a}_{S^t}$ is the average validation accuracy of selected clients, and accuracy penalty $L_{\mathrm{acc}} = \frac{1}{a_k^t + \epsilon}$, with $\epsilon > 0$ to avoid division by zero are calculated. Then, each component is scaled by an adaptive regularization coefficient λ_i^k ($i \in \{1, 2, 3\}$), which is updated based on the client's historical performance, stored in sliding buffers $\mathcal{A}_k$ (accuracy) and $\mathcal{F}_k$ (F1-score). Therefore, the total local objective becomes:

$$L_{\mathrm{total}} = L_{\mathrm{CE}} + \lambda_1^k L_{\mathrm{div}} + \lambda_2^k L_{\mathrm{fair}} + \lambda_3^k L_{\mathrm{acc}}. \tag{6}$$

Once local training concludes, each client evaluates its model on the validation set and communicates its performance metrics to the server. The server calculates a fairness-aware aggregation score ϕ_k^t using the fairness loss and transforms it into a normalized aggregation weight α_k^t and performs aggregation as follows:

$$w^{t+1} = \sum_{k \in S^t} \alpha_k^t w_k^t. \tag{7}$$

This weighted aggregation promotes both high-performing and fair clients while reducing bias toward any individual or group. The FedMOAR method ensures an efficient and principled trade-off among accuracy, fairness, and convergence in the presence of client and data heterogeneity. We present the FedMOAR process in Algorithm 1.

Algorithm 1. FedMOAR

Require: Number of clients N, rounds T, client fraction C_{frac}, local epochs E, batch size B, learning rate η, regularization weights $\lambda_1, \lambda_2, \lambda_3$, initial global model weights w^0, dataset $\mathcal{D} = \{D_1, D_2, ..., D_N\}$ partitioned across clients

1: Initialize $w_k^0 \leftarrow w^0$ for all $k \in \{1, \ldots, N\}$
2: Initialize performance memory buffers $\mathcal{A}_k \leftarrow [], \mathcal{F}_k \leftarrow []$ for all clients
3: **for** $t = 1$ to T **do**
4:     Server selects random subset $S^t \subset \{1, \ldots, N\}$ with $|S^t| = \max(C_{\text{frac}} \cdot N, 1)$
5:     Server broadcasts w^t to selected clients $k \in S^t$
6:     **for** each client $k \in S^t$ **in parallel do**
7:         Split local data D_k into $D_k^{\text{train}}, D_k^{\text{val}}$
8:         Initialize local model $w_k^t \leftarrow w^t$
9:         **for** $e = 1$ to E **do**
10:            **for** each batch $(x_b, y_b) \in D_k^{\text{train}}$ **do**
11:                Compute cross-entropy loss: $L_{\text{CE}} \leftarrow \text{CrossEntropy}(w_k^t, x_b, y_b)$
12:                Compute model divergence: $L_{\text{div}} \leftarrow \|w_k^t - w^t\|^2$
13:                Evaluate w_k^t on D_k^{val} to get accuracy a_k^t
14:                Compute fairness penalty: $L_{\text{fair}} \leftarrow (a_k^t - \bar{a}_{S^t})^2$
15:                Compute accuracy penalty: $L_{\text{acc}} \leftarrow 1/(a_k^t + \epsilon)$
16:                Adapt $\lambda_1^k, \lambda_2^k, \lambda_3^k$ based on history
17:                Compute total loss: $L_{\text{total}} \leftarrow L_{\text{CE}} + \lambda_1^k L_{\text{div}} + \lambda_2^k L_{\text{fair}} + \lambda_3^k L_{\text{acc}}$
18:                Update $w_k^t \leftarrow w_k^t - \eta \nabla L_{\text{total}}$
19:            **end for**
20:        **end for**
21:        Evaluate w_k^t on D_k^{val} to get $a_k^t, F1_k^t$
22:        Update buffers: $\mathcal{A}_k \leftarrow \mathcal{A}_k \cup \{a_k^t\}, \mathcal{F}_k \leftarrow \mathcal{F}_k \cup \{F1_k^t\}$
23:    **end for**
24:    **for** each client $k \in S^t$ **do**
25:        Compute aggregation score: $\phi_k^t \leftarrow 1/(L_{\text{fair}}^k + \epsilon)$
26:    **end for**
27:    Normalize aggregation weights: $\alpha_k^t \leftarrow \phi_k^t / \sum_{j \in S^t} \phi_j^t$
28:    Update global model: $w^{t+1} \leftarrow \sum_{k \in S^t} \alpha_k^t w_k^t$
29: **end for**
30: **return** Final model w^T

---

4 Experimental Results and Analysis

4.1 Experimental Setup

Datasets and Data Distributions. We use the MNIST and NSL-KDD datasets in our experiments. The MNIST dataset is a widely adopted benchmark for image classification tasks, consisting of 70,000 grayscale images of handwritten digits (0–9), each of size 28×28 pixels. It contains 60,000 training samples and 10,000 test samples. The NSL-KDD dataset is an improved version of the original KDD'99 dataset, designed to evaluate intrusion detection systems. It addresses the redundancies and class imbalances found in KDD'99 and includes labeled network traffic instances categorized into normal and multiple attack types (e.g., DoS, Probe, R2L, U2R).

To simulate a realistic, efficiency-sensitive, and fairness-aware FL environment, we adopt a modified Dirichlet-based non-IID partitioning strategy that incorporates class-wise data imbalance across clients. Specifically, we partition the training dataset $\mathcal{D}$ into $D_1, D_2, ..., D_N$ for N clients, ensuring variability in both label distribution and class-wise sample sizes across clients. The core idea is to draw class proportions for each client using a Dirichlet distribution as follows:

$$\mathbf{p}^{(c)} \sim \mathrm{Dir}(\alpha), \quad \forall c \in 1, ..., C, \tag{8}$$

where $\mathbf{p}^{(c)} = [p_1^{(c)}, ..., p_N^{(c)}]$ denotes the proportion of class c assigned to each client, and α is the Dirichlet concentration parameter. Lower values of α result in more skewed (i.e., non-IID) distributions. To further simulate real-world heterogeneity and fairness-related challenges, we introduce class-dependent sample imbalances such that the number of samples per class varies across clients, i.e., $|D_k^{(c)}| \neq |D_j^{(c)}|$ for some $k \neq j$. This setting reflects practical scenarios where certain clients have richer data for specific classes while others lack adequate representation. We enforce a minimum sample threshold, requiring each client to hold at least 50 data samples.

To ensure meaningful model training and stable evaluation, the partitioning strategy imposes two fairness-preserving constraints: (i) each client must contain at least $S_{\min}$ samples (i.e., $|D_k| \geq S_{\min}$), and (ii) each class c must be represented in the data of at least $M_{\min}$ clients (set to 5 for MNIST and 3 for NSL-KDD). This partitioning strategy produces a federated environment characterized by realistic data heterogeneity, class imbalance, and fairness trade-offs. It enables rigorous evaluation of FL algorithms designed to achieve both fairness and efficiency under non-IID conditions. The sample counts and class proportions of the MNIST and NSL-KDD datasets are illustrated in Fig. 1a, Fig. 1b, Fig. 1c, and Fig. 1d, respectively.

Neural Network Architecture. For local model training, we employ a lightweight fully connected feedforward neural network (NN) to ensure computational efficiency in our FL environment. The input data is first passed through a linear transformation followed by a ReLU activation function to produce the first hidden layer consisting of 64 neurons. This is followed by a second hidden layer with 32 neurons, also utilizing a linear transformation and ReLU activation. The final output layer maps the transformed features to the target space (e.g., class logits) using another linear transformation. The total number of trainable parameters is 52,650 for the MNIST dataset and 10,117 for the NSL-KDD dataset, reflecting the input dimensionality and output space of each task.

Hyperparameters. In the FedMOAR approach, we use a total of 100 clients, selecting either 20% or 100% of them per CR ($C_{\mathrm{frac}} = 0.2, 1.0$). The training runs for 500 CRs on the MNIST dataset and 200 CRs on the NSL-KDD dataset. Each participating client trains locally for 2 epochs ($E = 2$) using a batch size of 32 and a learning rate of $\eta = 0.01$. To simulate data heterogeneity, we apply a Dirichlet-based partitioning with concentration parameters $\alpha = 1.0, 0.5$, and 0.1, thereby generating varying levels of class distribution skew across clients. The

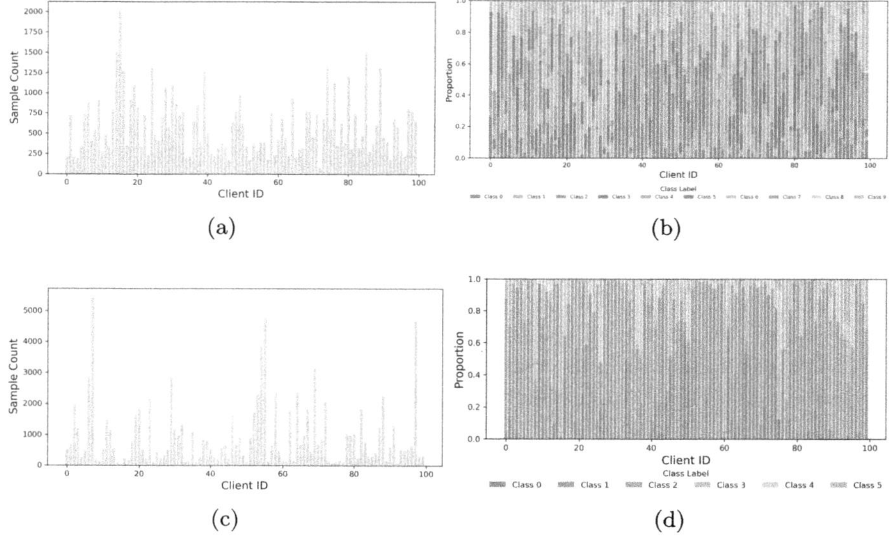

Fig. 1. Sample count and class proportion generated from our modified Dirichlet distributions for: (a) MNIST sample count, (b) MNIST class proportion, (c) NSLKDD sample count, (d) NSLKDD class proportion.

adaptive regularization weights λ_1, λ_2, and λ_3 are used to balance the penalties associated with model divergence, fairness deviation, and prediction accuracy, respectively.

Evaluation Metrics. We evaluate performance using several complementary metrics: training loss, test accuracy (TACC), F1 score (F1), MMAG, JFI, and CR. The MMAG quantifies performance disparity across clients, calculated as:

$$\text{MMAG} = \max_{k \in S^t} a_k^t - \min_{k \in S^t} a_k^t \tag{9}$$

where a_k^t is the accuracy of client k at round t.

JFI evaluates the equity of model performance across clients and is defined as:

$$\text{JFI} = \frac{\left(\sum_{k \in S^t} a_k^t\right)^2}{|S^t| \cdot \sum_{k \in S^t} (a_k^t)^2} \tag{10}$$

A JFI value closer to 1 indicates a more equitable distribution of accuracy among clients. In our evaluation, we predefine a target accuracy and measure the number of CRs required to reach this performance threshold.

4.2   Experimental Results

We conduct extensive experiments to evaluate the performance of existing FL methods and our proposed FedMOAR. The results are summarized in Table 1 and Table 2, which report key metrics on the MNIST and NSL-KDD datasets under varying non-IID conditions ($\alpha = 0.1, 0.5, 1.0$) and client participation settings ($C = 0.2$ for partial participation and $C = 1.0$ for full participation). Both tables present the key performance and efficiency indicators: test accuracy (TACC), F1 score (F1), MMAG, JFI, and the number of CRs, along with the relative speedup compared to FedAvg. This comprehensive setup enables a thorough evaluation of each method's ability to balance global model performance, fairness across clients, and communication efficiency.

Table 1. Performance of the different methods on the MNIST dataset under different levels of non-IIDness and client participation.

MNIST $\alpha = 0.1$									
Method	C = 0.2					C = 1.0			
	TACC	F1	MMAG	JFI	CR	TACC	F1	MMAG	JFI	CR
FedAvg	0.909	0.880	0.068	0.891	450	0.909	0.879	0.181	0.900	398
FedProx	0.902	0.875	0.082	0.879	405 (1.11×)	0.900	0.889	0.126	0.844	496 (0.80×)
FairFed	0.902	0.879	0.091	0.838	498 (0.90×)	0.904	0.872	0.115	0.860	420 (0.94×)
FedVal	0.898	0.876	0.123	0.786	–	0.888	0.866	0.224	0.870	–
FedMOAR	0.917	0.915	0.048	0.978	305 (1.47×)	0.926	0.925	0.096	0.977	194 (2.05×)
MNIST $\alpha = 0.5$									
FedAvg	0.912	0.890	0.071	0.929	363	0.911	0.897	0.076	0.915	328
FedProx	0.902	0.895	0.063	0.905	374 (0.97×)	0.902	0.889	0.095	0.880	469 (0.69×)
FairFed	0.909	0.887	0.101	0.865	320 (1.13×)	0.905	0.873	0.093	0.809	315 (1.04×)
FedVal	0.902	0.882	0.109	0.871	466 (0.77×)	0.901	0.869	0.099	0.700	483 (0.67×)
FedMOAR	0.933	0.932	0.051	0.976	165 (2.20×)	0.931	0.930	0.048	0.984	179 (1.83×)
MNIST $\alpha = 1.0$									
FedAvg	0.915	0.893	0.062	0.944	301	0.914	0.903	0.081	0.940	311
FedProx	0.904	0.892	0.066	0.933	325 (0.92×)	0.905	0.893	0.082	0.917	414 (0.75×)
FairFed	0.912	0.890	0.057	0.925	301 (1.00×)	0.908	0.881	0.064	0.845	390 (0.79×)
FedVal	0.907	0.895	0.126	0.943	411 (0.73×)	0.906	0.879	0.127	0.777	402 (0.77×)
FedMOAR	0.936	0.935	0.036	0.974	146 (2.06×)	0.935	0.934	0.039	0.985	159 (1.96×)

On the MNIST dataset (Table 1), the proposed FedMOAR consistently outperforms all baselines across all α values and client participation setups. While FedAvg achieves competitive test accuracy (e.g., 0.909 at $\alpha = 0.1$, $C = 0.2$), its corresponding F1-score is substantially lower (0.880), indicating inconsistency in performance across clients. This discrepancy likely results from class imbalance and unequal data volumes under non-IID conditions. Similar inconsistencies are observed for other baselines, highlighting their limited ability to ensure equitable client-level performance. In contrast, FedMOAR achieves both high TACC and high F1-score (0.917 and 0.915, respectively), reflecting not only strong

Table 2. Performance of the different methods on NSLKDD dataset under different levels of non-IIDness and client participation.

NSLKDD $\alpha = 0.1$										
Method	C = 0.2					C = 1.0				
	TACC	F1	MMAG	JFI	CR	TACC	F1	MMAG	JFI	CR
FedAvg	0.990	0.981	0.056	0.974	154	0.992	0.973	0.075	0.958	150
FedProx	0.992	0.972	0.060	0.972	137 (1.12×)	0.992	0.969	0.060	0.982	180 (0.83×)
FairFed	0.991	0.964	0.110	0.971	153 (1.01×)	0.992	0.969	0.099	0.976	149 (1.01×)
FedVal	0.992	0.969	0.067	0.951	187 (0.82×)	0.990	0.962	0.160	0.970	181 (0.82×)
FedMOAR	0.996	0.983	0.034	0.992	89 (1.73×)	0.996	0.988	0.041	0.990	95 (1.57×)
NSLKDD $\alpha = 0.5$										
FedAvg	0.992	0.970	0.045	0.992	153	0.990	0.955	0.047	0.982	144
FedProx	0.992	0.968	0.037	0.989	131 (1.17×)	0.992	0.967	0.056	0.983	153 (0.94×)
FairFed	0.992	0.966	0.032	0.985	141 (1.08×)	0.993	0.970	0.084	0.982	147 (0.98×)
FedVal	0.991	0.967	0.045	0.994	177 (0.86×)	0.990	0.958	0.049	0.987	175 (0.82×)
FedMOAR	0.996	0.985	0.021	0.998	67 (2.28×)	0.995	0.985	0.036	0.998	74 (1.95×)
NSLKDD $\alpha = 1.0$										
FedAvg	0.993	0.977	0.048	0.985	144	0.991	0.962	0.064	0.987	132
FedProx	0.992	0.970	0.045	0.983	132 (1.09×)	0.993	0.974	0.040	0.989	144 (0.92×)
FairFed	0.993	0.975	0.030	0.986	136 (1.06×)	0.991	0.965	0.061	0.984	132 (1.00×)
FedVal	0.992	0.971	0.090	0.982	163 (0.88×)	0.990	0.959	0.079	0.987	169 (0.78×)
FedMOAR	0.996	0.987	0.019	0.997	60 (2.40×)	0.996	0.985	0.023	0.998	69 (1.91×)

global performance but also balanced local effectiveness across clients. This performance trend continues at $\alpha = 0.5$ and $\alpha = 1.0$, where FedMOAR attains F1-scores of 0.932 and 0.935, respectively, surpassing all competing methods.

More importantly, FedMOAR achieves the lowest MMAG values 0.048, 0.051, and 0.036 as α increases, indicating more stable and consistent gradient updates from clients. This is crucial in FL, where unstable or divergent local updates can hinder global convergence. Additionally, FedMOAR yields near-optimal fairness, with JFI consistently above 0.97. This performance significantly surpasses baselines like FedProx (e.g., 0.879 at $\alpha = 0.1$, $C = 0.2$), demonstrating FedMOAR's capacity to ensure equitable treatment of all clients, even in highly heterogeneous environments. Furthermore, FedMOAR achieves the lowest number of required CRs as few as 305 rounds (1.47× faster than FedAvg) and even 146 rounds (2.06× speedup), highlighting its superior communication efficiency and faster convergence. In comparison, other methods require more CRs, and FedVal fails to converge under the most skewed setting ($\alpha = 0.1$).

On the NSL-KDD dataset (Table 2), the advantages of FedMOAR are even more pronounced. NSL-KDD presents a more complex challenge due to its class imbalance and categorical feature space, which heightens the difficulty of achieving generalization and fairness. Despite this, FedMOAR maintains the highest

TACC (0.996) across all α values and participation modes. More notably, it achieves significantly higher F1-scores, especially at $\alpha = 0.1$, $C = 1.0$, where it reaches 0.988 compared to FedAvg's 0.973. This performance gap underscores FedMOAR's ability to provide not just high accuracy but also consistent and fair predictions across clients, which is better captured by the F1-score in imbalanced settings.

FedMOAR's advantages in gradient stability and fairness are further exemplified on this dataset. It achieves extremely low MMAG values (as low as 0.019) and exceptionally high JFI scores (up to 0.998), reflecting near-perfect fairness among clients. In contrast, competing methods such as FairFed and FedVal exhibit greater variability in both MMAG and JFI, revealing inconsistency and reduced fairness under non-IID distributions. Moreover, FedMOAR demonstrates substantial convergence speed gains, requiring as few as 60 CRs at $\alpha = 1.0$, $C = 0.2$, achieving up to a 2.4× speedup compared to FedAvg. While some baselines show marginal improvements in certain cases, none maintain the same level of fairness or convergence efficiency as FedMOAR. The consistently higher MMAG values for these methods reflect weaker performance equity among clients with heterogeneous data distributions.

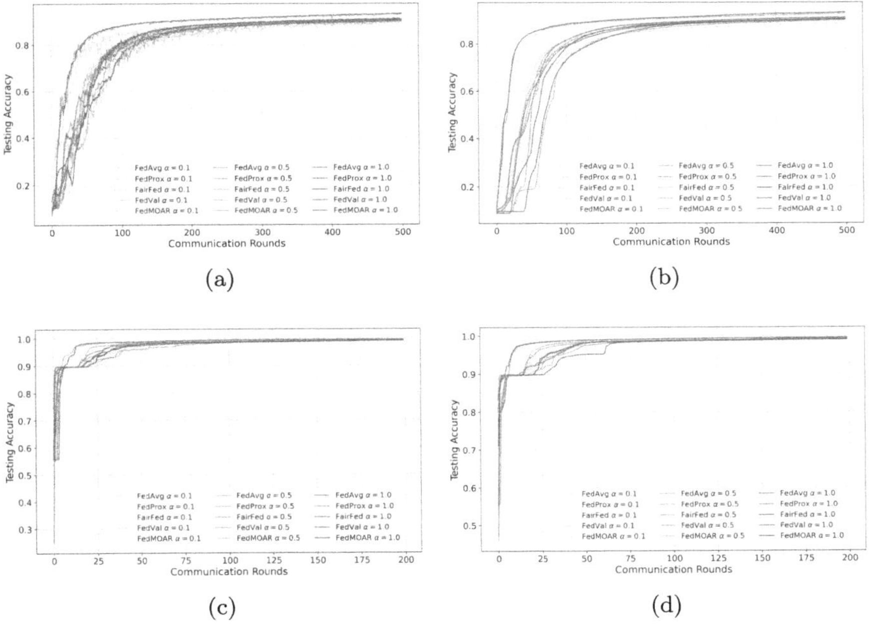

Fig. 2. Testing accuracy vs. CRs under non-IID settings ($\alpha = 0.1, 0.5, 1.0$) for: (a) MNIST Partial, (b) MNIST Full, (c) NSLKDD Partial, (d) NSLKDD Full client participation.

Across both MNIST and NSL-KDD, the proposed FedMOAR framework consistently outperforms all baselines across varying non-IID levels and client participation settings. Although some methods may match FedMOAR in terms of test accuracy, they fall short in terms of F1-score, fairness, or gradient stability, exposing their limitations in balancing global and client-level objectives. FedMOAR's multi-objective adaptive regularization enables it to jointly optimize accuracy, fairness, and communication efficiency. Its superior F1-scores indicate stronger per-client model generalization, while its low MMAG and high JFI values confirm equitable learning across clients. Combined with significantly reduced CRs, these results affirm FedMOAR's effectiveness in addressing key challenges in FL under heterogeneous conditions.

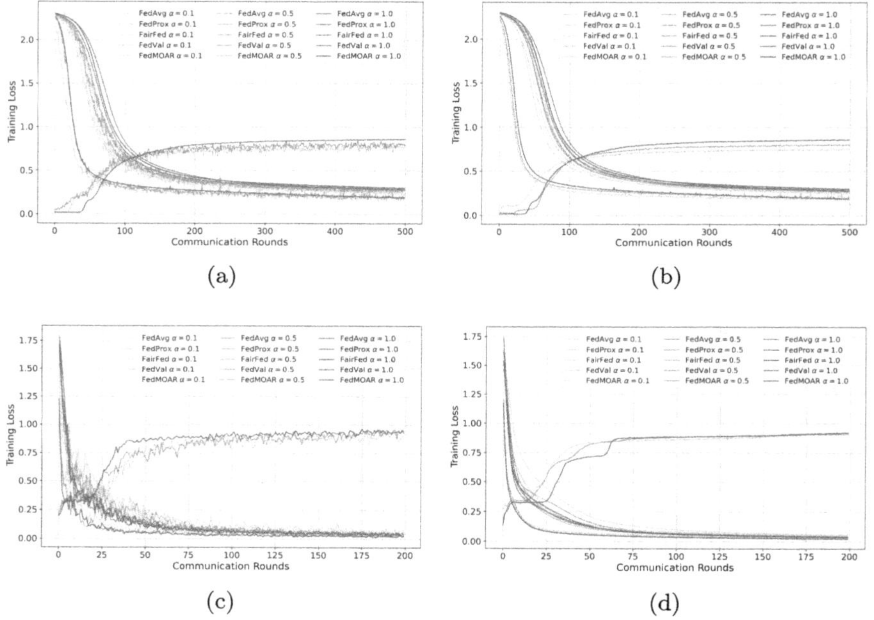

Fig. 3. Training loss vs. CRs under non-IID settings ($\alpha = 0.1, 0.5, 1.0$) for: (a) MNIST Partial, (b) MNIST Full, (c) NSLKDD Partial, (d) NSLKDD Full client participation.

4.3   Performance Visualizations

We present the testing accuracy and training loss trends for both the MNIST and NSL-KDD datasets in Fig. 2 and Fig. 3. The accuracy and loss curves across varying non-IID levels and client participation settings clearly demonstrate the superiority of FedMOAR over existing baselines. In terms of test accuracy, Fed-MOAR consistently outperforms FedAvg, FedProx, FairFed, and FedVal across all Dirichlet α values (0.1, 0.5, 1.0), under both partial and full client participation. Notably, FedMOAR not only achieves higher final accuracy but also converges substantially faster, indicating superior communication efficiency.

In the training loss plots shown in Fig. 3, FedMOAR exhibits the lowest and most stable loss trajectories, even under conditions of extreme data heterogeneity. This reflects its ability to maintain effective local optimization and minimize divergence from the global model. In contrast, baseline methods, particularly FairFed and FedVal, exhibit higher and more erratic loss curves, especially on the NSL-KDD dataset. These results further validate that FedMOAR is not only more accurate and communication-efficient but also robust and fair in highly heterogeneous federated environments.

5   Conclusion and Future Work

To enable fair and efficient FL, this work introduced FedMOAR, a Multi-Objective Adaptive Regularization approach that jointly optimizes global model accuracy, client-level fairness, and communication efficiency. Rather than relying solely on prediction loss (e.g., cross-entropy), FedMOAR incorporates additional objectives, such as model divergence regularization, fairness penalties, and accuracy compensation, to overcome the key limitations of conventional FL frameworks. These components help ensure that local updates remain aligned with the global model, promote equitable performance across clients, and enhance stability of participation. Through comprehensive experiments on the MNIST and NSL-KDD datasets under varying non-IID distributions and participation levels, FedMOAR consistently outperformed state-of-the-art baselines, including FedAvg, FedProx, FairFed, and FedVal. Specifically, FedMOAR achieved higher F1-scores, significantly lower MMAG, and near-optimal JFI, confirming its capability to deliver fair and efficient model performance across heterogeneous clients. Additionally, FedMOAR required up to 2.4× fewer CRs for convergence. Even in cases where test accuracy was comparable, FedMOAR demonstrated superior per-client performance, effectively balancing fairness, convergence, and communication efficiency.

Future work will explore the extension of FedMOAR to cross-domain and multi-modal FL scenarios. Moreover, integrating privacy-preserving techniques (e.g., differential privacy, secure aggregation) and adapting the framework to more complex real-world applications involving diverse data modalities and resource constraints could further enhance its scalability and applicability in practical deployments.

References

1. McMahan, B., Moore, E., Ramage, D., Hampson, S., y Arcas, B.A.: Communication-efficient learning of deep networks from decentralized data. In: Artificial Intelligence and Statistics, pp. 1273–1282. PMLR (2017)

2. Uddin, M.P., Xiang, Y., Hasan, M., Bai, J., Zhao, Y., Gao, L.: A systematic literature review of robust federated learning: issues, solutions, and future research directions. ACM Comput. Surv. **57**(10), 1–62 (2025)
3. Zhang, C., Xie, Y., Bai, H., Yu, B., Li, W., Gao, Y.: A survey on federated learning. Knowl.-Based Syst. **216**, 106775 (2021)
4. Wang, Z., Zhu, Y., Wang, D., Han, Z.: Federated analytics informed distributed industrial IoT learning with non-IID data. IEEE Trans. Netw. Sci. Eng. **10**(5), 2924–2939 (2022)
5. Nguyen, D.C., et al.: Federated learning for smart healthcare: a survey. ACM Comput. Surv. (CSUR) **55**(3), 1–37 (2022)
6. Li, T., Sahu, A.K., Zaheer, M., Sanjabi, M., Talwalkar, A., Smith, V.: Federated optimization in heterogeneous networks. In: Proceedings of Machine Learning and Systems, vol. 2, pp. 429–450 (2020)
7. Li, Q., He, B., Song, D.: Model-contrastive federated learning. In: Proceedings of the IEEE/CVF Conference on Computer Vision and Pattern Recognition, pp. 10713–10722 (2021)
8. Du, W., Xu, D., Wu, X., Tong, H.: Fairness-aware agnostic federated learning. In: Proceedings of the 2021 SIAM International Conference on Data Mining (SDM), pp. 181–189. SIAM (2021)
9. Papadaki, A., Martinez, N., Bertran, M., Sapiro, G., Rodrigues, M.: Minimax demographic group fairness in federated learning. In: Proceedings of the 2022 ACM Conference on Fairness, Accountability, and Transparency, pp. 142–159 (2022)
10. Wang, J., Liu, Q., Liang, H., Joshi, G., Poor, H.V.: Tackling the objective inconsistency problem in heterogeneous federated optimization. Adv. Neural. Inf. Process. Syst. **33**, 7611–7623 (2020)
11. Jin, C., Chen, X., Gu, Y., Li, Q.: FedDyn: a dynamic and efficient federated distillation approach on recommender system. In: 2022 IEEE 28th International Conference on Parallel and Distributed Systems (ICPADS), pp. 786–793. IEEE (2023)
12. Valadi, V., Qiu, X., de Gusmão, P.P.B., Lane, N.D., Alibeigi, M.: FedVal: different good or different bad in federated learning. In: 32nd USENIX Security Symposium (USENIX Security 2023), Anaheim, CA, pp. 6365–6380. USENIX Association (2023)
13. Zheng, H., Zhang, T., Chen, J.: Fedfair: a debiasing algorithm for federated learning systems. In: 2024 IEEE 23rd International Conference on Trust, Security and Privacy in Computing and Communications (TrustCom), pp. 2322–2326 (2024)
14. Li, Z., Lin, T., Shang, X., Wu, C.: Revisiting weighted aggregation in federated learning with neural networks. In: International Conference on Machine Learning, pp. 19767–19788. PMLR (2023)
15. Horvath, S., Laskaridis, S., Almeida, M., Leontiadis, I., Venieris, S., Lane, N.: FjORD: fair and accurate federated learning under heterogeneous targets with ordered dropout. Adv. Neural. Inf. Process. Syst. **34**, 12876–12889 (2021)
16. Li, T., Sanjabi, M., Beirami, A., Smith, V.: Fair resource allocation in federated learning. In: International Conference on Learning Representations (2020)
17. Karimireddy, S.P., Kale, S., Mohri, M., Reddi, S., Stich, S., Suresh, A.T.: Scaffold: stochastic controlled averaging for federated learning. In: International Conference on Machine Learning, pp. 5132–5143. PMLR (2020)
18. Hu, S., Wu, Z.S., Smith, V.: Fair federated learning via bounded group loss. In: 2024 IEEE Conference on Secure and Trustworthy Machine Learning (SaTML), pp. 140–160. IEEE (2024)

Unveiling Reliability in Multi-omics Classification: Fusion, Calibration, and Dynamic Scaling

Dang Khoa Nguyen[✉][iD], Duoyi Zhang[iD], Md Abul Bashar[iD], and Richi Nayak[iD]

Centre for Data Science, School of Computer Science, Queensland University of Technology, Brisbane, QLD 4000, Australia
{d98.nguyen,duoyi.zhang}@hdr.qut.edu.au, {m1.bashar,r.nayak}@qut.edu.au

Abstract. Multi-omics data, encompassing heterogeneous sources such as genomics, transcriptomics, and proteomics, has become a powerful resource for disease classification in bioinformatics. Fusion mechanisms combine features from multiple modalities into a single joint representation for predictive models to improve classification performance. However, while most studies focus on accuracy, model calibration - how well predicted probabilities match actual outcomes - remains underexplored in the multi-omics setting. Poor calibration can make even accurate models overconfident and unreliable. This study investigates the impact of three fusion strategies, early, intermediate and late, on three state-of-the-art post-hoc calibration techniques. We further introduce *dynamic scaling*, a regression-based calibration method that estimates instance-specific temperature parameters for more adaptive calibration than dataset-level techniques. Experiments on four benchmark omics datasets evaluate both classification and calibration metrics. Results show that data quality and class imbalance strongly influence performance, and while *dynamic scaling* often achieves the best calibration, challenges persist for small and imbalanced datasets.

Keywords: calibration · multi-omics data · multimodal fusion · deep learning

1 Introduction

Multi-omics data are widely available in real-world applications, especially in pharmacology, medication, and disease diagnosis [16]. Leveraging these datasets for bioinformatics classification has become a growing focus, as they enable models to capture complex and complementary information encoded in different biological layers, such as genomics, transcriptomics, epigenomics, and proteomics [11]. To fully explore the potential of these diverse omics layers, multimodal deep learning models have been developed, using fusion techniques to learn joint representations. In this context, multi-omics can be regarded as a specialised application of multimodal learning, where the "modalities" correspond to distinct biological data types.

Q. V. Nguyen et al. (Eds.): AusDM 2025, CCIS 2765, pp. 63–78, 2026.
https://doi.org/10.1007/978-981-95-6786-7_5

The availability of open-source multi-omics datasets such as TCGA, BRCA, ROSMAP and METABRIC [11] has opened new frontiers in biomedical research, enabling deeper insights into complex diseases like cancer and advancing precision medicine [16] [18]. They also facilitate rapid advances in deep learning for various omics-related tasks, including predicting patient survival [24], identifying cancer subtypes [15], and guiding treatment selection [26].

Despite the promise of deep learning as an analytical framework, extracting meaningful information from multi-omics data remains challenging. Individual data types, such as mRNA, CNV, DNA methylation, and miRNA, are already extremely high-dimensional. Their integration compounds complexity, making pattern detection more difficult and requiring advanced computational strategies [23]. Additionally, this challenge, often referred to as 'the curse of dimensionality', increases the risk of overfitting [1]. Furthermore, effective integration is hindered by the inherent heterogeneity of these datasets.

Undoubtedly, rapid advances in deep learning have led to more complex and expressive architectures for multi-omics integration. However, this advancement is accompanied by a critical, often-overlooked issue: the overconfidence of neural networks [7]. In biomedical domains, especially oncology and clinical diagnostics, model evaluation must go beyond accuracy to include reliability - the degree to which predictions can be trusted by human experts. Reliability is central to ensuring transparency, accountability, and safe decision-making. Model calibration, also referred to as confidence calibration, addresses this issue by aligning predicted probability scores with the true likelihood of correctness [7]. For example, if a model predicts that a patient has a 70% chance of surviving one year, then across many similar patients, about 70% should indeed survive that long. However, deep learning networks are often miscalibrated, typically overconfident. Even highly accurate models may be poorly calibrated, risking serious mistakes in clinical decisions. Factors contributing to miscalibration include over-parameterised architectures, poor regularisation, limited data, and class imbalance [7, 20]. Therefore, development of deep learning models must move beyond optimising accuracy alone, emphasising trustworthiness and interpretability in model outputs.

Model calibration methods can generally be grouped into four categories: post-hoc calibration, regularisation-based strategies, uncertainty estimation, and composition approaches [18]. In this paper, unlike prior studies that have examined mainly calibration in unimodal settings, we present a comprehensive analysis of three widely used post-hoc calibration techniques applied to multimodal fusion strategies. Our experiments span four benchmark multi-omics datasets: BRCA, TCGA, ROSMAP and METABRIC. Our contributions are as follows:

- We systematically evaluate three state-of-the-art post-hoc calibration techniques in multimodal settings, providing insights into their effectiveness beyond unimodal applications.
- We introduce *dynamic scaling*, a novel instance-based calibration approach that leverages a regression model to estimate a temperature parameter for

each instance, enabling finer-grained and more adaptive calibration than dataset-level approaches.
- We conduct a comprehensive evaluation of multimodal fusion methods (including various machine learning and deep learning models) for classification tasks on four widely used multi-omics datasets (BRCA, TCGA, ROSMAP, and METABRIC), with emphasis on calibration performance alongside predictive accuracy.

2   Related Work

Multimodal fusion. Multimodal learning has gained increasing attention in bioinformatics for its ability to improve predictive performance by integrating heterogeneous information sources. Unlike unimodal approaches that rely on a single data type, multimodal frameworks can combine complementary modalities, which may include omics layers (genomics, transcriptomics, proteomics) and non-omics data (e.g. demographic or clinical information) [11,25]. This integration enables models to capture complex and informative patterns and has shown effective in tasks such as Alzheimer's disease classification, breast cancer prognosis, and other disease-related studies [3,21].
A central challenge lies in how best to combine modalities. Fusion strategies are generally categorised into early, late, and intermediate approaches [11,25]. These strategies differ in terms of the stage at which modalities are integrated and how inter-modal relationships are captured. Early fusion concatenates modalities at the input level to form a joint representation processed by machine learning algorithms. For instance, in drug response prediction, an attention mechanism was applied on concatenated features to model inter-modal interactions [19]. However, combining modalities with different abstraction levels at the input stage can obscure within-modality features and hinder the learning of meaningful cross-modal interactions [14]. As an alternative, late fusion integrates modalities at the decision level by aggregating outputs from independently trained unimodal models. While this avoids abstraction-level mismatch and allows independent processing of each modality, it is prone to feature-level bias (FLB), where dominant modalities disproportionately affect predictions [22]. Adaptive weighting techniques have been proposed to mitigate FLB by adjusting contributions based on unimodal overfitting ratios [22], performance confidence gaps across modalities [12], or prototype distances [6]. Intermediate fusion has recently emerged as a promising compromise, allowing interactions across modalities at multiple stages. This approach flexibly captures both within and cross-modality relationships. For example, in mortality prediction, interactions between demographic factors (e.g. age and gender) and physiological measures (e.g. blood pressure) provide nuanced insights that neither modality alone can capture [13]. Such layered integration is well-suited for modeling complex dependencies in biomedical tasks [3,8,21].
Despite these advances, most multimodal studies in bioinformatics emphasise accuracy while neglecting calibration. In high-stakes applications, poorly calibrated models can yield overconfident and unreliable predictions, undermining

clinical decision-making. Thus, both accuracy and calibration must be evaluated to ensure trustworthy outcomes.

Calibration Techniques. In bioinformatics, calibration is critical for ensuring both the trustworthiness and interpretability of predictive models [10]. Many tasks, such as disease risk prediction or mortality prediction, require not only accurate classifications but also reliable estimates of prediction confidence [17]. Well-calibrated models output probabilities that closely match the true likelihood of outcomes, enabling clinicians to make informed, risk-aware decisions. Conversely, miscalibrated predictions can lead to overconfidence and potentially harmful outcomes in high-risk applications [7], making calibration a key requirement for deploying models in real-world bioinformatics settings.

Post-hoc calibration is the most commonly used strategy [5], relying on validation data to learn parameters that adjust model outputs. For example, temperature scaling [7] applies a single scalar to rescale logits before the softmax layer, assuming a global correction suffices for all predictions. Matrix scaling [7] generalises this approach by learning a linear transformation with a matrix and a bias vector to map the model's outputs to calibrated probabilities, still relying on validation data to optimise these parameters. While effective, these methods operate at the dataset level, implicitly assuming all test samples share similar miscalibration patterns. This assumption is often violated in bioinformatics, where patient profiles exhibit strong heterogeneity.

To address this, we propose a novel instance-level calibration method that generates a customised scaling factor for each input. This allows adaptive correction of sample-specific biases rather than applying a uniform transformation across all instances.

Other calibration strategies also exist, including regularisation-based methods, uncertainty estimation, and composition methods [18]. Regularisation integrates calibration into training, while uncertainty estimation can be applied post-hoc to provide confidence ranges but does not directly adjust probabilities. Composition methods combine multiple techniques but are more complex to implement. Since our goal is to evaluate and compare the calibration performance of existing models, we restrict our analysis to post-hoc calibration methods (Fig. 1).

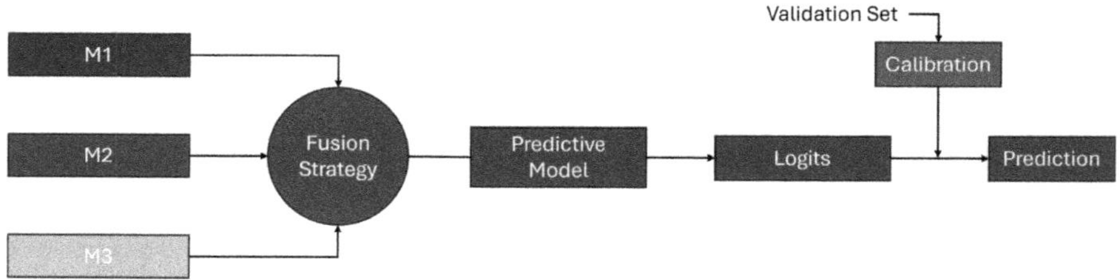

Fig. 1. Architecture of a multimodal calibration pipeline, where multiple modalities (e.g. M1, M2, M3) are combined through a fusion strategy, processed by a predictive model, and then passed to a calibration stage before generating the final prediction.

3   Methodology

The calibration of multimodal classification models in this study involves three main stages: data preprocessing, multimodal classification, and calibration. Below, we outline the techniques used for empirical analysis.

3.1   Data Preprocessing

We use the four datasets, preprocessed by their original authors. BRCA and ROSMAP were initially split into training and testing sets with a 0.3 ratio; we merged these splits and applied 5-fold cross-validation. TCGA and METABRIC were provided without predefined splits, so no additional preprocessing was required. The datasets with N instances can be represented as:

$$\mathcal{D} = \left\{ \left(\{\mathbf{x}_i^m\}_{m=1}^M, y_i \right) \right\}_{i=1}^N$$

where each instance i consists of M modality-specific inputs $\{\mathbf{x}_i^m\}_{m=1}^M$ with label $\{y_i\} \in \{1, ..., K\}$.

3.2   Multimodal Classification

After preprocessing, the modalities are integrated into a joint representation for classification.

We evaluate three commonly used fusion strategies: early, late, and intermediate fusion. The objective of each strategy is to transform the set of modality-specific inputs $\{\mathbf{x}_i^{(m)}\}_{m=1}^M$ into a joint representation $\mathbf{z}_i$, which is subsequently used to produce the classification output $\hat{y}_i$.

Early Fusion: In this method, processed data per modality are combined at the data level. i.e.: $\mathbf{z}_i = \left[\mathbf{x}_i^{(1)}; \ldots; \mathbf{x}_i^{(M)} \right]$ For M modalities with inputs $\mathbf{x}_i^{(m)}$, early fusion concatenates them into a single input $\mathbf{z}_i$ which is then passed to downstream classifiers, such as support vector machines, XGBoost, or Gaussian Process classifiers [4].

Late Fusion: In this method, processed data per modality are initially passed through unimodal models for feature extraction. i.e. $\mathbf{z}_i^{(m)} = h_m(\mathbf{x}_i^{(m)})$ where $h_m(\cdot)$ represents a unimodal encoder (e.g., a neural network) for converting x_m into $\mathbf{z}_i^{(m)}$. Advanced late fusion methods [6, 12, 22] assign different learning rates for each $h_m(\cdot)$, resulting in more balanced learning and improved classification performance. In late fusion, for M modalities with inputs $\mathbf{x}_i^{(m)}$, each modality is encoded independently and obtains its representations. These features are then averaged, and the integrated input is fed into the main classifier:

$$\mathbf{z}_i = \frac{1}{M} \sum_{m=1}^M \mathbf{z}_i^{(m)}$$

Intermediate Fusion: In this method, features are initially learned per-modality (i.e. $\tilde{\mathbf{z}}_i^{(m)} = f_m(\mathbf{x}_i^{(m)})$), where $f_m(\cdot)$ is the neural networks for unimodal feature extraction such as fully connected layers, then combines those features representation in a fusion block where the modalities will interact before the final prediction

$$\mathbf{z}_i = F\big(\tilde{\mathbf{z}}_i^{(1)}, \ldots, \tilde{\mathbf{z}}_i^{(M)}\big)$$

where F is the fusion block that combines unimodal embeddings into one joint representation, which can be implemented through approaches such as gating mechanisms [2,8], attention mechanisms [9], or other interaction modules.

Classification: The fused representation $\mathbf{z}_i$ is then passed to the classifier h_{clf}, producing logits $\mathbf{l}_i$:

$$\mathbf{l}_i = h_{clf}(\mathbf{z}_i)$$

where $h_{clf}(\cdot)$ is the classifier and $\mathbf{l}_i \in \mathcal{R}^K$ is the classification logits where K represents number of classes.

A Softmax function is then applied to convert the logits into a probability using the *Softmax* function:

$$\hat{p}_i = Softmax(\mathbf{l}_i)$$

where $\hat{p}_i \in \mathcal{R}^K$ represents the probability toward K classes, and $\sum_{k=1}^{K} \hat{p}_i = 1$. Finally, the predicted label is obtained as the highest value from $\hat{p}_i$:

$$\hat{y}_i = \arg\max_k(\hat{p}_i)$$

where $\hat{y}_i$ is a scalar value representing the predicted label.

A cross-entropy loss is then applied to optimise the multimodal classification networks.

3.3   Calibration

Model calibration measures how well a model's predicted probabilities reflect the correctness [7]. Given the predicted probability $\hat{p}_i \in \mathbb{R}^K$ and the predicted label $\hat{y}_i = \arg\max_k \hat{p}_{i,k}$, the prediction confidence and prediction correctness for instance i is defined as:

$$c_i = \max_k \hat{p}_{i,k}, \qquad o_i = \mathbb{1}\{\hat{y}_i = y_i\}.$$

While c_i and o_i are defined per instance, calibration is evaluated at the dataset level (e.g. on the test dataset) by grouping instances according to their confidence scores. To this end, partition $[0, 1]$ into N_B disjoint bins (intervals) $\{B_b\}_{b=1}^{N_B}$. For each bin, let

$$\mathcal{B}_b = \{i : c_i \in B_b\}$$

be the index set of instances whose prediction confidence lies in bin b, and denote its size by $n_b = |\mathcal{B}_b|$. The accuracy and average confidence for bin b are

$$\mathrm{acc}(b) = \frac{1}{n_b} \sum_{i \in \mathcal{B}_b} o_i, \qquad \mathrm{conf}(b) = \frac{1}{n_b} \sum_{i \in \mathcal{B}_b} c_i.$$

Let N_t be the total number of instances in the testing dataset. The Expected Calibration Error (ECE) and Maximum Calibration Error (MCE) are defined as:

$$\text{ECE} = \sum_{b=1}^{N_B} \frac{n_b}{N} \big| \operatorname{acc}(b) - \operatorname{conf}(b) \big|, \qquad \text{MCE} = \max_{1 \leq b \leq N_B} \big| \operatorname{acc}(b) - \operatorname{conf}(b) \big|.$$

ECE and MCE are the two metrics for evaluating the performance of calibration. Specifically, a lower ECE or MCE indicates a well-calibrated model, and vice versa.

Machine learning models—particularly deep learning models—are often poorly calibrated, prompting the development of various calibration techniques [7]. In this study, we focus on three widely used post-hoc methods: temperature scaling, matrix scaling, and vector scaling, each of which transforms the model logits $\mathbf{l}_i$ into calibrated probabilities $\hat{q}_i$. In addition, we introduce Dynamic Scaling, a variant of temperature scaling that assigns an instance-specific temperature.

Temperature Scaling: A single parameter applies to both binary and multiclass calibration. A singular scalar value T is introduced for all classes to soften the predicted probability of the model, where we will calculate calibrated probabilities $\hat{q}_i$ as follows:

$$\hat{q}_i = \max_k(Softmax(\mathbf{l}_{i,k}/T_k)), \quad T_k > 0$$

The following will happen to $\hat{q}_i$ if
(1) $T \to \infty \Rightarrow \hat{q}_i \to \frac{1}{K}$ (2) $T = 1 \Rightarrow \hat{q}_i = \hat{p}_i$ (3) $T \to 0 \Rightarrow \hat{q}_i = 1$
The optimal temperature is determined by minimising the non-negative log likelihood (NLL) loss on the validation set. Since the method's main goal is to soften the 'softmax', it will not alter the accuracy of the model.

Matrix Scaling: The method applies a linear transformation to the logits:

$$\hat{q}_i = \max_k(Softmax(W\mathbf{l}_{i,k} + b)) \quad W \in \mathbb{R}^{K \times K}, \quad b \in \mathbb{R}^K$$

where W is a $K \times K$ matrix and b is a vector length of K. This means that the model's accuracy changes according to:

$$\hat{\mathbf{y}}_i = \arg\max_k(Softmax(W\mathbf{l}_{i,k} + b))$$

Both parameters are optimized using NLL loss on the validation set. However, when the number of classes becomes too large, the method will involve many parameters, increasing the risk of overfitting, especially since the target set (e.g., testing dataset) is typically small.

Vector Scaling: Similar to matrix scaling, but here the weight matrix $W \in \mathbb{R}^{K \times K}$ is constrained to be diagonal, i.e., $W = \operatorname{diag}(v)$ with $v \in \mathbb{R}^K$.

Dynamic Scaling: In standard Temperature Scaling, logits $l_{i,k}$ for class k are scaled by a fixed temperature T. Dynamic Temperature Scaling extends this by predicting an instance-specific temperature $T(x)$ using a regression model $f(x)$. The calibrated probabilities become:

$$\hat{q}_i = \max_k(Softmax(l_{i,k}/T_k(l'_i))), \quad T_k > 0$$

where l'_i represents the logits of the validation set. Similarly to temperature scaling, the model is trained using the NLL loss. Since T is introduced to soften the 'softmax', it will not alter the accuracy of the model.

4   Empirical Evaluation

This paper investigates the calibration performance of deep learning and machine learning fusion models in multi-omics classification tasks. Further, it examines the impact of different calibration techniques on these models. We begin by introducing the datasets and describing the evaluation criteria.

4.1   Data Collection

We evaluated model performance on four multi-omics datasets spanning different applications. **Breast Cancer Gene (BRCA)** and **Religious Orders Study and Rush Memory and Aging Project (ROSMAP)** datasets contain 875 and 351 instances, respectively. Both include three types of omics data: mRNA expression (transcriptomics), DNA methylation (epigenomics), and miRNA expression (transcriptomics). The classification tasks are PAM50 subtype prediction for BRCA and Alzheimer's Disease vs. normal control for ROSMAP.
METABRIC and TCGA-BRCA (TCGA) datasets contain 1980 and 1080 trial data of breast cancer patients, respectively. Each patient record comprises three modalities: two omics layers (copy number alterations from genomics and gene expression from transcriptomics) and one non-omics layer (clinical profile). The classification task is survival prediction, distinguishing long-term from short-term survivors.

4.2   Baseline Models

To evaluate how well the confidence of machine learning and deep learning models aligns with actual outcomes, we select models from four families of methods: (1) Early Fusion: We concatenate the three modalities and feed the joint representation to SVM, Gaussian Process (GP), XGBoost, Logistic Regression (LR), and Neural Network (NN).
(2) Late Fusion: We calculate softmax scores per modality via a deep neural network and averaged across modalities. Besides, advanced late fusion strategies, including Gradient Blend (GB) [22], Gradient Modulation (GM) [12], and

PMR [6], are used.

(3) Intermediate fusion: We utilise GMU [2] to fully capture the interactions across all modalities. Other reliable approaches for multi-omics classification include DFTMC [8] and MMB [9]. MMB adapts a boosting-inspired strategy to improve the multimodal interactions.

4.3   Evaluation Criteria

Since TCGA and METABRIC are imbalanced, overall accuracy may be biased towards the majority class. To account for this, we evaluated models on accuracy, weighted F1, and macro-average F1. Due to 5-fold cross-validation, we report both the mean and standard deviation across the five runs. Additionally, Expected Calibration Error (ECE) and Maximum Calibration Error (MCE) are computed before and after applying three calibration techniques to assess how well the model's predicted confidences reflect actual outcomes.

4.4   Experimental Settings

All experiments were conducted on a Windows platform with an AMD Ryzen 5 CPU and a RTX 3070 GPU, except for MMB on the ROSMAP dataset, which was run on an HPC platform. Experiments were implemented in Python 3.11 using PyTorch 2.5.

The scikit-learn library was used to implement all machine learning models, and grid searching was applied to optimise accuracy. Details for the hyperparameter are provided in the Appendix. Deep learning models were trained and tuned with the following ranges: epochs in $\{1000, 2000, 3000, 4000, 5000\}$ to account for models with slower converge (i.e. GB), learning rates $\{1e\text{-}3, 1e\text{-}4, 1e\text{-}5\}$ as METABRIC and TCGA require smaller rates due to class imbalance, batch sizes $\{32, 64\}$ and sequence lengths $\{100, 200, 500, 1000\}$. Stratified 5-fold cross-validation was applied to all models, with evaluation metrics averaged across folds. The epoch achieving the highest accuracy was chosen for calibration. ECE, MCE, and accuracy were computed before and after calibration for side-by-side comparison.

5   Results and Discussion

In this section, we first report the classification performance of all methods on four datasets, followed by their calibration performance and the post-calibration accuracy. We then compare the effectiveness of our proposed Dynamic Scaling against existing calibration techniques.

5.1   Performance on Multiomics Classification Tasks

Due to space constraints, all tables and details are provided in the Appendix. Overall, machine learning models prove to be more useful in small sample settings such as Rosmap, while deep learning learning models are better in larger datasets like METABRIC and TCGA. In BRCA, where $p \gg n$, margin-based linear methods (e.g., SVM) outperform complex learners.

Regarding the fusion type, early fusion struggles with small and high-dimensional data, whereas late and intermediate fusion perform better by capturing modality-specific signals and balancing across modalities. Class imbalance remains a challenge across datasets; this results in high accuracy while lowering F1 macro.

Table 1. ECE across calibration methods on BRCA. Row-wise minima in **bold**.

Model	Type	Before Calibration	Matrix Scaling	Temperature Scaling	Vector Scaling	Dynamic Scaling
MMB	Intermediate	5.40 ± 2.13	10.23 ± 3.14	6.22 ± 1.79	4.98 ± 1.12	**4.88 ± 1.73**
DFTMC	Intermediate	4.35 ± 2.39	9.15 ± 4.73	4.98 ± 0.75	5.65 ± 1.06	**3.17 ± 1.41**
GMU	Intermediate	15.34 ± 2.74	10.74 ± 5.81	6.83 ± 1.78	8.34 ± 2.73	**5.49 ± 2.35**
GB	Late	9.99 ± 1.53	9.23 ± 4.01	10.75 ± 3.90	9.27 ± 3.98	**7.58 ± 2.68**
GM	Late	4.74 ± 2.85	8.61 ± 4.56	5.17 ± 3.11	**4.18 ± 2.88**	5.02 ± 0.47
PMR	Late	21.94 ± 5.53	8.50 ± 5.15	7.67 ± 3.71	**5.60 ± 0.93**	13.47 ± 2.22
DNN	Late	4.57 ± 0.97	8.43 ± 4.24	5.23 ± 2.29	**4.52 ± 2.11**	5.66 ± 1.72
DNN	Early	11.71 ± 2.42	10.51 ± 2.01	5.92 ± 1.86	6.09 ± 1.50	**5.05 ± 1.43**
Gausian Process	Early	13.27 ± 3.00	5.77 ± 2.62	5.65 ± 2.03	8.98 ± 3.00	**5.50 ± 2.74**
Logistic Regression	Early	7.82 ± 2.46	6.95 ± 2.76	**5.39 ± 1.10**	8.22 ± 3.38	6.14 ± 2.12
SVM	Early	7.49 ± 1.98	6.89 ± 3.24	**6.55 ± 1.77**	7.07 ± 1.04	7.50 ± 1.18
Xgboost	Early	44.55 ± 3.30	15.62 ± 14.88	7.38 ± 2.11	9.96 ± 3.69	**5.97 ± 2.78**

5.2   Calibration Performance

Tables 1, 2, 3 and 4 illustrate model calibration as measured by ECE (with $M = 5$ bins), before and after applying the three widely used calibration methods and our proposed method, Dynamic scaling. The detailed evaluation of MCE (with $M = 5$ bins) is provided in the Appendix. It is important to note that most datasets and models show some level of miscalibration, with ECE values ranging between 2% and 20 %, except for MMB on METABRIC and TCGA. This exception is likely due to the model fails to learn meaningful patterns, and its high accuracy is driven by class imbalance rather than learning. We found that miscalibration is not fusion strategies or model types specific (i.e. machine learning, deep learning, etc.), all types experience a level of deviation with no clear trend observed.

Matrix, Vector and Dynamic Scaling performs poorly on ROSMAP. This is expected as the goal of post-hoc calibration is to use validation data to learn a calibration mapping that transforms the model's logits into better calibrated outputs. For this to be effective requires sufficient sample size. This is evident

Table 2. ECE across calibration methods on ROSMAP. Row-wise minima in **bold**.

Model	Type	Before Calibration	Matrix Scaling	Temperature Scaling	Vector Scaling	Dynamic Scaling
MMB	Intermediate	14.12 ± 5.55	12.32 ± 4.51	$\mathbf{8.66 \pm 2.46}$	12.32 ± 4.51	11.97 ± 5.42
DFTMC	Intermediate	11.98 ± 10.98	9.89 ± 7.85	$\mathbf{8.14 \pm 6.59}$	8.28 ± 6.49	9.22 ± 7.61
GMU	Intermediate	20.83 ± 7.10	9.73 ± 3.64	$\mathbf{8.92 \pm 3.72}$	9.01 ± 2.95	9.59 ± 4.98
GB	Late	20.71 ± 5.75	9.32 ± 3.17	$\mathbf{8.80 \pm 4.09}$	9.32 ± 3.17	10.24 ± 4.24
GM	Late	$\mathbf{9.61 \pm 4.59}$	11.49 ± 2.92	10.66 ± 4.61	10.11 ± 5.84	9.98 ± 3.02
PMR	Late	13.11 ± 7.34	12.92 ± 5.35	10.49 ± 4.53	12.91 ± 5.34	$\mathbf{10.14 \pm 5.14}$
DNN	Late	13.05 ± 4.62	$\mathbf{10.25 \pm 4.12}$	13.16 ± 3.59	12.39 ± 2.69	12.42 ± 3.18
DNN	Early	$\mathbf{10.60 \pm 2.64}$	14.08 ± 3.25	13.66 ± 3.50	14.77 ± 2.21	11.69 ± 5.85
Gausian Process	Early	13.41 ± 4.03	13.78 ± 5.21	12.90 ± 5.10	13.78 ± 5.21	$\mathbf{12.52 \pm 6.24}$
Logistic Regression	Early	17.73 ± 9.46	10.00 ± 8.87	$\mathbf{6.54 \pm 5.11}$	10.00 ± 8.87	11.86 ± 8.10
SVM	Early	11.06 ± 4.98	9.77 ± 5.48	$\mathbf{9.26 \pm 7.61}$	9.77 ± 5.48	10.28 ± 7.24
Xgboost	Early	12.72 ± 4.31	13.84 ± 3.47	$\mathbf{11.70 \pm 6.15}$	13.84 ± 3.47	13.23 ± 6.63

in smaller sample datasets such as ROSMAP, where deep learning models may achieve higher accuracy but suffer from calibration overfitting due to the limited validation set. In such cases, simpler calibrators like temperature scaling are preferable, whereas more complex methods that require fitting additional parameters are prone to failure with small sample sizes.

Table 3. ECE across calibration methods on METABRIC. Row-wise minima in **bold**.

Model	Type	Before Calibration	Matrix Scaling	Temperature Scaling	Vector Scaling	Dynamic Scaling
MMB	Intermediate	66.26 ± 27.78	78.28 ± 1.96	$\mathbf{63.54 \pm 33.86}$	63.65 ± 33.61	78.28 ± 1.96
DFTMC	Intermediate	4.38 ± 1.30	3.13 ± 0.75	3.38 ± 0.95	3.07 ± 0.74	$\mathbf{2.98 \pm 1.37}$
GMU	Intermediate	9.05 ± 1.63	5.57 ± 2.37	4.67 ± 1.82	5.43 ± 2.19	$\mathbf{4.52 \pm 1.82}$
GB	Late	7.11 ± 3.60	6.46 ± 1.48	6.17 ± 2.19	6.66 ± 1.44	$\mathbf{5.98 \pm 1.44}$
GM	Late	3.46 ± 1.85	4.22 ± 1.74	4.17 ± 1.86	4.22 ± 1.74	$\mathbf{3.43 \pm 1.87}$
PMR	Late	$\mathbf{4.95 \pm 1.66}$	5.14 ± 2.67	5.60 ± 2.92	5.14 ± 2.67	5.46 ± 2.18
DNN	Late	$\mathbf{3.59 \pm 1.70}$	4.43 ± 2.54	5.02 ± 2.59	4.43 ± 2.54	4.11 ± 2.38
DNN	Early	$\mathbf{2.37 \pm 1.14}$	3.95 ± 1.25	4.28 ± 1.58	3.95 ± 1.25	4.05 ± 1.47
Gausian Process	Early	5.93 ± 1.78	$\mathbf{3.87 \pm 2.00}$	4.33 ± 1.83	$\mathbf{3.87 \pm 2.00}$	6.05 ± 2.10
Logistic Regression	Early	3.34 ± 1.30	3.76 ± 1.78	3.78 ± 1.83	3.76 ± 1.78	$\mathbf{2.79 \pm 2.20}$
SVM	Early	3.80 ± 1.66	$\mathbf{3.76 \pm 2.17}$	4.14 ± 1.70	$\mathbf{3.76 \pm 2.17}$	5.65 ± 2.49
Xgboost	Early	$\mathbf{3.53 \pm 1.62}$	4.58 ± 2.95	4.26 ± 2.88	4.58 ± 2.95	3.53 ± 2.90

Dynamic Scaling extends Temperature Scaling by introducing additional learnable parameters, enabling instance-wise adjustments. While this added risk of overfitting on small datasets, it is highly effective in larger datasets where sufficient validation set is available, which is evident in BRCA and METABRIC.
The only dataset where none of the methods able to calibrate is TCGA. In this case, only Matrix and Vector Scaling show slight improvement. Since this dataset is already well-calibrated (ECE $\leq\sim 5\%$), there is limited room for improvement, making post-processing largely unnecessary. It is also possible that the results are influenced by the dataset split or the chosen ECE hyperparameter. TCGA is

Table 4. ECE across calibration methods on TCGA. Row-wise minima in **bold**.

Model	Type	Before Calibration	Matrix Scaling	Temperature Scaling	Vector Scaling	Dynamic Scaling
MMB	Intermediate	44.79 ± 29.07	61.05 ± 30.72	35.71 ± 37.19	**34.70 ± 38.06**	61.07 ± 30.67
DFTMC	Intermediate	6.71 ± 3.36	6.20 ± 0.86	**5.14 ± 1.47**	6.48 ± 4.48	6.85 ± 2.65
GMU	Intermediate	**4.77 ± 1.05**	6.07 ± 5.66	4.97 ± 1.67	6.07 ± 5.66	5.28 ± 1.63
GB	Late	**8.97 ± 2.95**	9.11 ± 2.85	10.11 ± 2.60	9.75 ± 3.50	9.85 ± 1.43
GM	Late	**4.09 ± 1.45**	5.70 ± 5.06	4.61 ± 3.80	5.70 ± 5.06	5.88 ± 4.30
PMR	Late	7.59 ± 2.45	**3.63 ± 1.97**	4.51 ± 1.78	4.15 ± 2.08	5.18 ± 1.84
DNN	Late	5.49 ± 1.85	**5.20 ± 3.78**	5.22 ± 2.97	**5.20 ± 3.78**	6.76 ± 3.99
DNN	Early	**5.04 ± 0.58**	6.15 ± 4.55	6.30 ± 4.85	6.15 ± 4.55	7.85 ± 5.36
Gausian Process	Early	**2.57 ± 2.10**	5.19 ± 4.22	5.19 ± 4.22	5.19 ± 4.22	5.19 ± 4.22
Logistic Regression	Early	**5.64 ± 0.89**	6.36 ± 2.62	6.45 ± 2.82	6.36 ± 2.62	7.25 ± 2.58
SVM	Early	**3.60 ± 1.90**	7.19 ± 1.77	6.59 ± 2.29	7.19 ± 1.77	8.61 ± 3.32
Xgboost	Early	6.66 ± 2.33	**5.72 ± 2.26**	6.20 ± 2.63	**5.72 ± 2.26**	6.53 ± 2.86

highly imbalanced evident by a drastic dropped in F1 macro while the accuracies of most model stays at 80%. This highlights that imbalance inflates accuracy while damaging both fairness and calibration.

Another noteworthy observation is that high calibration doesn't imply high accuracy or vice versa. In BRCA, XGBoost achieved $\sim 80\%$ accuracy but was highly miscalibrated (ECE ≈ 44.5 before scaling). Conversely, Logistic Regression reached only $\sim 80\%$ accuracy but was much better calibrated (ECE ≈ 7.8). This shows that higher accuracy does not guarantee trustworthy probability estimates. Even after calibrated, a model that output well-calibrated probabilities can still misclassify frequently.

5.3 Classification Performance After Calibration

Tables 5, 6, 7 and 8 illustrate the classification performance after calibration. It is worth noting that the temperature scaling and dynamic scaling goal are to soften the probability. As a result, model accuracy remains unchanged under these methods. For this reason, we decided to not include them in the table and will use the accuracy before calibration when evaluating Temperature Scaling and Dynamic Scaling.

With high-dimensional dataset such as BRCA, calibration generally led to a drop in accuracy. This indicates that when models (i.e. GM, SVM) are already well-tuned, recalibration can disrupt the decision boundaries and reduce classification performance.

Table 5. Accuracy across calibration methods on BRCA. Row-wise maxima in **bold**.

Model	Type	Before Calibration	Matrix Scaling	Vector Scaling
MMB	Intermediate	**81.28 ± 1.95**	76.94 ± 4.50	80.82 ± 3.57
DFTMC	Intermediate	**83.79 ± 4.54**	77.83 ± 6.26	82.88 ± 4.20
GMU	Intermediate	**79.90 ± 3.46**	63.27 ± 30.62	79.89 ± 3.87
GB	Late	70.98 ± 7.63	73.49 ± 6.75	**74.86 ± 6.75**
GM	Late	**84.24 ± 1.74**	78.29 ± 5.19	82.87 ± 2.60
PMR	Late	78.06 ± 5.53	79.21 ± 4.73	**79.90 ± 3.85**
DNN	Late	**83.79 ± 3.94**	65.88 ± 30.00	83.10 ± 3.11
DNN	Early	**80.59 ± 2.63**	77.85 ± 3.45	79.22 ± 1.61
Gausian Process	Early	77.62 ± 4.54	77.83 ± 4.95	**78.07 ± 4.69**
Logistic Regression	Early	**80.36 ± 2.69**	78.76 ± 2.40	77.63 ± 3.41
SVM	Early	**84.93 ± 4.49**	83.56 ± 5.51	80.14 ± 3.37
Xgboost	Early	**79.68 ± 3.30**	78.08 ± 3.25	64.43 ± 27.85

Table 6. Accuracy across calibration methods on ROSMAP. Row-wise maxima in **bold**.

Model	Type	Before Calibration	Matrix Scaling	Vector Scaling
MMB	Intermediate	**71.43 ± 5.35**	67.43 ± 5.19	67.43 ± 5.19
DFTMC	Intermediate	**76.00 ± 9.82**	71.43 ± 11.61	**76.00 ± 9.82**
GMU	Intermediate	76.00 ± 6.26	76.57 ± 8.43	**77.14 ± 5.35**
GB	Late	**73.71 ± 3.73**	73.14 ± 2.56	73.14 ± 2.56
GM	Late	80.00 ± 6.70	79.43 ± 8.67	**80.57 ± 7.40**
PMR	Late	76.57 ± 10.58	**77.71 ± 9.13**	**77.71 ± 9.13**
DNN	Late	**82.86 ± 8.33**	81.14 ± 9.17	81.14 ± 8.71
DNN	Early	**77.14 ± 7.56**	76.00 ± 9.39	75.43 ± 9.60
Gausian Process	Early	**74.29 ± 8.08**	69.71 ± 6.58	69.71 ± 6.58
Logistic Regression	Early	**78.29 ± 9.60**	77.71 ± 8.67	77.71 ± 8.67
SVM	Early	73.71 ± 9.13	**74.86 ± 8.67**	**74.86 ± 8.67**
Xgboost	Early	**75.43 ± 8.94**	68.00 ± 5.50	68.00 ± 5.50

In ROSMAP, TCGA, and METABRIC, the effect of calibration on accuracy was less consistent. Some models experienced slight gains (e.g., GB in BRCA, PMR in TCGA), while others showed minor reductions. In ROSMAP, certain models show drastic drop in accuracy (i.e. MMB, Gaussian Process), likely due to overfitting the small validation set. In METABRIC, the largest dataset, accuracy was generally stable, with changes remaining within error margins. In these instances, calibration techniques such as Temperature Scaling and Dynamic Scal-

Table 7. Accuracy across calibration methods on METABRIC. Row-wise maxima in **bold**.

Model	Type	Before Calibration	Matrix Scaling	Vector Scaling
MMB	Intermediate	79.60 ± 2.73	78.28 ± 1.96	**79.90 ± 3.28**
DFTMC	Intermediate	**85.25 ± 1.73**	85.25 ± 1.61	84.95 ± 2.00
GMU	Intermediate	85.35 ± 1.86	**85.35 ± 2.58**	84.85 ± 2.11
GB	Late	78.28 ± 4.30	78.18 ± 4.47	**78.38 ± 4.38**
GM	Late	**84.04 ± 1.94**	83.94 ± 2.43	83.94 ± 2.43
PMR	Late	**85.86 ± 2.11**	85.86 ± 2.05	85.86 ± 2.05
DNN	Late	**85.66 ± 2.44**	85.35 ± 2.20	85.35 ± 2.20
DNN	Early	**86.26 ± 1.40**	85.66 ± 1.66	85.66 ± 1.66
Gausian Process	Early	82.73 ± 2.06	**83.13 ± 1.91**	**83.13 ± 1.91**
Logistic Regression	Early	**84.75 ± 1.53**	84.44 ± 1.69	84.44 ± 1.69
SVM	Early	83.43 ± 2.16	**83.64 ± 2.25**	**83.64 ± 2.25**
Xgboost	Early	**85.76 ± 1.80**	85.76 ± 1.53	85.76 ± 1.53

Table 8. Accuracy across calibration methods on TCGA. Row-wise maxima in **bold**.

Model	Type	Before Calibration	Matrix Scaling	Vector Scaling
MMB	Intermediate	76.30 ± 3.85	77.04 ± 5.38	**78.33 ± 5.26**
DFTMC	Intermediate	77.59 ± 4.06	75.00 ± 4.04	**77.78 ± 3.87**
GMU	Intermediate	**79.44 ± 3.84**	79.26 ± 5.01	79.26 ± 5.01
GB	Late	73.33 ± 4.87	73.52 ± 5.14	**73.70 ± 5.10**
GM	Late	**80.37 ± 3.49**	79.82 ± 2.40	79.82 ± 2.40
PMR	Late	79.44 ± 4.16	80.19 ± 4.47	**80.37 ± 4.21**
DNN	Late	**81.30 ± 3.30**	80.56 ± 2.78	80.56 ± 2.78
DNN	Early	**80.74 ± 3.03**	80.00 ± 3.85	80.00 ± 3.85
Gausian Process	Early	**76.11 ± 3.49**	**76.11 ± 3.49**	**76.11 ± 3.49**
Logistic Regression	Early	**80.56 ± 3.98**	79.63 ± 3.21	79.63 ± 3.21
SVM	Early	**80.19 ± 1.55**	79.63 ± 1.46	79.63 ± 1.46
Xgboost	Early	81.11 ± 3.31	**81.48 ± 3.46**	**81.48 ± 3.46**

ing are generally better as they only rescale the probabilities, therefore preserving the accuracy.

6   Conclusion and Future Direction

This study provides a comprehensive study of four post-hoc calibration methods and multimodal fusion strategies across four multi-omics datasets (BRCA, TCGA, ROSMAP and METABRIC). We found that calibration does not directly correlate with accuracy, and its performance is largely influenced by dataset characteristics rather than model type or fusion strategy. As this study focuses solely on multi-omics and calibration techniques, foundational models were not investigated and are left for future work.

We recognise that Dynamic Scaling is highly sensitive to small dataset sizes. Future work could address this by incorporating regularisation during training or by developing calibration methods that operate at the modality and instance-level.

7 Appendix

Appendix is accessible via https://github.com/dkav0206/Unveiling-Reliability-in-Multi-Omics-Classification-Fusion-Calibration-and-Dynamic-Scaling

Acknowledgment. Part of this research was supported by the Monash node of MAC-SYS (ARC Centre of Excellence for the Mathematical Analysis of Cellular Systems)

References

1. Acosta, J.N., Falcone, G.J., Rajpurkar, P., et al.: Multimodal biomedical AI. Nat. Med. **28**, 1773–1784 (2022)
2. Arevalo, J., Solorio, T., Montes-y Gómez, M., González, F.A.: Gated multimodal networks. Neural Comput. Appl. **3** (2020)
3. Arya, N., Saha, S.: Multi-modal advanced deep learning architectures for breast cancer survival prediction. Knowl.-Based Syst. **221**, 106965 (2021)
4. Bishop, C.M., Nasrabadi,N.M.: Pattern recognition and machine learning, volume 4. Springer (2006)
5. Chen, J., Su, B.: Transfer knowledge from head to tail: uncertainty calibration under long-tailed distribution. In: Proceedings of the IEEE/CVF Conference on Computer Vision and Pattern Recognition, pp. 19978–19987 (2023)
6. Fan, Y., Xu, W., Wang, H., Wang, J., Guo.: PMR: Prototypical Modal Rebalance for Multimodal Learning. In: Proceedings of the IEEE/CVF Conference on Computer Vision and Pattern Recognition (CVPR), pp. 20029–20038, vol. 6 (2023)
7. Guo, C., Pleiss, G., Sun, Y., Weinberger, K.Q.: On calibration of modern neural networks (2017)
8. Han, Z., Yang, F., Huang, J., Zhang, C., Yao, J.: Multimodal dynamics: dynamical fusion for trustworthy multimodal classification. In: Proceedings of the IEEE/CVF Conference on Computer Vision and Pattern Recognition (CVPR), pp. 3 (2022)
9. Mai, S., Sun, Y., Xiong, A., Zeng, Y., Hu, H.: Multimodal boosting: addressing noisy modalities and identifying modality contribution. IEEE Trans. Multimedia, pp. 1–16, vol. 1 (2023)
10. Montesinos-López, O.A., et al.: A new deep learning calibration method enhances genome-based prediction of continuous crop traits. Front. Genetics, **12**, 798840 (2021)
11. Nayak, R., Luong, K.: Subspace Learning for Multi-aspect Data. In: Nayak, R., Luong, K., eds., Multi-aspect Learning: Methods and Applications, pp. 77–101. Springer International Publishing, Cham (2023)

12. Peng, X., Wei, Y., Deng, A., Wang, D., Hu, D.: Balanced Multimodal Learning via On-the-Fly Gradient Modulation. In: Proceedings of the IEEE/CVF Conference on Computer Vision and Pattern Recognition (CVPR), pp. 8238–8247 vol. 6 (2022)
13. Pingi, S.T., Zhang, D., Bashar, M.A., Nayak, R.: Joint representation learning with generative adversarial imputation network for improved classification of longitudinal data. Data Sci. Eng. **9**(1), 5–25 (2024)
14. Stahlschmidt, S.R., Ulfenborg, B., Synnergren, J.: Multimodal deep learning for biomedical data fusion: a review. Briefings Bioinform. **23**(2), bbab569, vol. 01 (2022)
15. Sønderstrup, I.M.H., et al.: Subtypes in BRCA-mutated breast cancer. Human Pathol. **84**, 192–201 (2019)
16. Tabakhi, S., Suvon, M.N.I., Ahadian, P., Lu, H.: Multimodal learning for multi-omics: a survey. World Sci. Ann. Rev. Artif. Intell. **01** (2023)
17. Calster, B.V.,et al.: Evaluating diagnostic tests, and prediction models of the STRATOS initiative Bossuyt Patrick Collins Gary S. Macaskill Petra McLernon David J. Moons Karel GM Steyerberg Ewout W. Van Calster Ben van Smeden Maarten Vickers Andrew J. Calibration: the achilles heel of predictive analytics. BMC Med. **17**(1), 230 (2019)
18. Wang, C., Calibration in deep learning: a survey of the state-of-the-art (2024)
19. Wang, C., Lye, X., Kaalia, R., Kumar, P., Rajapakse, J.C.: Deep learning and multi-omics approach to predict drug responses in cancer. BMC Bioinform. **22**(10), 632 (2022)
20. Wang, D.-B., Feng, L., Zhang, M.-L.: Rethinking calibration of deep neural networks: Do not be afraid of overconfidence. In: Ranzato, M., Beygelzimer, A., Dauphin, Y., Liang, P.S., Wortman Vaughan, J., eds., Advances in Neural Information Processing Systems, volume 34, pp. 11809–11820. Curran Associates, Inc., (2021)
21. Wang, T., Shao, W., Huang, Z., Tang, H., Zhang, J., Ding, Z., Huang, K.: MOGONET integrates multi-omics data using graph convolutional networks allowing patient classification and biomarker identification. Nat. Commun. **12**(1), 3445 (2021)
22. WWang, J., Tran, D., MFeiszli, D.: What makes training multi-modal classification networks hard? In: 2020 IEEE/CVF Conference on Computer Vision and Pattern Recognition (CVPR). IEEE, **6**(2020)
23. Wu, C., Zhou, F., Ren, J., Li, X., Jiang, Y., Ma, S.: A selective review of multi-level omics data integration using variable selection. High Through. **8**(1), 4 (2019)
24. Zack, M., et al.: Artificial intelligence and multi-omics in pharmacogenomics: a new era of precision medicine. Mayo Clin. Proc. Digit. Health **3**(3), 100246 (2025)
25. Zhang, D., Nayak, R., Bashar, M.A.: Exploring fusion strategies in deep learning models for multi-modal classification. In: Data Mining (2021
26. Zhong, Z., Jiang, W., Zhang, J., et al.: Identification and validation of a novel 16-gene prognostic signature for patients with breast cancer. Sci. Rep. **12**(1), 12349 (2022)

Stability Evaluation of Clusterings Across Time

Sergej Korlakov[(✉)] [iD], Nina A. Liebrand [iD], and Stefan Conrad [iD]

Faculty of Mathematics and Natural Sciences, Department of Computer Science,
Heinrich Heine University, Universitätsstr. 1, 40225 Düsseldorf, Germany
{sergej.korlakov,nina.liebrand,stefan.conrad}@hhu.de

Abstract. Clustering is a fundamental unsupervised learning technique for grouping similar instances in unlabeled datasets. When data evolves over time, clustering can be applied at each time point, resulting in an over-time clustering (ot-clustering). Several approaches incorporate temporal context into ot-clustering in different ways, all aiming to uncover unknown patterns in multivariate time series data. However, many of these methods require tuning of a weighting parameter for temporal context, which must be set manually or determined through computationally intensive algorithms, depending on the method. More recently, a study introduced a concept of over-time stability for such ot-clusterings and proposed a method that measures changes in clusterings over time, eliminating the need for a weighting parameter for temporal context.

In this work, we propose an alternative to the existing definition of over-time stability and introduce a framework for evaluating ot-clusterings based on this definition. Beyond producing stability scores, our approach enables the visual analysis of ot-clusterings in a two-dimensional space, regardless of the number of dimensions in the original dataset.

Our experiments demonstrate that applying the proposed framework yields alternative yet meaningful, ot-clusterings compared to not taking temporal context into account. Compared to the only other ot-clustering stability evaluation method called CLOSE, our approach offers a greater flexibility in defining temporal context, produces ot-clusterings with less granularity but also reduced noise, and achieves better runtimes.

Keywords: Clustering over time · Clustering of multivariate time series · Over-time stability evaluation of clusterings

1 Introduction

Clustering has been used for decades for different purposes like customer and market segmentation [1], anomaly detection [2] and, more generally, data exploration [6]. As a technique within unsupervised learning, clustering is commonly applied to unlabeled datasets, relying solely on the intrinsic structure of the data [12]. State-of-the-art methods like K-means [10] and DBSCAN [5] exploit this structure, to group similar instances without supervision.

Q. V. Nguyen et al. (Eds.): AusDM 2025, CCIS 2765, pp. 79–93, 2026.
https://doi.org/10.1007/978-981-95-6786-7_6

When a dataset consists of multiple time series, the information available for clustering increases due to the evolving distribution of data points over time. Chakrabarti et al. [3] were the first introducing the clustering of such data, calling their method evolutionary clustering. Instead of treating each time point independently, evolutionary clustering incorporates the information about previous clustering with a certain weight into the clustering of the next time point. To achieve this, the authors utilize two criteria: snapshot quality and history cost. The snapshot quality at time t indicates how good a clustering is at t and can be assessed using metrics such as the silhouette coefficient [13]. In contrast, history cost quantifies the dissimilarity of two consecutive clusterings. These two criteria are combined into the following objective function:

$$sq_t - cp \cdot hc_t, \tag{1}$$

where sq_t is the snapshot quality at time t, hc_t is the history cost, and cp is a user-defined weight that balances the trade-off between snapshot quality and history cost.

Subsequent studies have either extended the objective function of evolutionary clustering to improve the over-time clustering quality [4,9], or pursued different methodological directions [11,14]. Some works left the issue of parameter selection unaddressed, as seen in [4], while others proposed specific strategies for determining the weight parameter within their methods [17,18]. However, approaches that involve automatic parameter determination encounter significant limitations, such as being NP-hard [17] or failing to incorporate temporal context [18]. Approaches that require the user to set the weight manually, typically involve experimenting with different weight values and analyzing the resulting over-time clusterings.

More recent works took a different direction. For instance, the approach by Korlakov et al. [8] focuses on a more intuitive weighting between snapshot quality and the incorporation of temporal information. They achieve this through a multi-objective optimization perspective on the relationship between the snapshot quality sq and temporal quality tq with latter defined as:

$$tq = 1 - hc, \tag{2}$$

where hc denotes the history cost. In contrast, Klassen et al. [7] pursued a different question: what characterizes a good or bad clustering over time? Their work led to the development of a stability concept for time series clustering across time. According to their definition, a stable clustering is one in which each object consistently remains with the same set of peers over time, and these peers continue to be assigned to the same clusters as the object itself. Based on their stability concept, the authors developed a method called CLOSE, which rates an over-time clustering based on its stability, thereby eliminating the need to explicitly weight the temporal context. However, CLOSE shows several limitations. First, its object-level perspective on temporal context is computationally intensive, leading to long runtimes. Second, the method employs a fixed definition of stability, which is tightly embedded in the underlying equations. This lack of flexibility may restrict users in exploring different notions of temporal stability, potentially limiting its applicability in broader data exploration scenarios.

Herein, we introduce our solution called SECAT **S**tability **E**valuation of **C**lusterings **A**cross **T**ime, a novel framework for assessing the stability to over-time clustered data. Similar to evolutionary clustering, SECAT quantifies temporal information by measuring differences between consecutive time point clusterings (tp-clusterings). However, instead of relying on a user-defined weight to balance between temporal-and snapshot quality, we extended the concept of stability of Klassen et al. [7] and incorporated it into SECAT. Specifically, we constructed a two-dimensional space defined by snapshot quality (sq) and temporal quality (tq), and evaluated both the fluctuation of the resulting (sq, tq)-tuples over-time and their proximity to the optimal values of sq and tq. Under our framework, a maximal stable over-time clustering (ot-clustering) achieves maximal snapshot- and temporal quality at every time point. When comparing two ot-clusterings with the same average (but not maximal) snapshot- and temporal quality, the one exhibiting smaller fluctuations in sq and tq over time is considered more stable. The design of our framework not only yields an interpretable stability score but also enables an intuitive visual analysis of snapshot- and temporal quality dynamics over time, as demonstrated in this work.

The main contributions of this paper to the over-time clustering of time series are:

- A new definition of clustering stability across time;
- A stability measure that is flexible with respect to the choice of snapshot- and temporal quality metrics;
- A method that avoids the need for weighting of the temporal context.

2 Preliminaries

Definition 1 (Time Series). *A time series T is an ordered sequence of observations measured at n discrete time points. In the multivariate case, each observation is a vector in $\mathbb{R}^D$:*

$$T = (o_{t_1}, o_{t_2}, \ldots, o_{t_n}), \quad o_{t_i} \in \mathbb{R}^D,$$

where $t_1 < t_2 < \cdots < t_n$ denote the time points.

Definition 2 (Dataset). *A dataset $\mathcal{D} = \{T_1, T_2, \ldots, T_m\}$ consists of m time series, each of equal length, with the assumption that no data is missing.*

Definition 3 (Time Point Clustering). *Let all observations of a single time point t_i from the m multivariate time series be given as a set $\mathcal{O}_{t_i} = \{o_{t_i,1}, \ldots, o_{t_i,l}\}$. The clustering $C_{t_i} = \{C_{t_i,1}, \ldots, C_{t_i,k}\}$ partitions $\mathcal{O}_{t_i}$ into k pairwise disjoint sets representing clusters. Each observation belongs to exactly one cluster. Each noise point is assigned to its own unique cluster with a negative index.*

Definition 4 (Over-Time Clustering). *An over-time clustering (ot-clustering) C^{ot} is an sequence of time point clusterings ordered by time.*

Definition 5 (Snapshot Quality). *The snapshot quality sq_i refers to the quality of a clustering C_{t_i} at a time point t_i. It can be quantified using bounded metrics for clustering evaluation like silhouette coefficient.*

Definition 6 (Temporal Quality). *The temporal quality tq_i indicates the difference of the time point clustering at time point t_i compared to the time point clustering at time point t_{i-1}. It can be quantified through bounded metrics which evaluate the similarity between two time point clusterings, e.g. Jaccard coefficient, the adjusted Rand index or the adjusted mutual information [16].*

Definition 7 (Distance). *Unless stated otherwise, the term "distance" in this work denotes the Euclidean distance.*

3   Method

Given an over-time clustering of time series (ot-clustering), SECAT assigns a score within the interval $[0, 1]$, enabling the comparison of different ot-clusterings with regard to their stability. A score of zero indicates minimum stability of an ot-clustering, while a score of one represents maximum stability. Without loss of generality, we assume in this work, that all time series begin at time point $t_{i=1}$.

To calculate the SECAT score for an ot-clustering of a set of time series of length n each, first, the snapshot quality sq_i must be calculated for each time point t_i with $i \in \{1, \ldots, n\}$ and the temporal quality $tq_{\hat{i}}$ for each time point $t_{\hat{i}}$ with $\hat{i} \in \{2, \ldots, n\}$. The differentiation of the sets of time points with regard to sq_i and $tq_{\hat{i}}$ is required, since temporal quality at time point $t_i = 1$ is undefined due to unavailable preceding information.

The calculation of snapshot- and temporal qualitys results in sets $SQ = \{sq_1, \ldots, sq_n\}$ and $TQ = \{tq_2, \ldots, tq_n\}$. For a clear identification of the elements from these sets, we define a total order $<_t$ as follows:

$$sq_a <_t sq_b \iff a < b \quad \text{and} \quad tq_a <_t tq_b \iff a < b \tag{3}$$

Both, the sets SQ, TQ and the total order $<_t$ are used for the definition of the equation of the SECAT score in the remainder of this section.

3.1   SECAT Score

The equation of the SECAT score consists of multiple components. The meaning of each component is illustrated in Fig. 1. First, the centroid c is calculated as:

$$c = \left(\frac{1}{|SQ|} \sum_{sq \in SQ} sq, \ \frac{1}{|TQ|} \sum_{tq \in TQ} tq \right) = (sq_c, tq_c), \tag{4}$$

where sq_c and tq_c represent the average snapshot- and temporal quality, respectively.

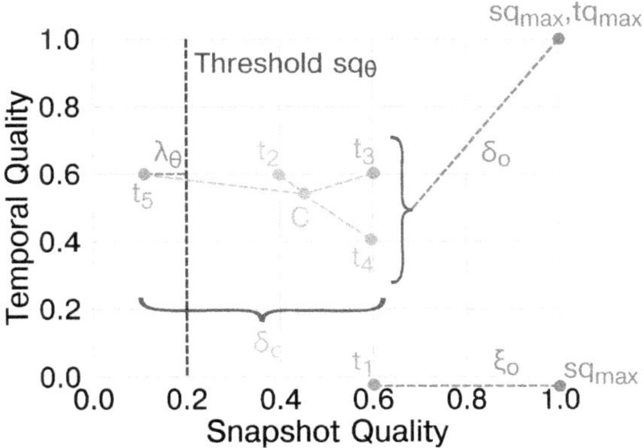

Fig. 1. Schematic representation of the components of the SECAT score within the space defined by snapshot- and temporal quality. Points marked t_1 to t_5 specify the time points of the corresponding (sq, tq)-tuples.

Next, ξ_c is calculated as the distance of snapshot quality at the time point $t_i = 1$ to the sq_c of the centroid c, normalized by the maximum achievable distance between $min_{<_t} SQ$ and sq_c:

$$\xi_c = \frac{|min_{<_t} SQ - sq_c|}{0.5 \cdot |sq_{max} - sq_{min}|}. \tag{5}$$

Here, sq_{max} and sq_{min} denote the upper and lower bound of the snapshot quality metric, respectively.

Subsequently, ξ_o, which is the normalized distance between maximum snapshot quality sq_{max} and the snapshot quality at time point $t_{i=1}$, is calculated as:

$$\xi_o = \frac{|min_{<_t} SQ - sq_{max}|}{|sq_{max} - sq_{min}|} \tag{6}$$

For all subsequent time points t_i with $i \geq 2$, the euclidean distance between each tuple $(sq_{<_t}^{(i)}, tq_{<_t}^{(i)})$ and the centroid $c = (sq_c, tq_c)$, normalized by the maximum possible distance, is calculated and summed as follows:

$$\delta_c = \sum_{i=2}^{n} \frac{\sqrt{(sq_{<_t}^{(i)} - sq_c)^2 + (tq_{<_t}^{(i)} - tq_c)^2}}{0.5 \cdot \sqrt{(sq_{max} - sq_{min})^2 + (tq_{max} - tq_{min})^2}}. \tag{7}$$

Next, for time points t_j with $j \geq 2$, the euclidean distance is calculated between each tuple $(sq_{<_t}^{(j)}, tq_{<_t}^{(j)})$ and the optimum $o = (sq_{max}, tq_{max})$, again normalized by the maximum possible distance and summed as follows:

$$\delta_o = \sum_{j=2}^{n} \frac{\sqrt{(sq_{<_t}^{(j)} - sq_{max})^2 + (tq_{<_t}^{(j)} - tq_{max})^2}}{\sqrt{(sq_{max} - sq_{min})^2 + (tq_{max} - tq_{min})^2}}. \tag{8}$$

Finally, the SECAT score is combined based on the Eqs. 5 to 8 as follows:

$$SECAT_{\xi,\delta} = \frac{(1 - \frac{\xi_c + \delta_c}{n}) + (1 - \frac{\xi_o + \delta_o}{n})}{2}.$$

(9)

Here, the distances to the centroid c at time points t_i with $i = 1$ and with $j \geq 2$ are summed, averaged by the number of given time points, subsequently substracted from one, to get a higher score for lower distances to c and vice versa. Same procedure is applied to the distances to the optimum o.

3.2 Penalty Term

In Eq. 9 snapshot- and temporal quality are treated symmetrically. As a result, certain edge cases can arise, where different ot-clusterings yield identical SECAT scores, despite having significantly different snapshot- and temporal quality. To solve this problem, we introduce a penalty factor λ_θ, which penalizes tp-clusterings where the snapshot quality is below the freely selectable threshold sq_θ. Given $\hat{SQ} = \{sq \in SQ | sq < sq_\theta\}$, the equation for λ_θ is as follows:

$$\lambda_\theta = \frac{1}{|\hat{SQ}|} \cdot \sum_{sq \in \hat{SQ}} |sq - sq_\theta|$$

(10)

Equation 7 extended by the penalty term λ_θ from Eq. 10 is given by:

$$SECAT = \frac{(1 - \frac{\xi_c + \delta_c}{n}) + (1 - \frac{\xi_o + \delta_o}{n})}{2} \cdot (|sq_{max} - sq_{min}| - \lambda_\theta)$$

(11)

4 Experiments

To evaluate SECAT, we compare ot-clusterings with the highest SECAT scores both against each other as well as against two other methods: the baseline and CLOSE. The baseline is defined as:

$$Baseline = \frac{1}{|SQ|} \sum_{sq \in SQ} sq$$

(12)

The authors of CLOSE propose two formulations of the method: one that applies quality measures to individual clusters and another that applies them to clusterings. In this work, we adopt the latter formulation, using the silhouette coefficient as the quality measure for clusterings. Additionally, we penalize noise in ot-clusterings if present, as described by Klassen et al. [7].

For the baseline, CLOSE, and SECAT, we use the silhouette coefficient as the snapshot quality measure. Since CLOSE requires clustering quality measures which return values between 0 and 1, we normalized the silhouette coefficient scores using the minâĂŞmax normalization. As the temporal quality measure for CLOSE and SECAT, we used the Jaccard coefficient, as it is natively supported by CLOSE. We also evaluate SECAT rankings with alternative temporal quality measures, namely adjusted mutual information (AMI) and adjusted Rand index (ARI). The SECAT penalty term for snapshot quality is set to 0.5, which corresponds to the normalized silhouette coefficient value indicating overlapping clusters.

In the course of this section, we also report the average runtimes with corresponding standard deviations for all three methods. All measurements were conducted on the same machine equipped with an Intel i7-14700K CPU, 32 GB RAM, and an M.2 SSD.

4.1   Generated Dataset and K-Means

In this experiment, we aimed to examine the rankings produced by the presented stability evaluation methods for ot-clusterings in a controlled setting. Therefore, we generated a two-dimensional time series dataset as follows: first, we defined the positions of three centroids and created normally distributed clusters of points around each of them. Then, to get data over-time, we defined three time points. At each of these time points, we moved two of the three clusters closer together to mimic a scenario in which clusters split up and merge over-time. The resulting dataset, along with the original cluster assignments, is shown in Fig. 2.

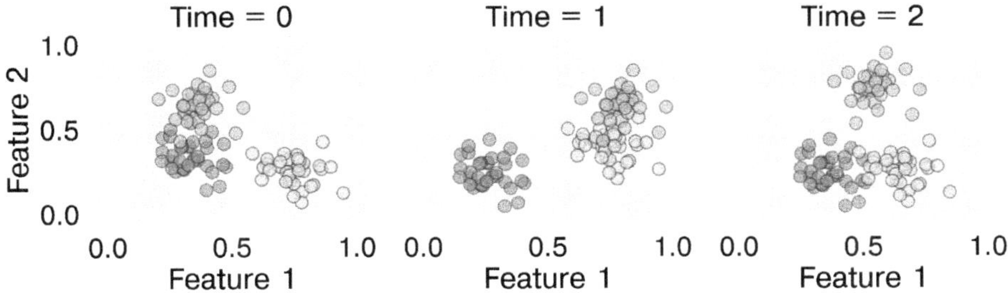

Fig. 2. Generated Dataset. The colors represent assignment of objects to the centroids which were created during the generation of objects.

In order to compare rankings based on presented stability evaluation methods, we removed the original cluster assignments of the generated dataset, clustered it per time point with K-means with $k \in [1, 10]$ and computed baseline, CLOSE and SECAT scores for each resulted over-time clustering. Table 1 displays the rankings and corresponding scores for each of the three methods.

Focusing first on the results of the baseline in Table 1, the highest scores were obtained for $k = 2$, $k = 3$ and $k = 4$, respectively. Since the baseline does not incorporate temporal context and uses the silhouette coefficient as the snapshot quality metric, this ranking shows, that the clustering with $k = 2$ yields the highest average silhouette coefficient.

Table 1. Comparison of ot-clustering rankings across evaluation methods of the generated dataset with regard to K-means.

Baseline			CLOSE			SECAT				
Rank	k	Score	Rank	k	Score	Rank	k	Score	Av. sq	Av. tq
1	2	0.809	1	10	0.026	1	3	0.803	0.790 ± 0.002	0.426 ± 0.039
2	3	0.790	2	9	0.021	2	2	0.776	0.809 ± 0.023	0.389 ± 0.057
3	4	0.737	3	8	0.019	3	4	0.770	0.737 ± 0.007	0.301 ± 0.015

In contrast to the baseline, CLOSE incorporates temporal context to calculate an over-time stability score. Here, the highest scores were achieved for $k = 10$, $k = 9$ and $k = 8$, respectively. The way temporal information is integrated into the CLOSE score results in a ranking, where ot-clusterings with a larger number of clusters receive higher scores, despite exhibiting much lower average snapshot quality. For instance, while the snapshot quality for $k = 2$ is 0.809 ± 0.023, it drops to 0.673 ± 0.006 for $k = 10$.

The highest SECAT scores were obtained for ot-clusterings with $k = 3$, $k = 2$ and $k = 4$, respectively. Among all methods, only SECAT identifies a top-ranked clustering whose number of clusters matches the number of centroids originally used to generate the dataset. A closer examination of snapshot- and temporal quality of the top three ot-clusterings reveals that the second-ranked ot-clustering exhibits a higher average snapshot quality but a lower average temporal quality, along with greater standard deviations for both metrics. This combination results in a lower overall SECAT score compared to the top-ranked ot-clustering.

To ensure a fair comparison between CLOSE and SECAT, and given that CLOSE supports Jaccard coefficient for quantification of temporal context, we also employed it in SECAT. The results presented above are based on this configuration. However, unlike CLOSE, SECAT is compatible with any bounded metric capable of comparing two clusterings. To explore the impact of different temporal quality metrics, we extended this experiment by evaluating SECAT rankings using adjusted Rand index (ARI) and adjusted mutual information (AMI). As result, the change of temporal quality to both, ARI and AMI, produced the similar top three rankings compared to the Jaccard coefficient: $k = 3$, $k = 4$, and $k = 5$, respectively. However, the ot-clustering with $k = 2$, which was second-ranked under the Jaccard coefficient, dropped to sixth place with ARI and ninth with AMI. This result highlights the sensitivity of stability rankings to the choice of temporal quality metric and suggests, that using multiple temporal quality metrics as an ensemble may lead to more robust and reliable evaluations.

In addition to comparing rankings and exploring different temporal quality metrics, we also measured the runtime of all three methods during the computation of stability scores. The baseline achieved a runtime of 0.002 ± 0.000 seconds, SECAT 0.009 ± 0.001 seconds, and CLOSE 0.690 ± 0.008 seconds. On average, SECAT runs slower than the baseline due to its incorporation of temporal information but is significantly faster than CLOSE. This difference arises from how temporal context is handled: CLOSE evaluates, for each time series, which peers remain grouped together across all time points, requiring extensive comparisons, whereas SECAT considers only the similarity between temporally consecutive tp-clusterings, resulting in fewer computations and improved runtime.

4.2 Generated Dataset and K-Means: Variation of Standard Deviation

As a follow-up experiment, we adopted the generation settings of the dataset shown in Fig. 2 and varied the standard deviation σ within the interval $[0.05, 0.15]$ in steps of 0.01. The random seed and initial centroids were kept constant to ensure clearer comparability. For each σ, we removed the original cluster labels, re-clustered the resulting dataset using different values of k with $k \in [1, 10]$, computed the stability scores using the baseline, CLOSE and SECAT, and ranked the resulting ot-clusterings accordingly. For the calculation of the stability scores, we used the Jaccard coefficient for tq and the silhouette coefficient for sq.

The highest scores together with their corresponding k-values are illustrated in Fig. 3. The x-axis indicates the standard deviation used for generating the dataset. The bars represent the number of clusters k of the ot-clusterings that achieved the highest scores according to the baseline, SECAT and CLOSE, corresponding to the left y-axis. The right y-axis denotes the maximum achieved score per method, visualized by the line plots.

The highest baseline scores range from 0.976 to 0.712 across all standard deviation values with $\sigma \in [0.05, 0.15]$ (step size 0.01) and decrease consistently with an increasing σ. Up to σ of 0.06, the baseline assigns the highest scores to ot-clusterings with $k = 3$. For higher values, including $\sigma = 0.15$, the highest scores are obtained for ot-clusterings with $k = 2$. This indicates that from $\sigma = 0.06$ onward, the average silhouette coefficient of ot-clusterings with $k = 3$ becomes lower than that of ot-clusterings with $k = 2$.

In contrast, the highest CLOSE scores range from 0.027 to 0.014, showing no recognizable trend. The corresponding k values vary between eight and ten, with no apparent dependence on the values of the standard deviation.

Compared to the baseline, the assigned highest SECAT scores range from 0.990 to 0.765 within the same interval of standard deviation values σ and step size, exhibiting a similar downward trend with increasing σ and having a slightly lower variation in assigned scores. Moreover, SECAT assigns the highest scores to ot-clusterings with $k = 3$ across a wider range of σ, up to and including $\sigma = 0.14$. Only from $\sigma = 0.15$ onward, the SECAT ranking shifts to a top-ranked ot-clustering with $k = 2$.

The decrease in baseline scores with increasing standard deviation occurs because the clusters at each time point increasingly overlap, which, on average, lowers the silhouette coefficient for a fixed number of clusters. SECAT's incorporation of temporal context mitigates this effect by reducing the influence of the snapshot quality metric on the stability score. At the same time, it increases

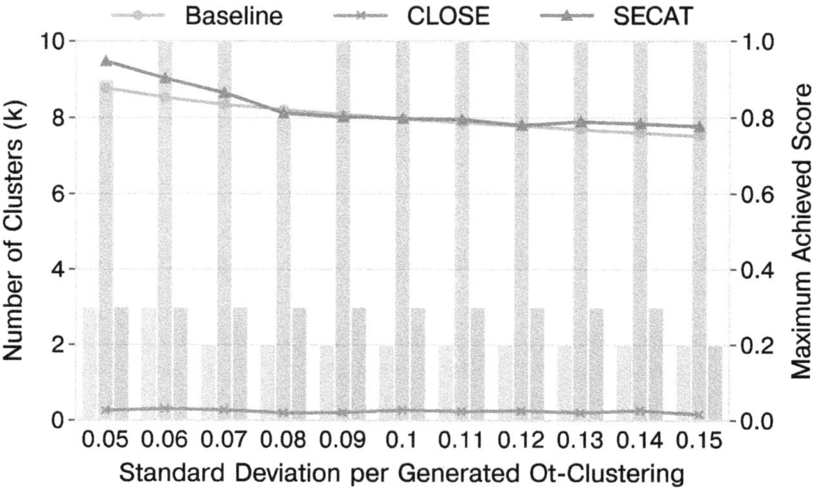

Fig. 3. Maximum achieved scores of the applied ot-clustering stability methods (right y-axis) along with the corresponding number of clusters of the top-ranked ot-clusterings (left y-axis) for each standard deviation value σ (x-axis).

the contribution of preceding ot-clusterings to the stability score through the corresponding temporal quality metric, leading to more stable ot-clusterings.

4.3   EIKON Dataset and DBSCAN

In this experiment, we use the dataset extracted by Tatusch et al. [15], which contains the features net sales and expected return for 30 randomly selected companies over the time period from 2008 to 2013. Each value for both features was first normalized by total assets of the corresponding company. Subsequently, the min-max normalization was applied by the authors.

We applied DBSCAN to cluster the dataset at each time point using a parameter range of $\epsilon \in [0.05, 1.0]$ in steps of 0.01, and $minPts \in \{2, 3, 4\}$. For each resulting ot-clustering, we computed stability scores using the baseline, CLOSE and SECAT, ranking the clusterings according to the resulting scores from each method. Table 2 presents the rankings, along with the corresponding clustering parameters. For SECAT, the table also includes the average snapshot- and temporal quality values as well as their standard deviations.

Starting with the baseline, the highest scores were obtained for the following DBSCAN parameter combinations $(\epsilon, minPts)$: $(0.15, 4)$, $(0.14, 4)$, and $(0.16, 4)$, respectively. Since the baseline computes solely the average silhouette coefficient for each ot-clustering, these ot-clusterings represent those with the highest average silhouette scores. The average number of clusters for the top-three ot-clusterings is $1,667 \pm 0.816$ for the first, 1.833 ± 0.983 for the second, and 1.500 ± 0.837 for the third.

Table 2. Ranking of ot-clusterings of the EIKON dataset with regard to DBSCAN. R stands for Ranking, MP for $MinPts$, av. sq for average snapshot quality and av. tq for average temporal quality.

Baseline				CLOSE				SECAT					
R.	ϵ	MP	Score	R.	ϵ	MP	Score	R.	ϵ	MP	Score	Av. sq	Av. tq
1	0.15	4	0.721	1	0.06	2	0.140	1	0.10	2	0.740	0.688 ± 0.030	0.364 ± 0.153
2	0.14	4	0.719	2	0.07	2	0.130	2	0.16	2	0.734	0.705 ± 0.038	0.440 ± 0.183
3	0.16	4	0.719	3	0.08	2	0.123	3	0.16	3	0.727	0.712 ± 0.033	0.660 ± 0.222

In contrast, the top three ot-clusterings ranked by CLOSE show significantly lower values of ϵ and $minPts$. The average numbers of clusters (6.333 ± 1.633, 5.5 ± 1.643, and 5.667 ± 1.366, respectively) are higher than those of all others ot-clusterings. This suggests that CLOSE favors ot-clusterings with finer granularity in the case of DBSCAN as well, consistent with the observations made for the generated dataset using K-means.

Compared to the baseline and CLOSE, SECAT produced a different ranking with the following parameter combinations ($\epsilon, minPts$): $(0.10, 2)$, $(0.16, 2)$ and $(0.16, 3)$, respectively. Here, the average number of clusters for the ranking shown in Table 2 is: 4.833 ± 1.169 for the first-ranked, 2.167 ± 0.753 for the second-ranked and 1.333 ± 0.516 for the third-ranked ot-clustering. These results indicate that the top-three ot-clusterings under SECAT tend to produce an average number of clusters that falls between those favored by the baseline and CLOSE, consistent with the findings from the previous experiment.

The average snapshot- and temporal qualities of the top three SECAT-ranked ot-clusterings reveal, that the first-ranked ot-clustering has lower average values for both, snapshot- and temporal quality, compared to the second- and third-ranked ot-clusterings. However, the first-ranked ot-clustering also exhibits the lowest standard deviations for both metrics, indicating lower changes of snapshot- and temporal quality over time. This observation is supported by the distribution of (sq, tq)-tuples shown in Fig. 4: in Fig. 4a, the tuples lie close together, whereas in Fig. 4b, they are more widely dispersed.

To analyze the influence of the selected temporal quality metric on the SECAT-ranking, we replaced it with adjusted Rand index (ARI) and the adjusted mutual information (AMI). When using ARI, the top three ot-clusterings changed to ot-clusterings with following combinations of ϵ and $MinPts$: $(0.13, 2)$, $(0.13, 4)$ and $(0.1, 2)$, respectively. In the case of AMI the ranking of the top-three ot-clusterings shifted to $(0.1, 2)$ for the first-ranked, $(0.08, 2)$ for the second-ranked and $(0.7, 2)$ for the third-ranked. Apart from the differing formulations of ARI and AMI, the variations in rankings can be attributed to a shared limitation: neither statistical measure accounts for noise. Despite this limitation, the clustering with $\epsilon = 0.1$ and $MinPts = 2$ consistently appears among the top three ot-clusterings for both alternative temporal quality measures.

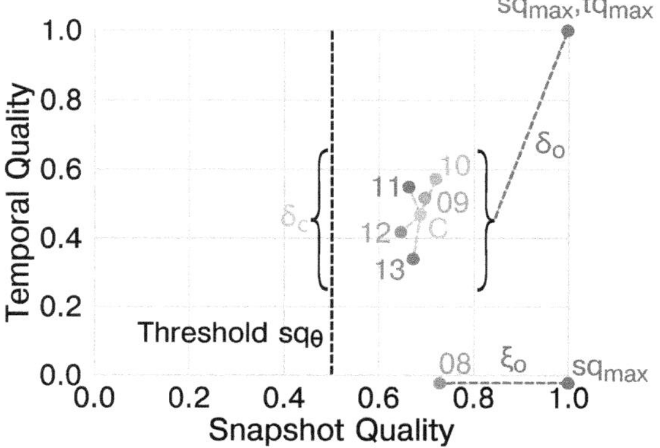

(a) Positions of the (sq, tq)-tuples for the ot-clustering with $\epsilon = 0.1$, $minPts = 2$.

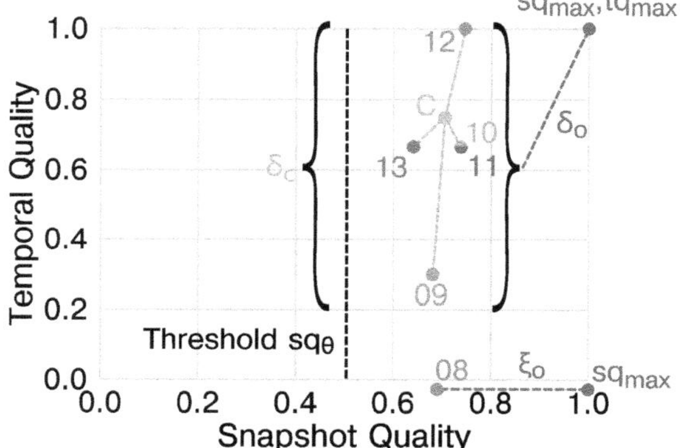

(b) Positions of the (sq, tq)-tuples for the ot-clustering with $\epsilon = 0.16$, $minPts = 2$.

Fig. 4. Illustration of the components of the SECAT score for top-two ot-clusterings with regard to SECAT. Points labeled *08* to *13* (last two digits of a year) are the time points of the corresponding (sq, tq)-tuples. The centroid c is computed from the positions of all other points.

The first-ranked ot-clustering for each method is shown in Fig. 5. To improve readability and facilitate easier visual comparison, the visualization is limited to the last three time points. The full clustering results for all time points are available in the corresponding GitHub repository[1]. The figure visually demonstrates, how the top-ranked ot-clustering can differ depending on whether and how temporal context is included in the assessment of clustering stability over time. While Fig. 5a presents the ot-clustering with the highest silhouette coefficient, the incorporation of temporal context leads to notably different top-ranked ot-

---

[1] https://github.com/YellowOfTheEgg/SECAT.

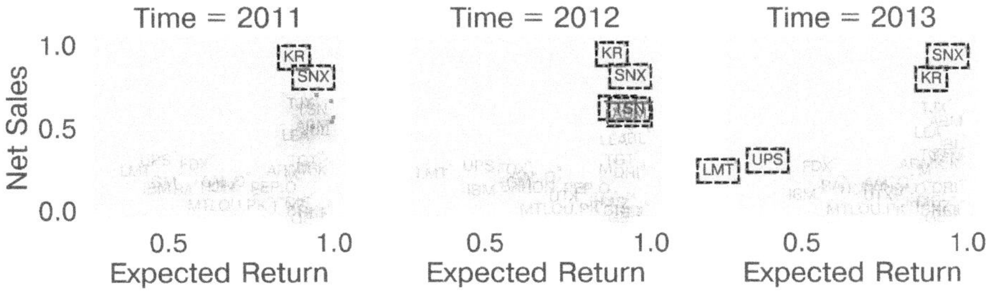

(a) Clustering that achieved the highest baseline score with $minPts = 4$ and $\epsilon = 0.15$.

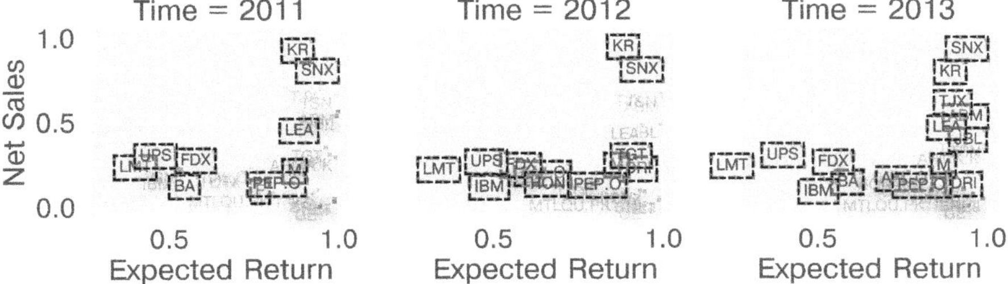

(b) Clustering that achieved the highest CLOSE score with $minPts = 2$ and $\epsilon = 0.06$.

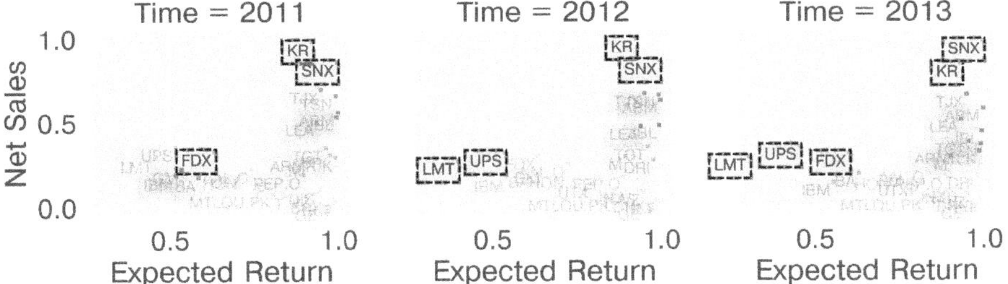

(c) Clustering that achieved the highest SECAT score with $minPts = 2$ and $\epsilon = 0.1$.

Fig. 5. DBSCAN Clusterings with the highest scores regarding the baseline, CLOSE and SECAT. Each square contains a label, with its color indicating cluster membership. Data points are positioned in the top-right corner of each square. Red dashed borders denote noise. (Color figure online)

clusterings. In the case of CLOSE, both the number of clusters and the amount of noise per time point increase significantly (Fig. 5b) compared to the baseline. In contrast, the way our approach incorporates temporal context results in an ot-clustering (Fig. 5c) with a higher number of clusters than the baseline and less noise than CLOSE, with respect to the dataset used in this experiment.

In addition to the points mentioned above, we also measured the runtimes of all three methods during the scoring of the ot-clusterings. In this setting, the baseline achieved a runtime of 0.002 ± 0.001 seconds, SECAT 0.019 ± 0.001

seconds, and CLOSE 0.848 ± 0.038 seconds. As observed with the generated dataset clustered over time using K-means, the baseline had the shortest runtime, followed by SECAT and then CLOSE.

5   Discussion and Conclusion

In this work, we introduced SECAT, a new method for quantifying the stability of ot-clusterings of time series. To ensure a broader applicability across diverse time series datasets, SECAT is designed to be highly flexible. It can be used with default parameters and also supports all bounded clustering quality metrics for snapshot quality, as well as bounded measures for comparing clusterings as the temporal quality measures.

In comparison to CLOSE, the only other existing approach with the same goal, SECAT relies on properties of Euclidean space, resulting in a highly interpretable and transparent solution. Our experimental evaluation on a generated and a real-world dataset demonstrates, that SECAT not only produces clusterings with high temporal coherence but also maintains strong snapshot quality. Compared to CLOSE, applying SECAT yields top-ranked ot-clusterings with lower granularity but also reduced noise, based on the datasets and metrics used. The runtime of SECAT is much closer to the baseline, which does not incorporate temporal context, than to CLOSE, highlighting SECAT's suitability for real-world scenarios.

Our work is also constrained by certain limitations. In the experimental setup, we fixed specific parameters of the data generation process to focus on merging and splitting clusters, which is a basic pattern commonly observed in ot-clusterings. Other temporal or structural patterns, however, were not considered. In addition, the current version of SECAT does not capture long-term temporal dependencies, as tq only reflects differences to the preceding tp-clustering. Incorporating such dependencies would require comparisons between non-adjacent tp-clusterings, an aspect that is reserved for future work.

Apart from the aspects mentioned above, SECAT provides a solid foundation for future research on the stability evaluation of ot-clusterings of time series and represents a promising step toward more interpretable and efficient analysis of temporal clustering behavior.

References

1. Afzal, A., et al.: Customer segmentation using hierarchical clustering. In: 2024 IEEE 9th International Conference for Convergence in Technology (I2CT), pp. 1–6. IEEE (2024). https://doi.org/10.1109/I2CT61223.2024.10543349
2. Ali, M., Scandurra, P., Moretti, F., Sherazi, H.H.R.: Anomaly detection in public street lighting data using unsupervised clustering. IEEE Trans. Consum. Electron. **70**(1), 4524–4535 (2024). https://doi.org/10.1109/TCE.2024.3354189
3. Chakrabarti, D., Kumar, R., Tomkins, A.: Evolutionary clustering. In: Proceedings of the 12th ACM SIGKDD International Conference on Knowledge Discovery and Data Mining, pp. 554–560 (2006). https://doi.org/10.1145/1150402.1150467

4. Chi, Y., Song, X., Zhou, D., Hino, K., Tseng, B.L.: On evolutionary spectral clustering. ACM Trans. Knowl. Discov. Data (TKDD) **3**(4), 1–30 (2009). https://doi.org/10.1145/1631162.1631165

5. Ester, M., Kriegel, H.P., Sander, J., Xu, X., et al.: A density-based algorithm for discovering clusters in large spatial databases with noise. In: KDD-96 Proceedings, pp. 226–231 (1996)

6. Ianni, M., Masciari, E., Mazzeo, G.M., Mezzanzanica, M., Zaniolo, C.: Fast and effective big data exploration by clustering. Futur. Gener. Comput. Syst. **102**, 84–94 (2020). https://doi.org/10.1016/j.future.2019.07.077

7. Klassen, G., Tatusch, M., Conrad, S.: Cluster-based stability evaluation in time series data sets. Appl. Intell. **53**(13), 16606–16629 (2023). https://doi.org/10.1007/s10489-022-04231-7

8. Korlakov, S., Klassen, G., Bauer, L.T., Conrad, S.: Multi-objective optimisation for the selection of clusterings across time. Eng. Proc. **68**(1), 48 (2024). https://doi.org/10.3390/engproc2024068048

9. Li, T., Chen, L., Jensen, C.S., Pedersen, T.B., Gao, Y., Hu, J.: Evolutionary clustering of moving objects. In: 2022 IEEE 38th International Conference on Data Engineering (ICDE). IEEE (2022). https://doi.org/10.1109/ICDE53745.2022.00225

10. Lloyd, S.: Least squares quantization in PCM. IEEE Trans. Inf. Theory **28**(2), 129–137 (1982). https://doi.org/10.1109/TIT.1982.1056489

11. Ma, X., Zhang, B., Ma, C., Ma, Z.: Co-regularized nonnegative matrix factorization for evolving community detection in dynamic networks. Inf. Sci. **528**, 265–279 (2020). https://doi.org/10.1016/j.ins.2020.04.031

12. Ranganathan, C.: Advanced Data Mining Techniques: Classification, Clustering, Regression and Prediction. Leilani Katie Publication (2024). https://books.google.de/books?id=W6kFEQAAQBAJ

13. Rousseeuw, P.J.: Silhouettes: a graphical aid to the interpretation and validation of cluster analysis. J. Comput. Appl. Math. **20**, 53–65 (1987). https://doi.org/10.1016/0377-0427(87)90125-7

14. Shankar, R., Kiran, G., Pudi, V.: Evolutionary clustering using frequent itemsets. In: Proceedings of the First International Workshop on Novel Data Stream Pattern Mining Techniques. ACM (2010). https://doi.org/10.1145/1833280.1833284

15. Tatusch, M., Klassen, G., Bravidor, M., Conrad, S.: Show me your friends and I'll tell you who you are. Finding anomalous time series by conspicuous cluster transitions. In: Le, T.D., et al. (eds.) AusDM 2019. CCIS, vol. 1127, pp. 91–103. Springer, Singapore (2019). https://doi.org/10.1007/978-981-15-1699-3_8

16. Vinh, N.X., Epps, J., Bailey, J.: Information theoretic measures for clusterings comparison: is a correction for chance necessary? In: Proceedings of the 26th Annual International Conference on Machine Learning, pp. 1073–1080 (2009). https://doi.org/10.1145/1553374.1553511

17. Xu, K.S., Kliger, M., Hero III, A.O.: Adaptive evolutionary clustering. Data Min. Knowl. Disc. **28**, 304–336 (2014). https://doi.org/10.1007/s10618-012-0302-x

18. Zhang, J., Song, Y., Chen, G., Zhang, C.: On-line evolutionary exponential family mixture. In: Twenty-First International Joint Conference on Artificial Intelligence. IJCAI (2009)

DriftSense: Adaptive Drift Detection
with Incremental Hoeffding Trees
for Real-Time Spatial Crowdsourcing

Md Mujibur Rahman[1]([⊠])[iD], Quazi Mamun[2][iD], Michael Bewong[1][iD],
and Md Zahidul Islam[3][iD]

[1] Charles Sturt University, Wagga Wagga, NSW, Australia
[2] Charles Sturt University, Albury, NSW, Australia
[3] Charles Sturt University, Bathurst, NSW, Australia
{mdmrahman,qmamun,mbewong,zislam}@csu.edu.au

Abstract. Spatial crowdsourcing (SC) platforms rely on predictive
models to allocate tasks to mobile workers. However, real-world dynamics such as traffic, weather, and user behaviour cause frequent data and
concept drift, which degrade model performance. Traditional drift detection methods are noise-sensitive, spatially agnostic, and computationally
expensive. We present DriftSense, a hybrid incremental learning approach for real-time drift detection and adaptation in SC systems. DriftSense introduces three innovations: (i) spatially localised entropy-based
drift detection, (ii) model-aware ADWIN (MA-ADWIN) that incorporates internal signals from Adaptive Hoeffding Trees (AHTs), and (iii) a
false-signal filtering mechanism for robust adaptation. Experiments on
real-world NYC Taxi and Yelp datasets, with injected abrupt, gradual,
and mixed drifts, show that DriftSense achieves up to 25% higher detection accuracy, reduces false alarms by over 8–15% points, and lowers
computational overhead by 20–25% compared to baselines. These results
demonstrate that DriftSense is both effective and lightweight, making it
suitable for deployment in dynamic SC platforms.

Keywords: Spatial crowdsourcing · Incremental learning · Concept
drift · Data drift · Adaptive Hoeffding Trees

1 Introduction

Spatial crowdsourcing (SC) systems, such as ride-hailing and food delivery platforms, assign real-time, location-based tasks using models trained on historical
spatiotemporal data [9,19]. These models, however, degrade quickly in dynamic
environments where user behaviour, demand distribution, and external conditions change continuously [5,12,15]. Traditional batch retraining is computationally expensive and fails to keep pace with real-time shifts [4,10], while most
approaches do not adequately distinguish between data drift (changes in input

Q. V. Nguyen et al. (Eds.): AusDM 2025, CCIS 2765, pp. 94–110, 2026.
https://doi.org/10.1007/978-981-95-6786-7_7

distribution) and concept drift (changes in the mapping from features to outcomes), both of which significantly reduce allocation accuracy if not detected and handled accordingly [1,11,19].

Incremental learning (IL) offers an efficient alternative to full retraining by enabling continuous model updates [11,18]. In SC research, IL methods integrate sliding-window-based drift detection with adaptive multi-armed bandit (MAB) strategies to optimise task allocation via partial updates [12]. Approaches, such as [5], also incorporate Markov model-based, temporally-aware drift detection to mitigate location drift, thereby preserving model stability and accuracy.

Popular drift detection techniques include DDM, EDDM, and ADWIN, which monitor data streams for distributional changes [7,11,18]. ADWIN, in particular, is widely adopted for its variable-length sliding window mechanism [4,14]. However, it lacks insight into a model's internal behaviour, limiting its ability to detect feature-level degradation and assess adaptation quality [2,6,20]. Moreover, existing methods often struggle to balance sensitivity with false positive suppression and lack structural adaptability [10,14,16]. As a result, most SC systems still rely on static heuristics or delayed retraining [9,12], making them slow to respond to real-time shifts, such as price surges during storms or behavioural changes after policy updates.

We present DriftSense, a lightweight approach specifically aimed at SC drift adaptation, advancing beyond standard detectors. DriftSense is innovative in three key aspects:

1. **DriftSense Approach:** We propose DriftSense, the first drift detection and adaptation method designed for SC, addressing its spatially localised and dynamic conditions.
2. **Hybrid Detection Mechanism:** We integrate entropy-based local drift detection, a model-aware extension of ADWIN with Adaptive Hoeffding Trees, and a false-signal filtering mechanism, enabling robust and efficient real-time adaptation under noisy conditions.
3. **Comprehensive Evaluation:** Using NYC Taxi and Yelp datasets with abrupt, gradual, and mixed drifts, DriftSense achieves up to 25% higher accuracy, 8–15 percentage points fewer false alarms, and 20–25% lower overhead compared to state-of-the-art baselines.

The remainder of this paper is organised as follows: Sect. 2 reviews related work relevant to SC. Section 3 formulates the problem. Section 4 outlines the proposed system architecture. Section 5 presents the experimental setup, results and evaluation. Section 6 concludes the paper.

2   Related Work

This section reviews research on the need for real-time drift adaptation in ridesharing and delivery systems. **Drift detection** in incremental learning (IL) is essential for handling classification and regression with streaming data [7,18,20]. *Concept drift* refers to changes in the conditional distribution $P(y|x)$, while *data* *drift* involves shifts in the marginal distribution $P(x)$ due to feature variations or environmental changes [2,11]. Drift in data streams occurs in two forms: abrupt and gradual. Abrupt drift involves sudden shifts from events, and gradual drift occurs slowly [6,11] (Table 1).

Table 1. Comparison of Key Drift Detection Methods in SC

Ref.	DT	SA	Strengths	Limitations
ADWIN [4]	Abrupt, Gradual Data	✗	Adaptive windowing, handles both drift types	Monitors only error rates, ignores internal model state; trigger false alarms
DDM [6]	Abrupt Data	✗	Lightweight and fast for abrupt concept drift	Sensitive to noise, weak on gradual drifts, spatially agnostic
QT [3]	Concept	✗	Area-based spatial segmentation, interpretable detection logic	Not real-time, struggles with high-dimensional data, lacks adaptiveness
HATs [14]	Concept	✗	IL, evolves structure, split criteria adapt over time	No built-in drift detector, reactive only
EM [16]	Data	✗	Lightweight, spatially agnostic, suitable for streaming	Needs careful threshold tuning, limited to data drift, less robust in concept drift scenarios
DDLP [19]	Concept, Data	✓	Privacy-aware drift detection, temporally aligned	Computationally heavy, specialised for location privacy, less generalisable to generic SC tasks
SD [12]	Concept	✓	Jointly models concept and data drift with spatio-temporal adaptation	Requires higher computation

Drift can occur abruptly due to events like system upgrades, making detection more straightforward [11], and gradually, as with shifting user preferences, complicating detection [6]. Statistical techniques like Maximum Mean Discrepancy (MMD), Kullback-Leibler (KL) divergence, and Population Stability Index (PSI) quantify shifts in feature distributions [2,17,22]. Entropy-based drift detectors offer a lightweight, non-parametric approach for high-frequency data streams, though further study is needed for spatial partitioning and real-time edge inference [10,14,16].

Adaptive Learning and Drift Adaptation in SC for Task Allocation: Real-time learning systems face challenges of temporal volatility and spatial heterogeneity. Research emphasises adaptive techniques like regional online classifiers [10] and mobility prediction systems [15]. Drift-aware algorithms are essential for managing concept drift [12], and detecting drift is crucial for privacy amid frequent location changes [19].

However, many systems require offline retraining or complex coordination, making them less suitable for real-time use [2,10]. Drift detection is often reactive rather than proactive. Hoeffding Trees (HT) can incrementally adjust to streaming data and concept drift [13,14]. Additionally, a Quad-Tree (QT)-based method detects drifts by checking overlaps between data classes [3].

Task allocation is challenging, typically using rule-based methods like nearest-neighbour heuristics or utility models [1]. Recent strategies leverage ML for predicting task success and response times. In dynamic settings, IL addresses drift detection with adaptive algorithms, and privacy-preserving techniques apply Markov models for stability [5,19]. Some work on context- or trust-aware allocation ignores drift, and many adaptive methods require centralized control and assume future rewards, limiting real-time use [1,9,12] (Fig. 1).

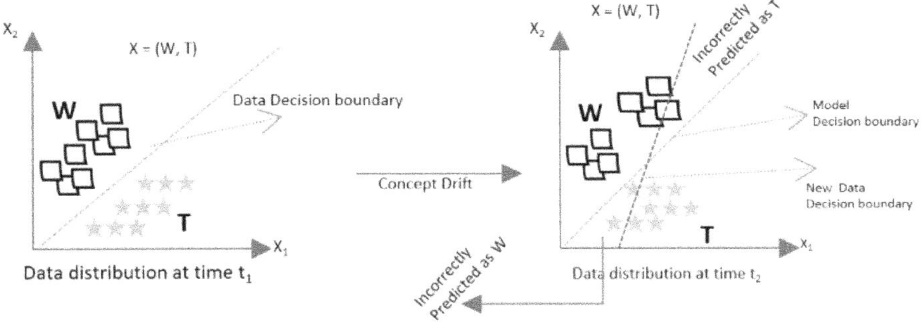

Fig. 1. Example of Drift Occurrence

Proposed DriftSense unifies entropy-based spatial data drift detection with ensemble-based concept drift detection using MA-ADWIN and Adaptive Hoeffding Trees (AHTs).

3   Problem Formulation

We examine a dynamic SC system where tasks, $T = \{t_1, t_2, \ldots\}$, are assigned in real-time to mobile workers, $W = \{w_1, w_2, \ldots\}$. Each task t_i has a feature vector $\mathbf{x}_i \in \mathbb{R}^d$, including attributes like location and user rating. Workers have profiles with availability and performance history. The goal is to learn a predictive model $f_t : \mathbb{R}^d \to \mathcal{Y}$ to decide on optimal worker assignments to tasks. This decision maximises the utility function $X(t_i, w_j)$, taking into account factors such as proximity and trust level.

At any time t, task-worker interactions follow an underlying *joint distribution* $P_t(\mathbf{x}, y)$, where: i) $P_t(\mathbf{x})$ is the **marginal distribution** of observed input features (e.g., spatial and temporal patterns), and ii) $P_t(y|\mathbf{x})$ is the **conditional distribution** describing the relationship between task features and outcomes (e.g., success or utility of assignment).

Over time, real-world dynamics can shift the underlying data distribution, leading to two main types of drift. *Data drift* occurs when $P_{t_1}(\mathbf{x}) \neq P_{t_2}(\mathbf{x})$, reflecting changes in inputs such as task density, spatial location patterns, or user demand. *Concept drift* arises when $P_{t_1}(y \mid \mathbf{x}) \neq P_{t_2}(y \mid \mathbf{x})$, indicating a change in the relationship between features and outcomes—for example, shifts in preferred assignment behaviour.

We formally define drift as,

$$\exists\, t_1, t_2 \text{ such that } P_{t_1}(\mathbf{x}, y) \neq P_{t_2}(\mathbf{x}, y) \Rightarrow f_{t_1}(\mathbf{x}) \neq f_{t_2}(\mathbf{x})$$

where f_t denotes the model used at time t, and D_t represents the data distribution at time t.

The key challenge arises from distributional shifts in the input features $\mathbf{x}_t$ (*data drift*) and changes in the predictive mapping f_t (*concept drift*) over time, driven by factors such as:

- Changes in worker behaviour and task dynamics,
- Spatio-temporal dynamics (e.g., traffic patterns, weather conditions, or public events),

Detecting data and concept drift in real time, DriftSense adjusts the task allocation model f_t for low latency and high accuracy, enabling efficient model adaptation in SC without full retraining.

4   Proposed Methods: The DriftSense

Spatially Localised Data Drift Detection: DriftSense, Fig. 2, is a modular method to detect and respond to drift in real-time SC systems, enhancing model stability and task allocation. To capture data drift with spatial granularity, DriftSense partitions the operational region using geohashing [19], a grid-based method. Each grid cell maintains a sliding window of recent data observations. We use Shannon entropy [10] to monitor variability in selected input features within each region,

$$H(R_i) = -\sum_{b=1}^{B} p_{ib} \log p_{ib} \tag{1}$$

where p_{ib} represents the empirical probability of a feature falling into a compartment b within cell R_i. An increase in entropy signals a shift in data distribution. DriftSense uses adaptive thresholding, such as the Page-Hinkley test [10], to compare current entropy with historical baselines, flagging significant changes for targeted model updates and localised retraining while avoiding unnecessary updates elsewhere.

Concept Drift Detection with Model-Aware ADWIN and AHT Methods: To handle concept drift (i.e., changes in the relationship between input features $\mathbf{x}_t$) and prediction target y_t, **DriftSense** employs an ensemble of Adaptive Hoeffding Trees (AHTs) in conjunction with a *model-aware variant of ADWIN (MA-ADWIN)*.

Each AHT node incrementally learns a predictive function $h_i : \mathbb{R}^d \to Y$, trained on streaming instances $(\mathbf{x}_t, y_t)$. The ensemble produces predictions via weighted majority vote:

$$\hat{y}_t = \arg\max_{y \in Y} \sum_{i=1}^{N} w_i \cdot \mathbb{I}[h_i(\mathbf{x}_t) = y], \tag{2}$$

where w_i is the weight of learner h_i, updated according to its recent accuracy.

ADWIN-Based Monitoring: Each learner maintains an error stream $E_i = \{e_1, e_2, \dots\}$, which is monitored by a variable-length window W. ADWIN splits W into subwindows W_1 and W_2 and tests for drift by computing the average error rate difference:

$$\Delta_{\text{err}} = |\mu_{W_1} - \mu_{W_2}|, \tag{3}$$

and comparing it against the Hoeffding bound:

$$\epsilon = \sqrt{\tfrac{1}{2} \cdot \ln\left(\tfrac{2}{\delta}\right) \cdot \left(\tfrac{1}{|W_1|} + \tfrac{1}{|W_2|}\right)}. \tag{4}$$

If $\Delta_{\text{err}} > \epsilon$, ADWIN signals a drift with confidence $1 - \delta$.

Extension to Model-Aware ADWIN (MA-ADWIN): While ADWIN relies solely on external error, MA-ADWIN enriches drift detection by incorporating *internal model signals* from AHT nodes. For a given node j, DriftSense monitors:

The node impurity change for feature j is defined as $\Delta_{\text{imp}}^{(j)} = \left| \hat{I}_j(W_1) - \hat{I}_j(W_2) \right|$. The split margin stability for feature j is computed as $\Delta_{\text{margin}}^{(j)} = \left| \hat{M}_j(W_1) - \hat{M}_j(W_2) \right|$. Finally, the feature distribution shift is measured using the divergence between the two distributions, $\Delta_{\text{dist}}^{(j)} = D(P_j(W_1) \| P_j(W_2))$.

Here, I_j denotes node impurity (e.g., entropy), M_j is the margin between the best and second-best split candidates, and $D(\cdot\|\cdot)$ is a KL divergence [17] measure. A drift is signalled only if both external error and at least one model-aware metric exceed their thresholds:

$$\Delta_{\text{err}} > \epsilon_1 \; \wedge \; \left(\Delta_{\text{imp}}^{(j)} > \epsilon_2 \; \vee \; \Delta_{\text{margin}}^{(j)} > \epsilon_3 \; \vee \; \Delta_{\text{dist}}^{(j)} > \epsilon_4 \right). \tag{5}$$

Drift Adaptation: In DriftSense, when drift is detected, the AHT is reset and retrained with recent samples, affecting only the subtrees $h_i^{(j)}$ while preserving stable regions. New subtrees are created in unstable nodes while learner weights w_i are updated using an exponentially weighted moving average (EWMA) [2],

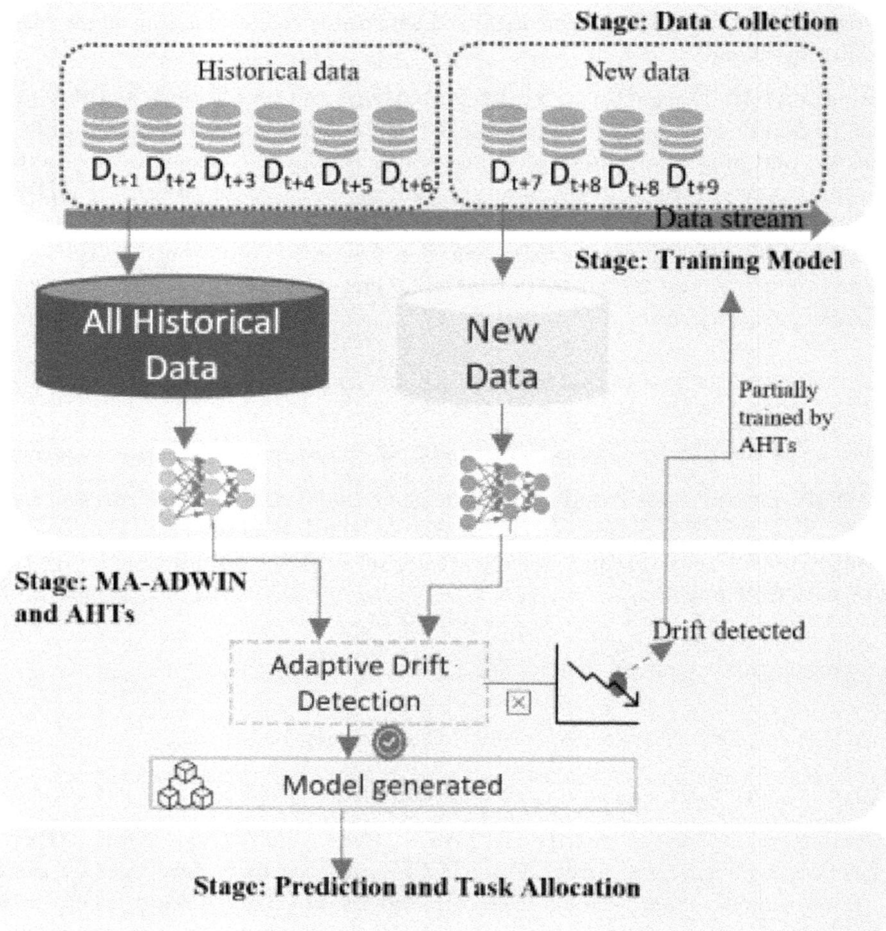

Fig. 2. DriftSense: Components and Connectivity Process

allowing DriftSense to quickly adapt to changing task-worker mappings in SC and minimise retraining.

False-Signal Filtering and Targeted Adaptation: To minimise false positives and overreactions caused by noise or transient drift signals [7], DriftSense implements a formal false-signal filtering module based on temporal and spatial statistics.

Let $s_r(t)$ represent the entropy or error-based signal strength observed in the region r at time t. The filtered drift score $\tilde{s}_r(t)$ is computed as:

$$\tilde{s}_r(t) = \alpha \cdot s_r(t) + (1 - \alpha) \cdot \tilde{s}_r(t - 1) \tag{6}$$

where α is a smoothing factor controlling temporal stability. A drift is only acted upon if:

$$\sum_{i=t-k}^{t} \mathbb{I}[\tilde{s}_r(i) > \theta] \geq m \tag{7}$$

The signal must exceed the decision threshold θ at least m times within a recent window of size k for persistence. To prevent global overreaction, DriftSense applies spatial containment, localising drift actions to a region r if $\tilde{s}_r(t) > \theta$, regardless of neighbouring cells. Once confirmed, targeted adaptation retrains affected AHT learners using data from region r, incorporating only recent W windows with a temporal decay weight $\omega_j = e^{-\lambda j}$, where j indexes window age. Our focus is on drift detection and adaptation, using the allocation algorithm from [9] to evaluate DriftSense's performance during task assignment.

5   Experimental Setup

We use two public datasets: the Yelp check-in data [1] with time-stamped check-ins and venue metadata, and the NYC taxi trips dataset [9,21] that simulates ride-hailing and delivery tasks with realistic spatio-temporal features.

We introduce synthetic drift types to reflect real-world dynamics: abrupt drift simulates sudden demand surges, gradual drift indicates slow changes in preferences, and mixed drift combines both. Controlled synthetic drifts mimic scenarios like weather conditions and holiday seasons (Table 2).

Table 2. NYC Taxi Trip Records for SC Modelling

PT	PU Long	PU Lat	DO Long	DO Lat	Dist	Fare	Tolls	Total
15/1/2015 19:05	−73.9939	40.7501	−73.9748	40.7506	1.59	12.00	5.05	17.05
10/1/2015 20:33	−74.0016	40.7242	−73.9944	40.7591	3.30	14.50	3.30	17.80
10/1/2015 20:33	−73.9633	40.8028	−73.9518	40.8244	1.80	9.50	1.30	10.80
10/1/2015 20:33	−74.0091	40.7138	−74.0043	40.7200	0.50	3.50	1.30	4.80
10/1/2015 20:33	−73.9712	40.7624	−74.0042	40.7427	3.00	15.00	1.20	16.20

PT = Pickup time and date, PU Long. = Pickup Longitude, PU Lat. = Pickup Latitude, DO = Drop Off, Dist. = Distance

Baselines and Metrics: We compare DriftSense with several state-of-the-art methods, including DDM [6], EDDM [19], and ADWIN [4] as global drift detection baselines. We also evaluate the QT [3] classifier, which uses static retraining for changes;

We use the following evaluation metrics, Precision/Recall [8]: Evaluated against injected or labelled drifts; F1-Score [8]: Harmonic mean of precision and recall over detected drifts; Drift Ratio (DR) [19]: To quantify drift in data and model learning: $DR(t) = \frac{d_t}{T_t}$, where d_t represents the number of timestamps with detected drift by time t, while T_t is the total observed timestamps. A higher DR signifies more frequent drift events and less model stability; False Alarm Rate (FAR) [7]: Frequency of false positives triggered by noise; Computational Cost [12]: Time and memory usage per window.

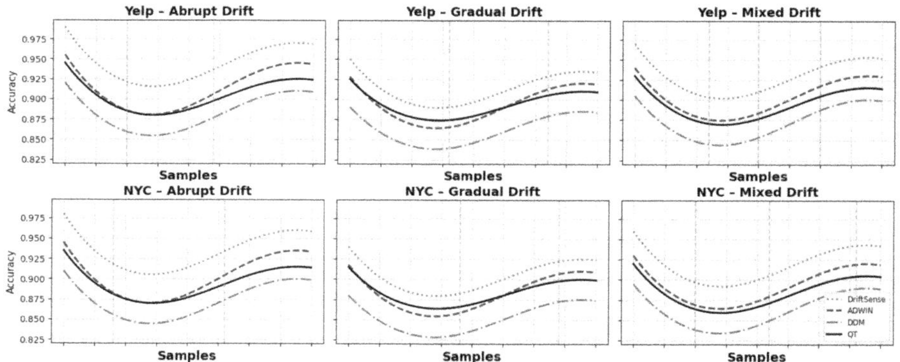

Fig. 3. Accuracy of DriftSense and baseline detectors over time

Hyper-parameter Settings: Experiments used Python with the scikit-multiflow library for IL methods (DDM, EDDM, ADWIN, Hoeffding Tree) and scikit-learn for baseline classifiers [9]. Conducted on an Intel i7 CPU with 32 GB of RAM running Ubuntu 22.04, each experiment was repeated 10 times with different random seeds to ensure reliability. We leverage the 100 attempts and achieve the best performance for the following hyperparameters: Window size w: 500 instances; ADWIN delta: 0.002; DDM warning threshold: 2 standard deviations; Drift threshold θ: 0.7 (normalised); Partial update window k: 3 most recent windows; Decay factor γ: 0.85; Number of samples set: 20,000.

5.1 Results and Analysis

In an analysis of real-world datasets like Yelp and NYC projected in Table 3, DriftSense outperformed all models, achieving the highest average accuracy, followed by ADWIN and QT. The difference between DriftSense and the lowest performer, DDM, EDDM, was moderate, highlighting the strong predictive capabilities of these models. DriftSense outperformed all baseline methods, focusing on metrics like F1-score, Recall, and correctly identified drifts. It achieved the highest Recall and maintained high precision due to effective filtering of false signals. QT ranked second in Recall and F1-score, but had slightly lower precision due to a higher number of detections. DriftSense scored 0.9288 on Yelp and 0.9104 on NYC, successfully identifying all six real drift events, while ADWIN and QT missed one, and DDM and EDDM missed two. This highlights DriftSense's superior ability to detect drifts in dynamic environments.

Table 3. Drift Detection Performance on Yelp and NYC Datasets

Dataset	Method	Accuracy(%)	Recall	Precision	F1-score	Detections	Drifts
Yelp	QT	91.84	0.7812	0.9251	0.8475	3	3/3
	ADWIN	92.45	**0.8279**	0.9110	0.8675	4	3/3
	DDM	91.73	0.6110	**0.9443**	0.7392	2	2/3
	EDDM	92.53	0.6830	0.9320	0.7102	2	2/3
	DriftSense	**94.38**	0.9370	0.9288	**0.9329**	3	3/3
NYC	QT	89.52	0.7603	0.9015	0.8245	3	3/3
	ADWIN	90.11	0.8097	0.8752	0.8412	5	3/3
	DDM	89.73	0.5982	**0.9211**	0.7217	2	2/3
	EDDM	91.13	0.6915	**0.9233**	0.7411	2	2/3
	DriftSense	**93.64**	**0.9270**	0.9104	**0.9186**	3	3/3

We evaluated four drift detection methods, DriftSense, ADWIN, DDM, and QT, across abrupt, gradual, and mixed concept drift using the Yelp and NYC datasets in Table 4. DriftSense consistently outperformed the others, achieving the highest accuracy, recall, and F1 score across all scenarios.

DriftSense achieved over 94% accuracy on both datasets, detecting all drift events with an F1-score above 0.92. While ADWIN and QT identified all drifts, they had lower recall and more false positives. DDM performed the weakest, missing several drifts. In gradual drift, DriftSense maintained high performance, although QT had higher precision but lower recall. For mixed drift, DriftSense achieved average F1-scores of 0.92 in NYC and 0.915 in Yelp. Notably, QT demonstrated substantial precision but poor recall, while DDM struggled with drift diversity.

Drift Detection Performance: We evaluate the ability of our hybrid drift detection mechanism to accurately and promptly identify concept drift (e.g., abrupt, gradual, and mixed) events injected into the task request streams. Figure 3 presents the drift detection accuracy over time for four methods—QT, ADWIN, DDM, and DriftSense—on the Yelp and NYC datasets. Across both datasets, DriftSense consistently maintained the highest accuracy throughout the data stream, demonstrating strong resilience to both abrupt and gradual drift events.

On the Yelp dataset, while all methods showed an initial performance rise, DriftSense exhibited minimal degradation during drift phases (between 5,000 and 20,000 examples), outperforming QT and ADWIN by 2–3% in midstream accuracy. On the NYC dataset, a similar pattern emerged: DriftSense began above 94% accuracy and stabilised around 92%, while ADWIN and QT followed closely, remaining 1–2% lower. DDM underperformed on both datasets, likely due to its limited sensitivity to fine-grained drift signals. Overall, DriftSense's combination of spatially localised entropy detection, adaptive ensemble updates,

Table 4. Performance of DriftSense and Baseline Methods on Yelp and NYC Datasets Across Drift Types

Dataset	Drift Type	Method	Accuracy (%)	Recall	Precision	F1-Score	Detections	True Drifts
Yelp	Abrupt	DriftSense	**94.25**	**0.91**	0.94	**0.925**	5	5/5
		ADWIN	91.60	0.86	0.92	0.89	6	5/5
		DDM	89.40	0.79	0.90	0.84	7	4/5
		QT	92.10	0.85	**0.95**	0.895	8	5/5
	Gradual	DriftSense	**93.10**	**0.88**	0.91	**0.895**	4	4/4
		ADWIN	90.75	0.84	0.89	0.865	5	4/4
		DDM	87.65	0.70	0.86	0.77	6	3/4
		QT	91.40	0.82	**0.93**	0.87	7	4/4
	Mixed	DriftSense	**93.70**	**0.90**	0.93	**0.915**	6	6/6
		ADWIN	90.30	0.83	0.91	0.87	7	6/6
		DDM	86.55	0.68	0.88	0.765	8	4/6
		QT	91.00	0.81	**0.94**	0.87	9	5/6
NYC	Abrupt	DriftSense	**95.00**	**0.92**	0.95	**0.935**	6	6/6
		ADWIN	92.50	0.87	0.94	0.905	7	6/6
		DDM	88.10	0.73	0.91	0.81	7	5/6
		QT	92.00	0.85	**0.96**	0.895	8	6/6
	Gradual	DriftSense	**93.60**	**0.89**	0.92	**0.905**	5	5/5
		ADWIN	91.00	0.84	0.89	0.865	5	5/5
		DDM	86.80	0.69	0.87	0.77	6	4/5
		QT	91.20	0.82	**0.94**	0.875	7	5/5
	Mixed	DriftSense	**94.10**	**0.91**	0.93	**0.92**	7	7/7
		ADWIN	91.70	0.85	0.91	0.88	7	6/7
		DDM	87.40	0.70	0.88	0.78	8	4/7
		QT	91.60	0.83	**0.95**	0.88	9	6/7

and false-signal filtering contributed to superior adaptation speed and retention of accuracy across highly dynamic task environments.

The hybrid method significantly reduces detection delay by approximately 30% compared to individual detectors, indicating faster reaction to drift onset. False positives and negatives are also reduced, demonstrating improved robustness against noise-induced alarms.

Concept Drift Ratio Over Time: Figure 5 shows the concept drift ratios over time across various detection methods in SC environments. From timestamps 0 to 25, the drift ratio shows significant variability due to sparse input distributions, leading to high sensitivity and potential false positives. The uniformity of prior spatial probabilities makes it challenging to differentiate early regions, leading techniques like ADWIN and QT to overreact to background noise rather than to actual changes.

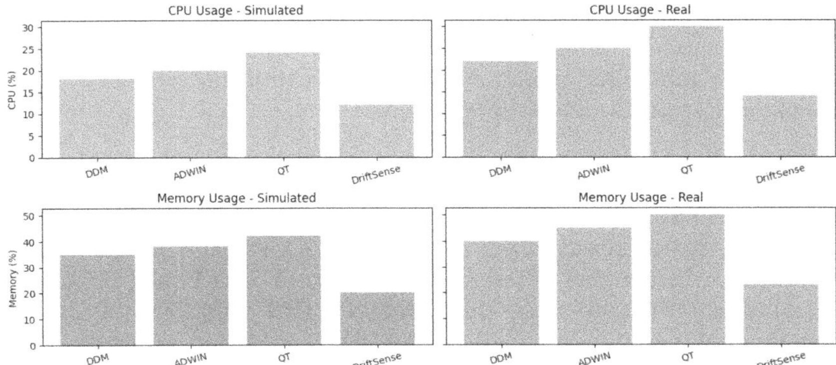

Fig. 4. Computational Overhead Comparison: Drift Detection Methods

As we adjust the timestamps from 25 to 50, a significant peak in the drift ratio for most detection methods is revealed, correlating with changes in spatial-task dynamics, such as worker availability and task density. Detection methods such as DDM, EDDM, and QT excel at identifying abrupt statistical changes during this phase. However, their increased sensitivity may result in missed noise as significant drift events.

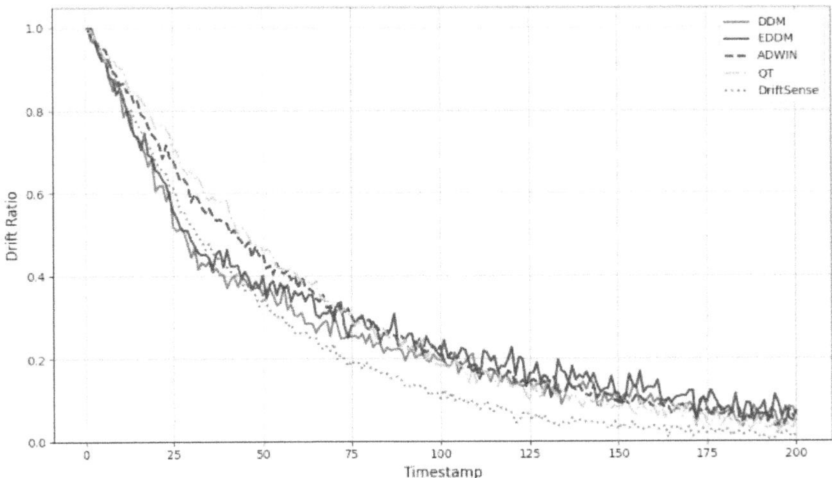

Fig. 5. Drift Ratio Over Time: DriftSense ensures balanced detection of true drifts with minimal false alarms, featuring smoother adaptation in zoomed-in areas during transitions

Once the environment stabilises and distinct patterns emerge, drift ratios decline across all detection methods, with DriftSense maintaining the lowest ratio, especially after timestamp 50. Its success is due to a hybrid scoring mechanism that combines various drift indicators and robust false-signal filtering. This approach minimises transient noise while capturing significant changes, enhancing concept drift detection performance.

Computational Overhead: Figure 4 compares the CPU and memory usage of four drift detection methods—DDM, ADWIN, QT, and DriftSense—across both simulated and real-world datasets. DriftSense consistently demonstrates the lowest computational overhead, using nearly half the CPU and memory of QT in both scenarios. In the real dataset, QT reached 30% CPU and 50% memory usage, whereas DriftSense maintained just 14% CPU and 23% memory, highlighting its lightweight design. DDM and ADWIN exhibited moderate overhead but lacked the adaptive filtering mechanisms that make DriftSense both efficient and robust. This efficiency stems from the gradual adaptation of AHTs and the integration of MA-ADWIN-based detection, which enables accurate responses to drift without requiring full retraining.

Table 5. False Alarm Rates (%) across Drift Types and Datasets

Method	Dataset	Abrupt	Gradual	Mixed
DriftSense	Yelp	8.3%	9.1%	10.5%
	NYC	7.2%	8.9%	11.0%
ADWIN	Yelp	13.5%	17.2%	20.1%
	NYC	12.1%	16.5%	21.7%
DDM	Yelp	21.0%	24.7%	28.9%
	NYC	19.4%	25.3%	30.1%
QT	Yelp	16.4%	19.2%	24.6%
	NYC	15.0%	18.5%	25.8%

Table 6. Accuracy Comparison Before and After Concept Drift

Model	Drift Accuracy (%)		
	Pre	*Post*	*Average*
Static Model	91.2	67.4	73.9
Scheduled Retraining	91.2	77.8	80.4
ADWIN	91.2	82.6	85.0
DDM	91.2	80.2	83.1
QT	91.2	79.5	82.6
DriftSense	**91.2**	**86.9**	**88.3**

Proportion of False Alarms (% of Total Detections): DriftSense consistently achieved the lowest false alarm rates across all datasets and drift types in Table 5, maintaining rates below 11% due to its hybrid scoring and robust filtering mechanisms. In contrast, DDM and QT exhibited higher sensitivity to noise, with false alarm rates exceeding 25% in mixed drift scenarios. While ADWIN averaged around 18% and performed moderately better than DDM and QT, it still generated more false detections than DriftSense. This means that DriftSense lowers false alarm rates by 8–15% points compared to baselines. These results emphasise its effectiveness in minimising unnecessary model updates while ensuring timely responses to drift.

Impact of Drift on Task Assignment Accuracy: Figure 3 illustrates allocation accuracy over a simulation period encompassing multiple drift events (sudden, gradual, and mixed). DriftSense achieves the highest task allocation accuracy among all evaluated models, maintaining post-drift performance within 4.3% points of the original pre-drift level. All models are initialised with the same pre-drift data, yielding a uniform baseline accuracy of 91.2% before drift onset. As shown in Table 6, DriftSense consistently adapts to concept and data drift more effectively than ADWIN, DDM, and QT, outperforming them by 3–6% on average in the post-drift phase. The static model, which does not adapt, suffers a 23.8% performance drop after drift. Scheduled retraining offers partial mitigation, but fails to respond promptly to unforeseen shifts due to its fixed update intervals. DriftSense's hybrid adaptation mechanism, which combines entropy-based detection and MA-ADWIN, responds to drift with AHT-targeted retraining, enabling it to maintain high task accuracy in dynamic conditions, thereby confirming its effectiveness in real-world SC scenarios.

Sensitivity Analysis: During sensitivity analysis, we adjusted the entropy threshold (θ_H) and found that values between 0.10 and 0.15 balanced early drift detection and noise tolerance. For concept drift, an ADWIN confidence level (δ) of 0.05 offered the best trade-off between precision and recall. Increasing the ensemble size beyond 5 yielded diminishing returns in accuracy, accompanied by increased computation. Additionally, applying a smoothing window of 3–4 time steps in the false-signal filter effectively suppressed noise-triggered updates without delaying adaptation. These results confirm that DriftSense maintains stable performance across a wide range of realistic settings. The decay factor γ impacts how quickly the model forgets outdated data; values around 0.85 provide the best trade-off.

Ablation Studies: The Table 7 shows that each module adds value to DriftSense. The ablation beyond single component removals to (i) quantify *module interactions* via a $2 \times 2 \times 2$ factorial (Entropy (E), MA-ADWIN (M), Filter (F)), 8 combinations of three modules were on or off to show individual and interaction effects. (ii) assess *hyperparameter sensitivity* for δ (MA-ADWIN), α (filter smoothing), (m, k) (persistence), and B (entropy), and (iii) evaluate *robustness* under label noise and spatial sparsity. Metrics include F1, Recall, Precision, False Alarm Rate (FAR), and Detection Delay (average samples from drift onset to alarm). All results are averaged over 10 seeds; we report an F1-score of 0.91 ± 0.02 (95% Confidence Interval) on the NYC Taxi dataset, showing stable performance.

Table 7. Ablation of DriftSense components on detection quality

Variant	Description	F1-Score	FAR
DriftSense (Full)	Entropy, MA-ADWIN & Filter	**0.92**	**10%**
w/o Entropy	Only MA-ADWIN & Filter	0.87	14%
w/o MA-ADWIN	Entropy & Filter	0.84	19%
w/o Filter	Entropy & MA-ADWIN	0.88	23%

6   Conclusion and Future Work

This paper presented DriftSense, an incremental learning framework for real-time drift detection and model adaptation in spatial crowdsourcing (SC). DriftSense integrates three key components: spatially localised entropy-based drift detection to capture regional dynamics, a model-aware extension of ADWIN with Adaptive Hoeffding Trees for more reliable concept drift detection, and a false-signal filtering mechanism to suppress noise while retaining sensitivity to true drifts. Experiments on real-world NYC Taxi and Yelp datasets under abrupt, gradual, and mixed drift scenarios demonstrate that DriftSense consistently outperforms baselines such as ADWIN, DDM, and QT. It achieves up to 25% higher detection accuracy, reduces false alarms by over 8–15% points, and lowers computational overhead by 20–25%, confirming its suitability for real-time SC deployment. Future work will explore extensions to federated learning and privacy-preserving settings, adaptive partitioning, and multimodal data integration to broaden its applicability in large-scale SC platforms.

References

1. AbuElHassan, S., Abo Alian, A., AbdelKader, T., Badr, N.: A review on privacy-preserving techniques for spatiotemporal data. Int. J. Data Sci. Anal. 1–14 (2025). https://doi.org/10.1007/s41060-025-00807-x
2. Aguiar, G.J., Cano, A.: A comprehensive analysis of concept drift locality in data streams. Knowl.-Based Syst. **289**, 111535 (2024). https://doi.org/10.1016/j.knosys.2024.111535
3. Amador Coelho, R., Bambirra Torres, L.C., Leite de Castro, C.: Concept drift detection with quadtree-based spatial mapping of streaming data. Inf. Sci. **625**, 578–592 (2023). https://doi.org/10.1016/j.ins.2022.12.085, https://www.sciencedirect.com/science/article/pii/S0020025522015808

4. Assis, D.N., Souza, V.M.: ADWIN-U: adaptive windowing for unsupervised drift detection on data streams: DN Assis, VMA Souza. Knowl. Inf. Syst. 1–30 (2025). https://doi.org/10.1007/s10115-025-02523-1

5. Bai, H., Hui, S.C.: A crowdsourcing-based incremental learning framework for automated essays scoring. Expert Syst. Appl. **238**, 121755 (2024)

6. Barros, R.S., Cabral, D.R., Gonçalves, P.M., Santos, S.G.: RDDM: reactive drift detection method. Expert Syst. Appl. **90**, 344–355 (2017)

7. Demšar, J., Bosnić, Z.: Detecting concept drift in data streams using model explanation. Expert Syst. Appl. **92**, 546–559 (2018). https://doi.org/10.1016/j.eswa.2017.10.003

8. Ferjani, I., Alsaif, S.A.: Dynamic road anomaly detection: harnessing smartphone accelerometer data with incremental concept drift detection and classification. Sensors **24**(24) (2024). https://doi.org/10.3390/s24248112, https://www.mdpi.com/1424-8220/24/24/8112

9. Ge, Y.F., Wang, H., Bertino, E., Cao, J., Zhang, Y.: Multiobjective privacy-preserving task assignment in spatial crowdsourcing. IEEE Trans. Cybern. **55**(8), 3584–3597 (2025). https://doi.org/10.1109/TCYB.2025.3573292

10. Gözüaçık, Ö., Can, F.: Concept learning using one-class classifiers for implicit drift detection in evolving data streams. Artif. Intell. Rev. **54**(5), 3725–3747 (2020). https://doi.org/10.1007/s10462-020-09939-x

11. Han, M., Chen, Z., Li, M., Wu, H., Zhang, X.: A survey of active and passive concept drift handling methods. Comput. Intell. **38**(4), 1492–1535 (2022). https://doi.org/10.1111/coin.12520

12. Jiao, Y., Lin, Z., Yu, L., Wu, X.: A fine-grain batching-based task allocation algorithm for spatial crowdsourcing. ISPRS Int. J. Geo-Inf. **11**(3) (2022). https://doi.org/10.3390/ijgi11030203, https://www.mdpi.com/2220-9964/11/3/203

13. Li, Z., Xiong, Y., Huang, W.: Drift-detection based incremental ensemble for reacting to different kinds of concept drift. In: 2019 5th International Conference on Big Data Computing and Communications (BIGCOM), pp. 107–114. IEEE (2019). https://doi.org/10.1109/BIGCOM.2019.00025

14. Liu, W., et al.: An adaptive hoeffding tree model based on differential entropy and relative entropy for concept drift detection. In: 2024 International Joint Conference on Neural Networks (IJCNN), pp. 1–8. IEEE (2024). https://doi.org/10.1109/IJCNN60899.2024.10650818

15. Qiu, G., Shen, Y.: Mobility-aware differentially private trajectory for privacy-preserving continual crowdsourcing. IEEE Access **9**, 26362–26376 (2021). https://doi.org/10.1109/ACCESS.2021.3058211

16. Sun, Y., Mi, J., Jin, C.: Entropy-based concept drift detection in information systems. Knowl.-Based Syst. **290**, 111596 (2024). https://doi.org/10.1016/j.knosys.2024.111596

17. Suryawanshi, S., Goswami, A., Patil, P.: Enhancing drift detection and model uncertainty handling in imbalanced streaming data using autoencoder-based approach. In: 2023 Second International Conference on Smart Technologies for Smart Nation (SmartTechCon), pp. 1265–1270 (2023). https://doi.org/10.1109/SmartTechCon57526.2023.10391432

18. Suárez-Cetrulo, A.L., Quintana, D., Cervantes, A.: A survey on machine learning for recurring concept drifting data streams. Expert Syst. Appl. **213**, 118934 (2023)

19. Wen, L., Xuebin, M., Xu, W.: DDLP: dynamic location data publishing with differential privacy in mobile crowdsensing. China Commun. **22**(5), 238–255 (2025). https://doi.org/10.23919/JCC.ja.2022-0734

20. Xiang, Q., Zi, L., Cong, X., Wang, Y.: Concept drift adaptation methods under the deep learning framework: a literature review. Appl. Sci. **13**(11) (2023). https://doi.org/10.3390/app13116515, https://www.mdpi.com/2076-3417/13/11/6515
21. Yang, Y., Zhu, M., Li, J., Wang, C., Fan, G.: Crowdsourcing regional coverage balancing method based on transfer learning in taxi service. IEEE Trans. Intell. Transp. Syst. **25**(11), 18063–18077 (2024). https://doi.org/10.1109/TITS.2024.3434561
22. Zhang, X., et al.: Continual learning with strategic selection and forgetting for network intrusion detection. In: IEEE INFOCOM 2025 - IEEE Conference on Computer Communications, pp. 1–10 (2025). https://doi.org/10.1109/INFOCOM55648.2025.11044615

Dynamic Meta-learning Ensemble with Pairwise Comparisons and Hierarchical Clustering for Financial Time Series Forecasting

Hanxue Yao$^{(\boxtimes)}$ and Irena Koprinska

School of Computer Science,The University of Sydney, Sydney, Australia
`hyao0763@uni.sydney.edu.au`, `irena.koprinska@sydney.edu.au`

Abstract. Forecasting stock market indices is challenging due to the dynamic and non-stationary nature of financial data. Traditional machine learning methods often struggle to capture these complexities, limiting their accuracy. In this paper, we propose two novel dynamic meta-learning ensembles—DME-PC and DME-HC, to address the critical challenge of ensemble member selection and adapting the prediction to each predicted data point, to enhance accuracy. To select ensemble members, DME-PC employs pairwise comparisons with statistical testing while DME-HC uses hierarchical clustering. To adapt the ensemble to the characteristics of the predicted data points, DME-PC and DME-HC combine dynamically the ensemble member predictions based on their performance on recent data and predicted future performance. Experiments on four financial datasets for 10 years and comparison with statistical, machine learning and deep learning methods showed the superior performance of the proposed dynamic ensembles. The best results were obtained with DME-HC (EWA), which utilizes hierarchical clustering for ensemble member selection and exponentially weighted average on most recent errors. Our results highlight the potential of the proposed dynamic ensembles to effectively capture temporal patterns in financial data and support better investment decisions.

Keywords: Dynamic Ensembles · Meta-Learning · Deep Learning · Pairwise Comparison · Hierarchical Clustering · Ranking · Time Series Forecasting · Stock Index

1 Introduction

Stock market indices are essential for assessing the overall state of financial markets, providing key data for investors and institutions to guide their strategic decisions. Forecasting these indices, however, presents significant challenges due to the dynamic, non-stationary nature of financial time series. These series are marked by non-linear fluctuations, abrupt market shifts, and external influences

Q. V. Nguyen et al. (Eds.): AusDM 2025, CCIS 2765, pp. 111–126, 2026.
https://doi.org/10.1007/978-981-95-6786-7_8

such as economic policies and market sentiment. Traditional machine learning techniques, while useful, often struggle to capture the complex and evolving patterns inherent in financial data, hindering their ability to reliably predict market trends.

To tackle these challenges, a widely adopted approach is the combination of multiple forecasting models, driven by evidence that different learning models exhibit varying strengths across different regions of the input space [3] and demonstrate fluctuating relative performance over time [1]. Dynamic ensemble techniques, in particular, have emerged as a promising solution. These techniques adaptively integrate the complementary advantages of multiple models for each predicted data point, enhancing forecasting accuracy and robustness, especially when dealing with complex and evolving time series data [4].

However, the effectiveness of dynamic ensembles depends critically on (1) the selection of ensemble members and (2) the combination of ensemble member prediction which should be adapted to the characteristics of the predicted data point and the changes in the time series. In this paper we propose novel methods to address these issues and create effective ensembles.

1.1 Selecting Ensemble Members

Selecting ensemble members requires a thorough understanding of the specific characteristics of the time series data, as different patterns often demand tailored modeling strategies. Determining the most suitable algorithms is rarely straightforward, making it a critical and ongoing challenge in time series forecasting.

In this paper, we propose two methods - based on pairwise comparison with statistical testing, and hierarchical clustering. The pairwise comparison method allows for a detailed assessment by comparing algorithms in pairs, ultimately producing a ranked list of candidates based on their relative performance. The top-ranked algorithms are then selected as ensemble members, serving as the foundation for further exploration and analysis.

However, pairwise comparisons, while effective, are not particularly efficient, especially as the number of prediction models b increases. The pairwise method necessitates the evaluation of $b(b-1)/2$ model pairs, leading to a computationally expensive process as b grows. To mitigate this issue, we propose an alternative method based on hierarchical clustering, which offers a more scalable solution. This method groups models based on their performance on the training set with models within the same cluster exhibiting similar strengths. From each cluster, only one model is selected, ensuring diversity. The final ensemble is composed of highly accurate models drawn from different clusters, capturing distinct patterns and enhancing forecasting accuracy.

Previous work on pairwise comparisons has explored psychological scaling methods, such as Thurstone's Law of Comparative Judgment [14], as well as statistical methodologies [11]. Building on these foundations, [6] proposed a method for label ranking by learning pairwise preferences. Methods for algorithm selection include using pairwise meta-rules [13] and active testing [9] to incrementally evaluate algorithm performance and optimize the selection process. Hierarchical clustering has been used to analyze the dynamics of stock market behavior, particularly during financial crises [7].

Pairwise comparison and hierarchical clustering methods have not yet been applied to selecting algorithms for dynamic ensembles in time series tasks.

1.2 Dynamic Ensembles for Time Series Forecasting

Most dynamic ensembles focus on weighted combination of the outputs of individual forecasters, where the weights are determined adaptively for new data points based on the forecaster's performance.

Cerqueira et al. [4] proposed ADE, a dynamic ensemble approach that adjusts each forecaster's contribution based on its expected future performance. This allows the model to adapt to varying forecasting challenges over time. Building on this, [5] refined the ADE methodology by introducing techniques to account for interdependencies among models during aggregation, including a blocked prequential procedure for generating out-of-bag predictions to improve data efficiency and a sequential re-weighting mechanism that dynamically adjusts model weights based on their correlations.

In [8], the ADE method was further enhanced by incorporating both past and predicted future performance metrics to optimize the ensemble for multi-step-ahead forecasting, thereby improving its robustness and adaptability. These studies underscore the growing effectiveness of dynamic ensemble methods, especially for complex and volatile time series such as solar power forecasting. Expanding on these advancements, [15] applied a dynamic meta-learning ensemble approach specifically to solar power forecasting and showed that leveraging multiple neural network based forecasters with dynamically adjusted weights led to significant improvements in forecasting accuracy.

In this paper, we build upon previous work and propose novel dynamic meta-learning ensemble methods for financial time series prediction that utilize two components to adaptively determine the contribution of ensemble members - based on most recent performance and predicted future performance.

1.3 Contribution

In summary, existing ensemble methods often select ensemble members directly, without systematically validating their suitability for specific tasks. In addition, the application of dynamic ensemble methods to financial time series forecasting remains relatively unexplored. Recent applications are predominantly driven by deep learning approaches [2, 10, 12, 17].

To address this gap, in this paper we introduce two novel dynamic meta-learning ensemble approaches and show their effectiveness for forecasting stock indices. The key contributions of this work are summarized as follows:

- We propose two novel dynamic meta-learning ensemble approaches, DME-PC and DME-HC. To select ensemble members, DME-PC employs pairwise comparisons with statistical testing and two ranking strategies (W and WL) and DME-HC uses hierarchical clustering for more efficient model comparison and selection.

- To dynamically determine appropriate weighting of ensemble members' contributions, DME-PC and DME-HC use a novel method leveraging two components: most recent performance and predicted future performance.
- We demonstrate the application of the methodology to stock price forecasting of global indices, using data from four datasets for 10 years. The results highlight the superior accuracy of the dynamic ensembles compared to existing statistical, machine learning and deep learning approaches, and their potential to support better investment decisions.

2   Proposed Methodology

This section outlines the core architecture of the proposed methodology, comprising two key modules: 1) Model selection to identify suitable ensemble members, using two methods - pairwise comparisons with statistical testing and hierarchical clustering (Fig. 1) and 2) Dynamic meta-learning ensemble (Fig. 2) that combines the predictions of the selected models, and utilizes both previous and predicted future performance.

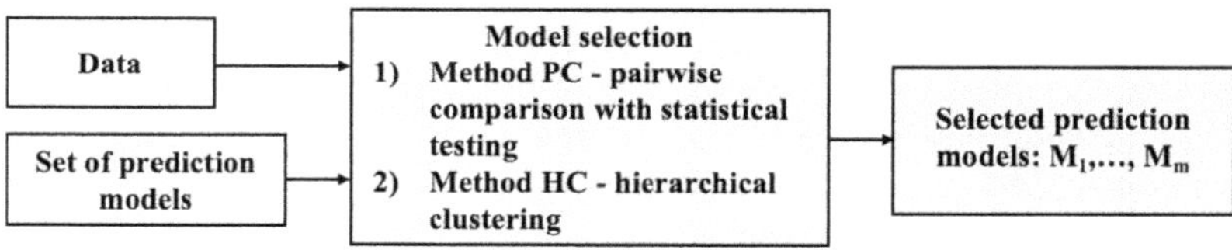

Fig. 1. Model selection methods PC and HC to select ensemble members.

2.1   Model Selection Method PC: Pairwise Comparison with Statistical Testing

This method assesses the model performance by using a time-series cross-validation and statistical testing. For each dataset and each pair of models (M_i and M_j), performance scores are computed using time-series cross-validation process with RMSE as the evaluation metric.

Then, a paired t-test is employed to determine whether the performance differences between the models are statistically significant, and a score is assigned as follows: $+1$ if model M_i significantly outperforms M_j; -1 if model M_j significantly outperforms M_i; and 0 if there is no statistically significant difference between the two models. The results are then aggregated across all datasets.

To identify the top-performing models, two *ranking strategies* are employed.

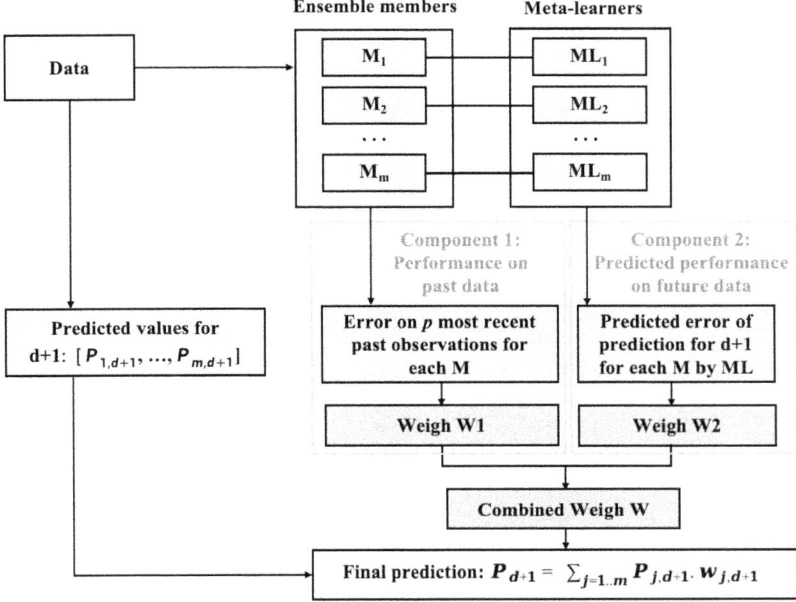

Fig. 2. Dynamic meta-learning ensemble approach DME.

Strategy WL. This strategy considers both wins and losses. The cumulative scores from wins and losses are computed and divided by the number of datasets to obtain the average score. Based on these average scores, the models are ranked, and the highest-performing models are selected as ensemble members.

Strategy W. In contrast, this strategy focuses exclusively on wins. The total win scores are aggregated and averaged to determine the final score for each model. The models are then ranked accordingly, and the highest-ranking models are chosen as the ensemble members.

2.2   Model Selection Method HC: Hierarchical Clustering

This is an alternative method for selecting predictive models, based on hierarchical clustering. It begins by assigning ranks to models based on their average RMSE, calculated through time-series cross-validation on the training set as in method PC. Models with smaller RMSE values are ranked higher, indicating superior predictive accuracy. The next step involves hierarchical clustering, which utilizes two input features: the average RMSE over the entire training set (spanning seven years) and the RMSE calculated for the seventh year. To compute the latter, the original training set is further subdivided, with the first six years used for training and the seventh year (comprising the final 252 data points of the original training set for all four datasets) used for testing. To ensure consistency across features, the input metrics are normalized prior to clustering.

The hierarchical clustering procedure begins with the construction of a distance matrix, where pairwise distances between models are computed using the Euclidean metric. Clustering is then performed using the Ward's linkage method, applied to assign models to distinct clusters based on their performance characteristics. The objective of this clustering process is to group models in such a way that those within the same cluster exhibit similar performance. For dynamic ensemble construction, models with high ranks and belonging to distinct clusters are selected, ensuring diversity within the ensemble.

2.3 Dynamic Meta-learning Ensemble DME

The dynamic ensemble with meta-learning is motivated by the observation that different models demonstrate varying strengths under different conditions, and their relative performance evolves over time. By adapting the ensemble to the characteristics of the predicted data point and changes in the time series, the performance may improve.

In line with this, we propose a method that involves the m selected ensemble members M and adaptively combines their prediction for the new day based on two error components: 1) M's error on previous recent data and 2) M's expected error for new days. Each M is trained to predict the target value for the new day and the two error components. The two error components are converted into weights W1 and W2, and the final prediction is a weighed average of the individual predictions of all models M using these weights. Hence, the ensemble incorporates both performance on past observations and predicted future performance, in determining the model weights for the final prediction. The prediction is dynamic as different weights are calculated for each predicted data point.

We formulate the task as a 1-step ahead prediction given the previous l observations. We employ a lag l of size 7. This means that the previous 7 d are used the input features, and we predict the value for the next day.

Past Observation Based Weighting Weigh W1. We evaluate the performance of each model M on the most recent previous days calculating the error. We used the last 5 d (previous window $p = 5$) and calculated MAE.

We propose two methods for calculating the combined MAE: Mean MAE, which averages the p MAEs, and EWA MAE which is an exponentially weighted average. EWA assigns progressively higher weights to more recent MAEs, thereby placing greater emphasis on most recent performance while still incorporating earlier MAEs. Hence, while Mean treats all past observations equally, EWA allows the model to adapt more responsively to the current trends in performance:

$$\text{Mean MAE} = \frac{1}{p} \sum_{t=0}^{p-1} \text{MAE}_t \tag{1}$$

$$\text{EWA MAE} = \sum_{t=0}^{p-1} w_t \cdot \text{MAE}_t \tag{2}$$

where MAE_t denotes the MAE observed on day t within the previous window p, and w_t is the corresponding weight calculated as:

$$w_t = \frac{\alpha^{p-1-t}}{\sum_{k=0}^{p-1} \alpha^{p-1-k}}, \quad t = 0, 1, \ldots, p-1 \tag{3}$$

where w_t represents the time weight assigned to the MAE of the t-th day within the previous window p, with $t = 0, 1, \ldots, p-1$ denoting the consecutive days in p, and α is the smoothing factor that governs the rate of weight decay, with more recent observations receiving greater emphasis. In this study, we used $p = 5$ and $\alpha = 0.8$, which can be adapted to different tasks.

Then, the error (Mean MAE or EWA MAE), is transformed into weigh W1 using the softmax method described below.

Future Prediction Based Weighting Weigh W2. This method employs meta-learners ML - one for each ensemble member M. Each ML forecasts the next day's MAE of M based on MAEs observed over the previous window p (we used $p = 5$ days). For the meta-learners, we utilize K-Nearest Neighbors regressors with $k = 5$ neighbors. The predicted MAE value for the new day is then converted into weigh W2 using the softmax method.

Converting Errors Into Weights. The MAE errors from 1) and 2) are transformed into weights to reflect the relative performance of the models.

We use the softmax function to calculate the weights of the ensemble members W1 and W2. The weights decrease exponentially as the error increases, hence more accurate ensemble members are given higher weights. The softmax function for predicting day $d + 1$ is formalised by the following equation:

$$w_{i,d+1} = \frac{e^{-\epsilon_i}}{\sum_{j=1}^{m} e^{-\epsilon_j}} \tag{4}$$

where $w_{i,d+1}$ represents the weight of the i-th ensemble member for day $d+1$, ϵ_i is the error for the i-th ensemble member (MAE on previous observations for W1 or predicted MAE error on future observations for W2), and m is the total number of models in the ensemble. This function transforms the predicted errors into weights using exponential decay, assigning smaller weights to models with higher errors and vice versa.

The final combined weight W for each ensemble member for each prediction day is calculated as the average of the two weights W1 and W2.

Final Prediction. The final prediction of the dynamic ensemble is calculated by the weighted average of the predictions of the individual ensemble members:

$$P_{d+1} = \sum_{j=1}^{m} w_{j,d+1} \cdot P_{j,d+1} \tag{5}$$

where P_{d+1} is the final prediction for day $d+1$, $w_{j,d+1}$ is the combined weight assigned to the j-th ensemble member at day $d+1$, $P_{j,d+1}$ is the prediction of the j-th ensemble member for day $d+1$, and m is the total number of ensemble members in the ensemble.

We developed two versions of the dynamic ensemble based on the model selection method - DME-PC which uses the pairwise comparison method (PC) and DME-HC which uses the hierarchical clustering method (HC).

3   Data and Experimental Setup

This section provides an overview of the four datasets used, outlines the data preprocessing procedures, defines the evaluation metrics, and details the prediction models used as ensemble members and baselines for comparison.

3.1   Data Description and Preprocessing

For the experiments, four prominent stock indices were selected: NASDAQ, S&P500, FTSE100, and DAX. They were chosen for their representation of key global financial markets, each serving as a benchmark for its respective economy. The raw data, consisting of daily closing prices from January 1, 2010, to December 31, 2019, were sourced from Yahoo Finance [16]. This 10-year-long dataset allows the model to train across various market phases, including bull, bear, and stagnant markets, providing a solid foundation for evaluating the performance of models under diverse market conditions. The four univariate financial time series, corresponding to the daily closing prices of stock indices are shown in Fig. 3 and summarized in Table 1.

Table 1. Statistics of four datasets

Dataset	Number of Instances	Mean Value	Standard Deviation	Min Value	Median Value	Max Value
NASDAQ	2515	4742.48	1877.28	2091.79	4620.37	9022.39
S&P500	2515	1962.10	588.48	1022.58	1986.45	3240.02
FTSE100	2524	6513.61	716.10	4805.80	6597.40	7877.50
DAX	2531	9530.66	2384.72	5072.33	9734.33	13559.60

For each dataset, non-trading days, including weekends and holidays, were removed to maintain consistent daily frequency. Missing values in the data were

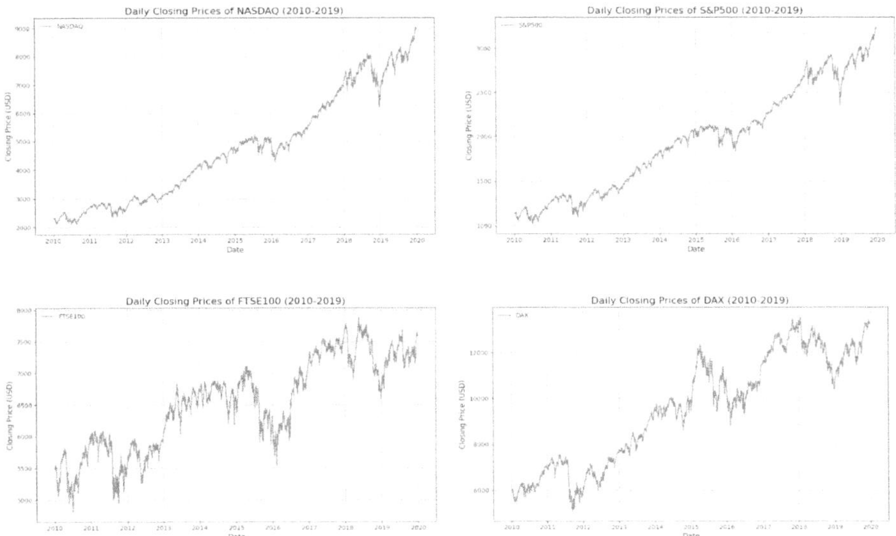

Fig. 3. Four financial time series datasets.

addressed using polynomial interpolation with a polynomial order of 2, thereby preserving the temporal integrity of the series. An outlier detection method based on the Interquartile Range method was applied.

The feature set consisted of the daily closing prices, while the target variable is the closing price for the following day. Each dataset was divided into training, validation, and test sets with a 7:1:2 ratio. These subsets are used for training, parameter tuning, and evaluation of the models, respectively.

Min-Max normalization to $[0, 1]$ is applied. To recover the original values after making predictions, the inverse scaling transformation is applied.

3.2 Evaluation Metrics

Both RMSE and MAE are used to provide a comprehensive evaluation of the models' predictive performance:

$$\text{RMSE} = \sqrt{\frac{1}{n} \sum_{i=1}^{n} (y_i - \hat{y}_i)^2}, \quad \text{MAE} = \frac{1}{n} \sum_{i=1}^{n} |y_i - \hat{y}_i| \tag{6}$$

where y_i is the actual value, $\hat{y}_i$ is the predicted value, and n is the number of observations.

3.3 Prediction Models Used in DME-PC and DME-HC

To introduce diversity, we select six different types of forecasting models - SARI-MAX, GB, RNN, LSTM, MLP, and CNN, representing statistical, machine learning and deep learning techniques. These algorithms have been shown to perform well in various time series forecasting applications and are also commonly used for stock index prediction, each bringing unique strengths.

For hyperparameter tuning, we employed grid search with 5-fold time series cross-validation on each dataset to minimize RMSE.

SARIMAX. The model was built using Seasonal Autoregressive Integrated Moving Average with Exogenous Variables (SARIMAX), configured to capture both seasonal trends and autoregressive components. In stock index prediction, it helps model linear temporal relationships and capture seasonal effects. Model selection was performed by tuning the parameters, order = [(1, 1, 0), (1, 1, 1)] along with seasonal order = [(0, 1, 1, 6), (1, 1, 1, 6), (1, 1, 1, 12)]. This approach allowed the model to effectively incorporate seasonality, trend, and noise, providing a robust forecasting framework for complex time-dependent data.

GB. The model was built using a Gradient Boosting (GB) Regressor, a widely used method that captures complex nonlinear relationships in financial datasets using multiple trees built sequentially to correct errors made by previous trees. For hyperparameter tuning, we explored: number of estimators = [50, 100, 200], learning rate = [0.01, 0.05, 0.1], and maximum tree depth = [3, 5].

RNN. The model was built using a Recurrent Neural Network (RNN), which is designed to capture temporal dependencies in sequential data and handle nonlinear patterns, making it suitable for stock index prediction. It consists of two SimpleRNN layers with ReLU activation to learn long-term dependencies. For hyperparameter tuning, we explored: number of neurons = [50, 100], epochs = [200, 500, 700], and batch size = [16, 32].

LSTM. The model was built using a Long Short-Term Memory (LSTM) network, a variant of RNN designed to handle long-range dependencies and mitigate the vanishing gradient problem. LSTMs are particularly effective in modeling sequential data with long-term dependencies, making them suitable for financial time series forecasting. With two LSTM layers featuring ReLU activation, this model is designed to capture complex temporal relationships in stock index data. The hyperparameter tuning explored: number of neurons = [50, 100], epochs = [200, 500, 700], and batch size = [16, 32].

MLP. The model was built using a Multi-Layer Perceptron (MLP), a fully connected feedforward neural network that is effective in learning complex patterns from structured data. MLPs are widely used in stock index prediction due to their ability to model nonlinear relationships between financial indicators. The MLP model we used consists of four fully connected layers designed to learn complex, non-linear relationships from structured stock index data. Hyperparameter tuning explored different number of neurons = [50, 100], epochs = [200, 500, 700], and batch size = [16, 32].

CNN. The model was built using a 1D Convolutional Neural Network (CNN), which utilizes convolutional layers to model sequential data and capture temporal dependencies. CNNs are well-suited for stock index prediction as they can efficiently detect patterns and trends within time series data by learning from local features. The architecture consists of two 1D convolutional layers followed by a flattening layer and a final dense layer for regression. The convolutional layers use ReLU activation functions, with the second layer having half the number of filters as the first. A dropout layer is added after the first convolutional layer to prevent overfitting. The hyperparameter tuning explored: number of neurons = [50, 100], epochs = [200, 500, 700], and batch size = [16, 32].

4   Results and Discussion

4.1   Model Selection

Table 2 presents the total counts of wins, losses, and ties achieved by each forecasting model across all datasets, as obtained using the pairwise comparison method for model selection.

Table 2. Model Selection Method PC: Total Wins, Losses and Ties

Model	Wins (+1)	Losses (-1)	Ties (0)
SARIMAX	14	0	6
GB	0	1	19
RNN	6	1	13
LSTM	5	5	10
MLP	0	13	7
CNN	3	8	9

To further assess the performance of the prediction models, we employed Strategy WL, which considers both wins and losses as evaluation metrics. Table 3 reports the average scores and rankings attained by each model across the four datasets. Models with the same average score are assigned the same rank. The results indicate that SARIMAX, RNN, and LSTM consistently outperformed the other models. Hence, these models were selected as ensemble members for subsequent analysis.

Table 3. Model Selection Method PC: Strategy WL - Average Score and Rank

Model	Average Score (WL)	Rank
SARIMAX	3.5	1
GB	-0.25	4
RNN	1.25	2
LSTM	0	3
MLP	-3.25	6
CNN	-1.25	5

Furthermore, Strategy W was employed to evaluate the predictive capabilities of the models from a win-based perspective. As presented in Table 4, this strategy identified SARIMAX, RNN, LSTM, and CNN as the top-performing models based on their average scores and rankings. Consequently, these models were selected as the ensemble members for further evaluation.

Table 4. Model Selection Method PC: Strategy W - Average Score and Rank

Model	Average Score (W)	Rank
SARIMAX	3.5	1
GB	0	5
RNN	1.5	2
LSTM	1.25	3
MLP	0	5
CNN	0.75	4

Table 5 presents the results of model selection method using hierarchical clustering across all four datasets. The findings demonstrate that RNN, LSTM, and MLP frequently achieve higher accuracy and belong to distinct clustering groups for the majority of the datasets. Based on these observations, these three models were selected as ensemble members to leverage their complementary strengths. CNN was excluded as it shares the same clustering group with MLP across all datasets, indicating similar performance.

4.2 Dynamic Meta-Learning Ensemble Performance

Table 6 shows the accuracy results of the proposed dynamic meta-learning ensemble methods DME-PC and DME-HC and the benchmark single prediction methods used for comparison, evaluated on the test set.

The main results can be summarized as follows:

- Overall performance: The proposed dynamic meta-learning ensemble DME was the best performing method on all datasets. All DME-PC and DME-HC versions outperformed all baseline methods, which are widely used for stock index prediction. The most accurate DME version across all datasets was DME-HC (EWA) which uses hierarchical clustering for model selection and exponential weighed average for weight calculation. After the DME methods, the best baseline methods were MLP and RNN, and the worst performing method was SARIMAX.
- Strategy WL vs. Strategy W for model selection in DME-PC: Strategy W outperformed Strategy WL on all datasets. W ranks and selects models based on wins only, while WL uses both wins and losses.

Table 5. Model Selection Method HC: Average RMSE, Rank and Cluster

(a) NASDAQ				(b) S&P500			
Model	Avg RMSE	Rank	Cluster	Model	Avg RMSE	Rank	Cluster
SARIMAX	312.89	6	3	SARIMAX	111.48	6	1
GB	265.09	5	2	GB	91.00	5	4
RNN	173.51	3	4	RNN	58.30	3	2
LSTM	256.40	4	2	LSTM	60.70	4	5
MLP	65.96	1	1	MLP	24.43	1	3
CNN	78.75	2	1	CNN	29.92	2	3

(c) FTSE100				(d) DAX			
Model	Avg RMSE	Rank	Cluster	Model	Avg RMSE	Rank	Cluster
SARIMAX	412.22	6	4	SARIMAX	1148.69	6	4
GB	107.49	2	1	GB	527.38	5	3
RNN	222.72	5	3	RNN	372.38	4	2
LSTM	161.78	4	2	LSTM	366.39	3	2
MLP	93.04	1	1	MLP	184.44	1	1
CNN	112.22	3	1	CNN	211.81	2	1

- DME-PC vs. DME-HC for model selection: Across all four datasets, DME-HC consistently outperformed DME-PC in terms of both forecasting accuracy and efficiency. Specifically, DME-HC which uses clustering for model selection, proved to be more time-efficient, while also achieving more accurate predictions, as it ensures greater diversity and captures more nuanced patterns within the data.
- Mean MAE vs. EWA MAE of previous window for determining weigh W1: The experimental results demonstrate that the EWA method consistently achieved better forecasting accuracy compared to Mean. This demonstrates that emphasizing recent performance while still accounting for earlier data is more effective for financial data prediction than treating all observations equally.

Table 6. Performance Evaluation Results (RMSE and MAE)

(a) NASDAQ				(b) S&P500			
Model	Avg RMSE	Rank	Cluster	Model	Avg RMSE	Rank	Cluster
SARIMAX	312.89	6	3	SARIMAX	111.48	6	1
GB	265.09	5	2	GB	91.00	5	4
RNN	173.51	3	4	RNN	58.30	3	2
LSTM	256.40	4	2	LSTM	60.70	4	5
MLP	65.96	1	1	MLP	24.43	1	3
CNN	78.75	2	1	CNN	29.92	2	3

(c) FTSE100				(d) DAX			
Model	Avg RMSE	Rank	Cluster	Model	Avg RMSE	Rank	Cluster
SARIMAX	412.22	6	4	SARIMAX	1148.69	6	4
GB	107.49	2	1	GB	527.38	5	3
RNN	222.72	5	3	RNN	372.38	4	2
LSTM	161.78	4	2	LSTM	366.39	3	2
MLP	93.04	1	1	MLP	184.44	1	1
CNN	112.22	3	1	CNN	211.81	2	1

5 Conclusion

This study focused on forecasting stock index values using historical daily closing prices, employing dynamic meta-learning ensemble techniques. We introduced two novel dynamic meta-learning ensemble approaches, DME-PC and DME-HC, which utilize different methods to select the most effective algorithms for ensemble membership - pairwise comparisons with statistical testing and hierarchical clustering, respectively. These methods address the critical challenge of ensemble member selection for dynamic ensembles. Two ranking strategies were explored for DME-PC: one that considers both wins and losses, and another that focuses solely on wins. Our results indicate that the win-based strategy yields better performance for stock index prediction. The clustering-based approach DME-HC obtained superior predictive accuracy compared to DME-PC, with the further advantage of greater time efficiency.

To adapt the ensemble to the characteristics of the predicted data points, DME-PC and DME-HC combine dynamically the ensemble member predictions based on two components. The first one is the ensemble member's previous performance - the mean or EWA of the most recent MAEs, and the second one is the predicted future performance - the predicted MAE for the upcoming time step, estimated by each associated meta-learner. The performance of DME-PC under the win-focused strategy and DME-HC were further evaluated using mean and EWA for Weigh 1 calculations.

Our comprehensive evaluation on four datasets using data for 10 years each, demonstrated that all dynamic ensembles performed superior than the baseline statistical, machine learning and deep learning methods used for comparison. DME-HC (EWA) most effectively captured the dynamics of stock index time series and achieved the best forecasting accuracy.

In future work, we plan to extend our proposed methodology to other financial datasets, including those from high-frequency markets. This will allow to evaluate the robustness and adaptability of the approach when applied to diverse and challenging forecasting scenarios in the financial domain.

References

1. Aiolfi, M., Timmermann, A.: Persistence in forecasting performance and conditional combination strategies. J. Econ. **135**(1–2), 31–53 (2006)
2. Bhambu, A., Gao, R., Suganthan, P.N.: Recurrent ensemble random vector functional link neural network for financial time series forecasting. Appl. Soft Comput. **161**
3. Brown, G., Kuncheva, L.I.: "Good" and "bad" diversity in majority vote ensembles. In: Proceedings of the 9th International Workshop on Multiple Classifier Systems, pp. 124–133 (2010)
4. Cerqueira, V., Torgo, L., Pinto, F., Soares, C.: Arbitrated ensemble for time series forecasting. In: Proceedings of the European Conference on Machine Learning and Knowledge Discovery in Databases (ECML PKDD), pp. 478–494. Springer (2017)
5. Cerqueira, V., Torgo, L., Pinto, F., Soares, C.: Arbitrage of forecasting experts. Mach. Learn. **108**, 913–944 (2019)
6. Hüllermeier, E., Fürnkranz, J., Cheng, W., Brinker, K.: Label ranking by learning pairwise preferences. Artif. Intell. **172**(16–17), 1897–1916 (2008)
7. Jaroonchokanan, N., Termsaithong, T., Suwanna, S.: Dynamics of hierarchical clustering in stocks market during financial crises. Phys. A: Stat. Mech. Appl. **607** (2022)
8. Koprinska, I., Rana, M., Rahman, A.: Dynamic ensemble using previous and predicted future performance for multi-step-ahead solar power forecasting. In: Proceedings of the 28th International Conference on Artificial Neural Networks (ICANN), pp. 436–449 (2019)
9. Leite, R., Brazdil, P.: Active testing strategy to predict the best classification algorithm via sampling and metalearning. In: Proceedings of the European Conference on Artificial Intelligence (ECAI), pp. 309–314 (2010)
10. Lin, Y., Yan, Y., Xu, J., Liao, Y., Ma, F.: Forecasting stock index price using the CEEMDAN-LSTM model. North Am. J. Econ. Finance **57** (2021)
11. Mosteller, F.: Remarks on the method of paired comparisons: I. the least squares solution assuming equal standard deviations and equal correlations. Selected Papers of Frederick Mosteller, pp. 157–162 (2006)
12. Srivastava, T., Mullick, I., Bedi, J.: Association mining based deep learning approach for financial time-series forecasting. Appl. Soft Comput. **155** (2024)
13. Sun, Q., Pfahringer, B.: Pairwise meta-rules for better meta-learning-based algorithm ranking. Mach. Learn. **93**(1), 141–161 (2013). https://doi.org/10.1007/s10994-013-5387-y
14. Thurstone, L.L.: A law of comparative judgment. In: Scaling, pp. 81–92 (2017)
15. Wang, Z., Koprinska, I.: Solar power forecasting using dynamic meta-learning ensemble of neural networks. In: Proceedings of the International Conference on Artificial Neural Networks (ICANN), pp. 528–537 (2018)

16. Yahoo Finance: Yahoo finance - Stock Market Live, Quotes, Business and Finance News. https://finance.yahoo.com, Accessed 10 Oct 2024
17. Zhang, C., Sjarif, N.N.A., Ibrahim, R.: 1D-CapsNet-LSTM: a deep learning-based model for multi-step stock index forecasting. J. King Saud Univer. Comput. Inf. Sci. **36**(2), 101959 (2024)

Application Track – Environment, Information Security and Productivity

Effective Missing-Data Imputation for Time Series with Seasonality and Causality

Devesh Bhogal[1]([✉]) [iD], Yanchang Zhao[1] [iD], Reena Kapoor[1] [iD], Tapas Biswas[1], Peter Toscas[1], and Klaus Joehnk[2]

[1] Data61, CSIRO, Canberra, Australia
bhogaldevesh@gmail.com, {yanchang.zhao,reena.kapoor,
tapas.biswas,peter.toscas}@csiro.au
[2] Environment, CSIRO, Canberra, Australia
klaus.joehnk@csiro.au

Abstract. Missing data is a problem commonly seen in most if not all real-world applications. Particularly for water quality monitoring systems, which are commonly plagued by sensor faults or network errors, missing or erroneous data pose a significant challenge in extracting accurate and meaningful insights. In this work, we investigate the problem of missing information in time-series data and propose a new method DISC - Data Imputation with Seasonality and Causality - which uses the concepts of seasonal decomposition and causal discovery to improve contextual accuracy of the imputations for time-series. DISC operates in two stages. First, it builds a causal relational graph representing inter-feature dependencies and uses this graph to impute missing values by adjusting estimates to the nearest-neighbour hourly data points. Second, it learns yearly, monthly, and daily seasonal patterns at an hourly resolution and imputes the remaining gaps. In scenarios where seasonal decomposition fails to fully resolve gaps, causal discovery exploits dependencies among time-series features to generate reference points that enhance the completeness of the seasonal pattern. The proposed method has been evaluated using real-world data from the Murray-Darling Basin and compared with multiple existing machine learning methods. The results validate the effectiveness of DISC, enabling accurate imputation of 14 consecutive days of missing hourly data with an R-squared of 80%, more than twice the performance of conventional approaches.

Keywords: Time Series · Data Imputation · Causal Discovery · Water Quality Measurement

1 Introduction

Time series data is information tracked over time about our environment and various other systems. This time lapse of reality allows us to gain a deeper

Q. V. Nguyen et al. (Eds.): AusDM 2025, CCIS 2765, pp. 129–144, 2026.
https://doi.org/10.1007/978-981-95-6786-7_9

understanding and to create models that can help predict the future. By recording data over time, we can predict stock prices in the field of finance [1], monitor the vitals of patients and predict adverse events in healthcare [2], forecast customer trends and user engagement in commerce [3,4] and track the climate or predict algal blooms in the field of environmental sciences [5,6]. In many of these fields, collecting data across all components simultaneously is challenging, and the data sources themselves are often unreliable. For example, water temperature sensors at dams and lakes can malfunction, be impacted by biofouling or tampering, leaving gaps or errors in the recorded datasets. As such, data ends up either discarded due to the errors or not recorded at all, resulting in datasets with missing information.

The missing information may degrade the performance of the model or make it unreliable. Even a small percentage can lead to major impact on model and analytical performance [7]. In cases where features in the dataset are entirely missing—such as when a light-and-temperature sensor with a damaged light component only records temperature and a limited portion of the light intensity history—the resulting data may introduce bias, as spatial or contextual factors such as sensor location are no longer properly represented. In addition, small datasets introduce further bias, as their limited sample size fails to capture edge cases, resulting in statistically weak models. In instances where the data recorded is sporadically missing, due to intermittent power failures, systematic skipping or similar issues, it often becomes unusable for certain machine learning models such as regression, where complete and continuous data is required. As a result, analysis and system understanding are compromised, with reduced model accuracy and distorted trends or correlations. Therefore, it becomes pertinent to impute the data and fill in gaps with meaningful and consistent data.

Imputing missing data is a known non-trivial and challenging problem. In time-series data, each data point is dependent on its neighbours, and irregularly missing values hinder accurate trend estimation. This, in turn, disrupts the cohesiveness of multivariate datasets, where interdependent features rely on complete temporal patterns. Simpler imputation approaches, such as forward/backward filling and linear interpolation, struggle with irregular gaps, because they fail to capture and preserve the underlying patterns. Deriving univariate or even multivariate relations using the remaining data is a greater challenge as the residual data often appears noisier, with outliers and incorrect readings giving rise to misleading assumptions and poor imputation accuracy. Machine learning and neural network models can handle non-stationary data, but at the cost of high data volumes and resource demands. Such requirements hinder their applicability in real-time operating environments, where imputations must be both accurate and computationally efficient.

To address the above problem, this paper presents a new method DISC – Data Imputation with Seasonality and Causality. It is a two-stage approach that combines seasonal decomposition with causal discovery to improve contextual accuracy in time-series imputations. In the first stage, causal discovery is employed to exploit inter-feature dependencies, generating additional reference points thereby allowing for a more complete seasonal representation. In the second stage, DISC learns yearly, monthly, and daily seasonal patterns at an hourly resolution taking into account the additional reference points. It fills missing values by aligning them with nearest-neighbour estimates of the corresponding hourly patterns. To alleviate the performance issues of forecasting models trained on sparse data, we conduct a step-by-step analysis of the components of DISC, comparing them against commonly used imputation approaches such as forward/backward filling, interpolation, and machine learning models. The proposed method has been evaluated using real-world data and compared with multiple existing methods, validating its superior performance. The experiments demonstrate that our method outperforms the standard approaches, maintaining high accuracy and correlation even when imputing long gaps of up to 14 days (336 hourly samples). Our method imputes a gap of 14 days of hourly data with an R^2 of 0.8 when compared to other approaches having an R^2 below 0.3.

Our paper offers the following contributions:

- a novel method for time-series data imputation that utilises causal modelling and seasonality decomposition trend analysis,
- a comprehensive experimental comparison with benchmark imputation models including machine learning approaches, and
- an application in a real-world scenario in the Murray-Darling Basin, Australia.

The rest of this paper is structured as follows. Section 2 reviews the related work, highlighting their applicability and limitations. Sections 3 and 4 present our proposed model, its framework, and components. Section 5 presents the experiment evaluation and discusses the results of different test cases, and Sect. 6 concludes the paper by discussing limitations and future work.

2   Related Work

Time-series data imputation has been tackled in a variety of ways in the past. Depending on the application, even something as simple as forward and backward filling can be used. The filling method is based on the concept of recency effect, involving either the repetition of the last known recent value for the missing segment or the closest value immediately after the missing segment of values. A more complex approach would be using interpolation [8], which computes estimates by assuming that the data forms a pattern representable by simple mathematical equations and uses it to predict the missing values. However, in real-world cases, the data cannot be fully modelled or explained through linear relations or even polynomial relations since the data relationships change over time both within variables and between dependent variables.

Seasonal decomposition is another technique used to break down a time-series into trend, seasonality, and residuals. The seasonality component gives us the underlying repetitive pattern in the time-series, the trend gives us the change in seasonality, and the residuals are the difference between the original and the combination of trend and seasonality. In 2022, Han et al. [9] extended previous work on seasonal decomposition by proposing a Univariate Imputation Method (UIM), which applies different imputation techniques to the individual components of the decomposition in order to reconstruct missing sensor data. However, using seasonal decomposition on its own gives a shallow model, which cannot capture the reasoning behind certain outliers, for example, water temperature depends on air temperature and rainfall amongst a few other parameters. Without the relational model between water temperature and rainfall, sudden drops in water temperature that coincide with heavy rainfall cannot be explained and therefore cannot be imputed by a univariate modelling approach such as seasonal decomposition.

Machine learning methods can be used to learn based on the data ingested, and then estimates from those models are used to fill in the gaps in time-series data. Various machine learning approaches have been used to impute missing data. A Python package, PyImpuyte [10], implements multiple machine learning methods, such as Bagging Regressor, Extra Trees Regressor, Random Forest Regressor and XGBoost, for imputing data with the help of multiple features. The KNN Imputation [11] takes the k-nearest neighbouring values and uses them to generate an estimate that then fills in the missing value. The EM Algorithm [12] is a two-step algorithm which iteratively converges towards the best estimates for missing data from the observed data and also best parameters for the maximising likelihood of observing the data. Deep-learning Imputation [13] captures the non-linear complexity of data by creating deep learning models using the observed data and then generating the estimates for missing data. The Random Forest approach [14] builds multiple regression trees, and the estimates produced work with non-linear data while reducing the chance of over-fitting. Machine learning approaches also come with drawbacks. They are very data hungry as they build relational graphs using as many data points as examples. They also come with computational challenges as they require larger resources to churn through the data multiple times. Their structuring is also complex, involving multiple stages and steps with iterative structures as in the case of deep learning models or neural networks.

Multiple related factors or features in time-series data vary with respect to each other over time. Causal modelling [15] can be used to identify the cause-effect relationship between the features, rather than a simple association between the different variables. Therefore, it can be used to boost predictive models and avoid over-fitting. In the past, causal discovery has been used alongside neural networks as an approach for filling in missing data, in observational studies and health data [16]. Almeida et al. [17] explored the use of graphical causal models built on completed datasets to impute other incomplete datasets. When causal models are used for generating estimates, the accuracy depends on the amount of data that is available synchronously for all variables. In a scenario where data is scarce, the causal graph is unable to accurately produce estimates. Moreover, in order to capture univariate relations, assumptions similar to interpolation are required.

In lieu of tackling the imputation problem of sparse multivariate data, our proposed method combines the benefits of univariate analysis from seasonal decomposition and multivariate relations from causal modelling and produces more robust imputation results.

3   Problem Statement

The target problem is to accurately imputing missing data from multivariate time series data in real-world applications. More specifically, we target the missing information in the water quality measurement data for hundreds of water monitoring stations (or sites) in the Murray-Darling Basin, Australia. The data includes many water quality measurements, such as pH, chloride, water temperature, nitrogen levels, dissolved oxygen, discharge rate, water level and turbidity. There are many missing values in the data, caused by equipment malfunction, data lost due to natural causes or vandalism, probe out of water, and other reasons. An effective imputation is essential for accurate water quality modelling and forecasting. Details of the application and data will be presented in Sect. 5.1. In the rest of this paper, we will use the above data as an example when presenting the details of our method.

4   Methodology

This section presents our proposed method and describes in detail the various components of the new time-series imputation method and the framework used to derive a better approach for imputation.

4.1   Framework

Our proposed method, Data Imputation with Seasonality and Causality (DISC), combines causal modelling with seasonal decomposition to fill in missing data in multivariate data. It follows a linear framework with two interchangeable stages as shown in Fig. 1. The data first passes through a Causal Imputation stage, and then through the Seasonal Imputation stage giving the final imputed result. The individual stages are described below.

4.2   Imputation Through Causal Modelling

The aim of a causal discovery stage in our approach is to extrapolate a relationship between the different features in the time-series. In our case we used the LiNGAM model [18] which generates a graph presenting the relationship between different features in a dataset as ratios. These ratios represent the coefficient of linear numerical relation between two features assuming that the features are linearly related. The LiNGAM causal graphs are generated assuming linearity between variables, hence providing us with a set of linear equations that we can

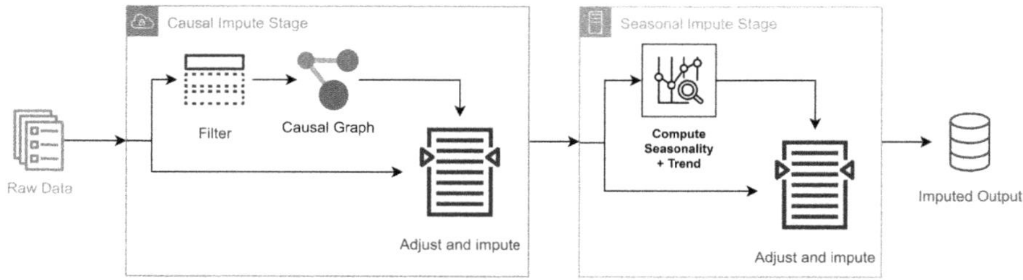

Fig. 1. The Framework of DISC – Data Imputation with Seasonality and Causality.

use. The graph represents the adjacency or coefficient matrix in a linear equation represented by $\mathbf{y} = \mathbf{B}x + e$, where x is the source variable, $\mathbf{y}$ is the output variable dependent on x, $\mathbf{B}$ is the coefficient and e is the error or residual from the conversion.

In order to get an accurate representation of the causal relationship, the model requires continuous data for all the variables. For this we needed to segregate the data into a pseudo-complete dataset and find a time segment where continuous data for all the features was available. Taking the water quality measurements data as an example, we segregated it by finding continuous data of about a year for all the features that were available with no missing data points for a given site. The data for each site is first filtered to isolate at least 2 years of continuous hourly data available for all features. This subset is then used to generate a casual graph, which can be used to generate the missing values for a feature from the other available features. However, not all sites have the required 2 years worth of continuous data. As an example, Fig. 2 shows how much data is available for the top 50 sites with the most data. The gap between the start date represented by blue and end date represented by red gives us the amount of data that we have access at each site, where it can be seen that most sites have less than 2 years worth of data without accounting for missing segments. Thus as a fallback, all the available data for such a site is used to generate the relational graph.

From the above prepared data, the LiNGAM algorithm is used to build causal graphs. Figure 3 is a causal graph generated for one such site, which gave us a conversion ratio to use when converting between the water temperature variable and all the other causally linked variables in the sites dataset. In Table 1 we have the variation of causal graph weights for all the sites, giving an insight into variables that are more reliable for generating the main focus variable which in our case is water temperature. The variation of weights is influenced by the data availability, with sparser datasets giving widely varying weights. As a result, the most reliable variable ended up being air temperature as its variance is not large across all the sites.

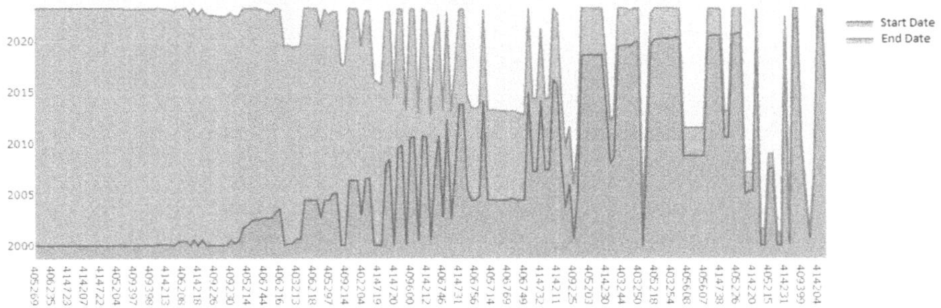

Fig. 2. Graph of available data from 50 sites in the dataset that have the most data. The x-axis shows water monitoring site IDs. The gap between the start date represented by blue and end date represented by red gives us the amount of data that we can use for each site. (Color figure online)

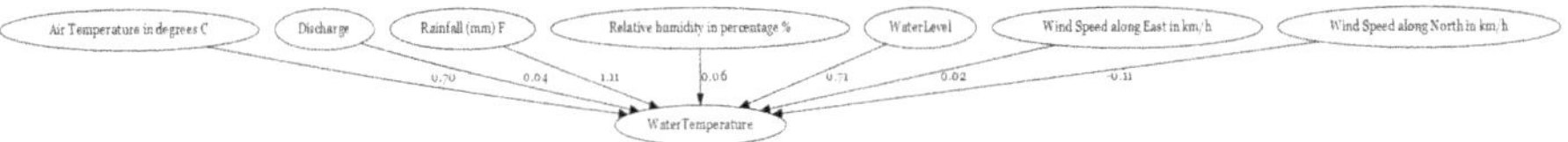

Fig. 3. Example Causal graph with water temperature as the focus variable and its relation to other variables for a selected site's data.

4.3   Imputation Through Seasonal Decomposition

In the Seasonal Imputation stage, the data is first processed into a one year average of all the data at the same frequency as the source dataset. This gives us an average trend for the dataset while retaining the seasonal patterns, for example water temperature data when averaged this way gives us a time-series that still has the yearly seasonal variation as well as the diurnal pattern.

Finally during the imputation steps in each stage, the built reference or the causal relationship is used depending on the stage. Each neighbour is compared with their own average mapped value or the causal estimate and then the average error of the nearest neighbours is used as an adjustment before imputation. Lets say that I is the estimate that we will use to impute, x is the estimate generated from causal graph for that time stamp, n is the number of neighbouring non-missing data points on each side of the value being imputed, b is the estimate generated and r is the ground truth for the neighbouring values suffixed by their distance from our chosen time stamp. The estimate I is calculated as,

$$I = x + \frac{\sum_{i=-n, i\neq 0}^{n} \frac{(r_i - b_i)}{|i|}}{\sum_{i=-n, i\neq 0}^{n} \left|\frac{1}{i}\right|}. \tag{1}$$

Table 1. Causal Graph Weight Statistics w.r.t Water Temperature

Variable	Stat		
	min	median	max
Air Temperature	−0.12	0.69	1.13
Discharge	−4.08	0.005	0.49
Rainfall	−11.98	0	4.69
Relative Humidity	−0.04	0.06	0.14
Water Level	−28.61	−1.19	33.14
Wind Speed along East	−0.05	0.03	0.14
Wind Speed along North	−0.23	−0.11	0

This allows for higher weight-age to be given to closer neighbours, while also factoring in data points on either side through a linear interpolation. Thus, it brings the estimate from each stage to be in line with the actual data around it.

5   Experimental Evaluation

We have tested and evaluated the proposed method in a real-world application of imputing water quality data from the Murray–Darling Basin in Australia, and the experiments and evaluation results are presented and discussed in this section.

5.1   Data

The water quality data is observational data from the Murray–Darling Basin, available publicly through the Victoria and NSW states' Water Measurement Information System. Presented in Table 2, we have the statistics of our dataset. The dataset consists of nearly 200 sites across the Basin, each with a set of time-dependent features. On average, about 20 years worth of data is available for the sites. As with all real-world observational data, this curated dataset has a fair amount of inconsistencies, errors or missing records. Each site has a varying amount of features as well as different duration of data collected.

5.2   Experiment Setup

For evaluating the different data imputation approaches, data was selected from 6 random sites ensuring that each site had at least 16 weeks worth of continuous hourly data for all recorded parameters. From the above data, we randomly withheld some entries and considered them as missing data for imputation purposes. After imputation, the withheld observations served as gold standard to validate the imputed values. Specifically, different time gap lengths or windows were created ranging from 1, 3, 6, 12, 24, 72 and 336 h gaps at random points

Table 2. Data Statistics.

Property	Stat		
	min	median	max
Start date (year)	2000	2004	2022
End date (year)	2001	2019	2023
Number of columns	4	4	7
Length of data (Hours)	8734	197031	210000
Percentage of missing values (%)	0	40.986	99.89

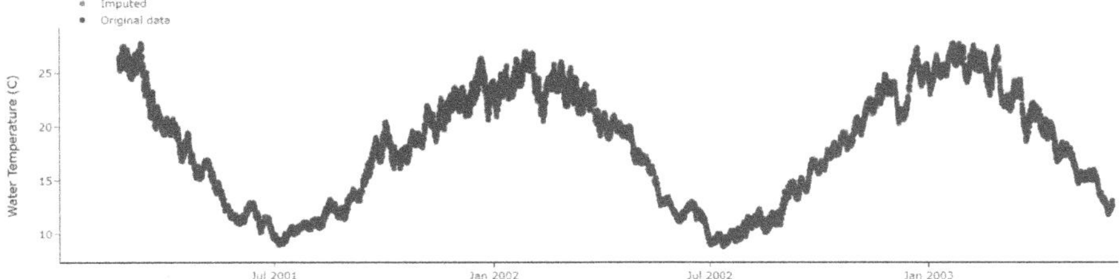

Fig. 4. DISC Seasonal Imputation stage estimates for water temperature. (Color figure online)

in the 16 week segment for each parameter, keeping a maximum limit of 40% missing data. Each test was repeated 10 times for each of these windows changing the pattern of missing data. Applicability of each approach was tested on different parameters. The parameter n in the Seasonal Imputation Stage was kept at a low value of 3 for computational simplicity.

The different imputation approaches are evaluated using Root Mean Squared Error (RMSE) and R^2. RMSE gives the magnitude of deviation or error in the estimates, while R^2 tells if the estimates vary in the same way as the expected values. A combination of both is necessary as the difference between models may not be as obvious in one metric, but more so in the other.

5.3   Experimental Results

This sections details the experiments conducted and their results.

Results for the DISC Seasonal Imputation Stage. An initial test was conducted to check if the Seasonal Imputation Stage would be able to estimate values and fill in the gaps of missing data without deviating too far from the neighbouring samples. No metric was calculated for this test, and it served as a visual representation of how the output estimates look. Two parameters were chosen to be imputed, turbidity and water temperature.

As can be seen in Fig. 4 the imputed results in red nearly completely overlap with the original data (blue). The slight deviations are visible as the red highlight

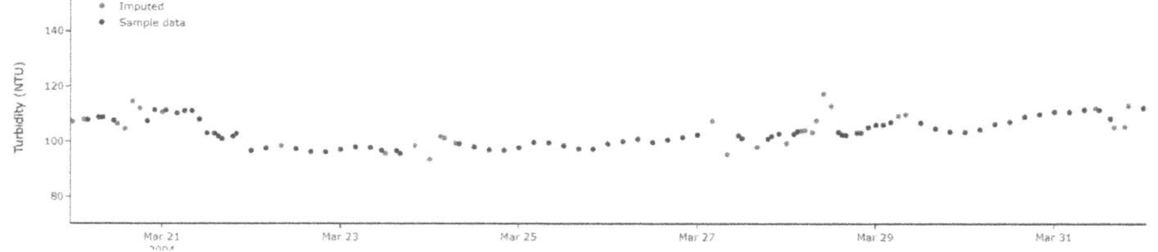

Fig. 5. DISC Seasonal Imputation Stage estimates for Turbidity.

Table 3. Smoothing rolling window size in hours, R^2 Comparison for Water Temperature.

Window size	Gap Length (in Hours)					
	1	3	6	12	24	72
Fill	**0.997**	**0.988**	**0.966**	**0.923**	**0.905**	**0.803**
3	0.759	−0.068	−1.850	−5.334	−4.866	−4.510
6	0.818	0.193	−1.311	−4.226	−3.858	−3.666
9	0.871	0.411	−0.687	−2.921	−2.653	−2.669
12	0.915	0.617	−0.113	−1.633	−1.498	−1.646
24	**0.997**	0.986	0.962	0.916	0.880	0.782

around the blue. In Fig. 5, we get a look at the results for imputing turbidity, the values are close to the neighbouring original data, aren't extremely noisy and follow an expected trend.

To reduce the noise on the estimates, the reference can be smoothened using a rolling window. Table 3 presents the R^2 results for the varying window sizes vs the different gaps to be imputed. The table puts the standard fill method as the most accurate across the board, however from Fig. 6 we can see that the standard fill gives a straight line, with no trend matching or even results that can be close to the expected values especially for larger gaps in data. This highlights that the closer the means of the estimates, the better the R^2 and thus, the flatter the estimate trend is, the better its supposed accuracy. Since we want to retain as much of the trend information, but still have a bit of smoothing applied to the data to reduce noise, we choose one of the rolling window options. For our tests, we chose to apply a window of length 6 as it preserves the diurnal variation just like a window length of 3, but is not erratic or computationally expensive.

In combination with the rolling window, interpolation is calculated using the mean of two nearest neighbouring errors, one in the past and one in the future, to adjust the rolling window estimates to further be in line with the expected data, as shown in Fig. 7.

Results for the Causal Discovery Stage. Figure 8 demonstrates that the estimates from causal discovery using the other available parameters effectively

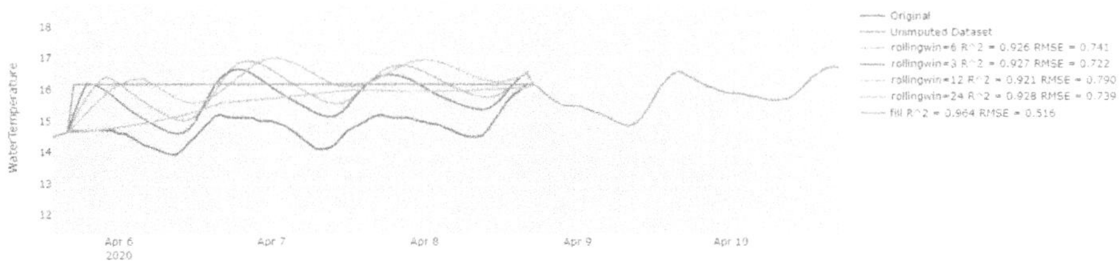

Fig. 6. Effects of rolling window on smoothing with water temperature as the parameter.

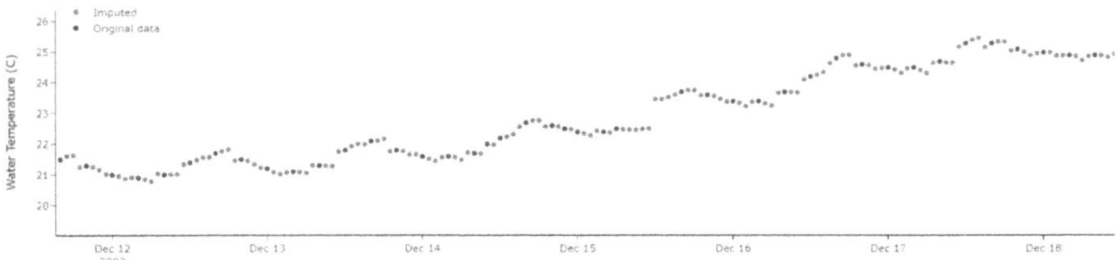

Fig. 7. DISC Seasonal Imputation stage estimates for water temperature with interpolation to correct for offset.

capture the trend of water temperature, matching the daily, monthly, and seasonal variations. However, the estimates are far off from the sample data. This is corrected by adjusting the estimates by calculating the error between the nearest neighbour values and the sample data it overlaps with. The improved results are shown in Fig. 9, where the estimates are now much closer to the original data. This approach does not fill all gaps, particularly when the relevant auxiliary parameters are unavailable for certain timestamps. Additionally, while the estimates remain noisy, they are more correlated to the other parameters linked with water temperature and thus act as good reference points that can help act as anchors when filling large gaps with the DISC method.

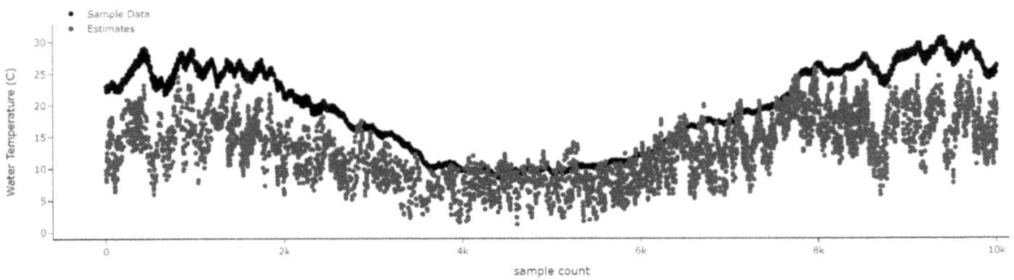

Fig. 8. Causal discovery estimates for water temperature.

Table 4. RMSE Pyimpuyte comparison for Water Temperature.

Model	Gap Length (in Hours)						
	1	3	6	12	24	72	336
Bagging	**0.951**	0.922	1.184	1.528	1.424	1.555	2.097
ExtraTrees	1.003	0.926	1.106	**1.385**	1.351	**1.442**	2.043
Random Forest	0.960	0.941	1.199	1.526	1.447	1.561	2.068
XG	1.025	**0.895**	**1.064**	1.562	**1.349**	1.555	**1.908**

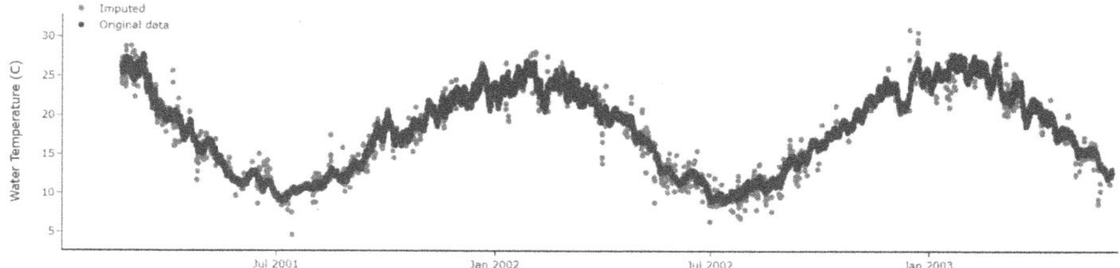

Fig. 9. Causal discovery corrected estimates for water temperature.

Results for the PyImpuyte Method. For comparative studies, the PyImpuyte library is used directly following its documentation [10], with default parameter settings for all models. Figure 10 compares performance of different machine learning algorithms within PyImpuyte to impute turbidity. XGBoost performs better for smaller gaps, while Bagging achieves better overall accuracy, slightly outperforming Random Forest. The same comparison was done with water temperature as the estimated parameter (see Table 4), where XGBoost turned out as the best model. Comparing Table 4 and Fig. 6, we see that the RMSE for PyImpuyte is much higher than Filling and DISC even for the 1 h imputation case. Thus, PyImpuyte was excluded from subsequent experiments. As with the Causal Discovery approach, since the estimates are dependent on other available parameters, some gaps remained unfilled in the absence of sufficient information.

Overall Results. Next, all the models are compared together with their best setup, with the results for imputing water temperature as tabulated in Table 5. The DISC method is presented without causal discovery and with causal discovery as a 2nd stage instead of as the 1st stage. We see that DISC performs the best for shorter gaps and reasonably well for longer gaps. Its performance is boosted significantly for the longer gaps when combined with Causal Discovery as the first stage. When Causal Discovery is used as the second stage, most of the gaps are

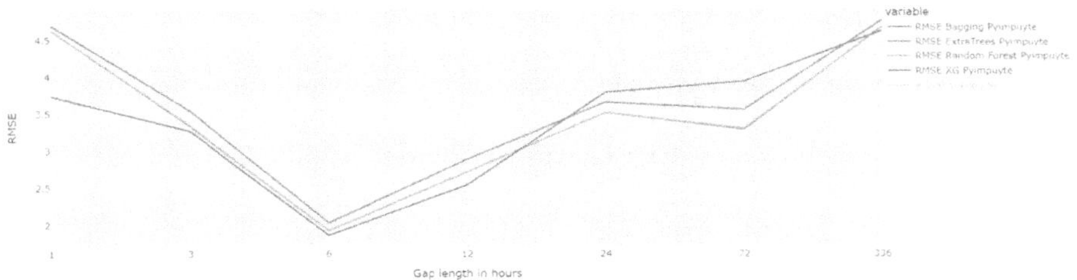

Fig. 10. Pyimpuyte performance for Turbidity with different imputation model settings.

Table 5. R^2 Comparison for Water Temperature. DISC – our proposed framework. DISC (SI+CI) – Seasonal Imputation in stage 1, followed by Causal Imputation in stage 2. DISC (CI+SI) – Causal Imputation in stage 1, followed by Seasonal Imputation in stage 2.

Model	Gap Length (in Hours)						
	1	3	6	12	24	72	336
Fill	0.994	0.976	0.934	0.856	0.832	0.772	0.332
Causal Imputation (CI)	0.166	0.166	0.166	0.165	0.162	0.160	0.129
Seasonal Imputation (SI)	**0.995**	**0.991**	**0.980**	0.944	0.898	0.870	0.628
DISC (SI+CI)	**0.995**	**0.991**	**0.980**	**0.945**	0.898	0.872	0.650
DISC (CI+SI)	0.663	0.988	0.947	0.799	**0.950**	**0.945**	**0.802**

already filled by Seasonal imputation stage and the few gaps are then addressed by Causal Discovery. In Table 6 are the results for imputing turbidity, a parameter that showcases yearly seasonality, typically high in spring-summer and low in autumn-winter [19]. Due to its seasonality, the Seasonal Imputation (SI) on its own ends up being the more consistent option compared to when adding the Causal Imputation (CI). The challenge here is due to the non-linearity of the relationship between turbidity and parameters such as temperature, the effects of which accumulate over time through microbial growth, altering the turbidity of the water. The DISC model performs better than simple filling. However, for features such as turbidity, which depend on non-linear cause–effect relationships a reduced performance observed for the 336-hour imputation. This is perhaps due to the linearity assumption imposed by LiNGAM, which if violated, may lead to incorrect causal ordering, making it difficult to distinguish between cause and effect features.

Table 6. R^2 Comparison for Turbidity. DISC – our proposed framework. DISC (SI+CI) – Seasonal Imputation in stage 1, followed by Causal Imputation in stage 2. DISC (CI+SI) – Causal Imputation in stage 1, followed by Seasonal Imputation in stage 2.

Model	Gap Length (in Hours)						
	1	3	6	12	24	72	336
Fill	0.858	0.754	0.698	0.237	0.406	−0.169	**−0.343**
Causal Imputation (CI)	**0.860**	0.774	0.534	0.489	0.077	−2.960	−20.591
Seasonal Imputation (SI)	0.855	0.719	**0.817**	**0.551**	**0.676**	**0.271**	−1.216
DISC (SI+CI)	0.857	**0.810**	0.813	0.519	0.412	0.119	−20.163
DISC (CI+SI)	**0.860**	0.768	0.446	0.459	0.318	−2.955	−20.181

6   Conclusions

This paper has provided a novel method - DISC (Data Imputation with Seasonality and Causality) and its framework for imputing multivariate time-series data, applying the concepts of causal modelling and seasonal decomposition. It has comprehensively evaluated the performance of the novel method in the field of water quality forecasting for a real-world dataset from the Murray-Darling Basin in Australia. The experiments showcase that the model is capable of imputing large time frame missing gaps, tested up to 14 days of hourly data or 336 missing data points. The model performs ahead of the benchmark approaches such as forward and backward filling, interpolation as well as machine learning techniques such as Causal Discovery, Bagging, ExtraTrees, Random Forest and XGBoost. The model imputes data with high accuracy and correlation with an R^2 accuracy of over 80% when compared to other approaches having an R^2 accuracy below 30% in the worst case scenario. From a computational perspective, DISC is a single-pass algorithm that eliminates the need for iterative model training as required by the machine learning methods. This makes it significantly faster while achieving comparable accuracy. Furthermore, it is worth noting that, though the method is demonstrated for the water domain, it is a generic method applicable to multivariate missing data imputation in other domains.

The method in our case is still very much limited to the data available. In many cases, data is only available for a few days or months within a year, severely limiting the average trend component. This lack of data can in some ways be supported by using data from neighbouring sites by directly adding them to the causal model. Remote sensing provides another source for the data, wherein the data points will first need to be extracted from the images and then fed to the causal model to act as anchors to guide the estimate of seasonality.

Given that the model can impute a fortnight of data easily while preserving the data pattern, its capability of imputing further than the 14-day period needs to be tested, for more than the limited scope of the water temperature parameter. Furthermore, the method can be used for forecasting highly seasonal variables similar to temperature as the seasonal component can be used on its own for expanding the time-series. Causal modelling presents an opportunity for further investigation in-terms of forecasting, where it can be used as a second step to correct and re-align the time-series data using the multivariate insights.

References

1. Chen, K., Zhou, Y., Dai, F.: A LSTM-based method for stock returns prediction: a case study of China stock market. In: 2015 IEEE International Conference on Big Data, pp. 2823–2824 (2015)
2. Morid, M.A., Liu Sheng, O.R., Dunbar, J.: Time series prediction using deep learning methods in healthcare (2022)
3. Abbasimehr, H., Shabani, M.: Forecasting of customer behavior using time series analysis. In: Data Science: From Research to Application, pp. 188–201. Springer, Cham (2020)
4. Tanhaei, H.G., et al.: Predictive analytics in customer behavior: anticipating trends and preferences. Results Control Optim. **17** (2024)
5. Bhogal, D., Zhao, Y., Kapoor, R., Toscas, P.: A comparative study of water temperature forecasting methods. In: 2023 IEEE/WIC International Conference on Web Intelligence and Intelligent Agent Technology (WI-IAT) (2023)
6. Demiray, B.Z., Mermer, O., Baydaroglu, O., Demir, I.: Predicting harmful algal blooms using explainable deep learning models: a comparative study. Water **17**(5) (2025)
7. Emmanuel, T., Maupong, T., Mpoeleng, D., Semong, T., Mphago, B., Tabona, O.: A survey on missing data in machine learning. J. Big Data **8**(1), 1–37 (2021). https://doi.org/10.1186/s40537-021-00516-9
8. Lee, Y.: Imputation method using local linear regression based on bidirectional k-nearest-components. J. lnf. Commun. Converg. Eng. **21**(1), 62–67 (2023)
9. Han, H., Sun, M., Han, H., Xiaolong, W., Qiao, J.: Univariate imputation method for recovering missing data in wastewater treatment process. Chin. J. Chem. Eng. **53**, 201–210 (2023)
10. Suresh, M., Taib, R., Zhao, Y., Jin, W.: Sharpening the BLADE: missing data imputation using supervised machine learning. In: Liu, J., Bailey, J. (eds.) AI 2019. LNCS (LNAI), vol. 11919, pp. 215–227. Springer, Cham (2019). https://doi.org/10.1007/978-3-030-35288-2_18
11. Troyanskaya, O., et al.: Missing value estimation methods for DNA microarrays. Bioinformatics **17**(6), 520–525 (2001)
12. Dempster, A.P., Laird, N.M., Rubin, D.B.: Maximum likelihood from incomplete data via the EM algorithm. J. Roy. Stat. Soc.: Ser. B (Methodol.) **39**(1), 1–22 (2018)
13. Sun, Y., Li, J., Yifan, X., Zhang, T., Wang, X.: Deep learning versus conventional methods for missing data imputation: a review and comparative study. Expert Syst. Appl. **227**, 120201 (2023)
14. Shah, A.D., Bartlett, J.W., Carpenter, J., Nicholas, O., Hemingway, H.: Comparison of random forest and parametric imputation models for imputing missing data using mice: a caliber study. Am. J. Epidemiol. **179**(6), 764–774 (2014)
15. Pearl, J.: Causality, 2nd edn. Cambridge University Press (2009)
16. White, I.R., Joseph, R., Best, N.: A causal modelling framework for reference-based imputation and tipping point analysis (2017)

17. Almeida, R.J., et al.: Graphical causal models and imputing missing data: a preliminary study. In: Information Processing and Management of Uncertainty in Knowledge-Based Systems, pp. 485–496 (2020)
18. Shimizu, S., et al.: A linear non-gaussian acyclic model for causal discovery. J. Mach. Learn. Res. **7**(72), 2003–2030 (2006)
19. Gayol, M., et al.: Temporal and spatial variability of turbidity in a highly productive and turbid shallow lake (chascomús, argentina) using a long time-series of landsat and sentinel-2 data. Hydrobiologia **851**, 1–23 (2024)

UniCausal: A Unified Approach to Causal Discovery from Hybrid Industrial Time Series and Events

Zhen Zhao[1,3]($\boxtimes$) (iD), Brian Kenneth Erickson[2] (iD), Shantanu Chakraborty[1] (iD), and Wei Liu[3] (iD)

[1] AVEVA Group plc (Australia), Sydney, NSW, Australia
`{zhen.zhao,shantanu.chakraborty}@aveva.com,`
`zhen.zhao-1@student.uts.edu.au`
[2] AVEVA Group plc (USA), Lake Forest, CA, USA
`brian.erickson@aveva.com`
[3] University of Technology Sydney, Sydney, NSW, Australia
`wei.liu@uts.edu.au`

Abstract. The discovery of temporal causality is essential for identifying the root causes of problems in real-world systems. In practice, causality can arise from two primary sources: time series (TS) data and discrete events. Although many studies have explored these two modalities independently, existing approaches typically focus on one or the other. To our knowledge, no existing framework unifies causal relationships derived from both TS and events. In this paper, we propose UniCausal, a unified framework for constructing hierarchical causal graphs from both TS and event data. Our approach integrates symbolic state-change events derived from raw TS with predefined discrete events, enabling multi-level causal inference across heterogeneous data sources. We further advance causal discovery by introducing a novel deep model that incorporates symbolic context conditions, enabling more interpretable and context-aware causal reasoning. To demonstrate the framework's practical utility, we present a case study in which UniCausal is combined with a Large Language Model (LLM) to perform interpretable root cause analysis. This case study emphasizes application-oriented reasoning in a real-world industrial scenario, highlighting the potential of UniCausal for transparent, human-in-the-loop diagnostics. While our case study focuses on industrial systems, the framework is broadly applicable to domains such as IoT, healthcare, and any other field where TS and event data co-exist.

Keywords: Temporal causality · Unified causal graph · Conditional causality · Time series representation learning · Symbolic event extraction · Industrial automation · Root cause reasoning

1 Introduction

Temporal causality discovery refers to the task of uncovering cause-and-effect relationships among variables that evolve over time using observational data [9].

Q. V. Nguyen et al. (Eds.): AusDM 2025, CCIS 2765, pp. 145–160, 2026.
https://doi.org/10.1007/978-981-95-6786-7_10

It plays a critical role in understanding complex dynamic systems, supporting root cause analysis, predictive control, and decision-making across a wide range of applications. Two primary types of temporal data are typically involved: continuous time series (TS) and discrete events. To represent causal relationships, Directed Acyclic Graphs (DAGs) are commonly employed, where nodes correspond to variables or events, and directed edges indicate causal influence. Beyond structural connections, temporal lag (the delay between a cause and its effect) is often explicitly modeled to capture not only what causes what, but also when the effect is likely to occur [7]. Accurate modeling of both causal structure and timing is essential for diagnosing and optimizing temporal processes in diverse domains.

1.1 Background and Motivation

In practice, understanding causal relationships has enabled impactful improvements across domains. In industrial automation, identifying that a valve malfunction causes downstream pressure drops can prevent system-wide failures and minimize downtime. In healthcare, recognizing that a specific medication leads to adverse interactions under certain physiological conditions informs safer treatment planning. In finance, uncovering causal links between macroeconomic indicators and market volatility supports more robust risk management strategies. These examples demonstrate how even partial causal knowledge can enhance diagnosis, planning, and proactive intervention, especially in systems where decisions are time-sensitive and consequences may propagate.

Despite its central importance, uncovering temporal causal relationships remains highly challenging. Traditional methods rely heavily on expert knowledge, which becomes less feasible in modern complex systems with numerous interacting variables. Furthermore, many influential causal factors are latent or non-obvious, requiring advanced computational techniques to extract them from raw observational data. These limitations motivate the development of scalable, data-driven approaches that can automatically infer causality across modalities.

1.2 Related Work and Limitations

Classical approaches to temporal causality discovery, such as Granger causality [13], form the foundation for many modern methods. Granger causality is a statistical technique for identifying causal relationships in TS data. It operates under the assumption that if the past values of one variable significantly improve the prediction of another, then the former is said to have a causal influence on the latter. Inspired by this principle, recent methods have leveraged deep learning to capture more complex and nonlinear causal dependencies. For example, the Temporal Causal Discovery Framework (TCDF) [10] trains multiple independent attention-based convolutional neural networks (CNNs), each of which predicts a target TS variable from all input TS variables. Causal relationships are inferred from the learned attention weights, along with mechanisms for causal validation and delay discovery. Another representative method is CausalFormer [8], which models multivariate TS using a deep architecture that integrates multi-kernel causal convolutions with scaled dot-product attention. After training the model to predict future values, Regression Relevance Propagation (RRP) is applied to extract interpretable causal influences between variables [2].

Deep learning is also widely applied to causal discovery from event sequences, commonly referred to as Event Causality Identification (ECI). A prominent example is Dr.ECI [6], which employs a multi-agent architecture. It includes specialized agents such as the Causal Explorer and Mediator Detector to capture direct and indirect causal relationships during the discovery stage, and Direct and Indirect Reasoners to apply causal priors for validating and refining results. CausalNET [18], another state-of-the-art (SOTA) ECI model, uncovers causal structures from categorized events using a topology-informed causal attention mechanism within a Transformer architecture.

Despite significant advances, existing methods face key limitations when applied to real-world scenarios. TS–based models focus exclusively on continuous numerical data, while ECI frameworks are limited to discrete event data. None provides a unified perspective that integrates both modalities. Moreover, most frameworks infer static causal graphs with fixed temporal lags, which fail to capture the conditional and context-dependent causal relationships frequently observed in complex operational systems. In industrial automation, for instance, machinery often operates under distinct modes such as Running, Maintenance, and Testing. The causal relationship between control variables (e.g., set points) and process variables (e.g., actual values) can vary significantly across these modes. A change in a set point may strongly influence the process value during Running mode, but have little to no effect in Maintenance mode when actuators are decoupled from the process. These limitations reduce the effectiveness of existing models in dynamic, multi-source environments.

1.3   Contributions

To address limitations of SOTA methods, we propose UniCausal, a unified framework for constructing hierarchical DAGs that capture context-aware causal relationships across both TS and event data. Unlike prior methods that focus exclusively on either TS (e.g., CausalFormer) or events (e.g., CausalNET), UniCausal unifies both modalities into a single, context-aware causal discovery framework. Our key contributions are as follows:

- **A unified causal discovery framework (UniCausal)** for constructing hierarchical DAGs that integrate continuous TS data with discrete event data, via a practical pipeline that fuses events derived from TS state transitions with predefined events (e.g., alarms, actions), which enables multilevel causal reasoning from raw signals to symbolic and semantic events.
- **A novel deep causal discovery model** that directly incorporates contextual information into causality relationships. This design allows causal dependencies and lags to be modulated by symbolic states of variables or system conditions, thereby improving both modeling accuracy and interpretability in real-world settings.

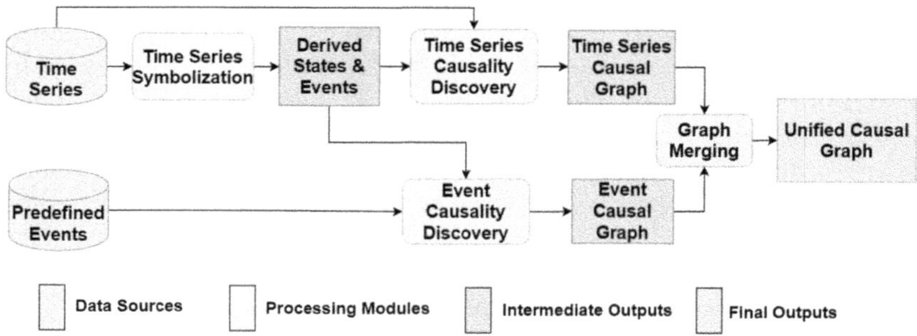

Fig. 1. Overview of the UniCausal framework. Raw TS and predefined events are processed in parallel: TS are symbolized to extract states and derive events, which are then used for both TS-based and event-based causal discovery. The resulting intermediate causal graphs are aligned and merged to produce a unified causal graph representing diverse dependencies.

- **A comprehensive set of experiments** that evaluate the effectiveness of UniCausal framework against SOTA methods on a diverse benchmark comprising both real-world and industrial-style datasets.
- **An application-oriented case study** demonstrating how UniCausal can be integrated with a large language model (LLM) to perform interpretable, context-aware root cause analysis in industrial systems. This use case highlights the potential of the framework to enable human-in-the-loop diagnostics by combining symbolic causal graphs with natural language reasoning.

2   UniCausal Framework Overview and Architecture

In this section, we present the UniCausal framework for constructing a unified causal graph from hybrid TS and event data. As illustrated in Fig. 1, the process begins with transforming raw TS into symbolic events via state identification and clustering. The derived states are passed to the **TS Causality Discovery** module, while the derived events, together with predefined system events, are fed into the **Event Causality Discovery** module.

The **TS Causality Discovery** module infers causal relationships from both the original TS and their symbolic representations, while the **Event Causality Discovery** module focuses on event-based causality. Although an integrated causal model is a direction worth exploring, we decouple the two modalities in the current version to enable modular benchmarking and improve scalability through parallel processing. The resulting intermediate graphs are then aligned using mappings between TS variables and event semantics, producing a unified causal graph that captures diverse dependencies in context-sensitive industrial environments.

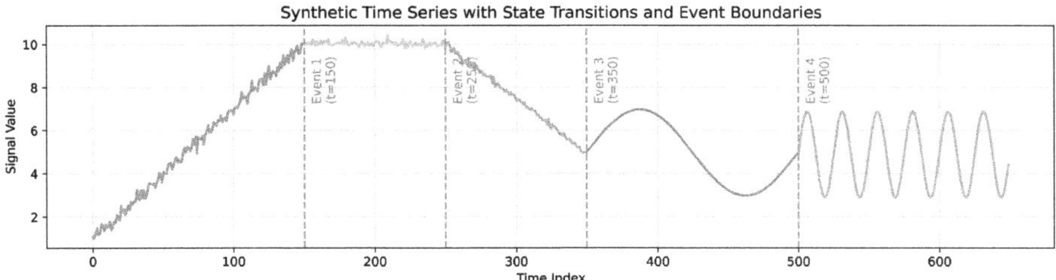

Fig. 2. Synthetic TS with state transitions and derived event boundaries. Each colored segment represents a distinct state: ▮ *Ramp Up,* ▮ *Steady High,* ▮ *Ramp Down,* ▮ *Low Frequency,* and ▮ *High Frequency.* Dashed vertical lines indicate state transitions, interpreted as derived events for ECI analysis. The value of t in parentheses denotes the time index.

2.1 Time Series Symbolization and Event Derivation

Symbolization serves as a foundational step in UniCausal, enabling both conditional causality modeling and the integration of TS and event-based representations. UniCausal adopts an unsupervised segmentation approach inspired by Time2State [15], which extracts latent state dynamics from raw signals and converts them into discrete symbolic events. These events provide a compact and interpretable abstraction of underlying temporal patterns, forming the bridge to event-level causal reasoning and conditional dependency estimation.

UniCausal employs an encoder network to learn latent representations from raw TS data. The encoder comprises a causal convolutional layer, a pooling layer, and a linear projection layer. A fixed-size sliding window is applied to the TS to extract overlapping segments, from which embeddings are generated. These embeddings are then clustered using a Dirichlet Process Gaussian Mixture Model (DPGMM) [4], a Bayesian clustering method that infers the number of clusters directly from data. Each cluster represents a latent state type, and consecutive windows assigned to the same cluster define a state instance. Transitions between different states are recorded as events. Figure 2 illustrates an example of the symbolic states and corresponding events derived from a TS signal.

Unlike the original Time2State design, which utilizes adaptive max pooling, UniCausal incorporates additive attention pooling [3] within the encoder, as our experiments show it yields more stable and accurate state representations, especially when handling long input windows.

2.2 Context-Aware Causal Modeling on Time Series and Events

Causal relationships identified by SOTA approaches are typically represented as

$$X_t \rightarrow Y_{t+\tau} \tag{1}$$

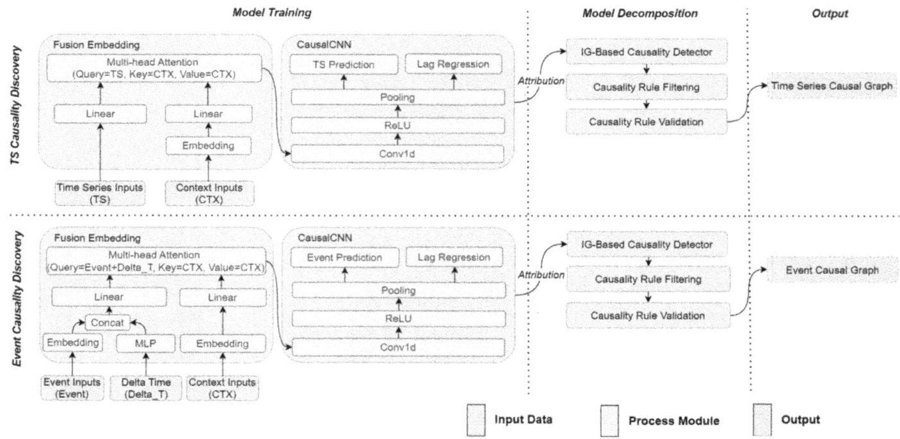

Fig. 3. Overview of the proposed context-aware causal discovery framework. The model is trained in a self-supervised manner to predict future time series or event outputs from past inputs and symbolic context, using attention-based fusion embeddings. After training, relevance-based model decomposition leverages Integrated Gradients (IG) to attribute predictions to input features. This enables the extraction and validation of condition-specific causal rules, ultimately yielding a context-aware causal graph.

where X denotes the cause variable, Y the effect variable, t the time index, and τ the temporal lag between cause and effect.

To account for system dynamics that vary under different operating conditions, we introduce a conditioning variable C, extending the formulation to:

$$X_t \xrightarrow{C} Y_{t+\tau} \tag{2}$$

where C represents the contextual or symbolic state under which the causal relationship holds. This denotes that the causal effect of X_t on $Y_{t+\tau}$ is valid only when the system is in context C.

Figure 3 presents the proposed framework for context-aware causal discovery from TS and event data. The framework comprises two main stages: *Model Training* and *Model Decomposition*.

In the *Model Training* phase, for **TS Causality Discovery**, raw TS inputs are augmented with symbolic context signals (e.g., TS-derived states or operational modes). These are embedded via a multi-head attention mechanism:

$$\text{Attention}(Q, K, V) = \text{softmax}\left(\frac{QK^{\top}}{\sqrt{d_k}}\right) V \tag{3}$$

where Q denotes the query vectors derived from the TS, and K, V represent the keys and values derived from the context inputs, respectively. This configuration allows the model to dynamically integrate contextual information that is most relevant to each TS segment, thereby improving the quality of causal inference.

For **Event Causality Discovery**, we consider an event sequence defined as $\mathcal{E} = \{(e_1, t_1), (e_2, t_2), \ldots, (e_n, t_n)\}$, where e_n denotes the n-th event class and t_n its corresponding timestamp. We begin by computing the delta time between each pair of consecutive events. Each event is passed through an embedding layer, and the corresponding delta time is processed by a sequential layer (e.g., a Multilayer Perceptron, or MLP). The resulting vectors are concatenated and projected via a linear layer to form the query representation for attention. As in the TS setting, the context inputs are embedded and used as the keys and values in the multi-head attention module to produce context-fused embeddings.

The resulting fusion embeddings derived from TS or event sequences are passed to a causal discovery model, implemented as a Causal Convolutional Neural Network (CausalCNN). A CausalCNN is a one-dimensional convolutional network that enforces temporal causality by restricting filters so that predictions at time t depend only on past inputs, never on future values. Our implementation consists of a Conv1D layer followed by a ReLU activation and pooling layer. To support both causality identification and lag estimation, the model includes two output heads: a prediction head to forecast future TS or event features and a regression head to estimate associated temporal lags. The model is trained in a self-supervised manner by minimizing the combined objective:

$$\mathcal{L}_{\text{total}} = \mathcal{L}_{\text{pred}} + W \cdot \mathcal{L}_{\text{lag}} \tag{4}$$

where $\mathcal{L}_{\text{pred}}$ and $\mathcal{L}_{\text{lag}}$ are the prediction and lag regression losses, respectively, and W is a tunable hyperparameter that balances their relative importance.

In the *Model Decomposition* stage, we apply Integrated Gradients (IG) [14] to interpret the trained model's predictions by computing relevance scores for each input feature. IG is a widely adopted attribution method that quantifies feature contributions by integrating the gradient of the model's output along a straight-line path from a baseline input (e.g., zero vector) to the actual input. The IG attribution for the i-th feature is given by:

$$\text{IG}_i(x) = (x_i - x_i') \times \int_{\alpha=0}^{1} \frac{\partial F\left(x' + \alpha \cdot (x - x')\right)}{\partial x_i} \, d\alpha \tag{5}$$

where x is the actual input, x' is the baseline input, F is the model output, and $\alpha \in [0, 1]$ is the interpolation parameter along the path from x' to x.

The IG-based causality detector identifies candidate causal relationships by evaluating the relevance of input variables across time and context. The *Causality Rule Filtering* module then selects the top-k most relevant rules based on these attributions. Finally, the *Causality Rule Validation* module associates the discovered causal patterns with specific contextual states and ensures that the resulting graph is acyclic. The final output is a context-aware causal graph that captures conditional dependencies grounded in system behavior.

2.3   Unified Causal Graph Construction

The final output of the UniCausal framework is a unified, hierarchical causal graph that integrates both TS and event-level dynamics under context-aware

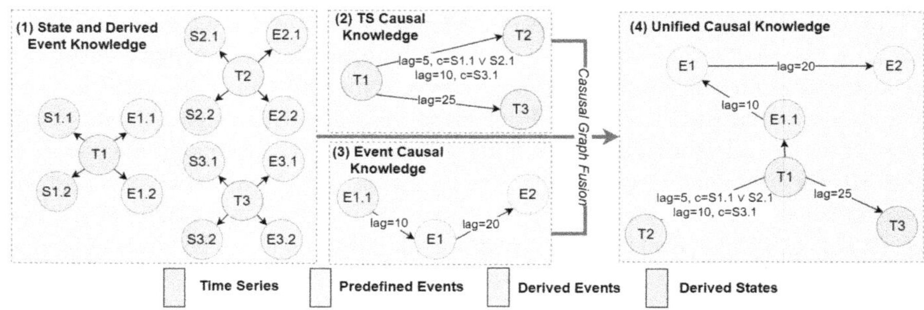

Fig. 4. Construction of a unified, context-aware causal graph. (1) Symbolic states and derived events are connected to TS variables via TS symbolization and event derivation; (2) TS causal graph with lag and condition annotations from TS causality discovery; (3) Event-level causal graph with lag and condition annotations from event causality discovery; (4) Final unified graph obtained by fusing TS and event causal knowledge, illustrating conditional causal edges across numeric and symbolic variables. Edge labels indicate the causal lag (e.g., `lag=5`) and, where applicable, the condition under which the causal relationship holds (e.g., `c=S1.1 ∨ S2.1`).

conditions. This graph is obtained by merging the TS and event causal graphs through shared intermediate structures, specifically, the derived events that connect TS behaviours to symbolic system events. As shown in Fig. 4, these derived events act as causal bridges: they originate from state transitions in TS variables (Fig. 4 (1)) and also participate as causes or effects within the event-level graph (Fig. 4 (3)).

Figure 4 (2) shows the TS causal subgraph annotated with temporal lags and the contextual conditions that qualify each causal relationship. Figure 4 (4) presents the fully merged hybrid graph, which captures causal links across numeric (TS) and symbolic (events) domains, while incorporating contextual information from states and condition-dependent relationships.

3   Experiments and Evaluation

In this section, we present a comprehensive set of experiments and analyze the results to evaluate the effectiveness of UniCausal framework. We begin by assessing the symbolization component, which plays a critical role in transforming raw TS data into meaningful symbolic states. The quality of this symbolization directly affects downstream tasks such as event derivation and condition detection for context-aware causal reasoning. Inaccurate or poorly segmented representations can obscure important patterns and compromise causal inference.

We then evaluate UniCausal's performance in discovering causal relationships across both TS and event modalities, as well as in hybrid settings that combine the two. Our experiments benchmark UniCausal against relevant baselines using detailed metrics for causality detection, lag estimation, and condition modeling.

Table 1. Performance comparison of symbolization methods using a window size of 512, evaluated with ARI and NMI metrics. The last row reports the mean and standard deviation across datasets.

Metric	Dataset	AdaMaxPool	Attn+FFT	Attn+Max	AddAttnPool	GaussMask+Reg	LearnAttnPool
ARI	ActRecTut	0.617	0.587	0.614	0.546	**0.669**	0.526
	MoCap	0.581	0.604	0.598	**0.714**	0.597	0.706
	PAMAP2	0.305	0.333	0.304	**0.352**	0.320	0.340
	USC-HAD	0.662	**0.816**	0.693	0.750	0.702	0.768
	Synthetic Data	0.613	0.658	0.626	0.755	0.659	**0.756**
NMI	ActRecTut	0.646	0.629	0.640	0.607	**0.668**	0.603
	MoCap	0.651	0.670	0.662	**0.749**	0.663	0.736
	PAMAP2	0.590	0.627	0.586	0.625	0.602	**0.630**
	USC-HAD	0.803	**0.878**	0.816	0.859	0.824	0.860
	Synthetic Data	0.686	0.716	0.689	**0.775**	0.712	0.773
	Average (mean, std)	0.675 ± 0.079	0.704 ± 0.104	0.679 ± 0.086	**0.723 ± 0.106**	0.694 ± 0.083	0.720 ± 0.105

3.1 Evaluation of Symbolization Module

The symbolization module is evaluated using four real-world human activity datasets (ActRecTut [5], MoCap [1], PAMAP2 [12], and USC-HAD [17]) and a synthetic dataset. To assess robustness across different temporal resolutions, each dataset is processed using multiple window sizes (128, 256, and 512). We build upon Time2State, an SOTA unsupervised symbolization method that serves as the foundation for UniCausal's symbolization component. Specifically, we replace its original adaptive max pooling (AdaMaxPool) layer with a set of alternative pooling and attention-based mechanisms. These include additive attention pooling (AddAttnPool), learned attention pooling (LearnAttnPool), probabilistic masking via Gaussian Mask with Regularization (GaussianMask+Reg), and hybrid strategies such as attention combined with adaptive max (Attn+Max) and attention combined with Fourier transforms (Attn+FFT). All variants share the same model architecture and training configuration, differing only in the pooling layer used for symbolization.

Performance is evaluated using the Adjusted Rand Index (ARI) and Normalized Mutual Information (NMI) to quantify segmentation quality. As shown in Table 1, AddAttnPool generally outperforms AdaMaxPool across most datasets (e.g., MoCap, PAMAP2, USC-HAD), though AdaMaxPool remains stronger on ActRecTut. Overall, AddAttnPool demonstrates stronger stability and domain generalization, making it the default symbolization module in UniCausal, particularly for long-duration industrial processes where extended temporal dependencies are common.

3.2 Evaluation of Context-Aware Causal Modeling

To assess UniCausal's performance across diverse data modalities and contextual scenarios, we conduct experiments on three categories of synthetic datasets: (1) multivariate TS with lag and context-aware causality, (2) event sequences with similar causal structures, and (3) hybrid datasets combining TS and event

Table 2. Accuracy and completeness comparison (mean ± standard deviation) of Uni-Causal with baseline methods across datasets of different modalities and ground truth configurations. Hybrid Causality denotes the overall ability to detect cross-modality causal links (i.e., from TS to events and vice versa). N/A indicates the dataset lacks ground truth for the metric; – indicates the method does not support the task (e.g., CausalFormer cannot handle event-based or hybrid causality). **Abbreviations:** Acc. = accuracy, Comp. = completeness, TS = time series.

Datasets	Method	TS Acc.	TS Comp.	Event Acc.	Event Comp.	Hybrid Acc.	Hybrid Comp.	Lag	Conditions
TS	CausalFormer	0.47 ± 0.13	0.33 ± 0.24	N/A	N/A	N/A	N/A	0.83 ± 0.19	–
	CausalNET	–	–	N/A	N/A	N/A	N/A	–	–
	UniCausal	**0.61 ± 0.23**	**0.72 ± 0.14**	N/A	N/A	N/A	N/A	**0.85 ± 0.22**	**0.65 ± 0.37**
Event	CausalFormer	N/A	N/A	–	–	N/A	N/A	–	–
	CausalNET	N/A	N/A	0.23 ± 0.06	0.31 ± 0.08	N/A	N/A	–	–
	UniCausal	N/A	N/A	**0.44 ± 0.18**	**0.92 ± 0.12**	N/A	N/A	**0.88 ± 0.13**	**0.40 ± 0.12**
Hybrid	CausalFormer	0.32 ± 0.32	0.28 ± 0.11	–	–	–	–	0.73 ± 0.17	–
	CausalNET	–	–	0.27 ± 0.23	0.44 ± 0.22	–	–	–	–
	UniCausal	**0.51 ± 0.13**	**0.88 ± 0.33**	**0.41± 0.15**	**0.78 ± 0.19**	**0.51 ± 0.32**	**0.73 ± 0.11**	**0.76 ± 0.12**	**0.42 ± 0.20**

data. Each category comprises 20 datasets generated from industrial-inspired templates, designed to capture distinct causal structures and support systematic evaluation of context-aware causal modeling.

The TS datasets consist of five process variables, each exhibiting state transitions among *increasing, decreasing,* and *stable* over time. Causal lags are randomly assigned within 1–8 timesteps. Causal relationships may be unconditional or conditioned on one or more TS states (e.g., "T1 increasing and T3 decreasing"). The event datasets comprise nine alarm types (six genuine, three noise), with conditions defined by operational modes—*Idle, Startup*, or *Shutdown*—or no condition, and lags ranging from 1–15 timesteps. Hybrid datasets support cross-modal causal links and mixed conditional structures.

For each dataset, we compare UniCausal with two baselines: CausalFormer for TS-based causality discovery and CausalNET for event causality inference. Evaluation is conducted along eight axes: TS causality accuracy and completeness, event causality accuracy and completeness, hybrid causality accuracy and completeness, lag prediction accuracy, and condition identification accuracy.

Evaluation metrics are as follows: **Causality accuracy** measures the proportion of predicted causal links that match the ground truth relationships; **Causality completeness** measures the proportion of ground truth causal relationships that are successfully identified by the model; **Lag accuracy** is the proportion of correctly identified causal pairs with an exact lag match; **Condition accuracy** is the proportion of correctly identified causal pairs with an exact condition match, including cases where no condition is specified.

The results presented in Table 2 show that **UniCausal** maintains consistent performance across all data modalities and ground truth configurations, including the challenging hybrid datasets that require inference across both TS and event variables. In contrast, the baseline methods are constrained by their architectural design: CausalFormer achieves TS causality accuracy of 0.47 ± 0.13 on TS datasets but cannot address event causality or hybrid tasks, while Causal-

NET is applicable only to event-based datasets and is unable to process TS data.

Across all evaluated scenarios, UniCausal attains higher accuracy and completeness compared to the baselines. In the most challenging hybrid datasets, UniCausal achieves 0.51 ± 0.13 for TS causality accuracy, 0.41 ± 0.15 for event causality accuracy, 0.51 ± 0.32 for hybrid causality accuracy, 0.76 ± 0.12 for lag prediction, and 0.42 ± 0.20 for condition accuracy, along with the highest completeness scores in each modality. These results indicate that UniCausal can effectively integrate and infer causal relationships across heterogeneous data sources, providing a unified and interpretable framework that generalizes to complex, real-world industrial scenarios.

4 Case Study: LLM-Augmented Causal Inference for Industrial Systems with Unified Causal Graph

We present a case study where UniCausal is integrated with an LLM to perform context-aware causal inference in natural language. This approach demonstrates how symbolic causal relationships, extracted from a Historian database (a specialized industrial system for storing historical sensor data) [16] and event logs, can be transformed into interpretable natural language explanations for root cause analysis.

4.1 System Design

In complex industrial systems, causal dependencies are often context-specific. For example, "Pump A causes a pressure increase in Tank B only when Valve C is open." UniCausal supports such conditional DAGs by ingesting event streams and logs, and generating a unified causal graph where edges are annotated with logical conditions and temporal lags. A JSON-like internal representation of the graph stores causal edges along with their source, target, condition, and lag metadata.

As illustrated in Fig. 5, the causal graph is used within an Industrial Copilot that answers operator queries. An Orchestrator determines which agent to invoke based on the query type—for example, root cause analysis requests are routed to the Causal Inference Agent. This agent contains three modules: Causal Knowledge Generation, which converts causal graph edges into templated natural language rules; Real-Time Data Access, which retrieves the current system state; and Prompt Composer, which merges causal knowledge, current observations, and the user query into a structured prompt for the LLM.

We adopt a structured template for causal reasoning, such as: "If $< Condition >$ is $< State >$, then $< Cause >$ has a cause effect on $< Effect >$ with a lag $< timestep >$." The condition part is optional; if present, $< Condition >$ may consist of one or more system states combined using logical conjunctions (e.g., "Valve C is open and System Mode is startup").

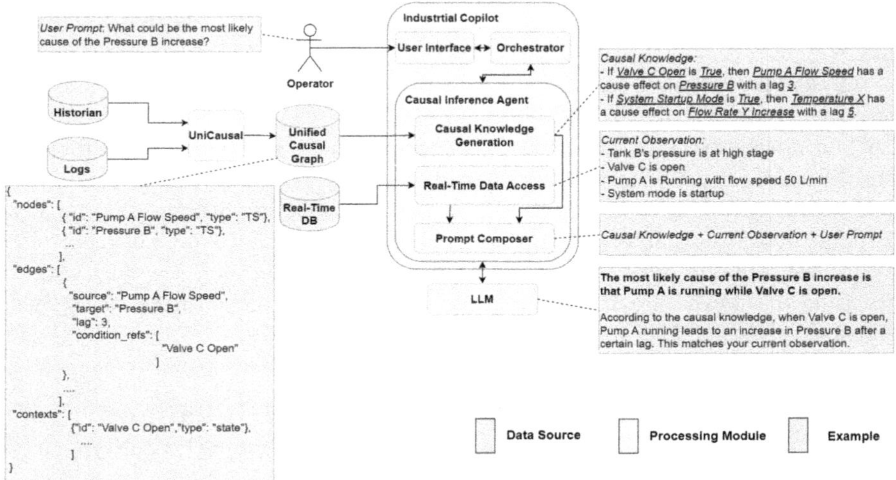

Fig. 5. System architecture of the Industrial Copilot integrating UniCausal and LLMs for context-aware causal reasoning. The figure depicts how historical logs and real-time sensor data are processed to build a unified causal graph, which is annotated with contextual conditions and temporal lags. The architecture shows the flow from data ingestion, causal inference, and state retrieval to prompt composition and natural language root cause explanation via LLM, supporting interpretable, operator-facing diagnostics.

These generated statements, combined with a user's prompt (e.g., "What is the most likely cause of the pressure B increase?"), are passed to a pre-trained LLM (e.g., GPT-4o [11]). The model returns a natural language explanation, such as: "The most likely cause of the Pressure B increase is that Pump A is running while Valve C is open."

This architecture enables interpretable, human-in-the-loop diagnosis in dynamic and condition-sensitive environments. While LLMs are inherently probabilistic, the templated causal input provides a structured scaffold that improves generalization and reproducibility across use cases of varying complexity. Preliminary testing in industrial-style scenarios indicated that the LLM's responses aligned well with the underlying causal graph, indicating promising potential for practical deployment. More comprehensive evaluation, including quantitative comparisons with baseline methods and validation against live-plant data, will be conducted as part of future work.

4.2   Evaluation of LLM-Augmented Causal Inference

As an exploratory assessment of the practical utility of the causal graph in downstream reasoning tasks, we conducted the experiment involving 20 simulated industrial scenarios, each centered around a triggered alarm. The objective is to assess whether GPT-4o can accurately infer the root cause of the alarm based

Table 3. Impact of Domain-Specific Causal Knowledge on GPT-4o Root Cause Inference (Case Study)

Aspect	Without Causal Knowledge	With Causal Knowledge
Prompt Input	Observation: - Tank 10 Level Low Alarm is On. - Valve 20 is Open. - Valve 30 is Closed.	Observation: - Tank 10 Level Low Alarm is On. - Valve 20 is Open. - Valve 30 is Closed. Causal Knowledge: - If Valve 30 is Closed, Valve 20 Flow Speed has a cause effect on Tank 10 Level with a lag 10. - Valve 20 Flow Speed Increasing has a cause effect on Tank 10 Level increasing with a lag 10.
LLM Inference Summary	Infers Valve 30 being closed prevents inflow; suggests it's the root cause for low level.	Infers insufficient flow through Valve 20 despite being open; provides lagged causal reasoning based on flow behavior.
Root Cause Identified	**Valve 30 is Closed** (blocks inflow)	**Insufficient Flow through Valve 20** (due to blockage or upstream issue)
Justification Quality	Moderately plausible but oversimplified; lacks context-specific reasoning about process dynamics.	More accurate and context-aware; applies lagged causal rules to interpret effect of flow behavior on tank level.
Inference Correctness	Partially correct—recognizes contributing factor but misses flow dependency.	More precise—aligns with system causality and identifies actual flow deficiency as root cause.
Conclusion	Inference is based on structural reasoning only.	Inference incorporates domain-specific dynamics, leading to a more accurate root cause analysis.

on system observations. We compare two experimental conditions: (1) the model is provided with both current observations and causal knowledge derived from the unified causal graph, and (2) the model is provided with the current observations only, without any causal knowledge. Each condition is evaluated using two criteria: **Correctness**, which assesses whether the model identifies the true root cause, and **Justification**, which evaluates whether the model provides a coherent and well-reasoned explanation for its inference.

To guide the model's behavior during the experiment, we used the following system prompt:

"You are an industrial AI assistant. You will be provided with alarms, system observations, and relevant causal knowledge about the process. Analyze all information and infer the most likely root cause of the current problem. Explain your reasoning based on the observations and the provided causal knowledge."

Our experiments indicate that GPT-4o tends to rely on general knowledge when explicit causal knowledge is not provided. While the model can occasionally produce correct inferences under such conditions, it often lacks the domain-specific understanding required for accurate industrial diagnostics. For instance,

Table 4. Effect of Causal Knowledge on GPT-4o Inference Accuracy and Justification in 20 Simulated Industrial Scenarios

Condition	Correctness	Justification	Notes
GPT-4o without Causal Knowledge	0.40	0.40	Model relies on general knowledge; often makes incorrect assumptions
GPT-4o with Causal Knowledge	**0.65**	**0.65**	Model leverages domain-specific knowledge for more accurate reasoning

it may be unable to differentiate between an inlet and an outlet valve when the observation merely states that a valve is open, or to determine whether a drop in tank level results from normal operation or an anomaly such as a leak or unintended valve activation. These limitations underscore the importance of integrating domain-specific causal knowledge, which improves the model's reasoning performance in complex and ambiguous industrial contexts. Table 3 presents a representative example comparing GPT-4o's performance with and without causal knowledge in one simulated alarm scenario.

Table 4 summarizes GPT-4o's performance across 20 simulated industrial alarm scenarios, showing improvements in both inference accuracy and justification quality when domain-specific causal knowledge is provided. Correctness was scored 1 if the predicted root cause matched the ground truth causal rule, and Justification was scored 1 if the explanation cited the correct condition/lag relationship. Reported values are averages across the 20 scenarios.

5   Conclusion and Future Work

In this paper, we presented UniCausal, a unified framework for causal inference from both TS and event data. UniCausal constructs context-aware causal graphs by associating each causal relationship with its relevant temporal lag and contextual conditions, enabling interpretable, multi-level reasoning across heterogeneous data sources.

To demonstrate its applicability, we conducted a case study integrating UniCausal with an LLM to enable human-interpretable root cause analysis. This integration supports flexible, context-aware diagnostics beyond rigid quantitative benchmarks. Our application-oriented prototype illustrates the potential of combining symbolic causal structures with natural language reasoning for practical use in industrial environments.

Like many causal discovery approaches, UniCausal relies on standard assumptions such as causal sufficiency (no unobserved confounders), approximate stationarity, and reliable lag estimation. These assumptions may limit identifiability in highly non-stationary or confounded systems. In addition, while our evaluation demonstrates promising results, further testing on larger and more diverse real-world industrial datasets will be an important next step to strengthen practical reliability.

Looking ahead, we plan to enhance and comprehensively evaluate the current causal inference application, as well as explore adaptive graph refinement, collaborative inference across multi-agent systems, and the extension of UniCausal to other industrial use cases, including alarm flood analysis, scheduled maintenance planning, process optimization, and anomaly explanation. We also aim to develop a unified and generic model capable of jointly processing time series, event sequences, and symbolic context data as inputs. This model will be trained to discover both standard and context-aware causal relationships across heterogeneous signal types, further expanding the applicability and generalization of our framework.

References

1. CMU motion capture database. http://mocap.cs.cmu.edu, Accessed June 2025
2. Bach, S., Binder, A., Montavon, G., Klauschen, F., Müller, K.R., Samek, W.: On pixel-wise explanations for non-linear classifier decisions by layer-wise relevance propagation. PLoS ONE **10**(7), e0130140 (2015)
3. Bahdanau, D., Cho, K., Bengio, Y.: Neural machine translation by jointly learning to align and translate. In: Proceedings of International Conference on Learning Representations (2015)
4. Blei, D.M., Jordan, M.I.: Variational inference for dirichlet process mixtures. Bayesian Anal. **1**(1), 121–143 (2006)
5. Bulling, A., Blanke, U., Schiele, B.: A tutorial on human activity recognition using body-worn inertial sensors. ACM Comput. Surveys **46**(3), 1–33 (2014)
6. Cai, R., Yu, S., Zhang, J., Chen, W., Xu, B., Zhang, K.: Dr.ECI: infusing large language models with causal knowledge for decomposed reasoning in event causality identification. In: Proceedings of the 31st International Conference on Computational Linguistics, pp. 9346–9375 (2025)
7. Gong, C., Zhang, C., Yao, D., Bi, J., Li, W., Xu, Y.: Causal discovery from temporal data: an overview and new perspectives. ACM Comput. Surv. **57**(4), 100:1–100:38 (2023)
8. Kong, L., Li, W., Yang, H., Zhang, Y., Guan, J., Zhou, S.: CausalFormer: an interpretable transformer for temporal causal discovery. IEEE Trans. Knowl. Data Eng. **37**(1), 102–115 (2025)
9. Moraffah, R., et al.: Causal inference for time series analysis: problems, methods and evaluation. Knowl. Inf. Syst. **63**(12), 3041–3085 (2021). https://doi.org/10.1007/s10115-021-01621-0
10. Nauta, M., Bucur, D., Seifert, C.: Causal discovery with attention-based convolutional neural networks. Mach. Learn. Knowl. Extraction **1**(1), 312–340 (2019)
11. OpenAI: Gpt-4o. https://openai.com/research/gpt-4o (2024)
12. Reiss, A., Stricker, D.: Introducing a new benchmarked dataset for activity monitoring (2012)
13. Shojaie, A., Fox, E.B.: Granger causality: a review and recent advances. Ann. Rev. Stat. Application **9**, 289–319 (2022)
14. Sundararajan, M., Taly, A., Yan, Q.: Axiomatic attribution for deep networks. In: International Conference on Machine Learning (ICML), pp. 3319–3328. PMLR (2017). https://arxiv.org/abs/1703.01365

15. Wang, C., Wu, K., Zhou, T., Cai, Z.: Time2State: an unsupervised framework for inferring the latent states in time series data. Proc. ACM Manag. Data **1**(1), 1–18 (2023)
16. Wang, J., Zhang, W., Shi, Y., Duan, S., Liu, J.: Industrial big data analytics: challenges, methodologies, and applications. arXiv preprint arXiv:1807.01016 (2018)
17. Zhang, M., Sawchuk, A.: USC-HAD: a daily activity dataset for ubiquitous activity recognition using wearable sensors, pp. 1036–1043 (09 2012)
18. Zhu, H., Huang, H., Yin, K., Fan, Z., Jin, H., Liu, B.: CausalNET: unveiling causal structures on event sequences by topology-informed causal attention. In: Proceedings of the Thirty-Third International Joint Conference on Artificial Intelligence, pp. 7144–7152 (2024)

Dynamic Source Code Vulnerability Characteristics Selection for Enhanced Vulnerability Discovery

Abdur Rehman Khan[1,2]($\boxtimes$) , Yue Xu[1] , and Yuefeng Li[1]

[1] School of Computer Science, Queensland University of Technology, Brisbane, QLD 4000, Australia
`{yue.xu,y2.li}@qut.edu.au`
[2] Centre for Data Science, Queensland University of Technology, Brisbane, QLD 4000, Australia
`aburrehman.khan@hdr.qut.edu.au`

Abstract. Software vulnerabilities are referred to as weaknesses in the source code. Hackers exploit these weaknesses to perform malicious actions, including accessing sensitive data and injecting a computer virus to hijack the computer system. Identifying these vulnerabilities is challenging. Even a perfectly functional program may have hidden vulnerable patterns. Contrastingly, locating these patterns manually by cybersecurity experts is onerous and time-consuming. Existing research uses deep learning algorithms to automate the process of finding vulnerable patterns. However, extracting relevant features from the source code is challenging, as source code can exceed hundreds of lines. Existing researchers either extract individual functions in a program or encapsulate the entire program, leading to under- or over-representation, respectively. A recent notion of using program slices has emerged, which only retains the most likely vulnerabilities causing statements. However, no significant research has been conducted to identify the most significant characteristics that are likely to cause vulnerabilities. Therefore, in this research, we investigate statistical heuristics to dynamically determine the most representative vulnerability characteristics and propose a deep neural network based on RoBERTa embeddings, which is fine-tuned on CodeBERT using multi-sample dropout to enhance generalizability. Our experimental results on real-world software source databases show up to 37% (F1 score) improvements in vulnerability prediction, with 61% reduced training time requirements when compared with the state-of-the-art baselines (VulDeePecker, SySeVR, and SlicedLocator).

Keywords: Software vulnerabilities · Cybersecurity · Deep Learning

1 Introduction

The exponential technological advancements and user adaptation have brought significant socio-technological challenges along with their benefits. Cybersecurity

Q. V. Nguyen et al. (Eds.): AusDM 2025, CCIS 2765, pp. 161–175, 2026.
https://doi.org/10.1007/978-981-95-6786-7_11

is one of the most emerging socio-technological challenges affecting all individuals and organizations. Cybersecurity incidents have increased tremendously in recent years. One of the leading causes of cybersecurity breaches is the weakness in software source code that hackers can exploit to perform malicious actions, leading to serious consequences such as unauthorized access, data breaches, and system compromises [10]. Consequently, the Massachusetts Institute of Technology Research and Engineering (MITRE) corporation made an effort to identify the most recurring Common Vulnerabilities and Exposures (CVEs) patterns to raise public awareness [6]. According to the recent survey, approximately 30% yearly expansion rate in cybersecurity incidents due to source code vulnerabilities has been reported [9]. In Australia, publicly reported CVEs increased by 31% in the financial year 2023–24[1].

These vulnerabilities can cause significant financial and reputation damage to organizations. For instance, in 2021, an improper authentication vulnerability in source code led to the *Kaseya Ransomware Attack*, compromising nearly 1,500 companies' data [1]. In 2022, a similar incident *Cyber Partisans* disrupted the Belarusian Railway ticketing system and traffic control [17]. In the same year, the Optus data breach incident, compromising 9.7 million user records, including sensitive information like driver license, Medicare, or passport numbers, caused the company $1.5 billion loss in brand value[2]. In 2023, the attackers exploited Google Authenticator to gain access to internal admin systems, specifically targeting accounts in the cryptocurrency, resulting in a loss of over $15 million [15]. In 2024, 26 billion user records were exposed due to an insufficient data protection vulnerability, resulting in the company incurring billions of dollars in compensation [16]. Moreover, since 48% of affected organizations did not report security incidents due to fear of repercussions, incidents could be twice the reported numbers [2].

While MITRE has created a public CVE database to raise public awareness, detecting and understanding vulnerabilities in source code remains a challenging and time-consuming process [6]. Moreover, manually identifying these vulnerabilities in source code is laborious, error-prone, and requires a high level of expertise, demanding the need to automate this process. Even a well-functioning software can have hidden vulnerability patterns. For instance, a common vulnerability known as *Improper Input Validation*, may allow an attacker to exploit this vulnerability by executing malicious database queries instead of entering a legitimate username and password into the input fields, retrieving sensitive data. Automated approaches to discovering vulnerabilities, including pattern matching and machine learning, have been widely explored. While pattern-matching methods rely on predefined rules to identify vulnerabilities, they often fail to generalize across varied and unseen code [11]. Machine learning methods, although more adaptive, do not adequately capture the deep structural nuances of source

---

[1] https://www.cyber.gov.au/about-us/view-all-content/reports-and-statistics/annual-cyber-threat-report-2023-2024.

[2] https://www.qld.gov.au/community/your-home-community/cyber-security/cyber-security-for-queenslanders/case-studies/optus-data-breach.

code, resulting in missed vulnerabilities. Deep learning, on the other hand, has demonstrated potential in modeling complex code structures and detecting vulnerabilities by leveraging the attention mechanism or low-level structural source code representations, including syntax, data, and control dependency graphs [9].

However, existing approaches suffer from the problem of representing source code programs, as they may contain hundreds of lines of code. As a result, most researchers consider either individual functions within a program or the program as a whole. The main challenge with considering individual function is the loss of crucial contextual information. A function may not be vulnerable itself; however, if used improperly, it could cause vulnerability (referred to as good source/bad sink), and vice versa. For instance, in a program, a bad source provides distrusted input that may be sanitized before reaching a good sink (Fig. 1 (right)), while a bad sink mishandles even trusted input from a good source (Fig. 1 (left)), potentially leading to vulnerabilities. This loss of context often results in a high false-positive/false-negative rate.

```python
def get_known_filename():
    # Trusted hardcoded path - a "good" source
    return "logs/server.log"
def open_file(filename):
    # Bad sink (vulnerable to path traversal if misused)
    with open("/var/app/" + filename, "r") as f:
        return f.read()
def main():
    safe_file = get_known_filename()   # good source
    content = open_file(safe_file)     # bad sink
    print(content)
```

```python
def get_user_supplied_path():
    return input("Enter the log file name: ")   # untrusted user input

def secure_open(user_path):
    base_path = os.path.abspath("logs/")
    target_path = os.path.abspath(os.path.join("logs/", user_path))

    # Good sink: validates that the resolved path stays within logs/
    if not target_path.startswith(base_path):
        raise ValueError("Invalid file path!")

    with open(target_path, "r") as f:
        return f.read()
def main():
    file_from_user = get_user_supplied_path()   # bad source
    content = secure_open(file_from_user)        # good sink
    print(content)
```

Fig. 1. Example of good source/bank sink (left) and bad source/good sink (right).

On the other hand, encapsulating a whole software program to train a neural model may become impractical due to challenges associated with input length and the introduction of irrelevant features, which limit the effectiveness of the deep neural model. Recent researchers have made an effort to identify the most likely vulnerability-causing characteristics in the source code. They argued that certain programming constructs, such as pointer usage, function calls, array usage, and arithmetic expressions, are more likely to lead to vulnerabilities in the program. For each characteristic, they extracted the dependent statements transitively, referred to as code slices. While the input length in code slices has decreased significantly, the number of generated slices remains nearly exponential. For instance, in a program, there could exist N number of statements. If each statement contains one of the vulnerability-causing characteristics, the transitively dependent statements are executed. However, a program may have conditional branches. Each B conditional branch results in the generation of 2^B slices perturbations, leading to an overall time complexity of $O(N.2^B)$. This may lead to redundant representation of features, consequently reducing model prediction performance and increasing model training time.

Therefore, to address this limitation, we employed the concept of renowned statistical heuristics to robustly determine the most significant slice characteristics, discarding the irrelevant slices. Our experimental results are based on the public repository of the source code vulnerability database (CVEFixes), which contains real-world projects. We further proposed a deep neural model that leverages the power of transformers in combination with RoBERTa embeddings and CodeBERT. When compared across various baselines (SySeVR [11], SlicedLocator [18], and VulDeePecker [12]), our approach demonstrates superior model prediction performance while requiring three times fewer model training time.

The rest of the paper is organized as follows. The next section provides a literature review. Section 3 summarizes the problems, motivations, and main contribution of our work. Section 4 explains the proposed approach. Section 5 describes the evaluation setup. Section 6 discusses the results. Finally, Sect. 7 concludes the discussion with future directions.

2   Literature Review

The discovery of source code vulnerability requires significant feature engineering before training a neural model. A source code can be represented as a sequence of tokens, low-level graphs containing data and instructions flow information, and hybrid techniques that combine both. In the subsequent subsection, we briefly explore source code representations, followed by a discussion of the notion of source code slices, and finally, deep neural models.

2.1   Source Code Representation

Source code is typically written in a programming language and can be viewed as a sequence of tokens, much like natural language text. Formally, a source code consists of programs (P). Each program has a set of functions, e.g., $P = \{f_1, f_2, ..., f_i\}$. Each function $f_i \in P$ in a program is a set of ordered statements, $f_i = \{s_{i,1}, s_{i,2}, \ldots, s_{i,j}\}$. These statements further encapsulates a set of ordered tokens (keywords, identifiers, operators, constants, etc.) in that function $s_{i,j} = \{t_{i,j,1}, t_{i,j,2}, \ldots, t_{i,j,k}\}$. A programming language is stricter in its syntax and has limited samples. This presents challenges in training a deep neural model. Moreover, source code tokens often exhibit high sparsity, particularly due to out-of-vocabulary (OOV) issues arising from the vast and diverse identifier space, such as variable, class, and function names, etc.

However, with advanced attention-based techniques such as transformers, researchers overcome this challenge by creating sub-pairs of each token via the Byte-Pair Encoding (BPE) technique [5]. An example of BPE could be the tokenization of a function name "unPremulSkImagePremul", which will be split into a list of subwords, i.e., ["un", "Prem", "ul", "Sk", "Image", "To", "Prem", "ul"]. The latent representation of each subword is passed to the training model.

Since a programming language is a sequential data, various dependencies exist in its tokens. These dependencies can be related to the interconnectivity of variables and/or functions of a source code. Therefore, to extract such structural information, existing researchers have used the graphical structure $G = (V, E)$ [7], where the vertices V represent the tokens and the edges E represent their relatedness. Usually, $V \in G$ are extracted from the abstract syntax tree AST, which is the underlying structure of the code, ignoring superficial details such as parentheses or semicolons. E in AST represents the hierarchy of the program flow, including syntactic relations, operator precedence, and scope. This G is usually enhanced with more semantic representations by modifying the edges $G = (V, E^\star)$. These extra edges incorporate various structural information, including the control flow graph (CFG) and dependency graph (PDG) [13] that describes the execution order and the flow of control (illustrated in Fig. 2). Providing graphs is another alternative to enriching the program semantics to enhance model prediction, especially with symbolic token representations.

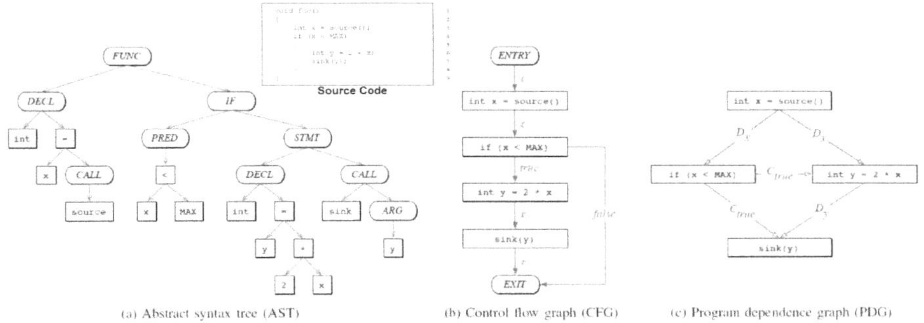

Fig. 2. Representation of source code taken and inherent graph structures.

2.2 Source Code Representation Granularity

The primary challenge associated with learning source code features is the proliferation of input space, such as OOV. This problem becomes even worse when representing a source code as a graph, as the number of nodes and edges increases manifoldly even for a simple program (Fig. 2). Therefore, existing researchers either consider the source code program as individual functions or encapsulate the entire program. In the former, crucial contextual information is lost. Since a program may not be vulnerable if handled safely (Bad Source/Good Sink), as represented in Fig. 1. In the latter, encapsulating the entire program may introduce unnecessary noise, which may reduce model efficiency and effectiveness.

Therefore, recent research focused on considering only the most likely causing source code Syntactic Vulnerability Characteristics (SyVC). In [12], researchers proposed *VulDeePecker*, assuming that most vulnerabilities are due to function calls. Consequently, they extracted only the transitive statements that depended on each function call in the program. They referred to this representation as *code gadgets*. Subsequently, researchers in [11] extended this notion in *SySeVR* to include more vulnerability characteristics such as array use, arithmetic expressions, and pointer use. Formally, let δ be a set of slices for P, $S_z = \{\delta_{ijz} | f_i \in P, s_{ij} \in f_i, z \in Z, \delta_{ijz} \in \delta\}$, where i represents i^{th} function, j represents j^{th} statement inside i^{th} function and $z \varepsilon SyVC$ represent z^{th} vulnerability characteristic in that statement. Where, SyVC is a set containing vulnerability characteristics such as pointer use (PU), Arithmetic Expressions (AE), Array Use (AU), or Function Calls (FC), given as $SyVC = \{PU, AE, PU, FC\}$. Each $\delta_{i,j,z}$ is derived via the forward and backward context from $PDG_p = \{V^p, E^p\}$, where $V^p = \bigcup AST(p)$ and $E^p = \bigcup_i E'_p \cup E_p^{interfunction}$, where $E^{interfunction} = \{(v_{i,j}, v_{k,1}) \mid v_{i,j} \text{ is a call to } f_k\} \cup \{(v_{k,m}, v_{i,j+1}) \mid v_{k,m} \text{ is return from } f_k\}$. The forward context is defined as $F_{i,j,z} = \{v \in V^p \mid v_{i,j} \prec v \wedge \exists \, path(v_{i,j} \rightsquigarrow v) \in E^p\}$ and the backward context is defined as $B_{i,j,z} = \{v \in V^p \mid v_{i,j} \prec v \wedge \exists \, path(\rightsquigarrow v_{i,j}) \in E^p\}$. The final slice representation is obtained as $\delta_{i,j,z} = order(F_{i,j,z} \cup B_{i,j,z})$, where $order(\cdot)$ preserves program semantics and removes redundant adjacent duplicates. A visual representation of the code slices, along with their characteristics, is illustrated in Fig. 3.

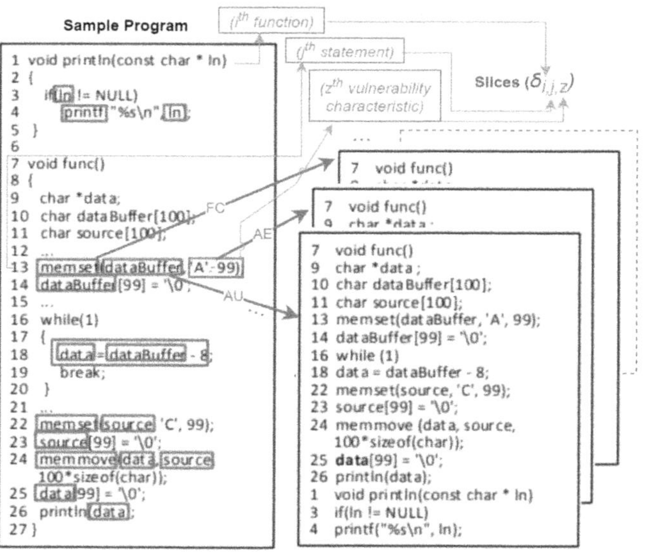

Fig. 3. A simplified example of multiple slices generated for each corresponding SyVC from one of the program statements seeded at red rectangles [11].

However, the number of slices generated via this approach is nearly exponential $O(N.2^B)$, as explained in the previous section. This makes the approach nearly impractical due to the time and space complexity required to train on advanced neural models, such as transformers. As a result, they trained their model on simple neural models. Researchers in [18] made an effort to incorporate semantic information by explicitly encoding graphs; however, they suffered from a similar problem.

2.3 Vulnerability Discovery Neural Models

Sequence-based models primarily rely on the tokens for model training. Notable examples include Bi-directional Long-Short Term Memory (BiLSTM) [12] using Word2Vec [11]. Such models are typically developed for generalized tasks that may require further fine-tuning to suit a specialized purpose, such as CodeBERT [5]. For instance, authors in [14] fine-tuned CodeBERT to discover vulnerabilities in the source code. In general, the transformer-based models have shown promising results in language understanding tasks.

Since the source code can also be represented as a graph (AST, CFG, PDG, etc.), neural networks that augment the graph as a structure have been extensively investigated in the literature. Most commonly, Graph Neural Network (GNN) [18], gated GNN [13], and graph convolution neural network [4] are prominent examples. These graph-based neural networks embed the program's node and edge information and attempt to learn the graphical latent space, as opposed to sequential learning. This enhances the deep structural and semantic understanding of the model and is proven to significantly enhance model learnability, especially in non-transformers-based models lacking semantics.

3   Problems and Motivations and Research Questions

Source code vulnerabilities have increased significantly in recent years. While significant research has been leveraged to propose various deep neural models and features, there exists a limited emphasis on finding the most suitable features that contribute positively to the deep neural model. Existing researchers either leverage individual functions or a whole program, leading to under- or over-representation, respectively. Although researchers proposed the notion of using program slices to retain only the dependent statements that are most likely to cause vulnerability characteristics in the program, these slice-based approaches suffered from an exponential number of generated slices and lacked robust heuristics to discard irrelevant vulnerability characteristics. As a result, training a transformer-based neural model, which is known for its promising prediction results, becomes nearly impractical due to time and space complexity [13]. To overcome aforementioned issues, in this research, we investigate:

- RQ1: How effective are transformers-based neural models in predicting source code slices' vulnerability?
- RQ2: What is the impact of removing the least representative vulnerability characteristics via statistical heuristic on model prediction performance?
- RQ3: How efficient can the model prediction performance be by training the neural model on the most representative vulnerability characteristics?

4　Proposed Approach

The proposed approach consists of three main phases. The first phase is loading the real-world software source code projects database containing the vulnerable and non-vulnerable software projects (Fig. 4(a)). Afterwards, source code slices are generated based on vulnerability characteristics. Contrary to selecting all SyVC or an arbitrary SyVC, we propose the notion of using statistical heuristics. These slices are further processed using statistical heuristics to eliminate redundant characteristics (Fig. 4(b)). Finally, based on optimum representative vulnerability characteristics, we propose a RoBERTa-CodeBERT-based deep neural model which is trained using layers of multi-sample dropout to enhance generalization (Fig. 4(c)). A discussion on each is provided in subsequent subsections.

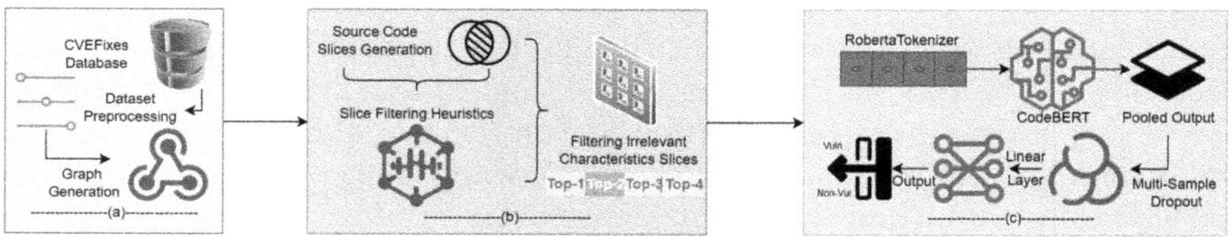

Fig. 4. Our approach workflow. (a) Source code database processing, (b) Slices filtering heuristics, and (c) Building neural model architecture.

4.1　Source Code Database Processing

We chose the CVEFixes database containing 5,320 real-world projects. This database contains both the vulnerable and non-vulnerable samples. From the source code, we performed preliminary preprocessing. This includes removing source code comments, tokenizing the source code into individual words, and generating PDG graphs based on vulnerability characteristics.

4.2　Slices Filtering Heuristics

From each program p both vulnerable PDG_p^{vul} and non-vulnerable PDG_p^{nonvul} samples code slices δ are extracted based on vulnerability characteristics $SyVC$, $z = \{AU, PE, AE, FC\}$:

$$\delta_{p,z}^{vul} \text{ from } PDG_p^{vul}, \qquad \delta_{p,z}^{nonvul} \text{ from } PDG_p^{nonvul}, \tag{1}$$

In total, we obtained eight distributions i.e., $D \in \delta_z^{vul|nonvul}$:

$$\mathcal{D} = \left\{ D_{AU}^{vul}, D_{AU}^{nonvul}, D_{PU}^{vul}, D_{PU}^{nonvul}, D_{FC}^{vul}, D_{FC}^{nonvul}, D_{AE}^{vul}, D_{AE}^{nonvul} \right\}, \tag{2}$$

To distinguish the most representative $SyVC$ from $\mathcal{D}$, we used cross-entropy. Formally, let $C(l, z)$ denote the count of code slices of type z with label $l \in \{vul, nonvul\}$ aggregated over all programs, i.e.:

$$C(l, z) = \sum_p |\delta_{p,z}^l|. \tag{3}$$

The count of slices for label l is:

$$N_l = \sum_{z \in SyVC} C(l, z), \tag{4}$$

and the total count over all labels and types is:

$$N = \sum_{l \in \{vul, nonvul\}} \sum_{z \in SyVC} C(l, z). \tag{5}$$

We then define the label-specific type distribution as:

$$P_l(z) = \frac{C(l, z)}{N_l}, \tag{6}$$

and the overall type distribution as:

$$Q(z) = \frac{\sum_l C(l, z)}{N}. \tag{7}$$

The cross-entropy of the label-specific distribution P_l with respect to the overall distribution Q is computed as:

$$H(P_l, Q) = - \sum_{z \in SyVC} P_l(z) \log Q(z), \tag{8}$$

which can be decomposed per type as:

$$H(P_l, Q) = \sum_{z \in SyVC} \underbrace{-P_l(z) \log Q(z)}_{\text{cross-entropy contribution of type } z}. \tag{9}$$

Lower values of $H(P_l, Q)$ indicate that the overall distribution Q effectively models the label-specific distribution P_l, suggesting that the corresponding $SyVC$ type z is representative within that label. Conversely, higher cross-entropy values highlight types that differ significantly from the overall distribution and may be more discriminative.

Based on the total cross-entropy $H(P_l, Q)$, we consider the individual per-type contributions as:

$$h_l(z) = -P_l(z) \log Q(z). \tag{10}$$

For each label l, we rank the characteristic types z by their per-type cross-entropy contributions $h_l(z)$ in ascending order and select the top-k types with the lowest values:

$$SyVC_l^{(k)} = \underset{S \subseteq SyVC, |S|=k}{\arg\min} \sum_{z \in S} h_l(z). \tag{11}$$

Choosing these types corresponds to identifying the most representative vulnerability characteristics for label l.

4.3 Building Neural Model Architecture

We design a neural model based on the CodeBERT architecture [5] to classify code slices $\delta^l_{p,z}$ extracted from program p, where $l \in \{\text{vul}, \text{nonvul}\}$ denotes the vulnerability label and $z \in SyVC$ the vulnerability characteristic type.

Each code slice is first tokenized using the RoBERTa tokenizer, which transforms the raw source code into a sequence of token IDs suitable for model input. Formally:

$$\mathbf{x} = Tokenizer_{\text{RoBERTa}}(\delta^l_{p,z}), \tag{12}$$

Our model, denoted as $\mathcal{M}$, builds upon the pre-trained RoBERTa-based CodeBERT model. It encodes the input token sequence $\mathbf{x}$ into a contextualized representation, using all 12 transformer layers of the pretrained CodeBERT model:

$$\mathbf{h} = Encoder_{\text{CodeBERT}}(\mathbf{x}) \in \mathbb{R}^d, \tag{13}$$

where d is the hidden size (i.e., 768). The pooled output $\mathbf{h}$ corresponds to the semantic embedding of the entire code slice.

To improve generalization and reduce overfitting, we employ multi-sample dropout [8], applying $\mathcal{K}$ parallel dropout masks $\{D_k\}_{k=1}^K$ with dropout rate r to the pooled representation:

$$\mathbf{h}_m = D_m(\mathbf{h}), \quad m = 1, \dots, M. \tag{14}$$

Each dropout sample is then passed through a shared linear classifier $f : \mathbb{R}^d \to \mathbb{R}$ producing logits as:

$$y_m = f(\mathbf{h}_m) = \mathbf{W}\mathbf{h}_m + b, \tag{15}$$

where $\mathbf{W}$ and b are learnable parameters.

The final output logit is computed as the average over all K dropout samples:

$$\hat{y} = \frac{1}{M} \sum_{m=1}^{M} y_m. \tag{16}$$

This logit $\hat{y}$ is used to compute the binary classification loss against the true label l via the binary cross-entropy with logits loss:

$$\mathcal{L} = \text{BCEWithLogits}(\hat{y}, l). \tag{17}$$

The model is trained using mini-batch stochastic gradient descent with the AdamW optimizer and evaluated through stratified $\mathcal{K}$-fold cross-validation. At each fold, the dataset $\{(\delta^l_{p,z}, l)\}$ is split into training and validation subsets, and model performance is assessed by standard classification metrics such as precision, recall, and F1-score.

5 Evaluation Setup

Our experimental setup for training the neural model consisted of an AMD Ryzen Threadripper workstation featuring a CPU model PRO 5955WX hexadeca-core processor with 32 GB of DDR4 RAM at 3200 MHz, along with an Nvidia RTX A4500 20 GB GPU. The epochs for the baseline were set to 10 iterations. The PyTorch 2.7 library was used to train the model in Python 3.12. All experimental results are validated using a 5-fold cross-validation.

5.1 Dataset

We used the most recent CVEFixes dataset to train the neural models. This dataset contains 5,320 real-world software projects. While the source code slice-based approaches train jointly on SARD and CVEFixes, we refrain from using the SARD dataset because it is synthetic and does not generalize well to real-world projects. It was criticized primarily for its inability to capture the complex and nuanced nature of vulnerabilities in real-world code [3]. The total number of slices generated from this dataset is 55,103. The complete distribution of slices generated is given in Table 1. The average slice length is 247 tokens.

5.2 Baseline Systems

We chose three state-of-the-art baseline systems: (i) VulDeePecker, (2) SySeVR, and (3) SlicedLocator. VulDeepecker proposed the notion of code gadgets that are most likely to cause dependent statements containing certain function calls. SySeVR extracts code slices based on statements that use arrays, contain arithmetic expressions, invoke functions, or involve pointer use. Both VulDeePecker and SySeVR are trained on a BiLSTM network. In input to their model was a latent representation of each token in the code slice using Word2Vec embeddings. SlicedLocator shares similar characteristics with SySeVR; however, it also encodes the instruction type and graph embeddings jointly on a GNN to enhance model prediction performance. To maintain a fair comparison, we reproduced their results on the CVEFixes dataset using their publicly available package.

5.3 Proposed Model Configurations

The proposed model fine-tunes CodeBERT (microsoft/codebert-base) for binary classification using multi-sample dropout (5 layers, p = 0.3) to improve generalization. It trains with a batch size of 16, over 5 epochs, keeping the original variable names. The model uses all 12 transformer layers (hidden size 768, 12 attention heads, vocab size 50,265), optimized with AdamW (LR $= 2 \times 10^{-5}$) and BCEWithLogitsLoss. This setup enhances semantic and syntactic understanding while reducing overfitting.

Table 1. CVEFixes dataset source code slices distribution.

Vulnerability Characteristics (SyVC)	Array Use (AU)	Pointer Use (PU)	Arithmetic Expression (AE)	Function Call (FC)	Total
Vulnerable	2708	5813	7222	7073	22816
Non-Vulnerable	3897	7624	10214	10552	32287
Total	6605	13437	17436	17625	55103

6   Results and Discussion

To answer our research questions, we compare the proposed architecture with our baseline systems in this section. Furthermore, we compare the impact of using statistical heuristics on the model prediction and training efficiency. A detailed discussion of each is provided as follows.

6.1   Model Prediction Performance

We compared the proposed neural model with the three renowned baselines: VulDeePecker, SySeVR, and Sliced Locator. VulDeePecker uses the notion of API calls to generate code gadgets, similar to code slices. SySeVR further uses three more vulnerability characteristics (pointer use, arithmetic expressions, and array use). SlicedLocator employs the same heuristics as SySeVR; however, it also incorporates the slice graph into the GNN to enhance model prediction performance. The comparative results in Table 2 demonstrate that the proposed approach consistently outperforms all baseline systems (Sliced Locator, VulDeePecker, and SySeVR) across most evaluation metrics. In terms of precision, it achieves the highest scores for both Non-Vul (0.72) and Vul (0.70) classes, improving the average precision by 17% over the best baseline. In the case of Vul recall, the proposed system also performed the best with (+37%) better performance. While recall for Non-Vul instances is slightly lower than VulDeePecker and SySeVR, the proposed approach maintains a balanced recall for both classes, resulting in the highest average recall (+17%). This indicates that the proposed model is less likely to be biased towards Non-Vul instances, thereby reducing false negatives, which is crucial in this domain. For instance, it is preferable for a model to classify vulnerability rather than misclassify a real vulnerability as benign. This balance between precision and recall translates into superior F1-scores, with notable gains of 25% in the average F1 compared to the next best system, indicating a better trade-off between false positives and false negatives. Overall, the proposed method achieves the highest accuracy (0.71), surpassing the baselines by up to 8%, showing its effectiveness in accurately identifying both vulnerable and non-vulnerable code segments while maintaining consistent performance across classes. Hence, to answer our research question 1, the proposed approach outperforms all baselines by 25% in terms of F1 scores.

6.2   Role of Statistical Heuristics

In this section, we discuss the role of statistical heuristics on the model prediction performance and training efficiency. We compared the performance and training

Table 2. Comparative model prediction performance for the proposed approach and across all the baseline systems.

Metric	Class	Sliced Locator	VulDeePecker	SySeVR	Proposed
Precision	Non-Vul	0.52	0.61	0.59	**0.72 (+11%)**
	Vul	0.24	0.46	0.32	**0.70 (+24%)**
	Avg	0.38	0.54	0.46	**0.71 (+17%)**
Recall	Non-Vul	0.65	0.87	**0.99**	0.83
	Vul	0.15	0.17	0.02	**0.54 (+37%)**
	Avg	0.40	0.52	0.50	**0.69 (+17%)**
F1-Score	Non-Vul	0.56	0.70	0.74	**0.77 (+3%)**
	Vul	0.19	0.17	0.04	**0.62 (+43%)**
	Avg	0.38	0.44	0.39	**0.69 (+25%)**
Accuracy	Overall	0.63	0.59	0.59	**0.71 (+8%)**

time efficiency by incrementally incorporating the most relevant vulnerability characteristics (SyVCs) for each class. The results discussed are anchored to the Top 4 (all SyVCs). According to cross-entropy, the following SyVCs were ranked for each class in ascending order: Non-Vul (AU, PU, AE, FC) and Vul (AU, FC, PU, AE). The comparative results are provided in Table 3. The Top-1 allowed the most to train the fastest (+88%) and produce results similar to those obtained by all SyVCs (Top-4). The best performance was achieved by incorporating only the top 2 SyVCs, with 10% better precision and 12% better recall. The overall model performance improved by 10% in terms of F1 scores. The accuracy also improved by 10%. After that, we observed that the model's performance gradually decreased. This could be due to the over-representation of SyVCs. Hence, to answer our research question 2, the model prediction performance improves significantly when only incorporating the most relevant SyVCs.

With each reduction in Top-k, the training time and steps required per epoch decreased significantly. By using only a fraction of SyVCs, the model training time per epoch was reduced on average by nearly three times compared to the Top-4 method. The model was able to train in 61% less time compared with Top-4. Therefore, incorporating the statistical heuristics serves three primary purposes. Firstly, researchers reported that datasets in source code vulnerability discovery contain significant impurities. Statistical heuristics can help discard irrelevant code samples. Secondly, by discarding irrelevant code samples, the model has a better chance of training on the most relevant samples, thereby increasing its performance. Finally, it can also potentially reduce the training dataset and model training by half at the same time. Hence, to answer our research question 3, using statistical heuristics to select more representative SyVCs not only increases the model's prediction performance but also significantly reduces the model's training time complexity.

Table 3. Effect of statistical heuristic on model prediction performance and training efficiency relative to Top-4 (all SyVCs).

SyVC	Precision	Recall	F1-score	Accuracy	Avg Time Per Epoch (m)	Avg Time Training (h)
Top-1	0.64 (-7%)	0.61 (-8%)	0.58 (-11%)	0.63 (-8%)	2.40	1 (-87.7%)
Top-2	**0.81 ($+10\%$)**	**0.81 ($+12\%$)**	**0.81 ($+12\%$)**	**0.81 ($+10\%$)**	**7.51**	**3.13 (-61.4%)**
Top-3	0.73 ($+2\%$)	0.72 ($+4\%$)	0.73 ($+4\%$)	0.73 ($+2\%$)	13.19	5.49 (-32.3%)
Top-4	0.71	0.69	0.69	0.71	19.46	8.11

7 Conclusion and Future Work

In this research, we proposed the notion of source code slices to discover vulnerabilities in real-world software projects. We used RoBERTa embeddings to fine-tune CodeBERT with multi-sample dropout to enhance generalizability on the source code vulnerability classification task. The proposed model achieved 25% improvements in terms of F1 scores, outperforming existing state-of-the-art baselines (VulDeePecker, SySeVR, and SlicedLocator). Furthermore, the notion of code slices depends on characteristics likely to cause vulnerability, which include array use, pointer use, function calls, and arithmetic expressions. Contrary to using either all or arbitrary vulnerability characteristics, we leveraged robust statistical heuristics to retain the most relevant and representative vulnerability characteristics. By incorporating only the fraction of slices, we observed further improvements in the model prediction performance ($+10\%$) in terms of the F1 score with 61% reduced time complexity. In the future, we look forward to extending our experimental setup across various datasets and extending the notion of using statistical heuristics to fine-grain prediction.

References

1. Aslan, Ö., Aktuğ, S.S., Ozkan-Okay, M., Yilmaz, A.A., Akin, E.: A comprehensive review of cyber security vulnerabilities, threats, attacks, and solutions. Electronics **12**(6), 1333 (2023). https://doi.org/10.3390/electronics12061333
2. Busetti, S., Scanni, F.M.: Evaluating incident reporting in cybersecurity. From threat detection to policy learning. Govern. Inf. Q. **42**(1), 102000 (2025). https://doi.org/10.1016/j.giq.2024.102000
3. Ding, Y., et al.: Vulnerability detection with code language models: how far are we? arXiv preprint arXiv:2403.18624 (2024)
4. Dong, Y., Tang, Y., Cheng, X., Yang, Y., Wang, S.: SedSVD: statement-level software vulnerability detection based on relational graph convolutional network with subgraph embedding. Inf. Softw. Technol. **158**, 107168 (2023)
5. Feng, Z., et al.: CodeBERT: a pre-trained model for programming and natural languages. arXiv preprint arXiv:2002.08155 (2020)
6. Haddad, A., Aaraj, N., Nakov, P., Mare, S.F.: Automated mapping of CVE vulnerability records to MITRE CWE weaknesses. arXiv preprint arXiv:2304.11130 (2023)

7. Hariharan, M., et al.: Proximal instance aggregator networks for explainable security vulnerability detection. Futur. Gener. Comput. Syst. **134**, 303–318 (2022)

8. Inoue, H.: Multi-sample dropout for accelerated training and better generalization. arXiv preprint arXiv:1905.09788 (2019)

9. Jiang, X., Xiao, Y., Wang, J., Zhang, W.: Multi-view pre-trained model for code vulnerability identification. In: Wang, L., Segal, M., Chen, J., Qiu, T. (eds.) WASA 2022. LNCS, vol. 13473, pp. 127–135. Springer, Cham (2022). https://doi.org/10.1007/978-3-031-19211-1_11

10. Li, D., Liu, Y., Huang, J.: Assessment of software vulnerability contributing factors by model-agnostic explainable ai. Mach. Learn. Knowl. Extract. **6**(2), 1087–1113 (2024)

11. Li, Z., Zou, D., Xu, S., Jin, H., Zhu, Y., Chen, Z.: SySeVR: a framework for using deep learning to detect software vulnerabilities. IEEE Trans. Dependable Secure Comput. **19**(4), 2244–2258 (2021)

12. Li, Z., et al.: VulDeePecker: a deep learning-based system for vulnerability detection. arXiv preprint arXiv:1801.01681 (2018)

13. McLaughlin, C., Lu, Y.: Multi-class vulnerability prediction using value flow and graph neural networks. Neural Comput. Appl. 1–23 (2024)

14. Pan, S., Bao, L., Xia, X., Lo, D., Li, S.: Fine-grained commit-level vulnerability type prediction by CWE tree structure. In: 2023 IEEE/ACM 45th International Conference on Software Engineering (ICSE), pp. 957–969. IEEE (2023)

15. Prajapat, S., Gaur, A.S., Soni, P., Nagar, N., Goswami, P.: Cyber attacks and review of recent data breach incidents their trends and measures. In: Naik, N., Jenkins, P., Prajapat, S., Grace, P. (eds.) C3AI 2024. LNNS, vol. 884, pp. 549–576. Springer, Cham (2024). https://doi.org/10.1007/978-3-031-74443-3_33

16. Rodrigues, G.A.P., Serrano, A.L.M., Vergara, G.F., Albuquerque, R.D.O., Nze, G.D.A.: Impact, compliance, and countermeasures in relation to data breaches in publicly traded us companies. Future Internet **16**(6), 201 (2024). https://doi.org/10.3390/fi16060201

17. Van Niekerk, B.: Vulnerability of South African commodity value chains to cyber incidents. Sci. Militaria: South Afr. J. Milit. Stud. **51**(3), 161–186 (2023). https://doi.org/10.5787/51-3-1345

18. Wu, B., Zou, F., Yi, P., Wu, Y., Zhang, L.: SlicedLocator: code vulnerability locator based on sliced dependence graph. Comput. Secur. **134**, 103469 (2023)

Modelling Financial Time Series of Returns and Covariance Matrices Using Time-Space Transformers

Xiaodan Dong[1]([✉])(iD), Kittituch Wongwatcharapaiboon[1](iD), Sugi Lee[1](iD),
Jennifer SK Chan[2](iD), and Weidong Huang[1](iD)

[1] University of Technology Sydney, Sydney, Australia
{Xiaodan.dong,Weidong.Huang}@uts.edu.au
[2] University of Sydney, Sydney, Australia
Jennifer.Chan@sydney.edu.au

Abstract. Recent advances in deep learning, particularly transformer-based models, have shown remarkable success in capturing both spatial and temporal dependencies in video and image data. Interestingly, financial time series data share analogous structures: asset returns at different time points form sequences, and the covariance matrix captures inter-asset relationships, similar to spatial correlations in images. Despite this analogy, existing financial forecasting models rarely leverage these spatio-temporal patterns to their full potential. This research addresses this gap by leveraging the Time-Space Transformer (TimeSformer) model, originally developed for video understanding, to analyze and predict financial time series data. Specifically, we transform the return features of portfolio stocks into image representations with a temporal structure. We experiment with four different TimeSformer architectures, employing both parametric and non-parametric loss functions to predict the subsequent image, which is then back-transformed into the corresponding returns and covariance matrices. Experimental results demonstrate that this approach greatly improves prediction accuracy, achieving lower RMSE for returns and reduced log-euclidean distance for covariance matrices compared to traditional methods.

Keywords: Time-Space Transformer · Covariance Matrix Time Series · Deep Learning · Distance Metrics · Negative Loglikelihood

1 Introduction

Portfolio optimization is the task of choosing asset weights to achieve a target balance between risk and return. In the classical mean-variance framework of Modern Portfolio Theory, investors either maximize expected return for a given level of risk or minimize risk for a given expected return, yielding an efficient frontier of optimal portfolios [10]. In practice, the quality of any optimized portfolio depends on two critical inputs: the expected returns of individual assets and the covariance matrix that captures their interdependencies. While

© The Author(s), under exclusive license to Springer Nature Singapore Pte Ltd. 2026
Q. V. Nguyen et al. (Eds.): AusDM 2025, CCIS 2765, pp. 176–191, 2026.
https://doi.org/10.1007/978-981-95-6786-7_12

expected returns provide an estimate of the likely average performance of assets, the covariance matrix plays an equally vital role by quantifying how assets co-move under dynamically changing economic conditions. Accurate estimation of both mean returns and the covariance matrix is essential for constructing efficient portfolios and managing risk. To minimize risk in today's rapidly changing world, forecasting the covariance matrix has become increasingly important. Although considerable research in recent decades has focused on improving the prediction of mean returns, challenges in covariance estimation remain critical. Leccadito et al. [7] employed the dynamic Gerber model (DGC) to assess its effectiveness in forecasting Value at Risk (VaR) and Expected Shortfall (ES).

Recent advancements in deep learning, particularly transformers [18], have attracted considerable attention for applications in video and image forecasting. Wang et al. [20] and Takahashi et al. [17] found that transformer-based approaches are effective in achieving high accuracy in their respective applications. Time-Space Transformer (TimeSformer or TSF) [1] is a convolution-free deep learning model specifically designed for video classification and action recognition tasks. It leverages the power of self-attention mechanisms, similar to those found in the original Transformer architecture, but extends them to handle both spatial and temporal dimensions within video data. Few papers apply transformer architectures directly to covariance matrix estimation. Cohen and Klein [4] derived an adaptive Kalman-informed transformer (A-KIT) designed to learn the varying process noise covariance online.

Video data and financial time series exhibit analogous structures. In videos, spatial patterns are captured by pixel values within each frame, while temporal dynamics arise across consecutive frames. Similarly, financial returns at different time points form sequences, with the covariance matrix representing relationships between assets, similar to spatial correlations in images. Consequently, transformer-based models, which effectively capture both spatial and temporal dependencies in video analysis, can be adapted to model inter-asset relationships and their evolution over time in financial data modeling.

1.1 Research Objectives and Questions

To our knowledge, prior work has not adapted video-based models, such as TimeSformer [1] for financial time series modeling. This research aims to address this gap by leveraging TimeSformer to jointly estimate both returns and covariance-based risks in high-dimensional financial time-series data. This research answers the following key research questions:

- Does this image-like encoding with TimeSformer improve joint forecasts of returns and covariance versus strong baseline?
- Does joint learning outperform training separate models on accuracy and stability?
- Which loss family gives the best accuracy and calibration for portfolio use?

2   Related Work

Modeling financial time series for portfolio optimization has been a prominent research domain. With the rapid advancement of machine learning and deep learning techniques, these methods have increasingly played a critical role in improving predictive performance for financial applications. The following section provides a summary of recent work in this domain, highlighting key developments and approaches that inform the context of the present study.

2.1   Return and Risk Forecast in Portfolio Optimization

In portfolio optimization, the return vector and covariance matrix are fundamental inputs that guide asset allocation decisions. The return vector contains information of the expected returns of each asset, representing the potential gains or losses over a given time horizon, while the covariance matrix captures the variances of individual assets and the covariances between asset pairs, reflecting both individual risk and inter-asset relationships. Accurate estimates of these quantities are essential: the return vector directs the portfolio toward maximizing expected gains, and the covariance matrix enables effective diversification to control overall risk. Sharpe Ratio [15] as a classic measure of risk-adjusted return of a portfolio requires both mean return vectors and covariance matrices. Together, they form the foundation of classical optimization frameworks, such as mean-variance optimization [10], and are central to more advanced modeling approaches explored in recent literature.

Traditionally, portfolio returns have been modeled using vector autoregressive (VAR) [16,25] and autoregressive moving average (ARIMA) models [3]. For risk modeling in portfolio optimization, approaches such as shrinkage estimators, factor models, Generalized Autoregressive Conditional Heteroskedasticity (GARCH) model [2], Dynamic Conditional Correlation GARCH (DCC-GARCH) model [5], DCC-GJR-GARCH [6] model and Kalman filters [19] have been widely employed to estimate volatility and capture dynamic correlations across assets. While these methods provide interpretable and grounded solutions, they often struggle to capture complex non-linear relationships and long-term dependencies in high-dimensional financial data.

On the other hand, Wang et al. [21] proposed the State Causality and Adaptive Covariance Decomposition (SCACD) method for time series forecasting. SCACD is designed to predict future values in complex, non-stationary time series, particularly when patterns are unclear or evolve over time, and is capable of handling long-term forecasts across large datasets. A notable limitation, however, is its reliance on the assumption that the data follows a Gaussian distribution, which may not hold in many real-world financial applications. Pienaar et al. [13] proposed a blended covariance matrix approach that leverages Mahalanobis distance to enhance portfolio optimization and risk management in volatile markets. Their method specifically addresses the issue of covariance instability during turbulent periods and demonstrates that using blended covariance inputs can reduce portfolio volatility and shift the efficient frontier leftward, indicating lower overall risk. However, the approach focuses solely on covariance estimation and does not address return prediction, highlighting a gap in the literature regarding joint modeling of returns and risk for more comprehensive portfolio optimization.

2.2 Deep Learning for Portfolio Optimization

Deep learning has been increasingly applied to portfolio optimization due to its ability to model complex, non-linear relationships in financial markets. Recent studies have employed Convolutional Neural Networks (CNNs) [12] to extract local temporal patterns from return series, while recurrent architectures such as Long Short-Term Memory networks (LSTMs) [11] and Gated Recurrent Units (GRUs) [14] are used to model sequential dependencies and temporal correlations across assets. Despite their successes, these architectures often struggle with long-range dependencies and high-dimensional data, which are critical for robust portfolio management.

Recently, the Transformer architecture, originally proposed by Vaswani et al. [18] for machine translation, has since inspired numerous adaptations for time-series forecasting because it demonstrated strong performance and promising results [22]. In particular, time-series Transformer variants have been developed to address forecasting, risk modeling, and other challenges in financial applications where modeling temporal dependencies is critical. The Temporal Fusion Transformer (TFT) [9], introduced in 2021, has been widely applied in financial forecasting tasks such as volatility, returns, and risk prediction. It integrates static covariates and time-varying input, while providing attention-based interpretability to enhance predictive accuracy. Frequency Enhanced Decomposed Transformer (FEDformer) [24] integrates the Transformer framework with seasonaltrend decomposition, where the decomposition component extracts the global temporal patterns of the series, while the Transformer layers model the finer-grained and more localized structures.

The TimeSpace Transformer (TimeSformer) was proposed by Bertasius et al. [1] as an extension of the Vision Transformer (ViT) to the temporal domain for video classification. In contrast to previous methods that relied heavily on 3D convolutional networks or recurrent neural architectures, TimeSformer employs a purely Transformer-based design. Its principal contribution, the divided space-time attention mechanism, facilitates the joint modeling of spatial and temporal dependencies within video sequences, thereby enabling efficient and interpretable representation learning.

In particular, we demonstrate that the proposed TimeSformer models can effectively leverage joint model for both return and covariance matrix series of financial time-series data. Furthermore, the proposed approach can be generalized to high-dimensional datasets by augmenting the data into multiple vector representations, enabling more comprehensive modeling of complex financial dependencies.

3   The Data

This study employs a dataset of assets carefully selected to address the research objectives. This section provides a detailed account of the dataset, including its source, structure, and the rationale for its selection. Additionally, the preprocessing steps, exploratory data analysis (EDA), and data augmentation strategies applied are described to ensure transparency and reproducibility.

3.1   Data Source and Preparation

We construct our dataset from a publicly available source containing OHLCV (open, high, low, close, volume) data obtained through the Yahoo Finance API [23]. The assets include individual stocks, indices, commodities, and fixed income instruments. This mix spans equities, commodities, bonds, and crypto are to capture a broad spectrum of market behavior. The timeline across instruments is aligned with the trading calendar of Standard and Poor's Depositary Receipts (SPDR) S&P500 Exchange-Traded Fund (ETF) (SPY) which removes weekends and exchange holidays to maintain consistency. This alignment is particularly important to assets in crypto category as Bitcoin (BTC-USD), which continuously trades, to avoid calendar mismatches and ensure comparability across assets.

For each asset, we extract eight features ($E = 8$) over the period from July 2019 to July 2025 using a 60-day rolling window. These include **log returns** (r_t), calculated from adjusted closing prices, and the 14-day Relative Strength Index (**RSI**), which captures momentum and signals overbought or oversold conditions. We also compute the Moving Average Convergence Divergence (MACD), represented by both the **MACD line**, defined as the difference between 12-day and 26-day exponential moving averages, and the 9-day **MACD signal line**, which smooths short-term fluctuations. To capture trend dynamics, we calculate two moving averages, **MA5** and **MA20**, representing short- and medium-term trends, along with their **ratio** (MA5/MA20) as a measure of short- to long-term trend comparison. Finally, trading activity is represented by log-transformed volume, $\log(1 + V_t)$, which stabilizes the scale while preserving market information.

3.2   Data Augmentation Strategies

TimeSformer, a video transformer, require input in the shape [B,F,C,H,W], corresponding to batch size, number of frames, channels, height, and width [1]. To adapt the dataset into the TimeSformer model, we develop two data augmentation strategies that differ in how features and assets are arranged spatially in one frame. This design reorganizes the asset and feature space into pseudo-image augmentation, with adjustments made to accommodate both the model input's requirement and empirical performance.

Features as Channels. In the first step (Fig. 1), features are stacked in the channel while assets are arranged in the height dimension and width fixed at one as before data augmentation. This configuration encodes features, multiple technical indicators, similar to how RGB channels represent pixel information in images. To further exploit the model's capacity for spatial attention, we extend this mapping by arranging it in a square grid across height and width dimensions. In practice, this is done by duplicating the width dimension to form a square (shown as the yellow blocks in the figure) which aligns with the input requirements of the TimeSformer, resulting in enhanced stability and consistency of the performance gains.

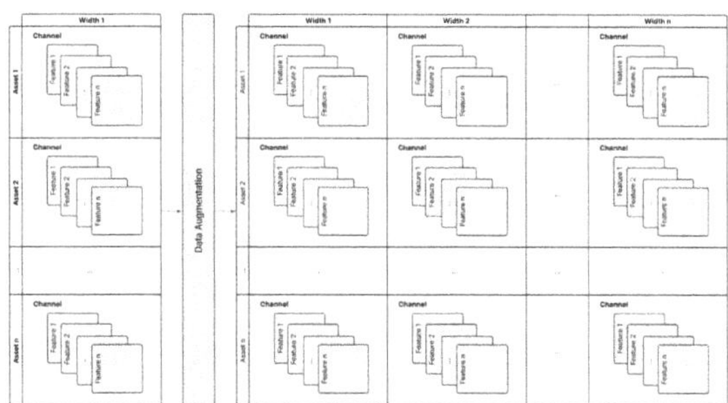

Fig. 1. Feature as Channel: features are stacked in the channel dimension and replicated across the width dimension.

Feature as Width. In this second strategy (Fig. 2), assets are organized along the height dimension. In contrast, features are arranged along the width dimension, resulting in a rectangular configuration (for instance, six assets by eight features). This setup emphasizes the cross-sectional interactions between assets and features, but deviates from the square input requirement of TimeSformer. To solve this issues, we implemented zero-padding to expand the grid into a square format. This ensures compatibility with TimeSformer's patch embedding mechanism which maintains the integrity of the original asset-feature information while enabling the model to leverage the benefits of a regularized square structure, which could lead to enhanced predictive performance.

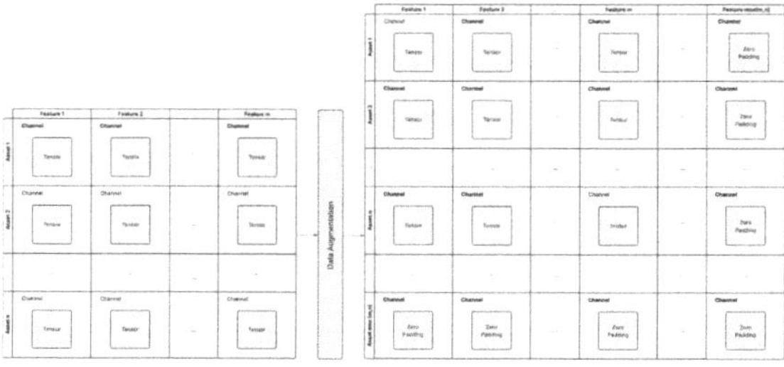

Fig. 2. Feature as Width: features are arranged along the width dimension, and zero-padding the grid when the size of asset and features are not the same.

4 Methodology

Let the time series of N asset returns be $\boldsymbol{r}_{1:T} = (r_{i,1:T}) \in \mathbb{R}^{T \times N}$ and their $N \times N$-dimensional sample covariance matrices $\boldsymbol{\Sigma}_{1:T} = (\sigma_{ij,1:T}) \in \mathbb{R}^{T \times N \times N}$ for day 1 to T in the training period. The feature sets are $\boldsymbol{X}_{0:(T-1),1:N,1:E} \in \mathbb{R}^{T \times N \times E}$. The $E = 8$ features for each asset at each day is described in Sect. 3.1.

We propose network models to train the mean return vector $\boldsymbol{\mu}_t$ using observed returns $\boldsymbol{r}_{1:T}$ as well as the covariance matrices $\hat{\boldsymbol{\Sigma}}_t$, $t \in 1:T$ using sample covariance matrices $\boldsymbol{\Sigma}_{1:T}$. We will describe various neural network architectures using transformer and propose various loss functions to estimate the network parameters. Since the time series of covariance matrices are not directly observed, we begin with a robust estimation of the covariance matrices.

4.1 Constructing Robust Sample Covariance Matrices

We first highlight the importance of the symmetric and positive definite (SPD) property of the sample $\boldsymbol{\Sigma}_t$ and modeled $\hat{\boldsymbol{\Sigma}}_t$ covariance matrices. To ensure that the covariance matrices are SPD, we employ the Cholesky decomposition.

To estimate the daily covariance matrix, we use a lookback window $\boldsymbol{W}_t$ of H log return vectors of N assets:

$$\boldsymbol{W}_t = \boldsymbol{r}_{(t-H+1):t} = (r_{t-H+1}, \ldots, r_t) \in \mathbb{R}^{N \times H} \tag{1}$$

with at least H days prior observations to estimate the sample covariance matrix on day t:

$$\boldsymbol{\Sigma}'_t = \frac{1}{H} \sum_{\tau=1}^{H} (\boldsymbol{r}_{t\tau} - \bar{r}_t)(\boldsymbol{r}_{t\tau} - \bar{r}_t)^{\top}. \tag{2}$$

However, $\boldsymbol{\Sigma}'_t$ can be very noisy and ill-conditioned if N is large relative to H. Hence, we initially build the daily labels for the covariance matrix of each asset using a rolling LedoitWolf estimator [8] and compute the LedoitWolf shrinkage covariance

$$\boldsymbol{\Sigma}_{\text{LW},t} = (1-\alpha)\boldsymbol{\Sigma}'_t + \alpha \boldsymbol{F}_t \approx \boldsymbol{\Sigma}'_t + \varepsilon \boldsymbol{I} = \boldsymbol{\Sigma}_t \tag{3}$$

where $\boldsymbol{F}_t$ is a structured, well-conditioned target matrix often taken as the scaled identity $\mathrm{tr}(\boldsymbol{\Sigma}_t')/N$, and the shrinkage intensity $\alpha \in [0,1]$. This blends the noisy sample covariance with a more stable target. Essentially, we add a small jitter $\varepsilon\boldsymbol{I}$ on the diagonal to ensure the sample covariance matrix is clearly positive definite for numerical robustness. We factor each $\boldsymbol{\Sigma}_t$ with a Cholesky decomposition:

$$\boldsymbol{\Sigma}_t = \boldsymbol{L}_t\boldsymbol{L}_t^{\top}. \tag{4}$$

where the lower-triangular entries of $\boldsymbol{L}_t$ are packed into one dimension matrix using row-major "vech" operator:

$$\boldsymbol{l}_t = \mathrm{vech}(\boldsymbol{L}_t) \in \mathbb{R}^M, \quad M = \frac{N(N+1)}{2}. \tag{5}$$

We then store both $\boldsymbol{L}_t$ and $\boldsymbol{l}_t$ by days for model training. This produces two tables: one with full SPD matrices $\boldsymbol{\Sigma}_t$ and one with their Cholesky vectors $\boldsymbol{l}_t$.

The covariance prediction tower of our TimeSformer maps encoded temporalspatial features $\boldsymbol{X}_t$ to a predicted Cholesky factor $\hat{\boldsymbol{L}}_t \in \mathbb{R}^M$. The covariance estimate is also factored as

$$\hat{\boldsymbol{\Sigma}}_t = \hat{\boldsymbol{L}}_t\hat{\boldsymbol{L}}_t^{\top} = \mathrm{lmat}(\hat{\boldsymbol{l}}_t)\mathrm{lmat}(\hat{\boldsymbol{l}}_t)^{\top}, \tag{6}$$

ensuring SPD by design where the "lmat" operator gives the inverse transform of "vech", transforming $\hat{\boldsymbol{l}}_t$ into the lower triangular matrix $\hat{\boldsymbol{L}}_t$.

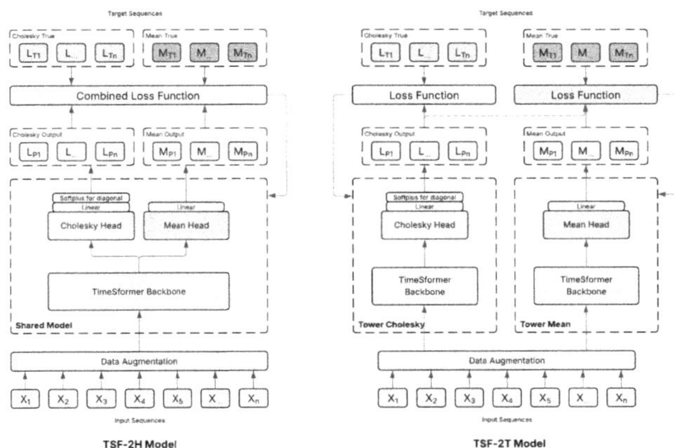

Fig. 3. Network Architectures for Model 3 (left) using combined loss functions and for Model 1 and 2 (right) using separate loss functions for the mean return vectors $\boldsymbol{\mu}_t$ and covariance matrices $\hat{\boldsymbol{\Sigma}}_t$ respectively.

4.2 Network Loss Function

To estimate the parameters of a transformer network, we define various loss functions so that network parameters are estimated to minimize these loss functions.

Model 1: Separate Networks Using WMSE Loss for Returns and Extended Log Euclidean Distance (log-ED) Loss for Covariance Matrices. For log returns, we train the *mean* vector $\boldsymbol{\mu}_t$ using the weighted mean square error (WMSE) loss function based on asset returns $\boldsymbol{r}_{1:T}$ and sample variances $\boldsymbol{\sigma}^2_{1:T} = (\sigma_{ii,1:T})$

$$\mathcal{L}_{\text{WMSE}} = \text{WMSE}(\boldsymbol{\mu}_{1:T}) = \sum_{t=1}^{T}\sum_{i=1}^{N} \frac{1}{\hat{\sigma}_{ii,t}}(r_{i,t} - \mu_{i,t})^2. \tag{7}$$

For covariance matrices, we train $\boldsymbol{\Sigma}_t$ using $\boldsymbol{S}_{1:T}$ and certain distance metric (DM) loss functions. Network parameter training optimizes a composite loss: (i) log-Euclidean distance between the observed $\boldsymbol{\Sigma}_t$ and predicted $\hat{\boldsymbol{\Sigma}}_t$ (log function is applied to reduce the effect of outlying variance and covariance particularly for the cryptocurrency market), (ii) a Frobenius penalty to reduce magnitude deviation, and (iii) a diagonal regularizer to prevent degenerate variances. We use mixed-precision training with gradient scaling for efficiency and stability. The loss function to estimate *covariance matrices* based on sample covariance matrices $\boldsymbol{\Sigma}_{1:T}$ is given by:

$$\mathcal{L}_{\text{cov}} = \mathcal{L}_{\text{logED}}(\hat{\boldsymbol{\Sigma}}_{1:T}|\boldsymbol{\Sigma}_{1:T}) + \mathcal{L}_{\text{frob}}(\hat{\boldsymbol{\Sigma}}_{1:T}|\boldsymbol{\Sigma}_{1:T}) + \mathcal{L}_{\text{diag}}(\hat{\boldsymbol{\Sigma}}_{1:T}|\boldsymbol{\Sigma}_{1:T}, \varepsilon) \tag{8}$$

where

$$\mathcal{L}_{\text{logED}}(\hat{\boldsymbol{\Sigma}}_{1:T}|\boldsymbol{\Sigma}_{1:T}) = \sum_{t=1}^{T}\sum_{i=1}^{N}\sum_{j=1}^{N} \left(\log(\sigma_{ij,t}) - \log(\hat{\sigma}_{ij,t}) \right)^2, \tag{9}$$

$$\mathcal{L}_{\text{frob}}(\hat{\boldsymbol{\Sigma}}_{1:T}|\boldsymbol{\Sigma}_{1:T}) = \sum_{t=1}^{T}\sum_{i=1}^{N}\sum_{j=1}^{N} \left(\sigma_{ij,t} - \hat{\sigma}_{t,ij} \right)^2, \tag{10}$$

$$\mathcal{L}_{\text{diag},t}(\hat{\boldsymbol{\Sigma}}_{1:T}|\boldsymbol{\Sigma}_{1:T}, \varepsilon) = \sum_{t=1}^{T}\sum_{i=1}^{N} \frac{1}{\max(\hat{\sigma}_{ii,t}, \varepsilon)}. \tag{11}$$

where $\varepsilon = 1 \times 10^{-8}$ and the three terms in Eq. (8) are equally weighted.

Model 2: Separate Networks Replacing WMSE with Multivariate Normal or T Density. This is a semi-parametric method as compared to the non-parametric method of Model 1. The covariance matrices $\hat{\boldsymbol{\Sigma}}_t$ are still separately trained using $\mathcal{L}_{\text{cov}}$ in Eq. (8) based on observed $\boldsymbol{\Sigma}_{1:T}$.

However, parametric models are applied to estimate the mean vector $\boldsymbol{\mu}_t$. Essentially, we replace the WMSE with the negative log-likelihood (NLL) of the multivariate normal or Student-t distributions. The loss function for covariance matrices remains the same as Model 1. The return vector $\boldsymbol{r}_t = \left[r_{1t}, r_{2t}, \ldots r_{Nt}\right]^{\top}$ is modeled as follows:

$$\boldsymbol{r}_t \sim \begin{cases} \text{MVT}\left(\boldsymbol{\mu}_t, \hat{\boldsymbol{\Sigma}}_t, \nu\right) & \text{(Multivariate } t\text{-distribution)} \\ \text{MVN}\left(\boldsymbol{\mu}_t, \hat{\boldsymbol{\Sigma}}_t\right) & \text{(Multivariate Gaussian distribution)} \end{cases} \tag{12}$$

where

$$\boldsymbol{\mu}_t = \begin{bmatrix} \mu_{1t}, \mu_{2t}, \ldots \mu_{Nt} \end{bmatrix}^\top \tag{13}$$

represents the $N \times 1$ mean returns vector of the assets at time t, the $N \times N$ variance-covariance matrix at time t is

$$\hat{\boldsymbol{\Sigma}}_t = \begin{bmatrix} \hat{\sigma}_{11t} & \hat{\sigma}_{12t} & \cdots & \hat{\sigma}_{1Nt} \\ \hat{\sigma}_{21t} & \hat{\sigma}_{22t} & \cdots & \hat{\sigma}_{2Nt} \\ \vdots & \vdots & \ddots & \vdots \\ \hat{\sigma}_{N1t} & \hat{\sigma}_{N2t} & \cdots & \hat{\sigma}_{NNt} \end{bmatrix}, \tag{14}$$

and ν represents the degrees of freedom (df) in the MVT distribution. However, to reduce the number of parameters estimated using $r_{1:T}$, the covariance matrix $\hat{\boldsymbol{\Sigma}}_t$ is not estimated from the model but instead, inputted from $\hat{\boldsymbol{\Sigma}}_t$ which minimizes the loss function $\mathcal{L}_{\text{cov}}(\hat{\boldsymbol{\Sigma}}_{1:T})$ in Eq. (8). The probability density function for MVT distribution is:

$$f_{\text{MVT}}(\boldsymbol{r}_t|\boldsymbol{\mu}_t, \hat{\boldsymbol{\Sigma}}_t) = \frac{\Gamma\left(\frac{\nu+N}{2}\right)}{\Gamma\left(\frac{\nu}{2}\right)\nu^{N/2}\pi^{N/2}|\hat{\boldsymbol{\Sigma}}_t|^{1/2}} \left(1 + \frac{1}{\nu}(\boldsymbol{r}_t - \boldsymbol{\mu}_t)^\top \hat{\boldsymbol{\Sigma}}_t^{-1}(\boldsymbol{r}_t - \boldsymbol{\mu}_t)\right)^{-\frac{\nu+N}{2}} \tag{15}$$

whereas the probability density function for MVN distribution is:

$$f_{\text{MVN}}(\boldsymbol{r}_t|\boldsymbol{\mu}_t, \hat{\boldsymbol{\Sigma}}_t) = \frac{1}{(2\pi)^{N/2}|\hat{\boldsymbol{\Sigma}}_t|^{1/2}} \exp\left(-\frac{1}{2}(\boldsymbol{r}_t - \boldsymbol{\mu}_t)^\top \hat{\boldsymbol{\Sigma}}_t^{-1}(\boldsymbol{r}_t - \boldsymbol{\mu}_t)\right). \tag{16}$$

In summary, the MVN or MVT models are separately trained by returns and inputed covariance matrices series. Essentially, this is not a MVN (or MVT) model but rather using them to define a loss function given by

$$\mathcal{L}_{\text{NLL}} = -\sum_{t=1}^{T} \ell(\boldsymbol{\mu}_t|\boldsymbol{r}_t, \hat{\boldsymbol{\Sigma}}_t) = -\sum_{t=1}^{T} \log f_{\text{MVN}}(\boldsymbol{r}_t|\boldsymbol{\mu}_t, \hat{\boldsymbol{\Sigma}}_t). \tag{17}$$

where $f_{\text{MVN}}(\cdot|\cdot)$ is given in Eq. (16). The density $f_{\text{MVN}}(\cdot|\cdot)$ can also be replaced by $f_{\text{MVT}}(\cdot|\cdot)$ in Eq. (15) in which ν can be obtained by a search method.

Model 3: Combined Network by Combining WMSE and Log-ED Into One Loss Function. We estimated both $\boldsymbol{\mu}_{1:N}$ and $\boldsymbol{\Sigma}_{1:T}$ together using the overall loss function defined as

$$\mathcal{L}_{\text{all}} = \mathcal{L}_{\text{WMSE}} + \mathcal{L}_{\text{logED}} \tag{18}$$

in Eq. (7) and (9) respectively.

4.3   Network Architecture

TimeSformer adapts the Vision Transformer to videos by splitting each frame into patches and using self-attention to model dependencies. The main difference from the case of videos is that it applies temporal attention and spatial attention separately within each block, called "divided space-time attention", so each patch first attends across time at the same spatial location and then cross space within the same frame. In TimeSformer, a video clip is shaped as $H \times W \times C \times F$ (height, width, channels, frames), then each frame is split into non-overlapping $P \times P$ patches, producing $\mathcal{N} = HW/P^2$ spatial locations per frame and temporal steps [1].

We propose four experimental configurations, the combined or separated losses (Sect. 4.2), by either the channel or width dimension ways of mapping features (Sect. 3.2). As illustrated in Fig. 3, the first configuration (left panel) relating to combined loss function employs a single TimeSformer model with two output heads.

The second configuration (right panel) adopts a dual-model design with separate loss functions, in which two independent TimeSformer models are trained separately, each dedicated to one of the outputs (referred to as the "Tower" architecture). For the Tower architecture, the network parameters for the predicted mean vector $\boldsymbol{\mu}_t$ are estimated to optimize three alternative loss functions WMSE, GNL, and TNLL, replacing $f_{\mathrm{MVN}}(\cdot|\cdot)$ by $f_{\mathrm{MVT}}(\cdot|\cdot)$. For the predicted covariance matrices $\hat{\boldsymbol{\Sigma}}_t$, the loss function is given by Eq. (8). This results in one experiment for the Head architecture and three for each loss type under the Tower architecture, for both channel and width placements, making a total of eight experiments.

Once the input sequences are transformed through the Data Augmentation process, the TimeSformer will take the input and extract the features through its divided space-time attention and pass the output to respective heads. For the Cholesky head, the output $\hat{\boldsymbol{L}}_t = \mathrm{lmat}(\hat{\boldsymbol{l}}_t)$ is the lower triangle of the Cholesky factor, and the softplus function is applied to its diagonal $\hat{l}_{ii}$, $i = 1 : N$ so that the transformed value is $\in \mathbb{R}$ in general.

5   Results

Table 1 presents a comparative evaluation of model performance using three metrics: WMSE, Frobenius Norm, and Log Euclidean distance. Our neural network models are compared to the traditional statistical models including the basic VAR model, serving as baselines and reporting the WMSE values, as well as the popular DCC-GARCH and DCC-GJR-GARCH models in which we report the Frobenius Norm and Log Euclidean results. In contrast, the TSF-based approaches are evaluated under multiple configurations: two-head models (TSF-2H) and two-tower models (TSF-2T), with features mapped to either the channel ("C") or width ("W") dimensions and further differentiated by the loss function used (WMSE, GNLL, and TNLL) for estimating the mean return vector $\boldsymbol{\mu}_t$. For models using TNLL, the degrees of freedom are set to (3,5,7,10) and 10 is chosen with the best performance according to WMSE, Frobenus Norm and Log Euclidean distance in Table 1.

Table 1. Model performance comparison across popular statistical models and proposed TimeSformer models using WMSE for the mean return vector and the Frobenius norm and log Euclidean distance for the covariance matrices

Model	WMSE	Frobenius Norm	Log Euclidean
VAR (1,0)	0.035304	0.009799	29.600006
DCC-GARCH (1,1)	–	0.004163	28.913357
DCC-GJR-GARCH (1,1,1)	–	0.004929	29.031575
TSF-2H-C	0.029397	0.001301	2.641179
TSF-2H-W	0.029140	0.001290	2.962452
TSF-2T-C-MSE	0.029325	0.001414	3.622555
TSF-2T-C-GNLL	0.029266	0.001207	2.865036
TSF-2T-C-TNLL (df=10)	0.028305	0.001219	2.668153
TSF-2T-W-MSE	0.030358	0.001281	3.159929
TSF-2T-W-GNLL	**0.028014**	0.001326	3.249065
TSF-2T-W-TNLL (df=10)	0.028446	**0.001165**	**2.462418**

TimeSformer variants show strong and often superior performance over traditional baselines. The width-based Student's t model (**TSF-2T-W-TNLL**) achieves the lowest Frobenius norm and log-Euclidean distance, indicating more accurate covariance estimation. The Gaussian version (**TSF-2T-W-GNLL**) records the lowest WMSE (**0.028014**) for mean returns. These results highlight that width-based feature mapping with probabilistic losses enhances predictive accuracy, making TimeSformer-based models strong contenders for multivariate financial forecasting.

The result suggests that a two-Towers configuration can learn better than sharing the backbone with two heads by allowing each tower to specialisze in learning the pattern. Similarly, modeling heterogeneous inputs as features along the width, rather than as channels, maintains clear feature boundaries. In our experiment, during the data augmentation for the feature as a channel, the data is replicated across the width (Fig. 1). This could make the model prones to overfitting the duplicated signals and impacting the generalization. From an optimization perspective, using TNLL as a loss function offers a better advantage over the MSE and GNLL in modeling heavy-tailed data that are evident in our dataset.

We perform validity check on predicted covariance matrices to ensure that they are SPD with eigenvalues $\lambda_i \geq 0$ for all i, and have finite numeric entries from our best model (**TSF-2T-W-TNLL**) indicated no asymmetry issues (maximum and mean symmetry error both 0.00e+00), and all matrices satisfy positive semi-definiteness, with the minimum eigenvalue across all matrices equal to 6.1806e-05. There are no asymmetric matrices above the 1e-08 tolerance (0 detected), no non-PSD matrices (0 detected), and no matrices contain NaN or Inf values (0 out of 250).

Figure 4 presents a comparison analysis of the predicted mean returns against the realized one-step-ahead returns across six representative assets. Across all assets, the model generally remains close to 0, a usual behavior of return series. For large-cap equites such as SPY and MSFT, the model predictions closely follow the observed returns, with some exceptions during Trump's tariff period. TSLA, which shows higher volatility, displays wider gaps between predicted and observed returns, though the model is able to capture some general trends over time well. For BTC-USD, it displays some sharp jumps that are typical in the more volatile cryptocurrency market. In contrast, the predicted mean returns for the low-volatility safe-haven assets GLD and IEF align well with the observed returns, indicating good model performance with more stable returns.

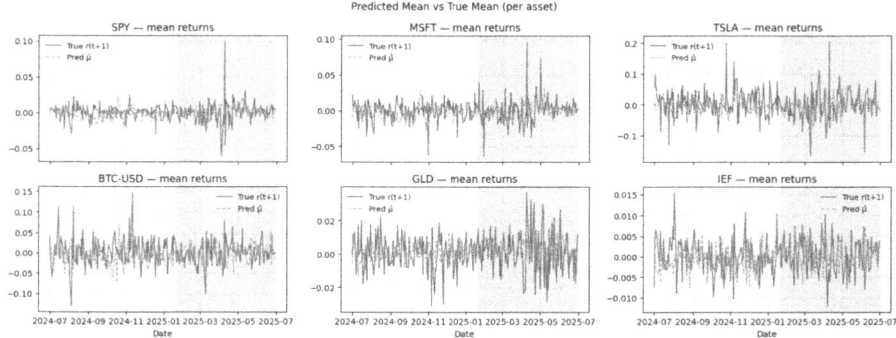

Fig. 4. Observed vs predicted mean returns (per asset) for TSF-2T-W-TNLL model. Trump's Tariff Stress Period is highlighted in light grey background.

Figure 5 compares realized (as observed; $\sigma_t^2 = \mathrm{diag}(\boldsymbol{\Sigma}_t)$) and predicted ($\hat{\sigma}_t^2 = \mathrm{diag}(\hat{\boldsymbol{\Sigma}}_t)$) variances for six assets. A sharp variance peak appears around April 2025, coinciding with Trump's tariff announcement, which triggered heightened volatility in equities such as SPY, MSFT, and TSLA. The model reflects this rise but underestimates the magnitude of the spike.

For BTC-USD, we observe the variance elevated through late 2024 to early 2025. As Bitcoin surged from 60k in October 2024 to nearly $100k by December, and new all-time highs above $100k in early 2025 before correcting. This rally and correction illustrates high variance, which model tracks but smooths relative to realized values.

In summary, the model captures regime shifts and broad variance trends across assets, though it tends to smooth extreme peaks, especially in high-volatility markets like Bitcoin and TSLA.

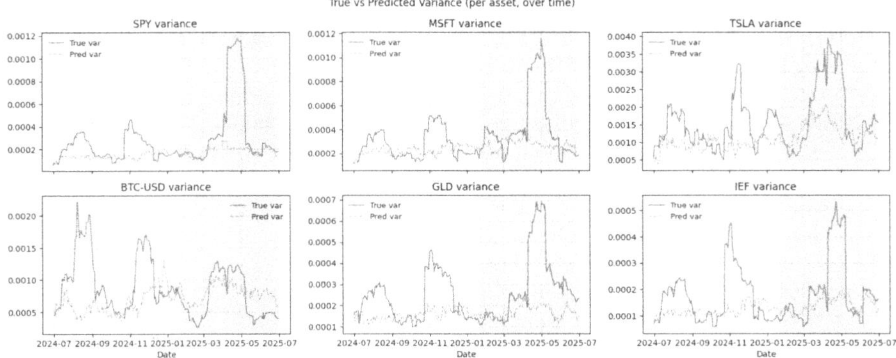

Fig. 5. Observed variance vs predicted variance (per asset) using TSF-2T-W-TNLL model. Trump's Tariff Stress Period is highlighted in light grey background.

6   Conclusion

This study introduces a new framework for modeling financial time series by transforming features like log returns into image-like vectors processed with TimeSformer. Two data representations and two architectures are tested: (1) separate TimeSformers for mean and covariance, and (2) a joint model with a combined loss. Mean returns use both non-parametric (WMSE) and parametric losses (Gaussian and Student's t), while covariances use non-parametric losses such as log-ED and extended log-ED.

Results show clear gains in predicting both mean and covariance. The Student's t model gives the most accurate covariance estimates, while the Gaussian model performs best for mean returns. These improvements enhance portfolio optimization by better balancing return and risk, showing the potential of combining representation learning with financial modeling.

However, the 20-day rolling window may over-smooth covariances due to overlap. Using intraday data or adding lagged returns could improve accuracy. The framework currently covers six assets; future work should scale it up. Further refinements include tuning the loss balance, testing Wishart-based covariance models, exploring hybrid losses, and using different learning rates for mean and covariance tasks to improve joint forecasting stability.

References

1. Bertasius, G., Wang, H., Torresani, L.: Is space-time attention all you need for video understanding? (2021). https://arxiv.org/abs/2102.05095
2. Bollerslev, T.: Generalized autoregressive conditional heteroskedasticity. J. Econom. **31**(3), 307–327 (1986)
3. Box, G.E.P., et al.: Time Series Analysis: Forecasting and Control, 5th edn. John Wiley & Sons, Inc., Hoboken (2016)

4. Cohen, N., Klein, I.: Adaptive kalman-informed transformer. Eng. Appl. Artif. Intell. **146**, 110221 (2025). https://doi.org/10.1016/j.engappai.2025.110221
5. Engle, R.F.: Dynamic conditional correlation: a simple class of multivariate generalized autoregressive conditional heteroskedasticity models. J. Bus. Econ. Stat. **20**(3), 339–350 (2002). http://www.jstor.org/stable/1392121
6. Glosten, L.R., et al.: On the relation between the expected value and the volatility of the nominal excess return on stocks. J. Financ. **48**(5), 1779–1801 (1993)
7. Leccadito, A., et al.: A novel robust method for estimating the covariance matrix of financial returns with applications to risk management. Financ. Innov. **10**(116), 1–24 (2024). https://doi.org/10.1186/s40854-024-00642-2
8. Ledoit, O., Wolf, M.: A well-conditioned estimator for large-dimensional covariance matrices. J. Multivar. Anal. **88**(2), 365–411 (2004)
9. Lim, B., et al.: Temporal fusion transformers for interpretable multi-horizon time series forecasting. Int. J. Forecast. **37**(4), 1748–1764 (2021). https://doi.org/10.1016/j.ijforecast.2021.03.012
10. Markowitz, H.M.: Portfolio selection, the journal of finance. 7 (1). N **1**, 71–91 (1952)
11. Moghar, A., Hamiche, M.: Stock market prediction using lstm recurrent neural network. Procedia Comput. Sci. **170**, 1168–1173 (2020). https://doi.org/10.1016/j.procs.2020.03.049
12. Nigam, S.: Forecasting time series using convolutional neural network with multiplicative neuron. Appl. Soft Comput. **174**, 112921 (2025). https://doi.org/10.1016/j.asoc.2025.112921
13. Pienaar, R., van Vuuren, G.: Enhancing portfolio asset allocation efficiency using blended covariance matrices. Int. J. Econ. Financ. Issues **15**(4), 343–355 (2025). https://doi.org/10.32479/ijefi.18883
14. Rahmadeyan, A.: Mustakim: long short-term memory and gated recurrent unit for stock price prediction. Procedia Comput. Sci. **234**, 204–212 (2024). https://doi.org/10.1016/j.procs.2024.02.167
15. Sharpe, W.F.: Mutual fund performance. J. Bus. **39**(1), 119–138 (1966)
16. Sims, C.A.: Macroeconomics and reality. Econometrica: J. Econom. Soc. 1–48 (1980)
17. Takahashi, S., et al.: Comparison of vision transformers and convolutional neural networks in medical image analysis: a systematic review. J. Med. Syst. **48**(1), 84 (2024). https://doi.org/10.1007/s10916-024-02105-8
18. Vaswani, A., et al.: Attention is all you need. In: Guyon, I., et al. (eds.) Advances in Neural Information Processing Systems, vol. 30. Curran Associates, Inc. (2017). https://proceedings.neurips.cc/paper_files/paper/2017/file/3f5ee243547dee91fbd053c1c4a845aa-Paper.pdf
19. Walker, D.: Kalman filtering of time series data. In: Soofi, A., Cao, L. (eds.) Modelling and Forecasting Financial Data, Studies in Computational Finance, vol. 2, pp. 127–148. Springer, Boston, MA (2002). https://doi.org/10.1007/978-1-4615-0931-8_7
20. Wang, J., et al.: A comprehensive transformer-based approach for high-accuracy gas adsorption predictions in metal-organic frameworks. Nat. Commun. **15**(1), 1904 (2024). https://doi.org/10.1038/s41467-024-46276-x
21. Wang, J., et al.: State causality and adaptive covariance decomposition based time series forecasting. Sensors **23**(2) (2023). https://doi.org/10.3390/s23020809
22. Wen, Q., et al.: Transformers in time series: a survey. arXiv preprint arXiv:2202.07125 (2022). https://doi.org/10.48550/arxiv.2202.07125

23. Yahoo Finance: Historical data for apple inc. (aapl). https://finance.yahoo.com/quote/AAPL/history. Accessed 23 July 2025

24. Zhou, T., et al.: Fedformer: frequency enhanced decomposed transformer for long-term series forecasting. arXiv preprint arXiv:2201.12740 (2022).https://doi.org/10.48550/arxiv.2201.12740

25. Zivot, E., Wang, J.: Vector autoregressive models for multivariate time series. In: Modeling Financial Time Series with S-Plus®. Springer, New York, NY (2003). https://doi.org/10.1007/978-0-387-21763-5_11

Temporal Fusion of Biophysical and Climate Data: A Data-Driven Hybrid Learning Approach for Short-Term Aboveground Biomass Forecasting

Han Xu$^{(\boxtimes)}$, Thirunavukarasu Balasubramaniam, and Richi Nayak

Centre for Data Science, School of Computer Science, Queensland University of Technology, Brisbane, QLD 4000, Australia
`x32.han@hdr.qut.edu.au`, `{thirunavukarasu.balas,r.nayak}@qut.edu.au`

Abstract. Accurate forecasting of aboveground biomass (AGB) is essential for sustainable pasture management. Existing approaches, such as traditional machine learning (ML) and biophysical models, have limitations: ML models fail to effectively capture temporal dependencies and underlying biophysical processes, whereas biophysical models are difficult to scale due to their requirement for extensive calibration. Although hybrid frameworks that integrate biophysical model outputs into ML models show potential, they still cannot effectively exploit temporal information. To address these issues, we propose a novel framework that combines a multi-view Gate-Controlled LSTM (GC-LSTM) with the biophysical model *Modvege* for short-term AGB forecasting. Our GC-LSTM employs dual branches to model both climate data and Modvege-derived biophysical variables, adaptively balancing their contributions through a gating mechanism. We evaluate GC-LSTM against nine baselines, including Modvege and eight ML and deep learning models that incorporate biophysical variables using three independent yearly datasets (2021, 2022, and 2023) of 56 Australian paddocks. Results demonstrate that GC-LSTM consistently achieves higher accuracy, highlighting its robustness and practical utility for AGB prediction.

Keywords: Multi-view Deep Fusion · Gate-Controlled LSTM · Biophysical Models · Aboveground Biomass Forecasting · Pasture Management

1 Introduction

Pasture - defined as land covered with grass and other low plants as forage for grazing animals - is a vital component of Australia's livestock industry [1]. Focusing on these grazing animals, statistics for the 2023–24 period indicate that the combined value of livestock disposals (animals sold for processing into red meats from cattle, calves, sheep, and lambs) and livestock products (wool and

Q. V. Nguyen et al. (Eds.): AusDM 2025, CCIS 2765, pp. 192–207, 2026.
https://doi.org/10.1007/978-981-95-6786-7_13

milk) reached approximately AUD 25.4 billion [2]. Accurate prediction of aboveground biomass (AGB) is crucial for effective pasture management, supporting decisions such as pasture scheduling or grazing intensity control to maintain the industry's profitability and sustainability.

Prior studies on AGB prediction encompass a diverse range of approaches. This study primarily reviews two major categories: traditional machine learning (ML) models and biophysical models. ML models such as Random Forest and XGBoost have shown promising predictive performance, particularly when integrating multiple data sources, including remote sensing, climate, and animal information [3,4]. However, these model remain limited in their ability to capture underlying biophysical processes and crucial temporal dependencies governing pasture growth. To represent biophysical processes more explicitly, biophysical models such as APSIM and Modvege simulate AGB or pasture growth based on environmental inputs and mechanistic equations [5,6]. Despite their interpretability, these models require extensive manual calibration [7] and rely on static predefined equations that cannot fully represent the complexity and variability of real-world conditions. Consequently, their scalability is restricted, particularly across Australia's vast and highly heterogeneous pasture regions [8].

To overcome these limitations, hybrid frameworks have been proposed to improve both prediction accuracy and scalability [7]. In such frameworks, biophysical model outputs serve as additional input features for ML models, combining the strengths of both approaches. Biophysical models provide mechanistic insights into growth processes, while ML models can learn nonlinear relationships between inputs and target variables [15], thereby capturing interactions not explicitly defined in the biophysical equations. For example, APSIM-derived variables have improved corn yield prediction in the United States, and DSSAT-derived features combined with remote sensing data have enhanced wheat yield prediction in India [9,10]. However, hybrid frameworks still face several challenges. Their performance can be degraded by noise originating from various data sources, such as inaccurate biophysical variables due to oversimplified assumptions or omission of external factors (e.g., animal loads or extreme climate) [11], errors in environmental tabular data [12], or contamination in remote sensing imagery caused by clouds and shadows [13]. Furthermore, these frameworks generally fail to capture temporal dependencies that reflect cumulative environmental effects on pasture growth [4].

Temporal Deep Learning (TDL) models, such as Long Short-Term Memory (LSTM), have shown strong potential in crop forecasting, with several studies reporting superior performance over traditional ML models in specific regions [14]. Their effectiveness comes from the ability to handle temporal dependencies and mitigate data noise by learning sequential patterns and suppressing random fluctuations within the input data [15]. Importantly, combining TDL models with biophysical models has been shown to improve predictive accuracy, robustness to noise, and generalisation to new locations [7]. These capabilities are particularly valuable for the heterogeneous pasture systems in Australia. Despite this potential, such hybrid frameworks remain underexplored.

To address this gap, we propose a novel framework that combines a multi-view gate-controlled LSTM (GC-LSTM) with Modvege [5] for short-term AGB forecasting using climate data. In this study, our contributions are threefold:

- We adapt Modvege [5], developed in France, to derive biophysical variables relevant to Australian pastures (Phalaris).
- We design GC-LSTM that jointly models climate data and Modvege-derived biophysical variables through dual branches, with an adaptive gating mechanism to fuse the two forecasts into a final prediction.
- We benchmark GC-LSTM against nine baselines, including Modvege and multiple traditional ML and DL models.

2   Methodology

The proposed GC-LSTM is a multi-view model that uses two distinct data streams, driven by climate and biophysical variables, to forecast AGB 15 days ahead. The model exploits three fusion strategies: early fusion (feature concatenation at the input level), intermediate fusion (concatenation of learned representations before the output layer), and late fusion (fusion of independent model forecasts). We first describe the problem context, followed by a detailed presentation of our proposed method.

2.1   The Problem Context: Short-term AGB Forecasting

Datasets. Phalaris aquatica, a perennial grass species, plays an important role in supporting grazing livestock in temperate regions of Australia [20]. To explore short-term AGB forecasting, we used a dataset covering 56 paddocks composed exclusively of *Phalaris* pastures. The paddock data was collected from Agri-Webb's farm management platform. The study of a single species helped eliminate the confounding effects of mixed pastures. A paddock refers to a fenced pasture area used for grazing. Each paddock is represented by three categories of data: daily climate records, daily Modvege-derived biophysical variables, and discrete AGB observations. A summary of the data is provided in Table 1.

Daily climate data were collected from the SILO database [16] for the period February 19, 2017, to September 1, 2023. Biophysical variables were generated using Modvege [5], an open-source biophysical model, selected for its alignment with AGB forecasting objectives. Originally developed in France, Modvege simulates daily AGB changes across four compartments: green vegetative (GV), green reproductive (GR), dead vegetative (DV), and dead reproductive (DR), based on climate inputs and grazing activities. To reduce complexity and isolate the impact of biophysical integration, we focused on the GV compartment and its derived variables (e.g., growth rate and senescence), which most strongly reflect leaf and sheath biomass dynamics. Discrete AGB measurements were sourced from CiboLabs [17], which estimates AGB using an ML model applied to Sentinel-2 satellite imagery. These observations were available at 5-day intervals, subject to cloud cover.

Table 1. Summary of daily climate data and biophysical variables, with discrete AGB.

Category	Feature	Description
Climate Data	Rainfall (mm)	–
	Maximum/ Minimum Temperature (°C)	–
	Maximum/ Minimum Relative Humidity (%)	Relative humidity at the times of maximum and minimum temperature
	ASCE Tall Crop Evapotranspiration (mm)	A measure of water lost to the atmosphere from crops [19]
	Incident Photosynthetically Active Radiation (PARi) (MJ/m^2)	Multiply SILO's solar radiation by a 0.45 conversion factor [18]
Biophysical Variables	Growth Rate (GRO) (kg DM/ha/day)	Actual AGB growth
	Potential Growth Rate (PGRO) (kg DM/ha/day)	Potential AGB growth
	Senescence (SEN) (kg DM/ha/day)	AGB lost due to senescence
	GV ageing (AGE) (kg DM/ha/day)	AGB lost due to GV ageing
	Sum of Thermal Time (SumT) (°C·day)	Accumulated daily temperature from 1 January
	Environmental Coefficients (ENV)	Environmental constraints affecting growth
Target Variable	AGB (kg DM/ha)	Aboveground Biomass

Modvege Calibration for Biophysical Variables. The calibration of the Modvege model [5] involves four key steps. **Step 1** focuses on data preparation. In this study, we only consider climate inputs. The simulation for each paddock requires daily average temperature (T), photosynthetically active radiation intercepted (PARi), precipitation (PP), and potential evapotranspiration (PET). Average temperature was derived as the mean of daily maximum and minimum temperatures, while PARi and PP were selected directly from the climate data. We used the ASCE Tall Crop Evapotranspiration for PET, as it is suitable for Phalaris, a perennial tall grass [19,20]. We set all other daily inputs, such as actual evapotranspiration (AET), leaf area index (LAI), cut height, and grazing animal count and weight, to zero. AET and LAI were later calculated using biophysical equations during simulation. Due to the lack of precise cutting and grazing records, we assumed that biomass dynamics were solely driven by natural growth and senescence.

Step 2 involves defining biophysical parameters and their search spaces. Modvege classifies pastures into four functional trait groups (A, B, C, D) [5]. Based on species traits such as the long leaf lifespan, Phalaris was assigned to Group B [20]. Initial Group B parameter values were used to establish calibration bounds. Site-specific parameters, water-holding capacity (WHC) and water reserve (WR), followed recommended ranges from [20], with WR constrained to be less than or equal to WHC [5].

Step 3 consists of initialising AGB and thermal time. Thermal time was calculated as the cumulative sum of daily mean temperatures. Our simulation begins on February 20, 2017, using the observed AGB from February 19 as the initial value. Since Modvege calculates thermal time from January 1, an offset corresponding to the cumulative thermal time from January 1 to February 19 was applied. For multi-year simulations, thermal time was reset at the start of each calendar year (January 1), and the initial AGB was updated using the simulated AGB from the last day of the preceding year (December 31).

Step 4 calibrates biophysical parameters for each paddock using Optuna [21] with multi-threading. Optuna, an open-source hyperparameter optimisation framework, samples parameter combinations from the defined search spaces (Step 3) and evaluates each through full Modvege simulations using paddock climate inputs. For each trial simulated AGB values, starting from March 6, 2017 (excluding the initial AGB on February 19) to September 1, 2023, were compared against satellite-estimated AGB observations, using evaluation metrics described in Sect. 3.4. Empirically, a minimum of 8,000 trials per paddock was required to find the optimal combination. Figure 1 illustrates the schematic flow of a paddock simulation, which tests different combinations to closely align the simulated curve with the satellite-estimated AGB curve.

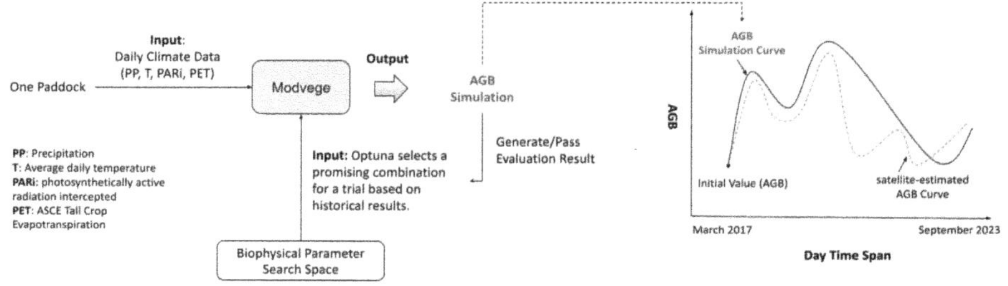

Fig. 1. Schematic workflow of a paddock calibration using Optuna.

These best-fitting parameters and climate inputs were then used to generate paddock-specific biophysical time series. During simulation, Modvege applies core biophysical formulations (Eqs. 1, 2 and 3), linking AGB dynamics with biophysical variables and climate inputs. This simulation incorporates calibrated parameters such as activation, optimal, and maximum growth temperatures (T0, T1, T2), maximum radiation-use efficiency (RUEmax), the PARi conversion coefficient for photosynthesis (alpha_PAR), two thermal time points (ST1 < ST2), and the nitrogen index (NI). In these equations, AGB change is governed by four environmental terms: seasonality (SEA controlled by ST1 and ST2), light availability (f(PARi), via alpha_PAR and PARi), temperature impact (f(T), calculated using T, T0, T1, and T2), and water stress (f(W), via WR and WHC).

$$\frac{dAGB}{dt} = GRO - SEN, \qquad GRO = PGRO \times ENV \times SEA \tag{1}$$

$$PGRO = RUEmax \times PARi \times \left[1 - \exp(-0.6 \times LAI)\right] \tag{2}$$

$$ENV = NI \times f(PARi) \times f(T) \times f(W) \tag{3}$$

Feature Aggregation and Merging. To capture the cumulative effects of short-term growth, we processed the daily climate data (Table 1) for each paddock using a 15-day sliding window spanning from day t to $t-14$. This procedure created data points at 15-day intervals, each representing 15-day averaged (seven) climate features. These were then merged with the corresponding daily (seven) climate feature values at day t, six biophysical variables, and the AGB recorded on day t. The resulting dataset comprises data points with a total of 21 features.

2.2   Multi-view Gate-Controlled LSTM

Our proposed model, GC-LSTM, is a temporal fusion network consisting of 5,886 trainable parameters. It introduces a gated control mechanism to fuse outputs from two distinct branches—one dedicated to climate inputs and another to biophysical variables—for AGB forecasting. As shown in Fig. 2, GC-LSTM processes sequential inputs through two separate LSTM encoders: a 24-neuron branch for capturing climate dependencies and a 10-neuron branch for biophysical dependencies. The final hidden state from each branch is then projected into a shared 24-dimensional embedding, followed by ELU activation and 0.1 dropout to enhance non-linearity and regularisation, respectively. A shared two-layer fully connected (FC) head generates individual forecasts from each embedding (FC1 and FC2). This head maps the 24-dimensional input to a 14-neuron hidden layer before producing a scalar output (FC3 and FC4). All hyperparameters were fine-tuned using Optuna to maximise averaged performance across validation sets. During training, the shared weights of the head are updated by adding gradients from both branches, using an optimiser, encouraging a unified mapping and improving convergence [15].

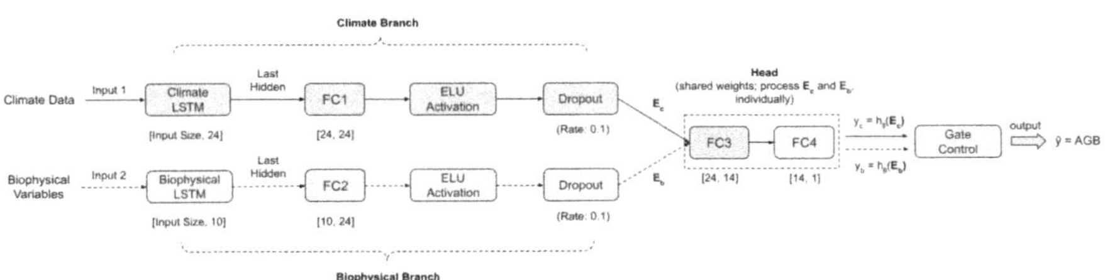

Fig. 2. Architecture of GC-LSTM, with the separate climate and biophysical branches, the head, and a gate control mechanism.

Mathematically, the head $h_\theta(\cdot)$ is defined as:

$$h_\theta(E) = W_2\left(W_1 E + b_1\right) + b_2, \tag{4}$$

where $W_1 \in \mathbb{R}^{d_h \times d_E}$, $b_1 \in \mathbb{R}^{d_h}$, $W_2 \in \mathbb{R}^{d_{out} \times d_h}$ and $b_2 \in \mathbb{R}^{d_{out}}$. The input $E \in \{E_c, E_b\}$ is a branch-specific embedding, with $E_c, E_b \in \mathbb{R}^{d_E}$. The set of head parameters is denoted by $\theta = \{W_1, b_1, W_2, b_2\}$ where d_E is the embedding dimension, d_h is the head's hidden layer dimension, and $d_{\text{out}} = 1$ is the scalar output dimension.

This head is applied independently to both branches, climate embedding E_c and biophysical embedding E_b, producing two forecasts:

$$y_c = h_\theta(E_c), \qquad y_b = h_\theta(E_b), \tag{5}$$

where $E_c = E_c(\cdot; \phi_c)$ and $E_b = E_b(\cdot; \phi_b)$ are embeddings produced by the climate and biophysical encoders, parameterised by ϕ_c and ϕ_b, respectively.

A gating mechanism then adaptively fuses the two forecasts:

$$\hat{y} = w \cdot y_c + (1 - w) \cdot y_b, \qquad w = \sigma(\alpha) = \frac{1}{1 + \exp(-\alpha)} \in (0, 1), \tag{6}$$

where $\hat{y}$ is the fused forecast and w is a learnable, sigmoid-based weight that assigns relative importance to each branch. This weight is controlled by a trainable scalar α, which is initialised in the logit space $\alpha_0 = \log(\frac{p}{1-p})$ with a prior $p \in (0, 1)$. This initialisation ensures that w is a valid probability while allowing stable optimisation. We define this sigmoid-based weighting as a gate control mechanism.

The model is trained with the Mean Absolute Error (MAE) loss:

$$\mathcal{L}_{\text{MAE}} = \frac{1}{N} \sum_{i=1}^{N} |y_i - \hat{y}_i|, \tag{7}$$

where y_i is the satellite-estimated AGB for the i-th sequence and $\hat{y}_i$ is the corresponding fused prediction.

3   Empirical Analysis

3.1   Experimental Setup

The proposed GC-LSTM and baseline models were implemented in Python 3.13.5 on an Apple Mac Pro (M2 chip, 10-core CPU, 16 GB RAM). DL models were implemented in PyTorch 2.7.1, data preprocessing and traditional ML models were implemented using scikit-learn 1.7.0, and XGBoost was implemented with the xgboost 3.0.3 library. For DL models, we used the Adam optimiser with a learning rate of 2e-4, a batch size of 128, and a maximum of 200 epochs. Early stopping (patience = 12) and L1 regularisation (1e-4) were applied to prevent overfitting. Training batches were not shuffled, and all models were trained using MAE loss, consistent with GC-LSTM.

3.2   Dataset Preparation

For the validation of GC-LSTM, we utilised the merged data (detailed in Sect. 2.1) to generate paddock-specific datasets and compare temporal and non-temporal model types. For temporal models, we used a six-point lookback window (90 days) to forecast the next 15-day AGB value, with each timestep containing 21 features. This setup helps the model capture seasonal and temporal dependencies. For non-temporal models, we flattened these six lookback steps into a single feature vector of 126 features (21 × 6). In both cases, we included historical AGB as an input to help capture growth dynamics without introducing data leakage. By default, this AGB is included in climate data. It is only included in biophysical variables when climate data is not used. As shown in Table 2, we created three datasets (D1–D3) using progressive training with rolling validation (used for hyperparameter tuning) and test years. The years 2021, 2022, and 2023 served as the test periods for datasets D1, D2, and D3, respectively. D3 dataset samples were collected until September 1, 2023. This strategy enables assessment under shifting climate conditions and evaluates the model's robustness to distribution changes [8]. For GC-LSTM, the optimal gate initialisation was obtained via Optuna, yielding prior probabilities of $p = [0.630, 0.050, 0.709]$ for D1, D2, and D3, respectively.

Table 2. Data Splits Overview.

Dataset	Train Period	Valid Period	Test Period
D1	2017.02–2020.09	2020.10–2020.12	2021
Samples	4,536	392	1,344
D2	2017.02–2021.09	2021.10–2021.12	2022
Samples	5,936	336	1,344
D3	2017.02–2022.09	2022.10–2022.12	2023
Samples	7,280	336	952

For normalisation, we applied a *Log1p* transformation to AGB in D1 and D2 to correct skewness; it was omitted in D3 as direct values yielded better model performance. All input features were scaled using Min-Max scaling [–1, 1], with scaling parameters computed from the training data and applied consistently to validation and test sets.

3.3 Baselines

We evaluated GC-LSTM against nine baselines spanning DL models, traditional ML models, and the biophysical model Modvege. The DL baselines include a custom Deep Neural Network (DNN) and three fusion-based GC-LSTM variants - Early, Intermediate, and Late Fusion—each designed to evaluate different integration strategies between climate and biophysical information. The traditional ML baselines are Random Forest, XGBoost, Bootstrapped Bagging, and Linear Regression. Except for Linear Regression, all DL and ML models had their hyperparameters fine-tuned with Optuna [21], and the final settings were selected based on performance across three validation sets.

Late Fusion LSTM was developed to evaluate a linear late-fusion strategy. We modified GC-LSTM by replacing the gated fusion with independent branch-specific heads. The climate branch uses a two-layer FC head (24 → 12 → output), while the biophysical branch uses a single FC layer (6 $\rightarrow$ output). The final forecast is the arithmetic mean of the two branch outputs.

Intermediate Fusion LSTM evaluates a linear feature concatenation strategy. It retains the separate LSTM encoders for climate and biophysical inputs, producing two separate 24-dimensional embeddings. These embeddings are then concatenated and fed into a forecast head consisting of one FC layer (14 neurons), followed by ELU activation and 0.1 dropout to generate the final prediction.

In Early Fusion LSTM, climate and biophysical variables are concatenated at the input level and processed jointly using a single LSTM layer (24 hidden neurons), followed by a dropout layer (rate 0.1) and an FC output layer to produce the forecast. To highlight the benefits of multi-view learning, we conducted two additional experiments using this architecture: (1) LSTM (Climate) trained using only climate variables; and (2) LSTM (Bio) trained using only biophysical variables. These serve as single-view baselines, enabling direct comparison to multi-view fusion strategies.

A DNN-based model [4] baseline evaluates performance without temporal dependencies. It consists of four FC layers: the first layer projects inputs to another layer of 32 neurons, followed by a 0.1 dropout layer. The second and third FC layers reduce to 24 and 16 neurons, respectively. An ELU activation is applied before the final FC layer.

Random Forest and XGBoost, ensembled decision tree methods, were chosen for their ability to capture non-linear interactions in climate-related prediction tasks with high accuracy and comprehensibility [9]. This implementation of Random Forest combined 300 decision trees with a maximum depth of 17 [4]. A minimum of 8 samples was required to split a node and 5 samples for a leaf. All features were considered at each split. This implementation of XGBoost fitted 1500 sequential trees with a maximum depth of 5, a learning rate of 0.05 and sampled 59% of the training data and 98% of the features for each tree. To prevent overfitting, it uses regularisations of L1 (1.05) and L2 (7.74).

Bootstrapped Bagging is an ensemble meta-estimator that fits base regressors on random subsets of the original dataset and then averages their predictions, which helps reduce variance and improve model robustness [4]. This model used 250 decision trees trained on bootstrap samples (59% of the data with replacement) with all features. Each tree has a maximum depth of 38, with a minimum of 7 samples per split and 5 samples per leaf.

As the simplest benchmark, Linear Regression assumes a linear relationship between input features and AGB, offering a baseline for assessing gains from non-linear, temporal and fusion models.

Modvege, configured with optimal biophysical parameters, serves as a process-based benchmark, providing a direct comparison with a biophysical simulation approach in Australian pasture systems.

3.4   Evaluation Metrics

We evaluated model performance using three commonly adopted regression metrics: Coefficient of Determination (R^2), Root Mean Squared Error (RMSE), and

Mean Absolute Error (MAE) [4]. R^2 measures the proportion of variance in the target variable explained by the model, with values approaching 1 indicating a better fit. RMSE penalises large errors more heavily, making it valuable for assessing the model's robustness against outliers. MAE provides an intuitive measure of average prediction error, expressing the average magnitude of deviations between predictions and actual values. Together, these metrics offer complementary insights into model accuracy, stability, and reliability.

3.5   Results and Discussions

This section presents a comparative analysis of all models evaluated on testing data from 2021, 2022, and 2023. Our objective is to demonstrate the effectiveness and robustness of GC-LSTM while addressing two key questions: (1) How effectively can multi-view TDL models, integrating climate and biophysical variables, improve short-term AGB forecasting accuracy; and (2) To what extent do temporal dependencies influence the forecasting accuracy of AGB? As the goal is not to study paddock-specific variability, we adopt macro-level evaluation, where predictions across all paddocks are aggregated before calculating performance metrics. To account for randomness in both DL and traditional ML models, we report the mean macro metrics performance across 10 independent runs, each using a different random seed. Standard deviations are included to indicate model stability. We also provide visual comparisons of forecasting curves across the three years to highlight the strengths and limitations of GC-LSTM relative to selected baselines.

Performance Comparison. All models were evaluated on normalised data to ensure fair comparison across paddocks with varying AGB ranges. Tables 3, 4, 5 summarise results for each test year (2021–2023) using R^2, RMSE, and MAE. For Modvege, we used paddock-specific climate data from 19 February 2017 to 1 September 2023, along with optimised biophysical parameters to generate continuous AGB simulations. The outputs were clipped to match each test year and aggregated paddock-wise for evaluation. As Modvege is deterministic, it was evaluated once and therefore does not include standard deviation values.

We demonstrate the superior performance and robustness of GC-LSTM through three key observations. First, GC-LSTM consistently achieves the highest R^2 and the lowest RMSE and MAE values across all three years. Second, it excels in more challenging forecasting conditions—particularly in 2022, when all models experienced performance degradation, likely due to extreme climate events—yet GC-LSTM retained a clear advantage. Third, although DL models generally exhibit higher variance than traditional ML models, GC-LSTM proves noticeably more stable than other DL models.

GC-LSTM's superiority is likely attributed to three primary factors. (1) Effective multi-view modelling: In addition to the high variability in climate data [8], Modvege's incomplete simulation (i.e. the GV compartment only) of biophysical variables introduces further noise. GC-LSTM's dual-branch architecture leverages the noise resistance of DL models [15] to embed and reconcile

Table 3. Performance comparison of all models in Year 2021. Mean ± std are reported.

Model	R^2	RMSE	MAE
GC-LSTM	**0.790 ± 0.009**	**0.130 ± 0.003**	**0.096 ± 0.003**
Late Fusion LSTM	0.784 ± 0.017	0.132 ± 0.005	0.099 ± 0.006
Intermediate Fusion LSTM	0.779 ± 0.014	0.134 ± 0.004	0.100 ± 0.005
Early Fusion LSTM	0.775 ± 0.015	0.135 ± 0.005	0.101 ± 0.005
LSTM (Climate)	0.781 ± 0.008	0.133 ± 0.002	0.099 ± 0.002
LSTM (Bio)	0.719 ± 0.022	0.151 ± 0.006	0.113 ± 0.006
DNN	0.732 ± 0.018	0.147 ± 0.005	0.109 ± 0.004
XGBoost	0.707 ± 0.000	0.154 ± 0.000	0.117 ± 0.000
Random Forest	0.750 ± 0.002	0.142 ± 0.001	0.107 ± 0.001
Bagging	0.750 ± 0.003	0.142 ± 0.001	0.106 ± 0.001
Linear Regression	0.642 ± 0.000	0.170 ± 0.000	0.134 ± 0.000
Modvege	0.207	0.253	0.172

Table 4. Performance comparison of all models in Year 2022. Mean ± std are reported.

Model	R^2	RMSE	MAE
GC-LSTM	**0.715 ± 0.008**	**0.099 ± 0.001**	**0.074 ± 0.001**
Late Fusion LSTM	0.699 ± 0.015	0.102 ± 0.002	0.076 ± 0.002
Intermediate Fusion LSTM	0.694 ± 0.015	0.103 ± 0.002	0.077 ± 0.002
Early Fusion LSTM	0.712 ± 0.012	0.100 ± 0.002	0.075 ± 0.001
LSTM (Climate)	0.681 ± 0.019	0.105 ± 0.003	0.078 ± 0.003
LSTM (Bio)	0.667 ± 0.020	0.107 ± 0.003	0.078 ± 0.002
DNN	0.645 ± 0.014	0.111 ± 0.002	0.085 ± 0.002
XGBoost	0.682 ± 0.000	0.105 ± 0.000	0.078 ± 0.000
Random Forest	0.684 ± 0.003	0.104 ± 0.000	0.077 ± 0.000
Bagging	0.688 ± 0.002	0.104 ± 0.000	0.077 ± 0.000
Linear Regression	0.361 ± 0.003	0.148 ± 0.000	0.120 ± 0.000
Modvege	0.064	0.179	0.141

signals from both views, enabling more reliable representation learning of AGB dynamics. (2) Shared head parameters: By applying a shared prediction head, the model learns a unified feature space, which supports more efficient convergence and improved generalisation. (3) Adaptive gated fusion: The learned gate dynamically weights the contributions of each branch, allowing the model to downweight noisy inputs and refine predictions. Initialising the gate with optimised prior probabilities ensures stable learning and mitigates the risk of local minima across datasets.

Table 5. Performance comparison of all models in Year 2023. Mean ± std are reported.

Model	R^2	RMSE	MAE
GC-LSTM	**0.814 ± 0.008**	**0.190 ± 0.004**	**0.145 ± 0.004**
Late Fusion LSTM	0.813 ± 0.009	**0.190 ± 0.005**	**0.145 ± 0.004**
Intermediate Fusion LSTM	0.811 ± 0.008	0.192 ± 0.004	**0.145 ± 0.003**
Early Fusion LSTM	0.780 ± 0.012	0.207 ± 0.005	0.156 ± 0.004
LSTM (Climate)	0.781 ± 0.014	0.206 ± 0.007	0.153 ± 0.004
LSTM (Bio)	0.737 ± 0.017	0.226 ± 0.007	0.165 ± 0.005
DNN	0.732 ± 0.010	0.228 ± 0.004	0.166 ± 0.003
XGBoost	0.809 ± 0.000	0.193 ± 0.000	0.147 ± 0.000
Random Forest	0.791 ± 0.002	0.202 ± 0.001	0.150 ± 0.001
Bagging	0.792 ± 0.001	0.201 ± 0.001	0.149 ± 0.001
Linear Regression	0.731 ± 0.000	0.228 ± 0.000	0.167 ± 0.000
Modvege	−0.021	0.445	0.340

Beyond GC-LSTM, our results demonstrate that the effectiveness of multi-view modelling is highly sensitive to the chosen fusion strategy, particularly when faced with low-quality biophysical variables. Despite inaccuracies, biophysical variables still carry relevant signals for AGB, evidenced by the consistent performance of LSTM(Bio) and the strong resilience of Early Fusion LSTM in 2022, a year with potential climate events. However, early fusion is not universally beneficial: in 2021 and 2023, LSTM(Climate) outperformed Early Fusion LSTM, suggesting that when climate data is dominant, naïvely combining noisy biophysical variables can impair performance. Notably, Late Fusion LSTM surpassed LSTM(Climate) in 2021, and both Late and Intermediate Fusion LSTMs outperformed LSTM(Climate) in 2023—reinforcing that even with noisy biophysical variables, an appropriate fusion strategy can unlock performance gains.

Finally, the role of temporal dependency is evident in AGB forecasting performance. Across most cases, temporal fusion models outperform non-temporal baselines (i.e. DNN and traditional ML models). An exception occurred with Early Fusion LSTM in 2023, where tree-based models (XGBoost, Random Forest, and Bootstrapped Bagging) outperform it. Two potential reasons: (1) Early Fusion LSTM may not effectively cope with noisy biophysical variables in each "look-back" step, and (2) Tree-based models can exploit strong climate signals through inherent feature selection, effectively filtering out noisy biophysical inputs. Nevertheless, temporal fusion models with more robust architectures—such as Late and Intermediate Fusion LSTMs and especially GC-LSTM—continue to deliver superior performance.

Forecasting Curve Visualisation. We conducted a comparative case study over three years using best- and worst-performing paddocks to visually assess the strengths and limitations of GC-LSTM against other representative models. To provide a meaningful comparison, one model from four distinct categories was chosen each year: a top-performing fusion model, a non-temporal DL model (DNN), a top-performing traditional ML model, and Modvege. Accordingly, the models compared were: GC-LSTM, Late Fusion LSTM, DNN, Bootstrapped Bagging, and Modvege in 2021; GC-LSTM, Early Fusion LSTM, DNN, Bootstrapped Bagging, and Modvege in 2022; and GC-LSTM, Late Fusion LSTM, DNN, XGBoost, and Modvege in 2023, along with the original AGB values (Satellite-estimated AGB).

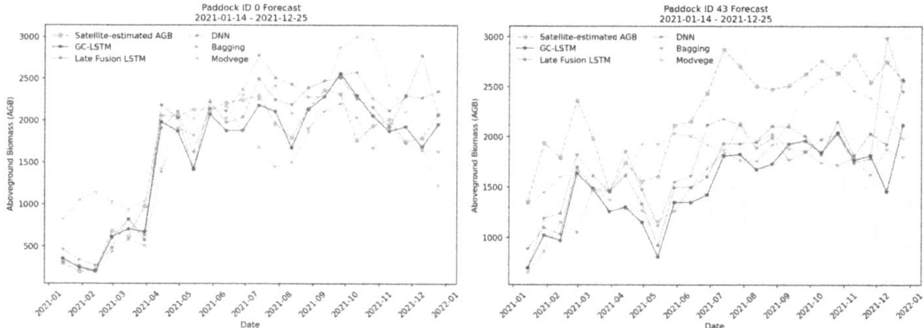

Fig. 3. Comparison of forecasting AGB for Paddock ID 0 (best-case) and Paddock ID 43 (worst-case) in 2021.

As shown in Figs. 3, 4, and 5, in the best-case scenarios, GC-LSTM consistently demonstrates strong robustness and higher performance across all three years. While temporal fusion models and traditional ML models perform similarly—consistent with their close evaluation metrics as reported in Tables 3, 4, 5, respectively—GC-LSTM more effectively captures subtle fluctuations in the data. For example, in 2021, its forecast for Paddock ID 0 closely follows the observed AGB from January to April and from mid-July to mid-September. Similarly, in 2022, for Paddock ID 62, GC-LSTM accurately models the growth trajectory from April to mid-June, including a key inflection in June missed by other models.

The worst-case analysis reveals two key limitations that suggest directions for improving GC-LSTM performance. First, a general decline in performance is observed across all models, including GC-LSTM, over the three years. Substantial deviations from observed AGB, such as those in Paddock ID 43 (2021) and Paddock ID 47 (2022), suggest that current climate and biophysical variables may be insufficient, potentially lacking crucial factors such as pasture-environment interactions or external factors like animal loads. Second, all models showed a short-term response lag, as evident in Paddock ID 74 (2023). While temporal models are designed to capture time dependencies [15], this lag suggests that current architectures may struggle to adapt to shifts in AGB immediately.

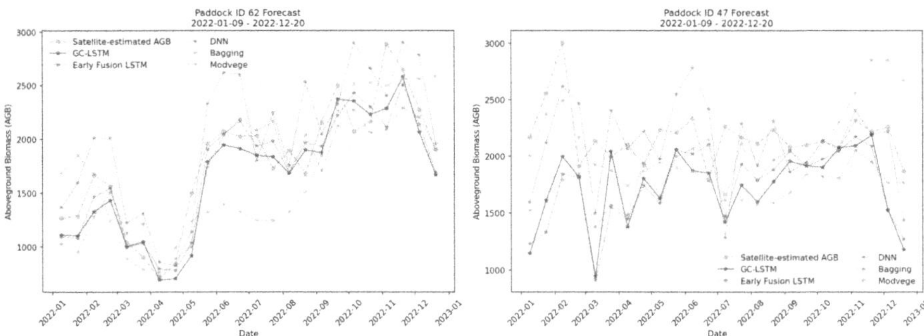

Fig. 4. Comparison of forecasting AGB for Paddock ID 62 (best-case) and Paddock ID 47 (worst-case) in 2022.

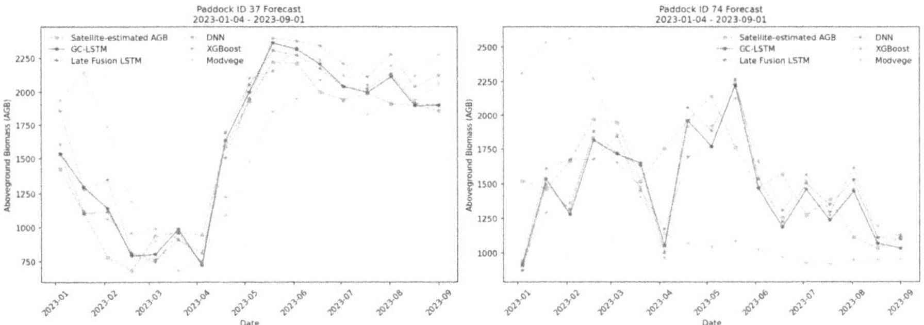

Fig. 5. Comparison of forecasting AGB for Paddock ID 37 (best-case) and Paddock ID 74 (worst-case) in 2023.

4 Conclusion

In this study, we addressed the limitations of existing AGB forecasting approaches by proposing GC-LSTM that effectively integrates climate data and Modvege-derived biophysical variables through an adaptive gating mechanism. Our results demonstrate that GC-LSTM delivers consistently superior performance and robustness across three independent test years (2021–2023) when compared to nine benchmark models. Its effectiveness is primarily attributed to its dual-branch architecture, which learns from distinct data streams, the use of shared weights for unified feature representation and a flexible gating mechanism that adapts to dynamic temporal patterns. Owing to its multi-year reliability and cost-efficient design, GC-LSTM presents a scalable and practical solution for operational pasture management across large numbers of paddocks.

Acknowledgments. This research was supported by funding from Food Agility CRC Ltd, Australia, funded under the Commonwealth Government CRC Program. The CRC Program supports industry-led collaborations between industry, researchers, and the community. This research was also partially supported by funding from Meat and Livestock Australia (MLA).

References

1. NSW Department of Primary Industries: Pastures and Rangelands. https://www.dpi.nsw.gov.au/agriculture/pastures-and-rangelands
2. Australian Bureau of Statistics: Agriculture and Livestock. https://www.abs.gov.au/statistics/industry/agriculture/australian-agriculture-livestock/latest-release
3. Defalque, G., Santos, R., Bungenstab, D., Echeverria, D., Dias, A., Defalque, C.: Machine learning models for dry matter and biomass estimates on cattle grazing systems. Comput. Electron. Agric. **216**, 108520 (2024)
4. Balasubramaniam, T., Mohotti, W.A., Sabir, K., Nayak, R.: Feature engineering on climate data with machine learning to understand time-lagging effects in pasture yield prediction. Eco. Inform. **86**, 103011 (2025)
5. Jouven, M., Carrère, P., Baumont, R.: Model predicting dynamics of biomass, structure and digestibility of herbage in managed permanent pastures. 1. Model description. Grass Forage Sci. **61**(2), 112–124 (2006)
6. Bosi, C., Huth, N.I., Sentelhas, P.C., Pezzopane, J.R.M.: APSIM model performance in simulating Piatã palisade grass growth and soil water in different positions of a silvopastoral system with eucalyptus. Agric. Syst. **195**, 103302 (2022)
7. Shi, Y., et al.: Deep learning meets process-based models: a hybrid approach to agricultural challenges. arXiv preprint arXiv:2504.16141 (2025)
8. Cobon, D.H., et al.: Evaluating the shifts in rainfall and pasture-growth variabilities across the pastoral zone of Australia during 1910–2010. Crop Pasture Sci. **70**(7), 634–647 (2019)
9. Shahhosseini, M., Hu, G., Huber, I., Archontoulis, S.V.: Coupling machine learning and crop modeling improves crop yield prediction in the US Corn Belt. Sci. Rep. **11**(1), 1606 (2021)
10. Kheir, A.M., et al.: Hybridization of process-based models, remote sensing, and machine learning for enhanced spatial predictions of wheat yield and quality. Comput. Electron. Agric. **234**, 110317 (2025)
11. Bai, H., Xiao, D., Tang, J., Liu, D.L.: Evaluation of wheat yield in North China Plain under extreme climate by coupling crop model with machine learning. Comput. Electron. Agric. **217**, 108651 (2024)
12. Sajid, S.S., Shahhosseini, M., Huber, I., Hu, G., Archontoulis, S.V.: County-scale crop yield prediction by integrating crop simulation with machine learning models. Front. Plant Sci. **13**, 1000224 (2022)
13. Dhillon, M.S., et al.: Integrating random forest and crop modeling improves crop yield prediction of winter wheat and oil seed rape. Front. Remote Sens. **3**, 1010978 (2022). https://doi.org/10.3389/frsen.2022.1010978
14. Shook, J., Gangopadhyay, T., Wu, L., Ganapathysubramanian, B., Sarkar, S., Singh, A.K.: Crop yield prediction integrating genotype and weather variables using deep learning. PLoS ONE **16**(6), e0252402 (2021)
15. Goodfellow, I., Bengio, Y., Courville, A.: Deep Learning. MIT Press, Cambridge, MA, USA (2016)
16. SILO: Australian climate data from 1889 to present. Queensland Government. https://www.longpaddock.qld.gov.au/silo/

17. Ogungbuyi, M.G., et al.: Enabling regenerative agriculture using remote sensing and machine learning. Land **12**(6), 1142 (2023)
18. Kiniry, J.R., Macdonald, J.D., Kemanian, A.R., Watson, B., Putz, G., Prepas, E.E.: Plant growth simulation for landscape-scale hydrological modelling. Hydrol. Sci. J. **53**(5), 1030–1042 (2008)
19. ASCE Task Committee on Standardization of Reference Evapotranspiration: ASCE's Standardized Reference Evapotranspiration Equation. In: Proceedings of the National Irrigation Symposium, pp. 1–10. Phoenix, USA (2000)
20. McLeod, M.K., MacLeod, D.A., Daniel, H.: Effect of degradation of phalaris+white clover pasture on soil water regimes of a Brown Chromosol in northern NSW, Australia. Agric. Water Manag. **82**(3), 318–342 (2006)
21. Akiba, T., Sano, S., Yanase, T., Ohta, T., Koyama, M.: Optuna: a next-generation hyperparameter optimization framework. In: Proceedings of the KDD 2019, pp. 2623–2631. ACM, New York (2019). https://doi.org/10.1145/3292500.3330701

Precision to Costing: Budgeted Modelling for Customer Contact Prediction

Ho Sing Tin[(✉)] [iD], Richi Nayak [iD], and Darren Wraith [iD]

Centre for Data Science, School of Computer Science, Queensland University of Technology, Brisbane, QLD 4000, Australia
`Hosing.tin@hdr.qut.edu.au`, `{r.nayak,d.wraith}@qut.edu.au`

Abstract. We predict weekly telephony contact for Centrelink customers using longitudinal administrative data aligned to a weekly grid, comparing statistical, machine learning, and deep learning models on 889,879 customers (150 features; 3.65M observations after class balancing; 5-fold CV). Feature-engineering regimes spanned from quick encodings to contact-history windows of 1, 4, 8, and 12 weeks; an 8-week history yielded the best trade-off, with diminishing returns beyond 8 weeks. Across 12 models, a Multilayer Perceptron attained the strongest performance on held-out data (AUC $\approx$ 0.84; accuracy $\approx$ 0.79). One-factor-at-a-time hyperparameter searches produced modest, non-monotonic gains, underscoring that additional tuning effort does not necessarily translate to uplift in model performance. To guide modelling investment and effort decisions, we introduced a cost-based budgeting equation associated with treatment applications of the predictive model. This framework approach encourages monetary costings to be considered alongside predictive performance for model development.

Keywords: Customer · Telephony · Prediction · Costing

1 Introduction

Customer contact is a pivotal part of service delivery for customer-facing industries. Customer experience (CX) hinges on the quality of the interaction between the customer and the business. Higher quality and shorter service wait-times can bolster brand loyalty [19], increase customer satisfaction [29] and lower churn rates [10]. In today's digital world, the telephony option remains a popular channel for contacting a business [6]. Telephony support offers the value of human connection, personalisation and higher levels of attention, which other digital channels often lack [11]. Emailing and online web chats often also redirect customers to an automated response [18]. Furthermore, complex and urgent issues often end up eventually requiring live human-to-human telephony resolutions [16]. As such, customer-facing businesses place great interest and strategic value in understanding and predicting customer telephony contact.

Services Australia (the Agency) is an Australian Federal Government Agency specialised in welfare payment delivery, child support payment, public health insurance (Medicare and healthcare) and aged care assistance [22]. The Agency is one of these

Q. V. Nguyen et al. (Eds.): AusDM 2025, CCIS 2765, pp. 208–223, 2026.
https://doi.org/10.1007/978-981-95-6786-7_14

businesses that depend greatly on analysing and understanding customer telephony contact. Receiving up to 55 million calls a year [23], improvements to the Agency's telephony business process to address traffic makes a substantial impact on the quality of service delivery. Knowing the likelihood of a customer calling can help the Agency identify business bottlenecks [31], optimise processes [10], create data-driven decision-making and enhance evidence-based outcomes [10]. From the customer perspective, streamlined telephony services can enrich their experience, increase satisfaction and boost public sentiment [19, 29] of the Australian Government. Predictive modelling is one of the tools that can help the Agency achieve these goals.

Predictive Models not purposed to provide a solution to a real-world business problem often do not have the added complexity of a monetary budget and tight time constraints. Whereas deployment predictive models prioritise the allowable cost budget and time constraints that the business is willing to allow for the modelling stages [24].

Model development and optimisation processes for uplifting model performance are well-documented in guides and academic literature. However, there is uncertainty that additional effort and time placed on the building stage of a predictive model can substantially improve predictive power or model performance. Few definitive, practical guides exist to help businesses rigorously assess how much money and time to invest into developing and optimising models for deployment [12] and when to stop. One such way of addressing this is taking into consideration the intended intervention (*treatment*) that the business intends to apply with the use of the predictive model. Business value can be generated by considering the cost-savings of utilising the predictive model through projected treatment application compared to a scenario without the model.

Consequently, decisions about how much time and budget to invest in improving a model's performance should weigh not only technical evaluation metrics, but also the business value. This paper explores the reality of predictive modelling on real administrative longitudinal data from the Agency with the following novel contributions:

(1) Presents data-based evidence that predicting a customer's likelihood to call in the following week with statistical, machine learning (ML) and deep learning (DL) predictive models can be achieved.
(2) Presents a series of empirical methods that show how a real-world predictive model's performance changes with feature engineering, model evaluation and hyperparameter tuning.
(3) Presents a novel quantifiable and reproducible equation for businesses to assess and evaluate how much time and money they should invest into further developing or optimising a model for more predictive power.

2 Related Work

2.1 Customer Contact Analytics

Predicting customer behaviour is a common data science problem in research within the business sector. To understand and anticipate the actions of customers, Kirui et al. [10] utilised probabilistic classifiers to predict customer churn in the telecommunications industry. Being able to pinpoint and identify customer churn pre-emptively, the business was able to effectively address sources of customer dissatisfaction and improve customer retention rate. Doing so allowed cost reductions with acquiring customers and retaining them in a more efficient and effective way.

Similarly, McKinsey [15] illustrated that advanced analytics can transform call centres into providing personalised and tailored customer service through predictive models. By predicting the behaviour of their customers, the staff were able to proactively address customer needs, resulting in reduced handling time, increased self-service initiatives, reduced employee costs and even further boosted the number of profitable deals being secured. The better that a business understands its customers and their behaviour, the better it can interpret and leverage complex interactions with its customers to their advantage [28].

2.2 Model Development and Optimisation

There are many documented methods to develop a predictive model and optimise it to uplift its predictive power. Feature engineering is the process of creating informative features by transforming, aggregating or manipulating the raw data. This process makes it easier for the model to identify and learn from the patterns. Effective feature engineering can drive most of the predictive power uplift in many domains [5, 15, 32].

Another well-researched approach is the evaluation of various types of predictive models, especially with longitudinal administrative data [5]. This is particularly true in the medical field due to the longitudinal nature and mass availability of EHR [30, 34]. For instance, Zhao et al. [35] discussed the prediction of survival in glioma patients using ML on available large-scale real-world clinical data. Using data over the course of 20 years, the study explored the effectiveness of three ML approaches (Cox proportional hazards model, support vector machine and random forest) in modelling survival outcomes. The study had found that all models had comparably good performance.

Hyperparameter tuning is another methodology that can improve the performance of classification predictive models. It achieves this by customising the parameters of a model towards the specific dataset. The practitioner specifies hyperparameters during the model training and validation stages. Hyperparameters are adjustments to the model, akin to changing the settings on a machine [25]. These adjustments are put in place before the model training, unlike parameters that are determined or learned during the model training phase such as the weights in an artificial neural network (ANN) or coefficients in regression models [3]. Model generalisability can be enhanced by different hyperparameter settings. This is seen in a basic logistic regression model in which the penalty strength (C parameter) can be controlled to improve the model's performance [3]. This practice is well-established and widely recognised by data professionals globally as a fundamental process in the field of data science [3, 34].

2.3 Cost-Based Analysis

Technical measurements are effective and provide clear results on model performances. However, in real world applications, especially in terms of customer servicing, considering the monetary costs of misclassification is equally as important for model deployment [9]. Cost-benefits are a major consideration in model deployment. Research into cost-savings which can be generated from modelling customer/client behaviour in enterprise settings is well-documented (even with real-world datasets [20, 34]), However, this area remains relatively underexplored in academic research compared to the extensive literature on patient outcomes in a healthcare/medical environment [9, 20]. The adoption of cost-benefit into the model build stage is also often hindered by the need for tailored approaches that consider the unique aspects and context of the business problem [9].

Weiner et al. [33] described that accuracy and technical measures are just one consideration and that the prevalence and cost of the undesirable outcome, and the effectiveness and cost of the planned interventions are also crucial elements for business value. Through these measures, the authors illustrated that despite a predictive model presenting with 0.8 AUC in identifying patients with Congestive Heart Failure (CHF), it was more financially favourable to simply apply the treatment to all patients admitted without the use of the predictive model at all. This idea of utilising the most preferable course of treatment application is expanded upon in this study in an administrative and project-based context.

3 Problem Formulation

With Services Australia's strong focus on service delivery and CX, early identification and prediction of repeat callers can support and streamline service delivery. We propose to investigate the capability of utilising statistical modelling, ML and DL in predicting the likelihood of a customer contacting the Agency via telephony services in the following week of the point-of-time of the prediction being made. We explore model evaluation, feature engineering and hyperparameter tuning to make this binary prediction. We assess whether using this model in applying the intended treatments can result in business value in the form of money savings for the Agency. This business problem can be framed in 2 parts: (1) How likely will a customer contact Services Australia by through the telephony channel in the following week? And (2) How can the Agency assess how much time and money to invest in developing and optimising this model?

4 The Proposed Solution

We follow the CRISP-DM methodology [24] to utilise the Agency's existing data assets to build predictive models for customer contact. The details of the data, methodologies and procedures are detailed in the sections below.

4.1 Business Understanding

The population of focus was Centrelink customers, a mix of ongoing and newly granted recipients. In collaboration with subject matter experts (SMEs) and data specialists from Services Australia, certain strong predictors of customer telephony contact were identified. The identified constructs were customer demographics, customer and Agency contact history, customer earnings, customer medical details, customer vulnerability details, customer benefit status, customer claimant history, customer urgent/crisis payment history and customer address history. Features related to these constructs were collected for modelling purposes.

4.2 Data Acquisition

Dataset extraction was performed internally within the Agency, where the process involved quality assurance and encrypted de-identification of customer details.

4.3 Data Preparation

The population contained 889,879 unique customers. Extraction also included 239 raw features across pertaining to each construct mentioned above. Each feature underwent correlation and independent analyses to determine its relevance and type. Throughout this process, a thorough understanding behind the properties of the features within each dataset were examined. Missing values were either removed or imputed based on the context and domain expertise. Distributions and skewness were rigorously examined through assessing the univariate statistics. Features were transformed using methods including log transformation, square-root transformation and Box-Cox transformations where necessary to normalise data distributions to ensure the robustness of the subsequent analyses and modelling results. Further guided by data experts and SMEs, nonsensical outliers were removed.

Further preprocessing included duplicate instances investigation, data transformation, discretisation, one-hot encoding, frequency encoding, label/binary encoding and collinearity/correlation analysis. Further data validity checks omitted customers who had deceased, were medically incapacitated, or went overseas within the timespan of the collected data. Feature interaction terms were created through aggregating certain contact types together and grouping them with the channel of contact (see Fig. 1). This feature engineering had resulted in thousands of newly engineered interaction features. Model stacking was used to select the top 150 features for the final model training data.

For further preparation of modelling, the data was transformed into a weekly view to align the data to a common time grid for the purpose of predicting a binary outcome (Y/N) of customer telephone contact in the following week. This was considered a suitable timeframe by SMEs for executive business decisions to be made.

An under-sampling approach to the data imbalance was also applied. The raw dataset had initially contained 35,854,451 (94.32%) instances of non-contact versus 2,157,809 (5.68%) instances of contact in the following week. Due to the largely imbalanced cohorts, the non-contacts group was under-sampled to match the number of contact group. After data-preprocessing and balancing, the final training dataset consisted of 150 features and 3,647,524 observations, of which 1,837,682 (50.38%) were non-contact target outcomes and 1,809,842 (49.62%) were contact target outcomes. The dataset was split into 80% training data and 20% test data and subjected to a 5-fold cross-validation.

4.4 Assumptions and Limitations

To address concept drift from policy and business procedural changes, domain knowledge from SMEs were used to update older variations of certain categories to newer terms and labels to match the more recent consistency in data labelling across the datasets.

4.5 Feature Engineering

We implemented 5 different feature engineering approaches where each subsequent approach requires a higher degree of effort, data manipulation and computing power. The 5 feature engineering approaches were:

1. Quick Feature Engineering – Simple one-hot encoding for the presence of certain features and where applicable a frequency count for all features in temporal windows of 3 monthly intervals.
1. In-depth Contact History Feature Engineering (1-week) – This approach included the quick feature engineering of all non-contact constructs and an in-depth feature engineering of the three constructs related to contacts. It involved the usage of previous internal research and collaboration with SMEs. With domain knowledge, feature interaction terms were created through aggregating certain contact types together and grouping them with the channel of contact. This resulted in 103 new features from just the contacts constructs with frequency counts of each of these features. In this feature engineering option, each instance was populated with a 1-week historical timeframe of these new features. A visual example of this feature engineering can be seen in Fig. 1.
2. In-depth Contact History Feature Engineering (4-weeks) – same as approach (2), except expanded to 4 weeks prior the target outcome.
3. In-depth Contact History Feature Engineering (8-weeks) – same as approach (2), except expanded to 8 weeks prior the target outcome.
4. In-depth Contact History Feature Engineering (12-weeks) – same as approach (2), except expanded to 12 weeks prior the target outcome.

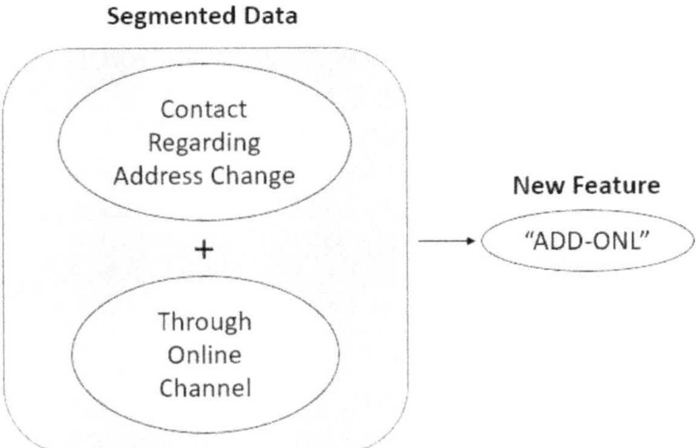

Fig. 1. An example of an interaction feature created from the feature engineering process

These time intervals for the feature engineering comparison were chosen due to point-biserial correlation analyses performed between the historical count of weekly contacts and the target outcome of customer contact in the following week. The strength of the correlations drops off significantly starting from week 13 onwards (see Fig. 2).

After feature engineering, feature selection was performed through model stacking where a random forest classification model was built first on all the features and the top 150 features based on Gini Coefficient were selected from each respective dataset for final modelling. This was repeated for all 5 feature engineering approaches.

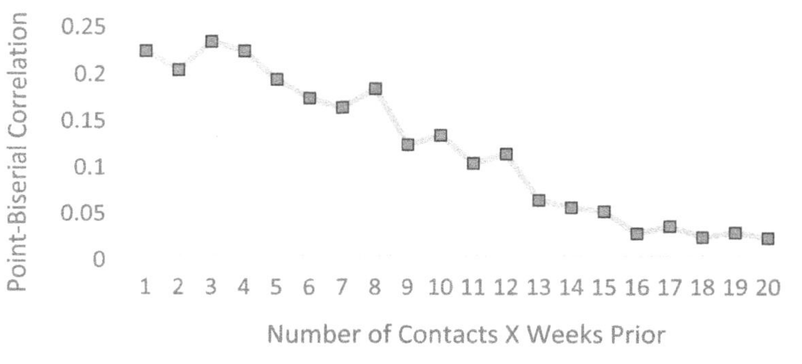

Fig. 2. Point-biserial correlations between the number of contacts that customer made in the prior weeks and the customer contacting the following week. The correlation tapers off as the number of contacts made by the customer is more impactful when recent. All p-values < 0.001.

4.6 Modelling

In this research, a comprehensive suite of models was evaluated spanning statistical, ML and DL models. Each was selected for its unique strengths in handling different aspects of predictive analytics. The statistical models chosen for evaluation were Naïve Bayes Classifier (NB), Logistic Regression (LR), Linear Discriminant Analysis (LDA) and Probit Regression (PR). The ML models chosen were Decision Tree (DT), Logistic Regression (LR), Neural Network (NN), Random Forest (RF), and XGBoost (XGB). And the DL models chosen were Multilayer Perceptron (MLP), Convolutional Neural Network (CNN) and Wide & Deep Learning (WDL). These models were chosen for a balance of their varying degrees of interpretability and complexity, reflecting a blend of traditional statistical inference and modern ML efficacy.

4.7 Hyperparameter Tuning

From the model evaluation, the highest AUC-performing model was chosen for hyperparameter tuning. Hyperparameters were manually grid searched with the intention to observe the impact that they have on the champion predictive model's AUC score on a test dataset.

5   Results and Evaluation

5.1   Feature Engineering

The modelling techniques that were used to assess the feature engineering efficacy were the 5 most used in the Agency: DT, LR, NN, RF and XGB. The mean AUC score from the cross-validation for each model for each of the 5 feature engineering scenarios can be examined and compared below in Table 1.

Table 1. Table of Feature Engineering Results

Feature Engineering	NN	XGB	RF	LR	DT
Quick Feature Engineering	0.7051	0.7041	0.6932	0.6840	0.5989
In-depth 1-Week History Contacts	0.7236	0.7255	0.7131	0.7044	0.6601
In-depth 4-Week History Contacts	0.8244	0.8163	0.8028	0.8041	0.6678
In-depth 8-Week History Contacts	0.8336	0.8378	0.8249	0.8151	0.6753
In-depth 12-Week History Contacts	0.8321	0.8333	0.8228	0.8137	0.6689

From the mean AUC scores, the best performing dataset was the one which underwent feature engineering for 8 weeks of historical contact data (see Fig. 3 for ROC curves).

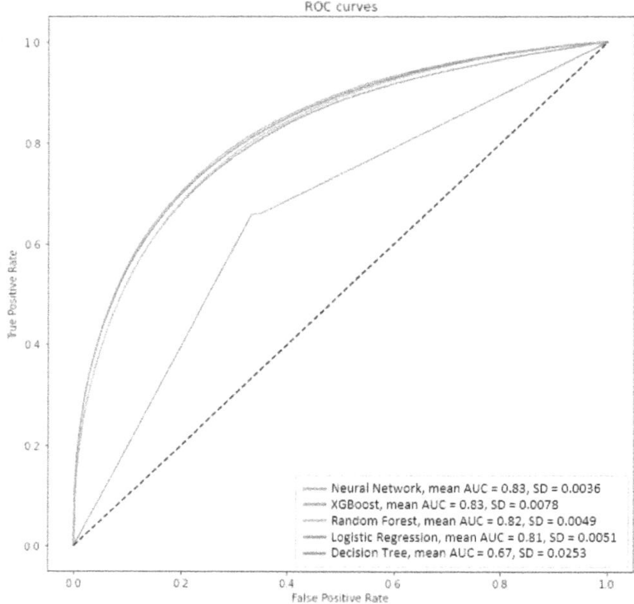

Fig. 3. ROC Curves – In-depth 8-week contact history feature engineering

A ceiling effect was observed where providing the model with a further 4 more historical weeks of contacts data (expanding to 12 weeks) did not further improve model performances. This is an indication that, at a certain point, the mere availability of more data may not equate to more accurate or better performing models. Subsequently, the dataset which was selected for the next model evaluation in the next step was the feature engineering approach option consisting of in-depth 8-weeks of contact history. This result matched with the correlation effect where contacts and feature approached a ceiling affect nearing 13 weeks (see Fig. 2).

5.2 Prediction Performance

Table 2 reports the performance of each model on an unseen test set of 199,422 customers: accuracy, ROC-AUC, precision, recall, and F1 score.

Table 2. Comparison Table of Statistical vs ML vs DL Models in Predicting Customer Contact

Type	Model	Accuracy	AUC	Precision	Recall	F1-Score
Statistical	NB	0.6193	0.6112	0.61	0.62	0.61
	LR	0.7344	0.8132	0.79	0.64	0.70
	LDA	0.6586	0.6481	0.67	0.60	0.64
	PR	0.6275	0.6363	0.62	0.63	0.62

(continued)

Table 2. (*continued*)

Type	Model	Accuracy	AUC	Precision	Recall	F1-Score
Machine Learning	DT	0.6612	0.6519	0.66	0.66	0.67
	RF	0.7422	0.8142	0.75	0.71	*0.75*
	SVM	0.7393	0.8082	0.73	0.72	0.73
	XGB	*0.7438*	*0.8241*	*0.78*	0.69	0.73
	NN	<u>0.7512</u>	<u>0.8281</u>	0.76	<u>0.74</u>	<u>0.75</u>
Deep Learning	MLP	**0.7910**	**0.8404**	**0.81**	**0.75**	**0.78**
	CNN	0.7108	0.7865	0.74	0.69	0.73
	WDL	0.7349	0.7942	0.76	*0.73*	0.75

Note. The top performing models in each metric are bolded, underlined and italicised in that order.

Overall, the DL model MLP performed better than the other models examined. Some of the statistical models provide for enhanced interpretability and transparency, whilst ML and DL techniques have the potential to utilise large datasets more efficiently, improving predictive capabilities [27]. Despite this, the complexity and vastness of Services Australia's data pose challenges for DL models, including computational intensity and a lack of minimal transparency, often required for bureaucratic purposes. Moreover, most DL models are not specifically designed for handling large-scale administrative longitudinal tabular data [7], as evidenced by the subpar performance of the CNN model compared to traditional ML models. Following model evaluation and selection conventions outlined above, the MLP model was selected to be the champion model for hyperparameter tuning experimentation.

5.3 Hyperparameter Tuning

The following hyperparameters for the MLP were taken into consideration for this manual grid search: batch size, hidden layers, hidden layer nodes, weight initialisation, regularisation, epoch count, learning rate and loss functions (see Table 3). All hyperparameter modifications were performed on the baseline default hyperparameter settings. When one hyperparameter was being investigated, the rest of the model hyperparameters were defaulted back to the original settings to observe the single effect of that one specific tweak. The interactions of the different hyperparameters may be reserved for future research due to a limited amount of computational power, current time-constraints and the current task not being an exhaustive purpose of searching for the very best model. It is evident that more time and effort expenditure onto this task may not necessarily produce higher modelling performance.

Table 3. Table of Hyperparameter Tuning Results on Model AUC

Hyper Parameter Type	Setting	AUC
Batch Size	32 (Default)	0.8397
	64	0.8387
	128	0.8355
	256	0.8341
Hidden Layers	1 (Default)	0.8397
	2	0.8409
	3	0.8439
	4	*0.8485*
	5	0.8401
	6	0.8382
Hidden Layer Nodes	10 (Default)	0.8397
	50	0.8412
	100	0.8402
	150	0.8356
Weight Initialisation	Zero/Constant Initialisation	0.8137
	Random Initialisation (Default)	0.8397
	Xavier Initialisation	0.8416
	LeCun Initialisation	0.8404
Regularisation	L1	0.8384
	L2 (Weight Decay) (Default)	0.8397
	Dropout	0.8399
	Early Epoch Stopping	0.8397
Epoch Count	5	0.8211
	20	0.8397
	100 (Default)	0.8399
	Early Epoch Stopping	0.8397
Learning Rate	0.1	0.8291
	0.01	0.8431
	0.001 (Default)	0.8397
	0.0001	0.8368
Loss Function	Binary Cross-Entropy (Default)	0.8397
	Logistic Loss	0.8388
	Hinge Loss	0.8302

(continued)

Table 3. (*continued*)

Hyper Parameter Type	Setting	AUC
	Squared Hinge Loss	0.8241

5.4 Business Effort and Investment Expenditure

These model development and optimisation processes conducted in the previous sections are well-documented, However, only partial answers are available for guiding businesses towards budgeted modelling, early-stopping of model development, cost-sensitive evaluations and monetary deployment analyses. The previous sections accentuated this view in which how various model development and optimisation activities can change and potentially improve the model performance, but there is no guarantee that more effort expended can result in these outcomes, especially hyperparameter tuning. To assess and evaluate continued monetary investment into a modelling project, we introduce the following equation for calculating recommended model cost budget.

Firstly, we define the following parameters:

- B: recommended model cost budget (currency)
- N_t: number of treatments applied (count)
- r: baseline/random rate of target event (fraction)
- p: model precision (fraction)
- c_e: cost of the target event occurrence (currency)
- M: model maintenance cost rate (currency)
- T_{sel}: the selected operating time-window (time-unit: days, weeks, etc.)
- T_{rec}: the max acceptable length of time to recover initial model build costs in units of T_{sel}

Then, to derive B (recommended model cost budget), we propose the equation:

$$((1 - r) * \left(\frac{N_t}{T_{sel}} \right) * c_e)) - ((1 - p) * \left(\frac{N_t}{T_{sel}} \right) * c\_e) - \left(\frac{M}{T_{sel}} \right)) * T_{rec} \qquad (1)$$

We assume in this scenario that the treatment is effective in nullifying the target event. We use the current case study in this equation as an example. For the purposes of this evaluation, assume that the Agency intends to apply a treatment where the customer's claim is escalated to immediate processing/finalisation. This is expected to avert any incoming phone calls (target event) in the following week from the customers who received this treatment. The Agency can use predictions from the predictive model to select who receives the treatment. Using the equation, we can calculate the savings from the treatment application compared with not having a model available. The champion MLP model from Sect. 5.2, where precision was 0.81, will be used in this example. The assumptions and business metrics for this scenario are given in Table 4 (see Appendix A for calculations).

Table 4. Table of Business Metrics & Assumptions

Item	Value
Selected Time-Frame	Weekly
Average Cost of Telephony Call (Target Event)	$9.33
Number of Treatments to be Applied/Selected Time-Frame	2000
Hourly Cost of Data Scientist	$66.07
Max Model Cost Recovery Time/Selected Time-Frame	8 (Weeks)
Maintenance Cost of Model/Selected Time Time-Frame	$495 (Weekly)

To derive the recommended model cost budget, we substitute the values into Eq. (1):

$$((1 - 0.4962) * (2000) * 9.33)) - ((1 - 0.81) * (2000) * 9.33) - (495)) * 8$$

Therefore, the recommended model cost budget for this model is $42,859.32. Assuming if the only expenditure in this scenario for the model build is only the hourly cost of allocating a data scientist to this task, we can then translate this into allowable time for model development and optimisation. If the quota of the cost of the data scientist is not depleted, then there is now a quantified allowable extra amount of time for the business to further develop the model and perform optimisation tasks. Additional savings generated by better model performance can also be used to inform the budget to allow for further optimisation (see Table 5).

In the event where the business wishes to invest in further model development and optimisation to uplift the current precision by 0.05 to 0.86, it will need to be accomplished within an extra $7460.69 budget or within 112.9 h of the data scientist's time to be cost-worthy based on the current assumptions. This allows the business to make a calculated and data-driven decision on whether this venture is feasible and worth attempting. Higher performing models allow more savings for the business and more time for further optimisation.

Table 5. Table of Precision-Informed Cost and Time Budget for Modelling

Model Precision	Recommended Model Cost Budget ($)	Translated time allowance for modelling (Hrs)
0.96	$65,241.41	987.50
0.91	$57,780.71	874.58
0.86	$50,320.01	761.65
0.81 (Current)	$42,859.32	648.73
0.76	$35,398.62	535.80
0.71	$27,937.92	422.87
0.66	$20,477.23	309.95
0.61	$13,016.53	197.02

This preliminary calculation can help guide the business to make educated and sensible cut-off decisions in terms of when to stop trying to uplift model performance. It helps the business to determine if there is any budget or leeway in further developing the model performance. This applied approach can also inform preliminary model build performance benchmarks. For instance, if the business is willing to only invest a max budget of $50,320. Then, in this current example, a precision of 0.86 would be needed for this model to be a worthwhile venture based on the current assumptions (See Table 5). Additionally, if the current budget has already exceeded the recommended model cost budget output, the business will need to increase their acceptable amount of time-in-loss expectations or decide to invest more into the project.

6 Conclusion

This study demonstrates that weekly prediction of customer telephony contact is both feasible and practically valuable. By transforming longitudinal administrative data into an aligned weekly grid, engineering interaction-rich features over an 8-week horizon and evaluating a broad family of statistical, ML and DL models, we found that a Multi-layer Perceptron achieved the strongest AUC (~0.84). This illustrated that model performance can vary greatly depending on the amount of time and effort invested into these model development and optimisation tasks. Additionally, predicting customer contact yielded diminishing returns beyond 8 weeks of contact data history, suggesting that recent contacts are a strong predictor of subsequent contacts.

We further illustrated that technical performance alone is not a stable guide for businesses to quantify model investment. We introduced a practical and reproducible applied approach quantify this aspect. The introduced approach connected the technical measure of precision into tangible business value informing the Agency of potential savings and cost budget scenarios.

The ultimate goal of this prediction remains to help the Agency understand and predict customer contact. Subsequently, this will allow for strategies addressing the telephony traffic volumes to be implemented to uplift business efficacy and customer satisfaction through enhanced treatment selection. Beyond the Agency, the findings in this paper offer actionable insights for sectors highly invested in customer contact.

6.1 Limitations and Future Work

In practicality, model deployment can hinge on manual resources available, technical resources available to maintain the model, staffing and personnel resources, business process bottle necks, volume of work, technical uptake and even in certain cases, ethical considerations of the model usage. None of which are considered in the scenarios within this paper. Assumptions are also made that the treatment with utilising the model will be effective. Additionally, transferring concepts from this paper into another context could potentially be challenging due to unique cost matrices with treatments. Furthermore, this paper only examined model expenditure simply as the cost of a data scientist hire, but in the real world there is infrastructure, data costs, platform-related costs and even unforeseen delays. Future work will include consideration of multiple treatments, differing treatment costs and success-rates of the intended treatment.

Disclosure of Interests. Author Ho Sing Tin is a PhD student at Queensland University of Technology. This research is also supported by Services Australia.

Appendix A

- Telephony Cost = avg hourly APS3/4 staff pay ($40.69) * avg call duration (0.23 h)
- Hourly cost of Data Scientist = avg hourly EL1 staff pay ($66.07)
- Maintenance Cost of Model = 0.2 FTE pay of hourly pay of Data Scientist (7.5 h * $66.07)
- Cost for each hour of Model Development = hourly pay of Data Scientist ($66.07)

References

1. Ascarza, E.: Retention futility: targeting high-risk customers might be less profitable than targeting customers with uncertain churn propensity. J. Mark. Res. **55**(1), 65–83 (2018)
2. Braha, D., Shmilovici, A.: Data mining for improving a cleaning process in the semiconductor industry. IEEE Trans. Semicond. Manuf. **15**(1), 91–101 (2002)
3. Machine Learning Mastery. https://machinelearningmastery.com/difference-between-a-par ameter-and-a-hyperparameter/. Accessed 19 Aug 2025
4. Carrasco-Ribelles, L.A., et al.: Prediction models using artificial intelligence and longitudinal data from electronic health records: a systematic methodological review. J. Am. Med. Inform. Assoc. **30**(12), 2072–2082 (2023)
5. Cheng, H.-T., et al.: Wide & deep learning for recommender systems. arXiv:1606.07792 (2016)
6. Genesys. https://www.genesys.com/en-gb/resources/contactbabel-report-the-uk-customer-experience-decision-makers-guide-2024-25. Accessed 12 Aug 2025
7. Gorishniy, Y., Rubachev, I., Khrulkov, V., Babenko, A.: Revisiting deep learning models for tabular data. arXiv:2106.11959 (2021)
8. Han, K.: Fixing the c parameter in the three-parameter logistic model. Pract. Assess. Res. Eval. **17** (2012)
9. He, H., Ma, Y.: Imbalanced Learning: Foundations, Algorithms, and Applications. Wiley-IEEE Press (2013)
10. Kirui, C., Hong, L., Cheruiyot, W., Kirui, H.: Predicting customer churn in mobile telephony industry. IJCSI Int. J. Comput. Sci. Issues **10**(2) (2013)
11. Kraus, M.W.: Voice-only communication enhances empathic accuracy. Am. Psychol. **72**(7), 644–654 (2017)
12. Kreuzberger, P., Kühl, N., Hirschl, S.: Machine learning operations (MLOps): a systematic literature review. J. Syst. Softw. (2023)
13. Kumar, V., Garg, M.: Predictive analytics: a review of trends and techniques. Int. J. Comput. Appl. **182**, 31–37 (2018)
14. Liu, B., et al.: AutoFIS: automatic feature interaction selection in factorization models for CTR prediction. In: Proceedings of the 26th ACM SIGKDD International Conference on Knowledge Discovery and Data Mining (KDD 2020) (2020)

15. McKinsey. https://www.mckinsey.com/capabilities/operations/our-insights/how-advanced-analytics-can-help-contact-centers-put-the-customer-first. Accessed 20 Aug 2025
16. Meuter, M.L., Ostrom, A.L., Roundtree, R.I., Bitner, M.J.: Self-service technologies: understanding customer satisfaction with technology-based service encounters. J. Mark. **64**(3), 50–64 (2000)
17. Moro, S., Cortez, P., Rita, P.: A data-driven approach to predict the success of bank telemarketing. Decis. Support. Syst. **62**, 22–31 (2014)
18. NTT Ltd. https://services.global.ntt/en-us/campaigns/2023-global-customer-experience-report. Accessed 12 Aug 2025
19. Pei, X.-L., Guo, J.-N., Wu, T.-J., Zhou, W.-X., Yeh, S.-P.: Does the effect of customer experience on customer satisfaction create a sustainable competitive advantage? a comparative study of different shopping situations. Sustainability **12**(18), 7436 (2020)
20. Pes, B., Lai, G.: Cost-sensitive learning strategies for high-dimensional and imbalanced data: a comparative study. PeerJ Comput. Sci. **7**, e832 (2021)
21. Reif, D.M., Motsinger, A.A., McKinney, B.A., et al.: Feature selection using a random forests classifier for the integrated analysis of multiple data types. In: 2006 IEEE Symposium on Computational Intelligence and Bioinformatics and Computational Biology (CIBCB) (2006)
22. Services Australia. https://www.servicesaustralia.gov.au/guide-to-australian-government-payments. Accessed 20 Aug 2025
23. Services Australia. https://www.servicesaustralia.gov.au/annual-report-2023-24. Accessed 20 Aug 2025
24. Shearer, C.: The CRISP-DM model: the new blueprint for data mining. J.Data Warehous. **5**(4), 13–22 (2000)
25. Suto, J.: The effect of hyperparameter search on artificial neural network in human activity recognition. Open Comput. Sci. **11**(1), 411–422 (2021)
26. Tan, M., Hatef, E., Taghipour, D., et al.: Including social and behavioral determinants in predictive models: trends, challenges, and opportunities. JMIR Med. Inform. (2020)
27. Taye, M.M.: Understanding of machine learning with deep learning: architectures, workflow, applications and future directions. Computers **12**, 91 (2023)
28. Taylor, P.N.: Customer contact journey prediction. In: Lecture Notes in Computer Science, pp. 278–290. Springer (2017)
29. Thompson, J.: Faster resolution, happier customers: the role of machine learning in customer service. HubSpot Blog. https://blog.hubspot.com/service/machine-learning-customer-service. Accessed 17 Aug 2025
30. Tian, Y.E., Cropley, V., Maier, A.B., Lautenschlager, N.T., Breakspear, M., Zalesky, A.: Heterogeneous aging across multiple organ systems and prediction of chronic disease and mortality. Nat. Med. **29**(5), 1221–1231 (2023)
31. Van Calster, B., Wynants, L., Timmerman, D., Steyerberg, E.W., Collins, G.S.: Predictive analytics in health care: how can we know it works? J. Am. Med. Inform. Assoc. **26**(12), 1651–1654 (2019)
32. Wang, R., et al.: DCN V2: Improved deep & cross network and practical lessons for web-scale learning to rank systems. In: Proceedings of the Web Conference 2021 (WWW 2021) (2021)
33. Weiner, M.G., Sheikh, W., Lehmann, H.P.: Interactive cost-benefit analysis: providing real-world financial context to predictive analytics. In: AMIA Annual Symposium Proceedings 2018, pp. 1076–1083 (2018)
34. Wong, J., Murray Horwitz, M., Zhou, L., Toh, S.: Using machine learning to identify health outcomes from electronic health record data. Curr. Epidemiol. Rep. **5**(4), 331–342 (2018)
35. Zhao, Y., Wong, Z.S.-Y., Tsui, K.L.: A framework of rebalancing imbalanced healthcare data for rare events' classification: a case of look-alike sound-alike mix-up incident detection. J.Healthc. Eng. **2018**, 1–11 (2018)

Defining Responsible AI: Contextual Insights Powered by LLMs

Lucky Atamhenwan[(✉)] ⓘ, Sangeetha Kutty ⓘ, and Lily D. Li ⓘ

Central Queensland University, Brisbane, QLD 4001, Australia
`ed.atamhenwan@outlook.com, {s.kutty,l.li}@cqu.edu.au`

Abstract. The increasing popularity of the concept of responsible AI has sparked discussions about it in various contexts. The terms accountability, ethicality, transparency and fairness have been repeatedly discussed as elements of responsible AI. Within the discussions, the definition of the concept of responsible AI has been broadly applied. Considering the variety of GenAI tools, their multifarious capabilities, and the different contexts in which they are used, there is a need for contextually relevant definitions of GenAI-related concepts, such as responsible AI. As what is responsible AI in one context or for one stakeholder may not be the same for another, this paper aims to provide a contextually relevant definition of responsible AI in the Australian Higher Education sector. Through the application of LLMs such as NotebookLM and AI-powered NVivo, the meanings attributed to responsible AI in datasets within the Tertiary Education Quality and Standards Agency (TEQSA) AI Knowledge Hub and those submitted to the Senate Inquiry on the Adoption of AI are analysed. Findings separate the meaning of accountability into developer responsibility, government responsibility, deployer responsibility and user responsibility. Ethical usage means alignment with Australian values, equitable access to GenAI tools and sustainable usage. Fairness is seen as AI safety for all, avoiding bias or discrimination by GenAI Tools, and ensuring a balance of interests for all parties impacted by GenAI decisions. The meaning attributed to transparency is providing clear information and full disclosure of GenAI usage and explaining processes of GenAI functionalities, capabilities and usage.

Keywords: Responsible AI · Accountability · Ethicality · Fairness · Transparency

1 Introduction

The concept of responsible AI is becoming increasingly popular in the discourse for AI regulation [1]. Both private and public sector organisations have contributed to the discourse on the responsible use of AI in different ways. Some of these organisations include the Australian Government Digital Transformation Agency (DTA), which released its policy on the responsible use of AI [2], and the Department of Industry, Science and Resources [3], which highlights the government's international partnerships on responsible AI [4]. Amongst the major private organisations, Google published its progress

Q. V. Nguyen et al. (Eds.): AusDM 2025, CCIS 2765, pp. 224–237, 2026.
https://doi.org/10.1007/978-981-95-6786-7_15

report on responsible AI in February 2025 [5], and Microsoft listed six core values as a show of commitment to responsible AI in its report [6]. Earlier, Meta published an article on its approach to responsible AI [7]. Considering the different activities and approaches towards responsible AI, there is a general idea that the concept of responsible AI refers to the development, deployment and use of GenAI tools in the right way. Consequently, broad meanings are attributed to the concept of responsible AI. While broad meanings give a general idea of the concept, they create challenges for clearer understanding, especially when a definition includes several terms. For example, within the Australian higher education sector, several organisations identify fairness as one of the elements of responsible AI [8–10]. This raises several questions. Does the term fairness mean the same thing in practice at the sector level, i.e., for all higher education organisations? Does it mean the same thing for every stakeholder within an organisation and in the sector? These questions could be asked for the other terms within the definitions of responsible AI. Considering these issues, there is a need for more contextually defined GenAI concepts.

Within the literature that discussed the concept of responsible AI in different contexts, the terms accountability, ethicality, transparency, and fairness appear repeatedly [8, 11–15]. A synthesis of the Australian Government AI governance solutions in the form of policies, frameworks, models, and tools identified the terms accountability, ethicality, fairness and transparency amongst others [13]. An explorative review of responsible and ethical AI practices to identify understanding and awareness gaps identified fairness, accountability and transparency as principles of responsible AI [12]. Similarly, other studies that discussed responsible AI in a variety of contexts identify these terms [8, 11, 14, 15].

While general definitions of responsible AI exist, sector-specific ones are lacking. Given the diverse contexts in which GenAI tools are deployed, context-specific definitions are necessary. The wide range of GenAI capabilities makes general definitions problematic, as interpretations of responsible AI can vary across contexts and stakeholders. For instance, using GenAI as personal tutors may benefit underserved learners but could negatively impact human tutors by reducing job opportunities.

Consequently, this paper will attempt to define responsible AI in the Australian Higher Education context. It will use the meanings and definitions attributed to the concept by Australian higher education organisations in datasets that are publicly available on the Tertiary Education Quality and Standards Agency (TEQSA) AI Knowledge Hub. Also, the meanings attributed to the concept in some of the datasets submitted to the Senate Inquiry on the Adoption of AI will be used. Considering the repeatability of the terms accountability, ethicality, fairness and transparency in the literature on responsible AI, the following research questions will be explored:

- RQ1: What does accountability mean within the context of responsible AI in the Australian Higher Education Sector?
- RQ2: How is ethical usage defined within the context of responsible AI in the Australian Higher Education Sector?
- RQ3: What is the meaning of fairness within the context of responsible AI in the Australian Higher Education Sector?

- RQ4: Which meaning is attributed to transparency within the context of responsible AI in the Australian Higher Education Sector?

The rest of this paper is structured as follows: Sect. 2 explains the methods used to apply NotebookLM and NVivo to extract meaning from the datasets. Section 3 presents the findings from the data analysis with summaries of the associated data items. Section 4 discusses the findings, Sect. 5 concludes the paper and Sect. 6 identifies the limitations of the paper.

2   Methods

This section discusses the design used and workflow process followed to extract meaning from the datasets and generating topics from the meanings attributed to the selected terms.

2.1   Stage One: Application of NotebookLM to Extract Meaning

This stage was done both manually and digitally. Digitally, we used Google's AI-powered research and note-taking tool called NotebookLM. Manually, some of the sources from the datasets that included discussion about each of the four responsible AI terms were identified. 26 suitable sources from TEQSA AI Knowledge Hub [16] were read, and the phrases and sentences that attributed definitions and meanings to the four terms were copied out. Digitally, 50 suitable sources from the publicly available 245 submitted to the Senate Inquiry on the adoption of AI (in which the four terms were discussed in different contexts) were uploaded to NotebookLM with the prompt to identify and highlight the definitions and meanings of the terms.

The prompt included directions such as: *"highlight and quote the sentences with definitions/meanings that are directly or indirectly similar to the following definitions* (definitions of the four terms) *for each of the sources"* and *"include sentences with the same or similar meanings but written differently"*. The definitions and meanings highlighted by NotebookLM were read manually, and the relevant phrases and sentences copied out. As summarised in Fig. 1, after identifying the concept that required a more specific definition from the literature (first step in the work flow), PDF/Doc files and links in which the concept was discussed were collated from the sources (step two) and uploaded into NotebookLM (step 3). After copying the suitable texts in line with the definitions of each of the four terms (step 4), the texts were organised using the four terms (step 5). Doing the steps manually and digitally enabled review and comparison for accuracy.

Among contemporary AI-assisted research tools, NotebookLM offers a distinct advantage in synthesising contextual information from a defined collection of documents, particularly when analysing specific terms. Unlike general-purpose models such as ChatGPT, or discovery-oriented platforms like Elicit and Perplexity AI, Notebook LM generates responses grounded exclusively in user-uploaded sources, thereby enhancing the accuracy and relevance of its outputs. In parallel, NVivo, especially when augmented with AI-assisted coding features, remains a robust tool for conducting systematic qualitative analysis through the categorisation and coding of textual data. While NVivo

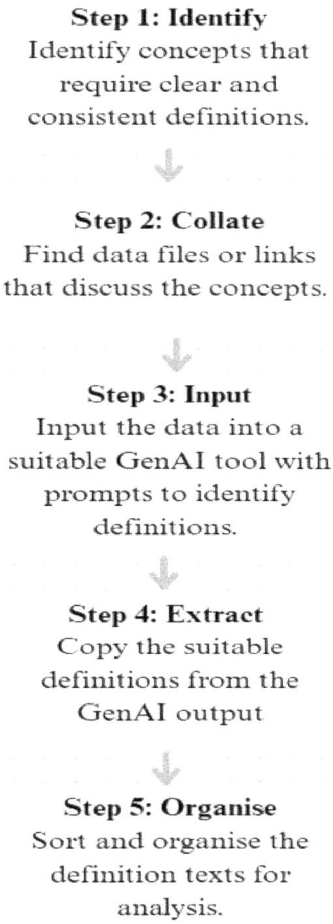

Step 1: Identify
Identify concepts that
require clear and
consistent definitions.

Step 2: Collate
Find data files or links
that discuss the concepts.

Step 3: Input
Input the data into a
suitable GenAI tool with
prompts to identify
definitions.

Step 4: Extract
Copy the suitable
definitions from the
GenAI output

Step 5: Organise
Sort and organise the
definition texts for
analysis.

Fig. 1. Text Extraction Workflow

facilitates the identification of patterns and thematic structures across large datasets, Notebook LM complements this process by providing coherent, context-sensitive interpretations of how particular concepts are articulated across the corpus. Together, these tools support a comprehensive and methodologically rigorous approach to defining concepts. Table 1 shows the number of times each term was referred to in the datasets. Table 2 summarises how the items were identified from the datasets by the tools used and selected.

Table 1. Terms and Items

Terms	No. of items
Accountability	53

(continued)

Table 1. (*continued*)

Terms	No. of items
Ethical Usage	22
Fairness	70
Transparency	28

Table 2. Selection Criteria and Tool

Inclusion Criteria	Exclusion Criteria	Tool Used	Tool Task
Definition/meaning is directly related to higher education in Australia Definition/meaning is related to another sector, but applicable to the higher education context in Australia	Definition/meaning is not directly or indirectly related to the higher education context in Australia	NotebookLM	Identify definitions/meanings of the terms in the dataset

2.2   Stage Two: Application of NVivo to Generate Topics

At this stage, keywords, phrases and sentences in the items identified were validated using a standard definition from the Oxford Learner's Dictionary for each term. That is, for accountability, the keywords and key phrases that were in line with the definition **"being responsible** for your decisions or actions and **expected to explain** them" were copied. For ethical usage, the definition "beliefs and principles about what is **right and wrong"** was used for validation. For fairness, **"treating people equally** or in a way that is **reasonable"** was used. For transparency, the definition followed was "the quality of something, such as a situation or an argument, that makes it **easy to understand"**.

The workflow is summarised in Fig. 2. The identified texts from stage one can be uploaded to NVivo, other software or a suitable GenAI tool for categorisation (step 1). The AI-powered tool called NVivo was more suitable on this occasion. The research topic, type and research questions helped to determine the suitable tool to use between the different tools available (with manual adjustments) for categorisation, generating topics (steps 2&3), and for determining relevance/accuracy during review (step 4). The validated phrases and sentences were categorised into suitable topics in NVivo. Items were added to the same category based on their similarity. The third heading titled "References" in Fig. 3 is the number of items (meanings/definitions) categorised under each topic after analysis.

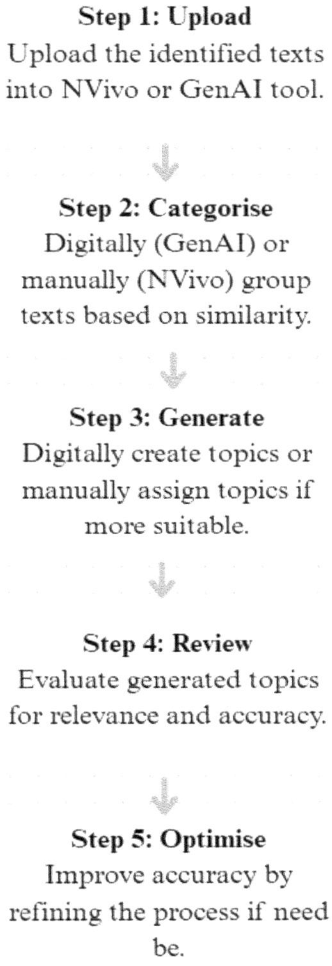

Step 1: Upload
Upload the identified texts
into NVivo or GenAI tool.

Step 2: Categorise
Digitally (GenAI) or
manually (NVivo) group
texts based on similarity.

Step 3: Generate
Digitally create topics or
manually assign topics if
more suitable.

Step 4: Review
Evaluate generated topics
for relevance and accuracy.

Step 5: Optimise
Improve accuracy by
refining the process if need
be.

Fig. 2. Topics Generation Workflow

⊕ Name	▲ Files	References
⊖ ○ Accountability in Responsible AI	3	78
○ Deployer Responsibilities	1	18
○ Developer Responsibilities	1	24
○ Government Responsibilities	3	22
○ User Responsibilities	1	14
⊖ ○ Ethical Usage in Responsible AI	1	16
○ Align with Australian Values	1	7
○ Equitable Access	1	3
○ Sustainable Usage	1	6
⊖ ○ Fairness in Responsible AI	2	50
○ AI Safety for All	2	13
○ Avoid Bias and Discrimination	1	15
○ Balanced Interest for All Parties	1	22
⊖ ○ Transparency	3	22
○ Clear Information	2	9
○ Explained Processes	2	8
○ Full Disclosure	2	5

Fig. 3. Topics and Meanings

During the categorisation stage, some of the items defined within a term aligned more practically with a topic in a different term. In Table 1, the number of items under the accountability term is 50, and it became 78 after categorisation under each topic. During categorisation, some definitions within the fairness term aligned with the responsibility of a key stakeholder. For example, developing a policy that requires a national ID to access a GenAI account that was originally defined under the fairness term aligned more practically with the topic of government responsibility under the accountability term, as this would require legislation. Although it could offer more safety to the general population, and consequently be seen as fair. Also, some items were separated to fit more than one topic and some items that did not align with the definitions of the terms and topics were omitted. This accounts for the differences in the number of items in Table 1 compared to the references in Fig. 3.

3 Findings

To determine the contextual meaning (for the Australian higher education context) of each topic within each term and answer the research questions, the items from the data, which have been categorised by similarity, were summarised, as shown in Tables 3, 4, 5 and 6.

3.1   Research Questions and Answers

Table 3. A summary of the data items under each accountability topic

Developer Responsibility	Government Responsibility	Deployer Responsibility	User Responsibility
Use suitable data to train GenAI models to avoid bias that could negatively impact student learning Train GenAI models to present accurate information without suppression/distortion to enable realistic learning Train GenAI models to avoid creating deepfake content for academic dishonesty or to offend fellow students and staff Provide info about GenAI's capabilities, functionalities, limitations and outcomes to deployers (universities) for risk and impact assessment	Regulations linked to the stakeholder responsible for redressing a given damage caused by AI Usage by learners and the academic community Regulatory governance and specific stakeholder obligations for GenAI's usage of data within the academic community Legal requirement to verify national identity before getting access to any GenAI account (like when buying a sim card)	Provide important information about proper GenAI usage and usage terms/conditions Ensure robust security systems to prevent cyber-attacks that would negatively impact learners and the academic community Educate learners and the academic community about the dangers of GenAI usage in addition to the benefits	Adhere to institutional and wider AI usage guidelines and regulations Take responsibility for the accuracy, ethicality and standard of created content for teaching, learning and assessment

RQ1: What does accountability mean within the context of responsible AI in the Australian Higher Education Sector?

From the data, accountability as an element of responsible AI in the Australian Higher Education Sector is shared among four major stakeholders. These are:

- Developers (including organisations that created the GenAI tools, such as OpenAI, Microsoft, Google, and their investors).
- Government (Legislature, Executive, and Judiciary).
- Deployers (Universities and other higher education organisations).
- Users (Students, lecturers, professors and other members of the academic community).

RQ2: How is ethical usage defined within the context of responsible AI in the Australian Higher Education Sector?

The meaning of ethical usage within the Australian Higher Education sector falls within the following three topics:

- Alignment with Australian values (freedom, respect, fairness, and equality of opportunity) [17].
- Equitable access to GenAI tools.

- Sustainable usage of GenAI tools

Table 4. A summary of the data items under each ethical-usage topic

Align with Australian Values	Equitable Access	Sustainable Usage
The use of GenAI tools by learners and the academic community should align with the values of freedom, respect, fairness, and equality of opportunity	The usage of GenAI tools should be equitably accessible to all learners and members of the academic community, not just some	GenAI tools should be used sustainably by learners and members of the academic community for economic benefit (e.g. more efficient learning), social good (e.g. group empowerment) and environmental preservation (e.g. fewer resources used)

RQ3: What is the meaning of fairness within the context of responsible AI in the Australian Higher Education Sector?

Fairness within the context of responsible AI in the Australian Higher Education Sector is seen as the following topics:

- Al safety for all stakeholders.
- Avoiding bias or discrimination by GenAI tools.
- Ensuring a balance of interests for all parties.

Table 5. A summary of the data items under each fairness topic

AI Safety for All	No Bias/Discrimination	Balanced Interest for All
Multi-stakeholder audit and risk assessment of GenAI models to ensure safe usage before release to learners and members of the academic community	Test GenAI models for bias against or discrimination towards some groups due to the data used to train them before usage in learning environments	Consider the interests of all parties impacted before GenAI-related decisions, e.g., existing jobs and job application process, students' admission screening and academic misconduct processes

RQ4: Which meaning is attributed to transparency within the context of responsible AI in the Australian Higher Education Sector?

The meaning attributed to transparency within the context of responsible AI in the Australian Higher Education Sector falls within the following topics:

- Clear information on GenAI usage.
- Explained processes of GenAI functionalities, capabilities and usage.
- Full disclosure of GenAI-created content.

Table 6. A summary of the data items under each transparency topic

Clear Information	Explained Processes	Full Disclosure
Clear information on how GenAI tools can and can't be used by learners and members of the academic community for teaching, learning, and assessment	The suitable respective stakeholder to provide information about the type of data used to train GenAI, GenAI's capability, actions, decision-making, limitations, purposes, outcomes and other usage process information	Embedding watermarks or other identifiers in content created by GenAI tools and voluntary disclosure by learners and members of the academic community

3.2 Proposed Responsible AI Framework for the Australian Higher Education Sector

The proposed Responsible AI Framework in Fig. 4 identifies the key areas that can be adapted to suite a specific higher education organisation. Within the accountability quadrant, developers include the organisations or individuals that created and/or own the GenAI tools, including their investors and other parties that contributed to creating it. The government includes the legislative, executive, and judicial arms of government with their respective components. Deployer includes the organisation that makes a GenAI tool directly available for usage by employees, customers/clients and other stakeholders. Within the Australian higher education context, deployers are universities and other higher education providers. Users are mainly students and employees. While the number of factors within each quadrant was determined by the dataset studied, this may change if a different dataset is explored and when the framework is applied to different sectors. For example, ethicality within the financial sector and transparency within the public sector may include more than three factors, respectively.

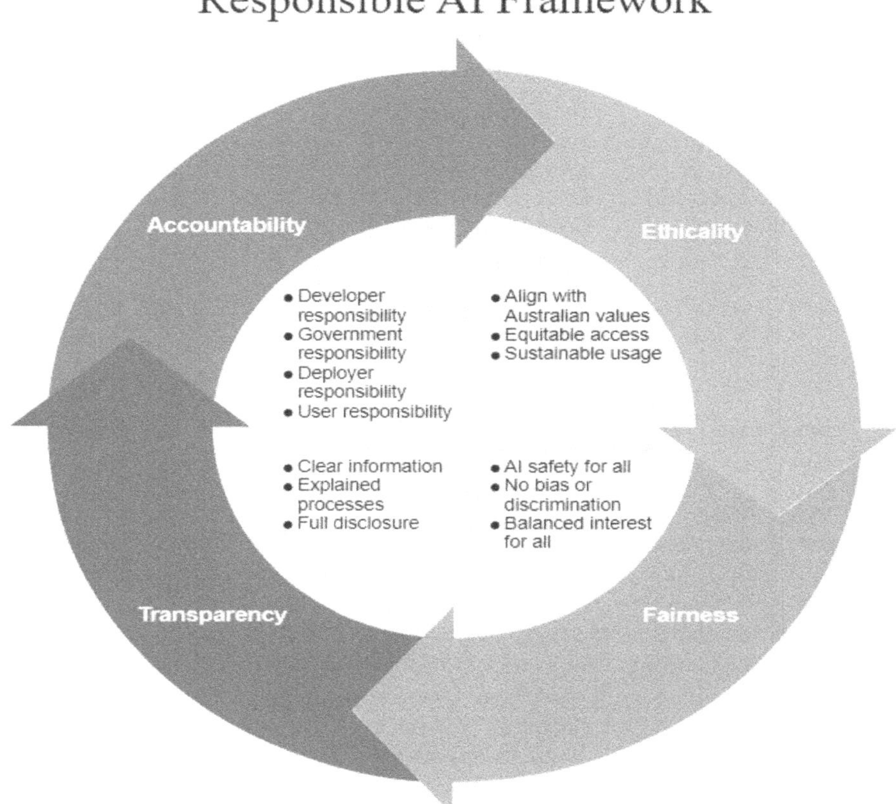

Fig. 4. Responsible AI Framework

4   Discussion

The definition of Responsible AI in the Australian higher education sector separates accountability into the responsibilities of respective stakeholders (see Table 3). Ethical usage should align with Australian values, enable equitable access to GenAI tools and ensure sustainable usage. Fairness means providing AI safety for all, avoiding bias or discrimination by GenAI tools, and ensuring a balance of interests for all parties impacted by GenAI decisions. Transparency emphasises the clarity of information and full disclosure of GenAI usage, as well as explaining the processes, functionalities, and capabilities of GenAI tools.

Within the Australian higher education sector, being responsible by using correct datasets to train GenAI models is a right step towards ensuring that learners are exposed to the correct information when they use GenAI tools as personal tutors, considering existing fears about hallucination by GenAI tools [18]. This raises the question of how quickly training data can be updated when there are changes in reality, to reflect the outputs by GenAI tools and thus remain responsible. In line with suggestions by the Business for Social Responsibility (BSR), having regulations that identify the specific stakeholder responsible for a given type of harm done due to the use of GenAI within the academic community may encourage respective stakeholders to take preventative measures [19]. However, while taking preventative measures is responsible, it may be challenging when the nature of the harm can't be traced to a specific stakeholder.

AI literacy is responsible because it will contribute to a safer use of GenAI tools and prepare learners for the workplace where GenAI is increasingly being used. While AI literacy should be fully embraced, it is important to be mindful that it would equally empower intentionally unethical users for unethical usage. For example, as shown in previous work [20], it requires a certain level of AI literacy to know the right AI humaniser platform that will successfully edit AI-created content to mirror human-created one. However, this wouldn't be the case when learners and members of the academic community take responsibility for the content created using GenAI and adhere to usage guidelines and regulations.

A responsible way to use GenAI tools in academia is to use them in line with Australian values, but inequitable access to GenAI tools is contrary to the values of fairness and equality. Thus, equitable access for learners and members of the academic community is at the heart of ethical GenAI usage and towards sustainable usage. However, the reverse is the case for intentionally unethical users. Multi-stakeholder audit, risk assessment and testing GenAI models to avoid bias and discrimination before release is both fair and responsible for stakeholders. As job loss due to automation, such as the use of GenAI, is never seen as fair by some stakeholders [21], considering the interests of the parties involved before GenAI-related job decisions is both fair and responsible. The same can be said for GenAI-related decisions that negatively impact stakeholders, such as the use of GenAI tools for student admission screening.

5 Conclusion

The need to contextualise the definitions and meanings of GenAI-related concepts is an urgent one. Considering the variety of tasks that GenAI tools are capable of, the heterogeneity of sectors that they are deployed in and the differences in contexts within which they are used, broad definitions and meanings are no longer suitable for clear understanding. This paper is a contribution towards clearer definitions and meanings. It proposed a contextual definition for the concept of responsible AI within the Australian Higher Education sector using the many broad meanings attributed to the concept in publicly available datasets. Accountability, ethical usage, fairness and transparency were identified as the main elements of the responsible AI concept within the Australian Higher Education sector. While these elements of the responsible AI concept (which could be more) may be similar to those in other sectors, their specific meaning won't necessarily be the same. It is when contextually related factors are embedded within the definitions of these elements that their meaning can be captured in a specific sector or context, as shown in Tables 3, 4, 5 and 6. Thus, the proposed responsible AI framework should be seen as an introductory contribution which should be adapted to suit other contexts and further developed.

6 Limitations

A limitation of NotebookLM is that it could not specifically identify only the definitions and meanings of the concepts, but included other information. This limitation was manually resolved by reading the output to copy out only the meanings and definitions of the concepts.

The findings are limited to the extent to which the concepts focused on were discussed in the datasets used. This means that, there could be more meaning when bigger datasets are studied. For example, accountability in the datasets was shared among four stakeholders, but there could be more stakeholders. Also, studying the definitions attributed to responsible AI in different sectors may reveal fewer or more than the four main concepts identified in this paper.

Disclosure of Interests.   The authors have no competing interests to declare that are relevant to the content of this article.

References

1. Centre of the Public Square (CPS). Submission to the Proposals Paper for Introducing Mandatory Guardrails for AI in High-risk Settings. https://percapita.org.au/wp-content/uploads/2024/10/Per-Capita-Submission_-AI-in-high-risk-settings_September-2024.pdf. Accessed 12 Aug 2025
2. Digital Transformation Agency. Policy for responsible use of AI in government, Digital Transformation Agency. digital.gov.au/policy/ai/policy. Accessed 25 Aug 2025
3. Department of Industry, Science and Resources. Artificial intelligence: We are committed to ensuring all Australians share the benefits of artificial intelligence (AI). https://www.industry.gov.au/science-technology-and-innovation/technology/artificial-intelligence. Accessed 23 Aug 2025
4. Department of Industry, Science and Resources. Artificial intelligence: Partnering internationally on responsible AI. https://www.industry.gov.au/science-technology-and-innovation/technology/artificial-intelligence#partnering-internationally-on-responsible-ai. Accessed 23 Aug 2025
5. Google LLC. Our AI Principles. https://ai.google/principles/#our-ai-principles-in-action. Accessed 23 Aug 2025
6. Microsoft Corporation. Responsible AI at Microsoft: Explore the tools, practices, and policies we've created to uphold our responsible AI principles. https://www.microsoft.com/en-us/ai/responsible-ai. Accessed 23 Aug 2025
7. Meta Platforms, Inc. Responsible AI: Connect 2024: The responsible approach we're taking to generative AI. https://ai.meta.com/blog/responsible-ai-connect-2024/. Accessed 23 Aug 2025
8. Australian Academy of Technological Sciences and Engineering (ATSE), and the Australian Institute for Machine Learning (AIML) at The University of Adelaide. Responsible AI: Your questions answered. https://ai.uq.edu.au/files/6726/Responsible%20AI_Your%20questions%20answered.pdf. Accessed 02 Oct 2025
9. Macquarie University. Responsible and Ethical Use of Artificial Intelligence Policy. https://policies.mq.edu.au/document/view.php?id=394. Accessed 02 Oct 2025
10. RMIT University. Responsible Artificial Intelligence (AI) Procedure. https://policies.rmit.edu.au/document/view.php?id=305. Accessed 02 Oct 2025

11. Zhu, L., Xu, X., Lu, Q., Governatori, G., Whittle, J.: AI and ethics—operationalizing responsible AI. In: Chen, F., Zhou, J. (eds.) Humanity Driven AI. Springer, Cham (2021). https://doi.org/10.1007/978-3-030-72188-6_2
12. Bano, M., Zowghi, D., Shea, P., Ibarra, G.: Investigating responsible AI for scientific research: an empirical study. arXiv preprint arXiv:2312.09561 (2023)
13. Batool, A., Zowghi, D. and Bano, M.: Responsible AI governance: a systematic literature review. arXiv preprint arXiv:2401.10896, (2023)
14. Greenleaf, G., Mowbray, A. and Chung, P.: Regulation of (generative) AI requires continuous oversight (AustLII submission on the 'Safe and responsible AI in Australia' Discussion Paper). UNSW Law Research Paper (23–49) (2023)
15. Kao, K.T.: From Robodebt to responsible AI: sociotechnical imaginaries of AI in Australia. Commun. Res. Pract. **10**(3), 387–397 (2024)
16. Tertiary Education Quality Standards Agency: Gen AI knowledge hub. https://www.teqsa.gov.au/guides-resources/higher-education-good-practice-hub/gen-ai-knowledge-hub. Accessed 01 Oct 2025
17. The Department of Home Affairs. Australian values. https://www.homeaffairs.gov.au/about-us/our-portfolios/social-cohesion/australian-values. Accessed 03 Oct 2025
18. McGrath, C., Pargman, T.C., Juth, N., Palmgren, P.J.: University teachers' perceptions of responsibility and artificial intelligence in higher education-an experimental philosophical study. Comput. Educ.: Artif. Intell. **4**, 100139 (2023)
19. Business for Social Responsibility. Remedy for Generative AI-Related Harms: Guide 8 of the Responsible AI Practitioner Guides for Taking a Human Rights-Based Approach to Generative AI, Business for Social Responsibility, San Francisco (2025)
20. Baron, P.: Are AI detection and plagiarism similarity scores worthwhile in the age of ChatGPT and other Generative AI?. In: Scholarship of Teaching and Learning in the South (SOTL) in the South, vol. 8, no. 2, pp.151–179 (2024)
21. Shetty, G.: Job loss due to automation technologies: perceptions of fairness in the information technology sector of India. OPUS: HR J. **12**(2), 2021 (2022)

Research Track – Deep Learning Fusion and Vision

Fusing Deep Object Detectors via Spatial Heatmap-Based Relevance Modelling

Anqi Xiong[1(✉)], Yuefeng Li[1], Yanming Feng[1], Xiaohui Tao[2], and Jianming Yong[2]

[1] Faculty of Science, Queensland University of Technology (QUT), Brisbane, Australia
anqi.xiong@hdr.qut.edu.au, {y2.li,y.feng}@qut.edu.au
[2] University of Southern Queensland (UniSQ), Toowoomba, Australia
{Xiaohui.Tao,Jianming.Yong}@unisq.edu.au

Abstract. This study presents a Spatial Heatmap-Based Relevance Model that fuses the outputs of deep learning-based object detectors (such as YOLOv5 and Faster R-CNN). To address the limitations of single-model detectors, such as spatial ambiguity, occlusion sensitivity, and inconsistent confidence calibration, the model integrates spatial relevance modelling, confidence-adjusted fusion, and structured matching logic. A relevance heatmap is constructed from the training dataset to estimate class-specific spatial likelihoods across the image plane. During inference, each detected bounding box is assigned a relevance score based on its spatial alignment with the pre-computed heatmap, capturing the expected spatial distribution of object classes. These relevance scores are then used to rescale the original confidence values produced by each detector, allowing a more context-aware interpretation of the detection certainty. Subsequently, a Top N confidence product rule is applied to select the final detection outputs from both models. Evaluation in a road object detection dataset demonstrates consistent improvements in precision and F1 score compared to individual detectors and baseline fusion strategies. The proposed model remains modular and model-agnostic, making it compatible with various detection architectures that support spatial priors. The results highlight the benefits of incorporating spatial relevance learned into detection pipelines to enhance robustness in real-world scenarios.

Keywords: Object Detection · Model Fusion · Deep Learning · Spatial Relevance Model · Confidence Calibration

1 Introduction

Object detection is a fundamental task in computer vision, enabling machines to identify and localise objects in diverse environments. Applications range from autonomous driving, where vehicles must detect pedestrians, signs, and other cars in real time, to industrial robotics, where accurate localisation supports tasks such as object grasping and obstacle avoidance. Broadly, detectors are categorised into one-stage and two-stage architectures. One-stage models, such as YOLO, directly predict bounding boxes and class probabilities in a single pass, offering high speed but often producing more false positives. In contrast, two-stage models, such as Faster R-CNN, first generate candidate regions and then refine predictions, typically achieving higher precision but at greater computational cost. These complementary strengths and weaknesses form the motivation for research on combining detectors.

Despite their advances, individual detectors remain vulnerable in real-world deployments. Environmental factors such as lighting, occlusion, and motion blur often degrade performance. More critically, each architecture exhibits distinct error patterns: one-stage models may fail due to overly aggressive non-maximum suppression and fixed anchor priors, while two-stage models may miss objects if the region proposal stage fails. As no single detector provides uniformly reliable performance across conditions, fusing multiple models has emerged as a promising strategy. Traditional fusion methods, however, remain simplistic. Common approaches, such as averaging scores, majority voting, or applying NMS on pooled predictions, provide some gains but disregard statistical knowledge from training data. They assume all spatial regions and detections are equally reliable, while ignoring contextual signals, which often results in poor calibration and suboptimal box selection in cluttered or ambiguous scenes.

Although ensemble methods and post-processing refinements have been explored, several gaps remain. Most existing fusion strategies are heuristic, relying on static weighting or confidence averaging without leveraging spatial priors. Confidence calibration techniques have largely been developed for single models, failing to consider cross-model consistency or contextual reasoning. Addressing these gaps requires a unified framework that integrates spatial statistics into the fusion stage, dynamically adjusts confidence, and produces interpretable outputs. This paper therefore investigates three core questions: how to encode class-specific spatial relevance for context-aware scoring, how to design a principled fusion mechanism that integrates heterogeneous detectors, and how to balance precision and recall under diverse conditions through relevance-aware ranking.

In this study, we propose a spatial heatmapâĂŞbased relevance model for object detection fusion. Instead of relying solely on model-internal confidence scores, our approach redefines confidence as a context-aware belief value shaped by spatial priors learned from training data. This belief assignment enables detections to be evaluated not only on the model's raw output but also on their spatial plausibility. Building on this representation, we introduce a two-stage fusion strategy: the first stage recalibrates confidence through spatial relevance maps, while the second stage resolves detector disagreements via consensus-based boosting and global ranking. We evaluate the framework on a road vehicle dataset. Results show consistent improvements in precision and F1 score compared to individual detectors and baseline fusion methods. This strategy is modular, requiring no retraining of base detectors, and demonstrates both interpretability and generalisability. These contributions highlight the practical value of embedding statistical spatial reasoning into post-processing pipelines for robust and adaptive object detection.

2   Related Work

This review of the literature offers a systematic summary of current advancements in relevance modelling methodologies, postprocessing and score refinement tactics, model fusion techniques, and foundational structures. In this paper, a new spatially informed fusion paradigm is suggested after each component critically analyses the assumptions, advantages, and disadvantages of existing approaches.

One-stage and two-stage techniques are the two most prevalent classifications for object detection architecture. To achieve more accuracy at the expense of slower inference, two-stage detectors, such as Faster R-CNN, first produce region proposals before classifying each proposal. One-stage models, such as YOLO (You Only Look Once), on the other hand, greatly increase speed by performing detection in a single pass without the proposal step [1]. In a comparative study, YOLO, SSD, and Faster R-CNN were evaluated under identical settings. It was discovered that YOLO performed exceptionally well in speed-critical tasks, although Faster R-CNN attained greater mean average precision (mAP) [2]. These trade-offs are supported by systematic reviews. In this respect, it is proposed that YOLO is more appropriate for real-time deployment in edge devices and video analysis, while Faster R-CNN is better suited for applications requiring high localisation precision, such as security surveillance or medical imaging [3]. Similarly, it has been highlighted that Faster R-CNN remains beneficial in scenarios with dense or overlapping objects, even when YOLO is computationally efficient [4].

Non-Maximum Suppression (NMS) is a commonly used post-processing step in object detection pipelines. Soft-NMS improves detection accuracy without retraining the model by altering the score decay instead of removing overlapping boxes [5]. However, calibration distortions may be introduced by post-processing techniques like NMS. By comparing pre- and post-NMS outputs, one study examined this phenomenon and found that well-calibrated raw predictions frequently became overconfident following suppression [6]. Furthermore, a transformer-based object detection calibration framework called Cal-DETR was presented to improve both in-domain and out-of-domain dependability, aligning projected confidence with genuine correctness likelihood via logit mixing regularization and uncertainty-guided logit modulation [7]. Moreover, contextual signals can greatly enhance post-processing and confidence assignment, according to recent research. To achieve greater average precision, a contextual rescoring framework was proposed that uses RNNs and attention processes to update detection scores based on the spatial and semantic association of surrounding objects [8].

To reduce missed detections and minimise false positives, a detection framework was proposed that makes use of opinion pooling and redundant detectors

[9]. Additionally, the advantages of multi-model diversity were demonstrated by fusing outputs from YOLOv7 and DETR, using weighted averaging and machine learning stacking approaches to adjust to shifting weather conditions in satellite data [10]. The three main categories of fusion techniques are early, intermediate, and late fusion. In detection applications, late-stage decision-level fusion is very common. An optimisation-based method was introduced that balances various detector outputs by adjusting model confidence scores using a weighted framework [11]. Yayambo is an iterative probabilistic output fusion technique with theoretical assurances of convergence and resilience against error variance [12]. After that, learning-based fusion has become more popular. A bilevel integrated model was presented that blends data-driven regularisation with multi-modality characteristics to adaptively adjust fusion weights [13]. A co-enhancement framework was developed, where image fusion and object detection are jointly trained through shared feature representations and gradient flow, forming a bi-directional feedback loop between the two tasks [14].

Modelling the distribution of object classes in real-world scenes provides crucial priors for improving detection robustness and cross-view consistency. A simultaneous localisation and mapping (SLAM) system that integrates stereo vision, IMU data, and semantic segmentation can leverage geometric and semantic priors to handle dynamic environments with improved localisation accuracy [15]. These applications emphasise that modelling class-specific spatial tendencies can provide informative biases, such as height-dependent target occurrence in aerial views or lane-specific car placements. Additionally, the failure of large vision-language models (LVLMs) in spatial tasks has highlighted the limitations of weak or missing spatial priors. Analysis shows that such models often misinterpret relative positioning or co-occurrence, failing in tasks requiring geometric logic [16]. Integrating spatial relevance mechanisms directly into detection pipelines has proven effective across 2D, 3D, and multi-modal contexts. The Spatial Attention Frustum method enhances 3D object detection in occluded regions by constructing frustum-shaped attention zones derived from 2D proposals and projecting anchor boxes into 3D space [17]. Region- and class-specific spatial priors support improved detection precision and semantic alignment. The Visual-Linguistic Agent (VLA) incorporates object detection with large language models, refining YOLO outputs via spatial-linguistic feedback loops to ensure contextual plausibility and relational accuracy [18]. Graph-based feedback mechanisms like SGScore evaluate layout correctness by comparing generated scene graphs to annotated ground truth, enabling spatial consistency enforcement during training and inference [19]. In multimodal VQA systems, spatial heatmaps are not only used for attention but are also directly embedded in structured representations.

Existing literature demonstrates significant progress in object detection architectures, post-processing and confidence calibration, and fusion techniques. However, notable gaps remain that motivate the proposed research. Firstly, while advanced confidence calibration methods have been explored (e.g., Soft-NMS, learnable NMS, contextual calibration), most approaches focus on improving single-detector performance and rarely consider calibration in the context of multi-model fusion. The interaction between spatial priors and confidence calibration in fused detection pipelines is under-explored. Secondly, although fusion methods are increasingly adopted to leverage complementary strengths of detectors, existing techniques are dominated by score averaging, voting, or heuristic post-processing. Few studies incorporate statistical spatial relevance modelling into fusion logic to guide matching and confidence adjustment. Thirdly, the modelling of class-specific spatial priors is well-studied in domains such as SLAM or VQA, but its integration into 2D object detection pipelines remains limited. Current object detectors often ignore spatial context, leading to reduced robustness in complex or cluttered scenes. Lastly, failure cases and robustness under real-world conditions, such as occlusion, domain shifts, and spatial distribution changes, are seldom systematically addressed in existing fusion frameworks. Addressing these gaps, this research proposes a relevance-aware, confidence-adjusted fusion framework that systematically integrates spatial priors and dynamic fusion strategies, aiming to advance both theoretical understanding and practical performance in adaptive object detection (Fig. 1).

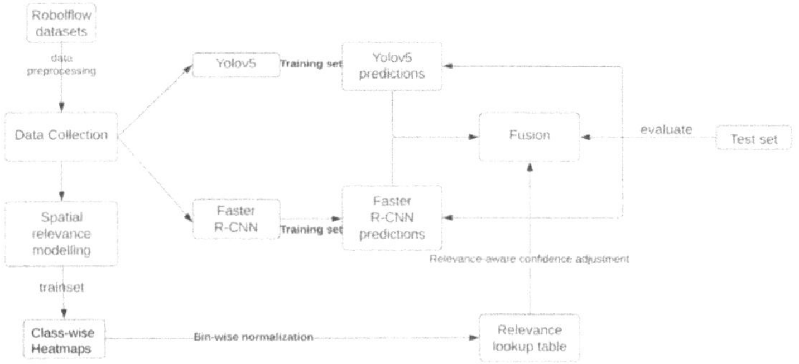

Fig. 1. Experimental Workflow Diagram.

3 Methodology

The proposed framework proceeds in three steps. Step 1: Training data are used to independently train YOLOv5 and Faster R-CNN, while preprocessing ensures consistent input resolution. Step 2: A spatial relevance model is constructed from the training annotations, producing class-specific heatmaps stored in a normalised lookup table. Step 3: During inference, predictions from both detectors are recalibrated using these relevance scores and then fused with IoU-based matching and relevance-aware ranking. The fused results are finally evaluated on a held-out test set.

3.1 Motivation for Confidence Adjustment in Multi-Model Fusion

Each projected bounding box in object detection tasks is usually given a confidence score that represents the probability that an object of a particular class is present within it. However, such confidence scores are computed internally and vary significantly across detection architectures, both in formulation and numerical scale. In the original YOLOv1 formulation [20], the objectness confidence is supervised using a separate loss component that penalises incorrect confidence predictions for both object and non-object grid cells. The confidence loss is defined as:

$$L_{confidence} = \sum_{i=0}^{S^2} \sum_{j=0}^{B} e_i^{obj} \left(C_i - \hat{C}_i \right)^2 \\ + \lambda_{noobj} \sum_{i=0}^{S^2} \sum_{j=0}^{B} e_i^{noobj} \left(C_i - \hat{C}_i \right)^2 \tag{1}$$

In this equation, C_i and $\hat{C}_i$ represent the ground-truth and predicted objectness confidence scores for grid cell i, while e_i^{obj} and e_i^{noobj} are binary indicators denoting whether an object is present or absent in the corresponding cell. The hyperparameter λ_{noobj} is used to balance the contribution of false positives from non-object cells. This formulation enables the model to penalise confident predictions in background regions while still emphasising accurate confidence regression in object regions.

On the other hand, the overall training objective of Faster R-CNN [21] jointly optimises the classification and regression branches of the Region Proposal Network (RPN). The multi-task loss is defined as:

$$L(\{p_i\}, \{t_i\}) = \frac{1}{N_{cls}} \sum_i L_{cls}(p_i, p_i^*) + \lambda \frac{1}{N_{reg}} \sum_i p_i^* L_{reg}(t_i, t_i^*) \tag{2}$$

Here, p_i is the predicted probability of anchor i being an object, and p_i^* is the ground-truth label (1 for positive, 0 for negative). t_i and t_i^* are the predicted and ground-truth bounding box coordinates, respectively. The classification loss L_{cls} is typically a log loss over two classes (object vs. background), while the regression loss L_{reg} is the smooth L_1 loss. The term p_i^* acts as a mask to include only positive anchors in the regression term. λ balances the two loss components.

In addition to architectural differences, raw confidence scores lack spatial awareness. This reflects the internal belief of the model, but does not explain whether detection occurs in a statistically plausible region of the image. For example, a model may assign high confidence to a detection in a location where the object class rarely appears in training data. This leads to false positives of high confidence and false negatives of low confidence, especially with domain occlusion, clutter, or domain shifts. To address these issues, a relevance-guided confidence adjustment strategy is proposed. Specifically, spatial relevance maps from training data represent the likelihood of each object class appearing at each location. During inference, each predicted bounding box is mapped to its corresponding spatial bin, and a relevance score is retrieved based on its class and position. The final belief score is computed by:

$$\text{Belief Score} = \text{Confidence} \times \text{ScaledRelevance} \tag{3}$$

Only when a detection is both extremely confident and spatially realistic is it deemed reliable, thanks to this multiplicative correction. From a probabilistic standpoint, this can be viewed as approximating the joint probability by combining the model's posterior with a learned spatial prior. This transformation of confidence into a context-aware belief measure lays the foundation for our spatially-informed fusion framework.

3.2   Fusion Design and Grid Mapping

We also experimented with using the KL-divergence as a measure of spatial uncertainty. For each class c, the divergence between the training prior $q_c(x, y)$ and the empirical prediction distribution $p_c(x, y)$ was computed with Laplace smoothing [22]:

$$D_{\text{KL}}(p_c \parallel q_c) = \sum_{x=1}^{W} \sum_{y=1}^{H} p_c(x, y) \log \frac{p_c(x, y)}{q_c(x, y)} . \tag{4}$$

However, this global score does not provide localised box-level adjustment and tends to penalise rare but correct detections. As a result, KL-divergence was retained only as a diagnostic metric and not integrated into the final fusion score. To integrate predictions from YOLOv5 and Faster R-CNN into a unified output, the method adopts a spatially informed fusion strategy based on grid mapping and relevance-guided confidence adjustment. Bounding boxes from both detectors are first transformed by mapping their centre coordinates into a 100×100 spatial grid. This discretisation enables structured analysis of training-time spatial distributions and provides a resolution-independent, perturbation-tolerant representation that is particularly useful when heterogeneous detectors produce slightly misaligned coordinates.

While intersection over union is standard for measuring overlap, in multi-detector fusion, it is sensitive to small spatial shifts, often yields sub-threshold overlaps for small or elongated objects, and lacks a single, scale- and class-invariant threshold. Heterogeneous detectors can also propose semantically identical yet geometrically misaligned boxes. Therefore, the approach combines light IoU filtering with grid-centre positional reasoning, where IoU removes clearly non-overlapping candidates, whereas the grid prior prevents semantically consistent detections from being discarded and improves matching robustness. Beyond geometric alignment, a class-specific relevance map is constructed from the training distribution. For each class, occurrences per grid cell are counted and normalised to produce a relevance score in $[0, 1]$. At inference, each predicted box retrieves the relevance value of the cell containing its centre; this value is linearly mapped to a scaling range $[a, b]$ (e.g., $[0.8, 1.2]$) and multiplied with the detector's confidence to obtain a certainty-aware score. If a box falls into a grid cell unseen during training, a neutral relevance of 1.0 is assigned to preserve recall.

3.3 Spatial Priors and Confidence Adjustment

To support effective fusion between YOLOv5 and Faster R-CNN, this research proposes a relevance-aware confidence adjustment method grounded in spatial distribution statistics. The fundamental premise is that particular item classes are more likely to be found in particular areas of an image. The relevance modelling process begins by applying spatial binning to the training set. To model this, each 640×640 training image is divided into a 100×100 spatial grid. For each object, the normalised bounding box centre coordinates are mapped to bin indices using:

$$x_{\text{bin}} = \lfloor x_{\text{center}} \times 99 \rfloor \quad \text{and} \quad y_{\text{bin}} = \lfloor y_{\text{center}} \times 99 \rfloor \tag{5}$$

where `x_center`, `y_center` are the normalised bounding box centre coordinates, with the range from 0 to 1. And `x_bin`, `y_bin` are the integer indices of the corresponding spatial bin, ranging from 0 to 99. Each class accumulates a 2D histogram of occurrence frequencies across these bins, resulting in a tensor of shape `[number_of_classes,100,100]`, where each value reflects the raw occurrence count of a given class in a particular spatial bin. These frequency maps are normalised so that the sum of all values for each class equals one, converting raw counts into probabilistic spatial priors.

Normalised relevance scores were used in the original setup and applied immediately without scaling. An intermediate experiment allowed for up to ±50% adjustment in detection confidence by introducing a wider scaling range of [0.5, 1.5] to amplify spatial influence. However, this setup sometimes suppressed correct detections in rare places by overemphasising spatial priors. A reasonable range of [0.8, 1.2], about ±20% adjustability, was chosen for the final setup in order to lessen this. This equilibrium maintained overall confidence calibration while preserving the advantages of spatial context. The following formula is used to apply the scaling:

$$\text{ScaledScore} = a + \frac{\text{RawScore} - \text{MinScore}}{\text{MaxScore} - \text{MinScore}} \times (b - a) \tag{6}$$

where `raw_score` is the normalised relevance score for a specific bin, `min_score` and `max_score` represent the minimum and maximum scores across the class-specific relevance lookup table. `a`, `b` specify the scaling bounds, and `Scaled_score` is the adjusted relevance score used in the fusion process.

3.4 Fusion Algorithm

The fusion strategy consists of two stages: Belief Assignment and Belief Fusion. In the Belief Assignment, each predicted bounding box is mapped to a 100×100

grid cell, and the relevance value from the training-derived lookup table is multiplied with the detector's confidence score to produce a belief score. Detections in unseen cells receive a neutral factor of 1.0 to preserve recall. Belief Fusion then merges the outputs from both detectors. For boxes of the same class with IoU above a threshold (e.g., 0.5), the detection with the higher belief score is retained. Unmatched detections are preserved and globally ranked, and a Top N selection is applied to balance precision and recall.

Algorithm 1. Belief Assignment

Require: Training set D (for building the relevance model); `Yolo_conf` (list of boxes from YOLOv5: ($imageID, class, x\_c, y\_c, w, h, confidence$)); `Rcnn_conf` (list of boxes from Faster R-CNN in the same format); `Grid` (a 100×100 spatial grid over the image).

Ensure: `Yolo_bel`, `Rcnn_bel` (lists of boxes with assigned belief scores).

    Method:

  1: **Build relevance lookup** LT from D $\triangleright$ LT stores tuples
     ($class, x\_bin, y\_bin, relevance\_score$)
        // For each object centre from D votes its bin; per-class counts are normalised to scores.

  2: **Assign belief to YOLO boxes**
  3: **for each** box $b \in$ `Yolo_conf` **do**
  4:     $(i, j) \leftarrow \mathrm{bin}(b.x\_c, b.y\_c, \mathtt{Grid})$ $\triangleright$ map center to 100×100 bin
  5:     $r \leftarrow LT[b.class, i, j]$ $\triangleright$ lookup relevance score
  6:     $b.belief \leftarrow b.confidence \times r$
  7:     push b into `Yolo_bel`
  8: **end for**

  9: **Assign belief to Faster R-CNN boxes**
10: **for each** box $b \in$ `Rcnn_conf` **do**
11:     $(i, j) \leftarrow \mathrm{bin}(b.x\_c, b.y\_c, \mathtt{Grid})$
12:     $r \leftarrow LT[b.class, i, j]$
13:     $b.belief \leftarrow b.confidence \times r$
14:     push b into `Rcnn_bel`
15: **end for**
16: **return** `Yolo_bel`, `Rcnn_bel`

Subsequently, Belief Fusion incorporates the adjusted forecasts from both models. The technique compares all bounding box pairings from Faster R-CNN and YOLOv5 that are in the same class for every image. The box with the higher belief score is chosen and added to the matched set if the IoU between two boxes is greater than a predetermined threshold (for example, 0.5). Unless they are already matched, both boxes are kept and added to the unmatched set if the IoU drops below the cutoff. Furthermore, the unmatched list also retains boxes that don't match any counterpart.

Algorithm 2. Belief Fusion

Require: `Yolo_bel`, `Rcnn_bel` (boxes with belief); τ (IoU threshold).
Ensure: `Fusion_matched`, `Fusion_unmatched`.
 1: `Fusion_matched` $\leftarrow \emptyset$, `Fusion_unmatched` $\leftarrow \emptyset$
 2: **for each** *imageID* in the test set **do**
 3:     **for each** pair (y, f) of the same class, where $y \in$`Yolo_bel`, $f \in$`Rcnn_bel` **do**
 4:         $u \leftarrow \mathrm{IoU}(y, f)$
 5:         **if** $u \geq \tau$ **then**
 6:             $b^\star \leftarrow \arg\max\{y.belief, f.belief\}$
 7:             add $b^\star$ to `Fusion_matched`
 8:         **else**
 9:             add y and f to `Fusion_unmatched` $\triangleright$ if not already present
10:         **end if**
11:     **end for**
12:     add any box with no counterpart to `Fusion_unmatched`
13: **end for**
14: **return** `Fusion_matched`, `Fusion_unmatched`

3.5   Summary of Methodology

This study proposes a modular fusion framework that integrates YOLOv5 and Faster R-CNN using relevance-guided confidence adjustment and spatial-aware post-processing. A 100×100 grid-based relevance map is constructed from training annotations to encode class-specific spatial priors. During inference, each predicted bounding box is treated as an ROI and mapped to the corresponding grid cell to retrieve its relevance weight. The original confidence score is recalibrated by multiplying it with this relevance factor, yielding a belief score. Box-level fusion is then performed based on IoU thresholds to match detections from both models. Matched boxes are boosted to emphasise cross-model agreement, while unmatched boxes are retained and globally ranked. A Top N selection strategy is applied to balance precision and recall. This approach enhances detection robustness by leveraging spatial distribution, belief-based ranking, and class-aware ROI modelling, without requiring retraining of base detectors.

4   Experiments and Results

This section reports experimental results obtained from different configurations of the proposed fusion framework. The goal is to evaluate how integrating spatial relevance with prediction confidence impacts object detection performance. Experiments are conducted on YOLOv5 and Faster R-CNN outputs under varying IoU thresholds, confidence adjustment ranges, boost factors, and Top N filtering strategies. We establish baselines on two datasets: the Roboflow road-vehicle dataset, which contains 2772 annotated bounding boxes across six classes. Performance is consistently measured using precision, recall, and F1 score.

4.1   Dataset Preparation and Models Training

For the road-vehicle experiments, a publicly available dataset from Roboflow is used, pre-split into training, validation, and test subsets with YOLO annotations. All images are resized to 640×640 pixels for consistency across detectors, and common augmentations (brightness change, random rotation, horizontal flip) are applied during preprocessing.

YOLOv5s, representing one-stage detectors, is trained for 50 epochs with batch size 16 using the official PyTorch implementation (v6.2). Stochastic Gradient Descent (SGD) with a learning rate of 0.01, momentum of 0.937, and weight decay 5×10^{-4} is applied. The multi-task loss combines CIoU, classification, and confidence terms to jointly predict objectness, class probability, and box coordinates.

Faster R-CNN with a ResNet-50 backbone is trained under the same input resolution and batch size for 50 epochs, using Adam with a learning rate of 10^{-4}. Its multi-task objective includes classification loss and Smooth L_1 regression. Both models are trained on the same training split and evaluated on the same fixed test set to ensure a fair comparison in later fusion experiments.

4.2   Baseline Model Performance

We first establish single-detector baselines on two datasets. On the Roboflow road-vehicle dataset, which contains 2772 annotated bounding boxes across six transportation classes, YOLOv5 and Faster R-CNN are evaluated separately to provide a benchmark for fusion. YOLOv5 achieves its best balance at a confidence threshold of 0.25 with an F1-score of 0.816, while the average score for both models is 0.767. These results establish reliable baselines on the road-vehicle dataset for assessing the proposed fusion framework.

As baselines for fusion, we also include a traditional method. These consist of confidence-voting (retaining the higher-confidence box when IoU > 0.5), Weighted Boxes Fusion (WBF), and probabilistic score multiplication. Such strategies rely purely on overlapping-box rules and score combination, without considering spatial priors. When the confidence threshold is set as 0.01, the F1-score achieves the highest as 0.790. While they sometimes yield small improvements, their performance is inconsistent across categories, highlighting the need for the proposed relevance-aware fusion framework.

4.3   Integration of Relevance-Based Confidence Adjustment

This section evaluates the effectiveness of integrating spatial relevance scores into the model confidence outputs. All predictions from YOLOv5 and Faster R-CNN are retained without thresholding to allow a fair post-hoc application of spatial relevance adjustment. Relevance scores derived from class-specific spatial priors are multiplied by the original confidence to yield adjusted scores. Among the tested configurations, the moderate scaling range of [0.8, 1.2] shows the best balance between recall and precision. In contrast, the wider range of [0.5, 1.5] often suppresses valid predictions in less populated areas, leading to performance degradation (Fig. 2).

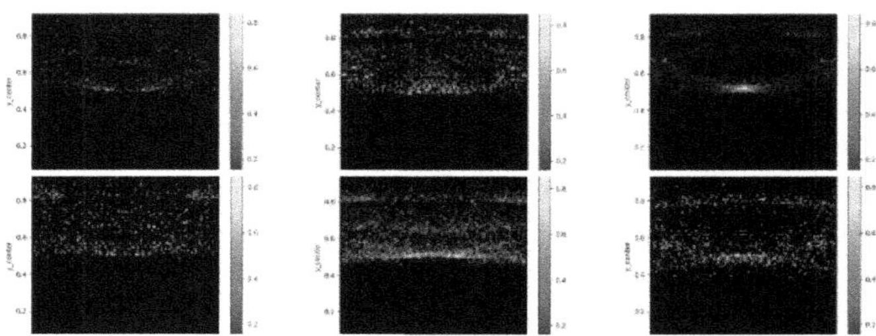

Fig. 2. Normalized heatmap for 6 classes.

The relevance-adjusted outputs yield consistent improvements in F1-score across both detectors, with YOLOv5 showing a more notable gain due to its denser prediction set. The relevance heatmaps are used for adjustment, highlighting the spatial priors learned from training data. These results confirm that relevance modelling enhances detection robustness by introducing lightweight, context-aware calibration without modifying model weights. The [0.8, 1.2] configuration is adopted as the default in the following fusion stages.

4.4 Analysis and Interpretation

The relative effectiveness of different fusion strategies is evaluated across precision, recall, and F1 score. As a baseline, we include a traditional late-fusion method, such as confidence voting, where overlapping boxes with the same class label are compared and the one with the higher confidence is retained. While simple, this approach is sensitive to calibration differences between detectors. Building upon this baseline, our proposed relevance-aware fusion introduces spatially guided confidence adjustment with a moderate scaling range of [0.8, 1.2], followed by consensus-based boosting and global Top N ranking. On the road-vehicle dataset, the traditional baseline achieves an F1 score of 0.816. With relevance-based adjustment, performance increases to 0.820, and the final optimised configuration achieves the best F1 score of 0.8225, using relevance scaling, boosting with a factor of 2.0, and Top 2500 filtering. This demonstrates that detection quality is enhanced by integrating spatial priors and structured ranking rather than relying solely on raw confidence outputs.

Further experiments with Top N filtering confirm that different cutoffs trade precision against recall. Without any limit, the system produces nearly 6000 detections, improving recall but significantly lowering precision. A stricter threshold of 0.5 yields high precision (0.977), but recall drops to 0.525, reducing detections to around 1500. The optimal balance is found in the range 2000âĂŞ3000, particularly at the Top 2500, which provides stable gains. Overall, compared to the best YOLOv5 single-model baseline (F1 = 0.816), the proposed fusion framework improves F1 by about 1%, and around 7.5% over the average single-detector baseline, validating the effectiveness of relevance-aware fusion for real-world object detection tasks.

4.5 Limitations

This method is a post-processing scheme that depends on fixed, training-derived spatial priors. As such, it can underweight valid but rare layouts under domain shift; centre-based binning oversimplifies object extents and may fail to capture boundary or shape cues; uniform agreement boosting can amplify correlated false positives; and performance is sensitive to hand-set hyperparameters. The approach does not correct systematic biases in the base detectors or the training data.

5 Conclusion and Future Work

This study presents a structured and interpretable fusion framework that improves object detection performance by combining predictions from YOLOv5 and Faster R-CNN through relevance-guided confidence adjustment. The approach integrates spatial priors, consensus-based boosting, and Top N filtering into a modular post-processing pipeline. Without requiring retraining, the framework enhances precision and recall by leveraging statistical object occurrence patterns derived from training annotations. Experimental results confirm the effectiveness of this method: the best-performing configuration achieved an F1 score of 0.8225, with precision of 0.8672 and recall of 0.7821, outperforming both YOLOv5 and Faster R-CNN standalone baselines. These improvements validate the central hypothesis that spatial relevance, when appropriately scaled and integrated, can guide more reliable confidence estimation and fusion decisions. The proposed strategy is lightweight and architecture-independent, making it suitable for scenarios where retraining the model is not feasible but enhanced model robustness is required. Furthermore, it can be extended to applications in security, compliance, and privacy, such as identifying and handling sensitive content, data deletion, data retention policies, or blocking specific objects, pornographic content, or inappropriate outputs in certain domains.

However, limitations remain. The spatial relevance model depends on training data distributions, which may introduce bias or suppress valid but rare detections in underrepresented regions. Confidence boosting can also amplify errors when both models agree on incorrect predictions. Moreover, reliance on fixed parameters such as scaling range and boost factors introduces sensitivity and limits adaptability across domains. To address these challenges, future work will focus on three key areas. First, we plan to enhance the spatial prior representation by replacing centre-point binning with overlap-based relevance grids, allowing finer-grained alignment with detection regions. Second, KL divergence will be revisited not only as a global evaluation metric but also as an adaptive modulator of relevance reliability per image or per class. In order to generalise across spatial layouts and semantic classes, we ultimately aim to develop learnable, context-aware relevance estimators—essentially attributed networks ([23]) in which the identified objects exhibit both structural cohesiveness and attribute homogeneity, possibly using transformer-based modules or attention methods. Longer-term intentions include improving the technique for real-time deployment on edge devices, expanding its application to more demanding multi-class detection tasks, and comparing it to more contemporary fusion-based designs like stacking ensembles or attention-driven late fusion. By embedding statistical reasoning into post-processing, this framework offers a practical and extensible foundation for robust object detection in intelligent transportation and other vision-driven applications.

Acknowledgment. This article was partially supported by the Grant DP220101360 from the Australian Research Council.

References

1. Cao, Y., Wang, H.: Object detection: algorithms and prospects. In: 2022 Int. Conf. on Data Analytics, Computing and Artificial Intelligence (ICDACAI), pp. 1–4. IEEE, Zakopane (2022)
2. Aboyomi, D.D., Daniel, C.: A comparative analysis of modern object detection algorithms: YOLO vs. SSD vs. faster R-CNN. ITEJ **8**(2), 96–106 (2023)
3. Reswara, E., Suakanto, S., Putra, S.A.: Comparison of object detection algorithm using YOLO vs faster R-CNN: a systematic literature review. In: Proc. 2023 6th International Conference on Big Data Technologies, pp. 419–424. ACM, Qingdao (2023)
4. Joiya, F.: Object detection: YOLO vs faster R-CNN. IRJMETS **4**(9) (2022)
5. Bodla, N., Singh, B., Chellappa, R., Davis, L.S.: Soft-NMS — improving object detection with one line of code. In: 2017 IEEE Int. Conf. on Computer Vision (ICCV), pp. 5562–5570. IEEE, Venice (2017)
6. Schwaiger, F., et al.: From black-box to white-box: examining confidence calibration under different conditions. arXiv preprint: arXiv:2101.02971 (2021)
7. Munir, M.A., Khan, S., Khan, M.H., Ali, M., Khan, F.S.: Cal-DETR: calibrated detection transformer. arXiv preprint: arXiv:2305.14691 (2023)
8. Pato, L.V., Negrinho, R., Aguiar, P.M.Q.: Seeing without looking: contextual rescoring of object detections for AP maximization. In: 2020 IEEE/CVF Conf. on Computer Vision and Pattern Recognition (CVPR), pp. 14598–14606. IEEE, Seattle (2020)
9. Boschmann, W., Bakhshande, F., Söffker, D.: Complementary object detection: improving reliability of object candidates using redundant detection approaches. In: Proc. 33rd European Safety and Reliability Conference, pp. 225–232. Research Publishing Services (2023)
10. Ahmed, M., El-Sheimy, N., Leung, H.: Optimizing satellite imagery object detection in challenging weather conditions using IoT-driven fusion strategies. In: 2024 IEEE 10th World Forum on Internet of Things (WF-IoT), pp. 450–456. IEEE, Ottawa (2024)

11. Teng, Z., Zhang, B.: An optimization method of fusing multiple decisions in object detection. In: Peng, W.-C., et al. (eds.) Trends and Applications in Knowledge Discovery and Data Mining, Lecture Notes in Computer Science, vol. 8643, pp. 29–35. Springer, Cham (2014)

12. Masakuna, J.F., Kafunda, P.K., Kayembe, M.L.: On the theoretical convergence and error sensitivity analysis of Yayambo for fusion of probabilistic classifier outputs. In: 2022 25th Int. Conf. on Information Fusion (FUSION), pp. 01–08. IEEE, Linköping (2022)

13. Liu, R., Liu, J., Jiang, Z., Fan, X., Luo, Z.: A bilevel integrated model with data-driven layer ensemble for multi-modality image fusion. IEEE Trans. Image Process. **30**, 1261–1274 (2021)

14. Dong, A., et al.: Co-enhancement of multi-modality image fusion and object detection via feature adaptation. IEEE Trans. Circuits Syst. Video Technol. **34**(12), 12624–12637 (2024)

15. Wen, S., Tao, S., Liu, X., Babiarz, A., Yu, F.R.: CD-SLAM: a real-time stereo visual-inertial slam for complex dynamic environments with semantic and geometric information. IEEE Trans. Instrum. Meas. **73**, 1–8 (2024)

16. Jiang, Y., et al.: Effectiveness assessment of recent large vision-language models. Visual Intell. **2**(1), 17 (2024)

17. He, X., et al.: Spatial attention frustum: a 3D object detection method focusing on occluded objects. Sensors **22**(6), 2366 (2022)

18. Yang, J., et al.: Visual-linguistic agent: towards collaborative contextual object reasoning. arXiv preprint: arXiv:2411.10252 (2024)

19. Chen, Z., Wu, J., Lei, Z., Chen, C.W.: What makes a scene? scene graph-based evaluation and feedback for controllable generation. arXiv preprint: arXiv:2411.15435 (2024)

20. Redmon, J., Divvala, S., Girshick, R., Farhadi, A.: You only look once: unified, real-time object detection. In: 2016 IEEE Conf. on Computer Vision and Pattern Recognition (CVPR), pp. 779–788. IEEE (2016)

21. Ren, S., He, K., Girshick, R., Sun, J.: Faster R-CNN: towards real-time object detection with region proposal networks. In: Advances in Neural Information Processing Systems, vol. 28 (2015)

22. Kullback, S., Leibler, R.A.: On information and sufficiency. Ann. Math. Stat. **22**(1), 79–86 (1951)

23. Berahmand, K., Nasiri, E., Pir mohammadiani, R., Li, Y.: Spectral clustering on protein-protein interaction networks via constructing affinity matrix using attributed graph embedding. Comput. Biol.d Med. **138**, 104933 (2021)

CarDamageEval: Benchmark Evaluation of Car Damage Assessment Using Vision Language Models

Md Jahid Hasan(✉) [iD], Cong Kha Nguyen [iD], Yee Ling Boo [iD],
Hamed Jahani [iD], and Kok-Leong Ong [iD]

Department of Business Information Systems, School of Accounting, Information
Systems and Supply Chain, RMIT University, Melbourne, VIC 3000, Australia
`S3952007@student.rmit.edu.au,`
`{cong.kha.nguyen,yeeling.boo,hamed.jahani2,kok-leong.ong2}@rmit.edu.au`

Abstract. Accurate and interpretable evaluation of vision language
models (VLMs) is crucial for applications where accuracy and transparency are essential, such as automotive damage assessment, where
structured outputs support reliable decision making. This paper presents
CarDamageEval, a dual-layer evaluation framework designed to measure
both the structural accuracy and semantic quality of VLM outputs. The
framework enforces a predefined structured output format comprising
explicit tuples of damage type, vehicle body part and severity level,
enabling rigorous quantitative assessment through pair-matching metrics such as precision, recall and F1 score. Complementing this, semantic
quality is assessed using the Holistic Description Score (HDS), which
captures correctness, completeness, coherence and relevance. These two
perspectives are unified through the hybrid CarDD Score, providing a
balanced metric that rewards factual accuracy and descriptive clarity. To
support structured evaluation, we compiled a dataset integrating public
vehicle images, each annotated with bounding boxes and linked damage
type, body part and severity labels. This baseline comparison further
demonstrates the framework's ability to distinguish model performance
across configurations, highlighting the effectiveness of fine-tuning in generating accurate and structured damage descriptions. While developed
for automotive assessment, the principles underlying CarDamageEval
also apply to other structured vision language tasks, such as lesion detection in medical imaging or defect localisation in industrial inspection,
making it a versatile and reusable evaluation standard.

Keywords: Automotive Damage Assessment · Vision-Language
Models · Structured Output Evaluation · VLM Evaluation

1 Introduction

Vision Language Models (VLMs) have shown significant progress in jointly interpreting visual content and corresponding textual descriptions. Recent advances

Q. V. Nguyen et al. (Eds.): AusDM 2025, CCIS 2765, pp. 256–270, 2026.
https://doi.org/10.1007/978-981-95-6786-7_17

in multimodal learning have enabled models such as CLIP [1], Qwen [2], and LLaMA [3] to excel in tasks including image captioning, visual question answering, scene understanding and multimodal retrieval. Their ability to produce fluent and semantically rich narratives has created opportunities for diverse applications, including e commerce, autonomous vehicles and healthcare.

However, despite these advances, the use of VLMs in domains requiring structured and interpretable outputs remains limited. In this context, structured output refers to predictions that adhere to a predefined schema, such as explicitly pairing attributes and objects (for example, damage type, affected vehicle body part and severity level), rather than producing unconstrained free form text. Many real world applications demand outputs that are not only semantically accurate but also conform to predefined schemas. For instance, in automotive damage assessment, relevant to vehicle resale platforms, insurance claims and fleet management, the system must identify explicit attributes such as damage type (for example dent or scratch), damage location (for example left front door or rear bumper) and severity level (for example minor, moderate or severe). Free form outputs (for example "the car appears to have some scratches on the side") lack structural consistency and therefore limit their reliability for automated downstream processes. Structured outputs enable more transparent reporting, support decision making and enhance trust for end users.

In the automotive domain, accurate and standardised vehicle damage assessment has direct commercial relevance, underpinning applications such as insurance claims processing, resale valuation and e commerce vehicle listing verification. While traditional computer vision approaches, such as Faster R-CNN [4] or YOLO [5], have been effective for localising damages and classifying predefined damage types, these models typically generate categorical labels or bounding boxes but offer limited contextual reasoning and descriptive capability, motivating the exploration of VLMs. VLMs have the potential to bridge this gap by combining fine grained visual understanding with natural language reasoning, enabling outputs that capture both the type and location of damage as well as its severity in human readable form. However, the generative nature of VLM outputs often produces unstructured text, which complicates rigorous and reproducible evaluation.

Existing evaluation metrics for multimodal tasks, such as BLEU, METEOR, CIDEr and CLIPScore [6], are designed to measure semantic similarity between generated captions and reference text. These metrics tolerate variation in wording, which is advantageous for open ended tasks but unsuitable for precision critical domains. As a result, comparing models for tasks that demand structured outputs such as mapping vehicle damage type, location and severity remains inconsistent and often requires costly manual inspection.

To address these challenges, this paper proposes CarDamageEval, a dual layer evaluation framework designed for structured vision language tasks. The first layer focuses on structured pair matching, where model outputs are represented as sets of (damage type, car body part, severity level) tuples, enabling rigorous quantitative benchmarking using classical metrics such as precision, recall and F1 score. The second layer incorporates semantic quality evaluation through the Holistic Description Score (HDS), adapted from the DeepEval [7] paradigm, which assesses four aspects of textual output: correctness, completeness, coherence and relevance.

To unify structural accuracy and semantic quality, we propose the CarDD Score, a hybrid metric computed as a weighted harmonic mean of the structured F1 score and the semantic HDS score. This ensures that models are rewarded for factual accuracy while maintaining narrative clarity, a critical property for real-world deployments where both structured and human-readable outputs are required. Although this work focuses on automotive damage detection, the principles underlying CarDamageEval are generalisable to other structured vision-language tasks, including medical imaging and industrial defect detection.

This paper makes the following key contributions:

- **CarDamageEval Framework:** We introduce a novel evaluation framework combining structured pair-matching metrics and semantic quality assessment for VLMs.
- **Benchmark Dataset:** We present a fine-grained automotive damage dataset linking damage types to specific vehicle body parts and severity levels, enabling structured model training and evaluation.
- **Hybrid Evaluation Metric:** We propose the CarDD Score, a hybrid metric harmonising structural correctness with descriptive quality.

2   Related Studies

Automotive damage detection has long been a focus of computer vision research, driven by applications in insurance assessment, resale inspection, and automated vehicle maintenance. Early methods predominantly utilised convolutional neural networks (CNNs) [8] for classification and detection, or employed segmentation frameworks such as YOLO [5] and Mask R-CNN [4] to identify damages including scratches, dents and cracks. Several datasets, such as the Car Damage Dataset(carDD) [9] and VehiDE [10], supported these approaches by providing bounding box or segmentation annotations. While these datasets have advanced detection accuracy, they primarily emphasise geometric localisation and class labels, lacking structured textual descriptions that link each damage type with the relevant vehicle body part. This limitation restricts their suitability for evaluating VLMs designed to produce semantically rich, structured outputs.

Evaluation of VLMs has traditionally focused on open-ended tasks such as image captioning, visual question answering (VQA), and referring expression comprehension. Benchmarks including COCO Captions [11], and VQA [12] typically assess model outputs using n-gram overlap (BLEU, METEOR, CIDEr) or answer accuracy [6]. These metrics are effective for flexible, free-form text but accommodate semantic variation and often fail to penalise critical attribute-object mismatches, such as reporting a "scratch on the bonnet" instead of the "rear door". More structured reasoning benchmarks, such as GQA [13] and VQA

2.0 [14], have introduced tasks requiring attribute-level or spatial reasoning. However, these benchmarks remain domain-agnostic and do not evaluate tasks that rely on specialised taxonomies, such as vehicle damage types and body parts.

Structured output evaluation benchmarks demonstrate that constraining outputs enables more objective assessment. GQA, for example, represents scene understanding via graph-based question answering, while RefCOCO and Ref-COCOg focus on localised object identification using natural language [15]. These tasks evaluate attribute-object relationships and spatial reference resolution but are not tailored to automotive contexts. In particular, no existing benchmark evaluates VLM outputs that jointly require precise structured prediction of (damage type, body part) pairs and narrative quality. CarDamageEval addresses this gap by combining pair-level quantitative matching with semantic scoring, providing a unified evaluation protocol suitable for precision-critical automotive scenarios. A comparison with existing datasets and benchmarks is presented in Table 1.

Table 1. Comparison of CarDamageEval with existing datasets and benchmarks.

Benchmark	Annotation Type	Output Format	Evaluation Metrics	Remarks
CarDD [9]	Bounding boxes	Class labels	Accuracy, IoU	Lacks structured text alignment
GQA [13]	Scene graphs	Free-form answers	BLEU, Answer correctness	Domain-agnostic reasoning
CarDamageEval	Text & bounding boxes	Structured pairs	F1, HDS, Hybrid Score	Combines structural precision and semantic quality

3 Dataset

Accurate evaluation of vehicle damage using VLMs requires a dataset that represents both the visual appearance of the damage and its structured semantic context. Existing automotive damage datasets often provide only bounding box annotations or simple class labels, which are insufficient for tasks that demand structured outputs linking damage types with specific vehicle components. To address this limitation, our evaluation framework adopts the publicly available CarDD [9] dataset and adapts it for structured pair-based evaluation.

The CarDD dataset provides high-quality bounding box annotations identifying regions of visible damage along with corresponding damage type labels. For our structured evaluation task, these annotations are mapped into a pair-based format that links each damage type to the corresponding vehicle body part, represented as (damage type, car body part). This structured format enables direct comparison between model predictions and ground-truth references using quantitative metrics. In addition to bounding box verification, each damage instance is annotated with a severity level (minor, moderate, severe), supporting fine-grained assessment of both detection accuracy and severity reasoning.

To enrich the dataset with text-based semantic information, we leverage ChatGPT-4o API to generate initial structured textual descriptions for each image, particularly focusing on precise body part naming and severity details. These automatically generated descriptions are manually reviewed and refined by human annotators to ensure accuracy, eliminate hallucinated content, and maintain consistency with the structured taxonomy. The resulting dataset thus combines precise visual localisation, structured semantic labels, and descriptive text aligned to each damage instance. Figure 1 illustrates the annotation pipeline, from using existing bounding box labels to automated description generation and human refinement.

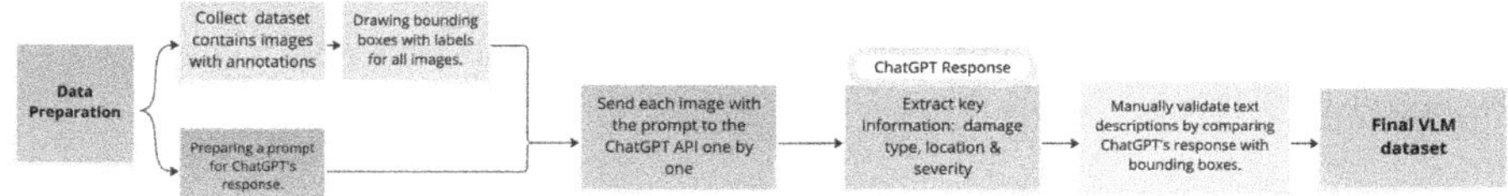

Fig. 1. The annotation pipeline used in CarDamageEval. Existing bounding box annotations from CarDD are mapped to a structured taxonomy, followed by automated textual description generation using ChatGPT-4o and human refinement to produce high-quality annotations for structured VLM evaluation.

To support reproducible benchmarking, we followed the same split ratio as CarDD, dividing the dataset into three non-overlapping subsets for training, validation and testing. This split ensures that model training, hyperparameter tuning, and final evaluation are conducted on mutually exclusive image sets, minimising overfitting and improving the reliability of reported performance. The resulting dataset captures a diverse range of vehicles, environmental conditions, and damage severities, providing a strong foundation for evaluating models designed to produce structured outputs. Although applied here to automotive damage, the preparation approach is generalisable to other structured vision language tasks such as industrial defect detection or medical imaging analysis.

4   Problem Formulation

The accurate assessment of vehicle damage requires not only the identification of damage types and their locations but also an understanding of severity levels, which directly influence repair costs and decision-making. While VLMs have shown strong capabilities in generating descriptive narratives of visual scenes, their unconstrained, free-form outputs pose challenges for tasks that demand structured, quantifiable, and comparable information. CarDamageEval addresses this limitation by defining the problem as a structured prediction task and establishing an evaluation protocol that integrates both quantitative and semantic assessments.

The CarDamageEval formalises vehicle damage description as the generation of structured outputs in the form of triplets:

$$\hat{y} = \{(d_i, p_i, s_i)\}_{i=1}^n \tag{1}$$

where d_i represents the damage type (e.g., scratch, dent, crack), p_i denotes the affected car body part (e.g., front left door, rear bumper), and s_i indicates the severity(e.g., minor, moderate, severe). This formulation ensures that the model's predictions can be aligned directly with human-verified ground truth, enabling precise quantitative evaluation metrics such as precision, recall, and F1 score.

In addition to structural evaluation, CarDamageEval integrates semantic scoring to capture narrative quality, rewarding outputs that are factually correct and contextually coherent. The combination of structured triplet matching, semantic assessment, and high-quality human-verified annotations creates a reproducible and objective framework for evaluating VLMs on vehicle damage tasks and other domains requiring precise, attribute-focused visual understanding.

5   Methodology of Evaluation Framework

CarDamageEval establishes a systematic methodology for evaluating VLMs on structured vehicle damage assessment tasks. The framework integrates two complementary evaluation layers: a structured matching process that quantitatively compares predicted damage attributes with ground truth and a semantic scoring component that assesses the descriptive quality of generated narratives. These layers are combined through a hybrid metric, enabling balanced assessment of factual accuracy and language quality. This methodology provides a reproducible protocol that is agnostic to model architecture and applicable beyond automotive damage assessment to other structured visual understanding domains.

5.1   Structured Damage Descriptor Prediction Task

To begin, we formalise the structured prediction task targeted by CarDamageEval. Given an input image I from the space of images $\mathcal{I}$, the model M must output a finite set of structured descriptors:

$$\hat{y} = M(I) = \{(\hat{d}_j, \hat{p}_j, \hat{s}_j)\}_{j=1}^m, \quad \hat{d}_j \in D, \ \hat{p}_j \in P, \ \hat{s}_j \in S. \tag{2}$$

where $\hat{d}_j$ represents the predicted damage type, $\hat{p}_j$ the affected vehicle body part, and $\hat{s}_j \in \{\text{minor, moderate, severe}\}$ the severity level.

The corresponding ground truth annotations are similarly structured:

$$y = \{(d_i, p_i, s_i)\}_{i=1}^n, \tag{3}$$

where d_i represents the annotated damage type, p_i the associated vehicle body part, and s_i its severity level.

Thus, the model operates in a structured output space:

$$\mathcal{Y} = \mathcal{P}(D \times P \times S), \tag{4}$$

where $\mathcal{P}(\cdot)$ denotes the set of all possible subsets. In this context, it represents every possible combination of damage types, body parts, and severity levels that the model might predict.

5.2 Evaluation Operator and Matching Logic

To ensure objective and reproducible comparison between model predictions and reference annotations, CarDamageEval defines a formal evaluation operator. This operator is built upon a binary matching function that determines whether each predicted structured damage descriptor exactly corresponds to a ground-truth descriptor, thereby forming the basis for subsequent metric computation.

The matching function is defined as follows:

$$Match((d_i, p_i, s_i), (\hat{d}_j, \hat{p}_j, \hat{s}_j)) = \begin{cases} 1, & \text{if } d_i = \hat{d}_j \wedge p_i = \hat{p}_j \wedge s_i = \hat{s}_j, \\ 0, & \text{otherwise.} \end{cases} \tag{5}$$

The evaluation operator $\mathcal{E}$ maps structured prediction–ground truth descriptor sets to a quantitative score:

$$\mathcal{E} : (y, \hat{y}) \longrightarrow \mathbb{R}. \tag{6}$$

True positives (TP) count all correctly predicted descriptors:

$$TP = \sum_{(d_i, p_i, s_i) \in y} \mathbf{1}\big(\exists(\hat{d}_j, \hat{p}_j, \hat{s}_j) \in \hat{y} : Match((d_i, p_i, s_i), (\hat{d}_j, \hat{p}_j, \hat{s}_j)) = 1\big), \tag{7}$$

while false negatives (FN) and false positives (FP) are given by:

$$FN = |y| - TP, \quad FP = |\hat{y}| - TP, \tag{8}$$

where $|\cdot|$ denotes set cardinality and $\mathbf{1}(\cdot)$ is the indicator function.

5.3 Quantitative Metrics for Structured Outputs

Building on the defined true positives (TP), false positives (FP), and false negatives (FN), CarDamageEval applies standard metrics to quantify model performance on structured damage descriptors. These metrics capture not only the specificity and coverage of predictions but also their overall balance and accuracy.

Precision measures the proportion of correctly predicted descriptors among all predicted descriptors:

$$Precision = \frac{TP}{TP + FP} \tag{9}$$

Recall measures the proportion of correctly predicted descriptors relative to all ground–truth descriptors:

$$\text{Recall} = \frac{TP}{TP + FN} \tag{10}$$

The **F1 score**, as the harmonic mean of precision and recall, provides a single balanced measure of predictive performance:

$$F1 = 2 \times \frac{\text{Precision} \times \text{Recall}}{\text{Precision} + \text{Recall}} \tag{11}$$

Additionally, **Accuracy** measures the proportion of correct predictions with respect to the ground truth:

$$\text{Accuracy} = \frac{TP}{TP + FN} \tag{12}$$

These metrics are computed on a per–image basis and then averaged across the evaluation dataset to obtain robust model–level scores.

5.4   Semantic Quality Assessment via HDS

While quantitative metrics capture factual correctness, they do not assess the quality, readability, or completeness of the generated textual descriptions. To address this limitation, CarDamageEval introduces a semantic scoring process known as the HDS, which evaluates the extent to which a model's outputs provide comprehensive and coherent information about detected damages.

This semantic HDS assessment is performed using the DeepEval framework, an LLM-as-a-Judge approach that leverages a large language model to evaluate descriptive quality across multiple dimensions. The HDS is composed of four equally weighted dimensions:

– **Correctness** (S_c): Alignment of descriptions with ground–truth annotations.
– **Completeness** (S_m): Inclusion of all relevant damage details.
– **Coherence** (S_o): Logical structure and grammatical clarity of the description.
– **Relevance** (S_r): Focus on damage–related content without unnecessary information.

Each dimension is rated on a scale from 0 to 1 using a large language model as an evaluator, and the final HDS is computed as:

$$HDS = \frac{1}{4}(S_c + S_m + S_o + S_r) \tag{13}$$

This semantic evaluation complements structured metrics by ensuring that model outputs are not only factually correct but also human–readable and contextually relevant, a key requirement for practical deployment in vehicle inspection tasks.

5.5   Unified Performance Metric: CarDD_Score

Building on the structured evaluation metrics and semantic quality assessment, CarDamageEval introduces a unified metric, termed CarDD_Score, which integrates both levels of evaluation. This score combines the factual correctness captured by the F1 score with the descriptive quality measured by the HDS, producing a single value that reflects a model's overall capability.

The CarDD_Score is computed using the harmonic mean of F1 and HDS, ensuring sensitivity to weaknesses in either component:

$$\text{CarDD\_Score} = \frac{2 \cdot \text{F1} \cdot \text{HDS}}{\text{F1} + \text{HDS}}$$

This formulation penalises models that excel in only one dimension, thereby promoting balanced performance across structured damage detection and natural language description. In practical automotive assessment scenarios, where both precise damage localisation and clear explanation are critical, the CarDD_Score provides a rigorous and interpretable indicator of overall model quality.

6   Experimental Setup

The experimental pipeline consists of two main components, an object detection model and a VLM for structured car damage understanding. The object detection stage is based on YOLOv9, which localises visible damages on the vehicle and outputs bounding boxes with associated damage labels. These outputs are then used as input for the second stage, where a vision-language model generates structured textual descriptions linking each detected damage to its corresponding vehicle body part. Three VLM variants, namely LLaVA, LLaMA and Qwen, are fine-tuned on the CarDamageEval dataset to ensure that the outputs follow the required structured format of damage type and car body part pairs. Fine-tuning is applied to encourage structured outputs, since generic VLMs often generate free-form captions that do not fully adhere to the triplet format. By adapting the models with task-specific pairs of images and structured annotations, the outputs are better aligned with evaluation requirements and demonstrate improved consistency. This two-stage configuration allows spatial localisation to be combined with semantically grounded textual reasoning. Figure 2 illustrates the experimental pipeline combining YOLOv9-based damage detection with a fine-tuned VLM for structured description generation.

For training, YOLOv9 is trained directly on the bounding box annotations using carDD dataset. The VLMs are fine-tuned on paired image and text samples from the same dataset so that the textual outputs are aligned with the detection results. To enhance efficiency, fine-tuning is performed using Low-Rank Adaptation (LoRA) [16], which inserts trainable low-rank adapter modules into the query and value projection layers of the transformer blocks. This method allows the models to adapt to task-specific behaviours while keeping the majority of

Fig. 2. Overview of experimental pipeline for car damage detection and structured description generation.

pre-trained weights frozen. Gradient checkpointing is applied to minimise memory consumption, and early stopping is employed to reduce the risk of overfitting. Each VLM model is trained for 20 epochs using the AdamW optimiser with a learning rate of 5×10^{-5}. The complete pipeline is evaluated on the CarDamageEval test set of 375 images. Model performance is measured using structured pair-matching metrics including precision, recall and F1 score, as well as semantic narrative quality scoring through the HDS. The final CarDD Score, computed as the harmonic mean of F1 and the HDS, provides a balanced measure of both structured correctness and descriptive quality. Table 2 presents the details of the models used in the experimental settings.

Table 2. Models used in the experimental pipeline and their primary roles.

Name	Architecture	Training Data	Purpose
YOLOv9 [17]	Object detection CNN	Bbox annotations	Damage localisation
LLaVA(13B) [18]	CLIP-ViT-L + LLaMA2	Image–text pairs	Structured output
LLaMA(11B) [3]	CLIP-ViT-L + LLaMA3.2	Image–text pairs	Structured output
Qwen(7B) [2]	Qwen2-VL-Chat	Image–text pairs	Structured output

7 Results

The evaluation of the proposed CarDamageEval framework was performed using the 375 image test set, ensuring a balanced representation in all six types of damage and multiple body parts. The results reported here assess both the detection and description stages of the pipeline, highlighting the importance of structured output evaluation. The section begins with an analysis of quantitative and semantic results, proceeds to the unified CarDD_Score metric, and concludes with a baseline comparison of fine-tuned and non-fine-tuned VLMs.

7.1 Quantitative Results

The first stage of the proposed pipeline employs YOLOv9 to localise vehicle damage with high accuracy. This detection backbone achieves a recall of 71.9%

and an average precision (AP) of 61.7%, producing bounding boxes that precisely delineate damaged regions. These localisations form the basis for generating structured descriptions in the subsequent stage.

In the second stage, the pipeline uses fine-tuned LLaVA, LLaMA, and Qwen models to produce structured outputs. These outputs are evaluated using a binary matching function that compares predicted tuples of damage type, body part, and severity with ground–truth annotations. Performance is assessed through precision, recall, F1 score, and overall accuracy, providing a comprehensive measure of the quality of structured descriptions. The evaluation results, presented in Table 3, show that the fine-tuned models effectively interpret the spatial damage cues provided by YOLOv9, yielding semantically meaningful structured outputs. Among the models, LLaVA demonstrates the highest performance across all metrics, followed by LLaMA and Qwen.

Table 3. Quantitative Performance of Fine-Tuned VLMs

Model	Precision	Recall	F1 Score	Accuracy
LLaVA	0.90	0.87	0.88	0.85
LLaMA	0.84	0.80	0.82	0.79
Qwen	0.82	0.78	0.80	0.77

7.2 Semantic Quality Assessment

Quantitative metrics measure factual accuracy but do not fully capture the quality of generated narratives. To address this limitation, semantic quality is assessed using the HDS, which aggregates four dimensions: correctness, completeness, coherence and relevance. Each dimension is scored between 0 and 1 using an LLM as a Judge framework, providing a more comprehensive assessment of descriptive quality.

Table 4. Breakdown of HDS Components for Fine-Tuned VLMs

Model	Sc	Sm	So	Sr	HDS
LLaVA	0.92	0.90	0.91	0.93	0.92
LLaMA	0.87	0.85	0.86	0.86	0.86
Qwen	0.85	0.83	0.84	0.84	0.84

The results from Table 4 indicate that LLaVA again achieves the highest performance, producing descriptions that are highly coherent and contextually relevant. LLaMA performs competitively but scores slightly lower for completeness, while Qwen shows reduced narrative flow despite maintaining accurate pair associations.

7.3   Unified Performance: CarDD_Score

To provide a single metric that reflects both structured correctness and semantic quality, we adopt the CarDD_Score, defined as the harmonic mean of the F1 Score and HDS. This formulation penalises weaknesses in either structural or semantic aspects, ensuring balanced evaluation.

Table 5. Unified CarDD_Score for Fine-Tuned VLMs

Model	F1 Score	HDS	CarDD_Score
LLaVA	0.88	0.92	0.90
LLaMA	0.82	0.86	0.84
Qwen	0.80	0.84	0.82

The results, summarised in Table 5, show that LLaVA achieves the highest overall performance with an F1 Score of 0.88 and an HDS of 0.92, resulting in a CarDD_Score of 0.89. LLaMA ranks second with an F1 Score of 0.82 and an HDS of 0.86 (CarDD_Score 0.83), followed by Qwen with an F1 Score of 0.80 and an HDS of 0.84 (CarDD_Score 0.81). These findings confirm that fine-tuned VLMs are capable of generating structured descriptions that combine precise attribute detection with coherent and contextually appropriate language. Figure 3 presents examples of these outputs, illustrating how the models effectively capture damage type, location and severity, and express this information through concise and contextually relevant descriptions.

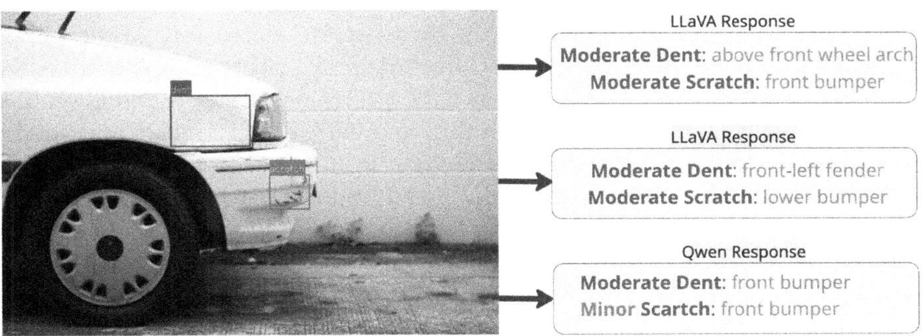

Fig. 3. Qualitative Comparison of Structured Outputs from Fine-Tuned VLMs. Green text denotes correct predictions, whereas red text marks incorrect outputs. (Color figure online)

7.4 Baseline Comparison: Fine-Tuned Vs Non-Fine-Tuned VLMs

To highlight the effect of domain-specific fine-tuning, we compare the proposed pipeline, which integrates YOLOv9 with fine-tuned VLMs, against an alternative configuration using the same detection backbone but non-fine-tuned VLMs. This comparison isolates the influence of fine-tuning on structured description generation by ensuring that both pipelines share identical detection outputs. Table 6 presents the detailed performance metrics for each model configuration.

As shown in Table 6, fine-tuned models consistently outperform their pretrained counterparts across all evaluation metrics. LLaVA (fine-tuned) achieves the highest scores, with an F1 Score of 0.88, HDS of 0.92, and a CarDD_Score of 0.90. Similarly, fine-tuned LLaMA and Qwen attain F1 Scores of 0.82 and 0.80, and HDS values of 0.86 and 0.84, resulting in CarDD_Scores of 0.84 and 0.82, respectively. These results confirm that fine-tuning significantly enhances both structured correctness and semantic coherence.

Table 6. Baseline Comparison using YOLOv9 with Fine-Tuned and Original VLMs

VLM Configuration	F1 Score	HDS	CarDD_Score
LLaVA (Fine-tuned)	0.88	0.92	0.90
LLaVA (Pretrained)	0.67	0.78	0.72
LLaMA (Fine-tuned)	0.82	0.86	0.84
LLaMA (Pretrained)	0.65	0.75	0.70
Qwen (Fine-tuned)	0.80	0.84	0.82
Qwen (Pretrained)	0.62	0.72	0.67

By contrast, non-fine-tuned VLMs demonstrate noticeably lower performance. For example, pretrained LLaVA records an F1 Score of 0.67 and HDS of 0.78, leading to a CarDD_Score of only 0.72. Pretrained LLaMA and Qwen perform similarly, with CarDD_Scores of 0.70 and 0.67, respectively. These models often generate free-form outputs that deviate from the required (damage type, body part, severity) structure, necessitating additional post-processing to extract structured information. This added complexity results in frequent omissions or incorrect output, which contribute to lower recall, reduced F1 Scores, and diminished semantic quality.

8 Conclusion

This paper has presented CarDamageEval, a dual-layer evaluation framework specifically designed to address the limitations of existing metrics for assessing VLMs in structured automotive damage analysis. The framework enforces a triplet output format linking damage type, vehicle body part, and severity level, enabling objective pair-level matching through established quantitative metrics such as precision, recall, and F1 score. To complement structural correctness, CarDamageEval incorporates the HDS, a semantic quality measure assessing correctness, completeness, coherence, and relevance, with the hybrid CarDD_Score harmonically uniting these two perspectives into a single balanced indicator. Experimental evaluation on fine-tuned and baseline VLMs demonstrated that the framework reliably differentiates model performance across diverse damage scenarios, including visible defects, spatially challenging locations, and subtle surface anomalies, with fine-tuned LLaVA achieving the highest overall scores. These findings highlight the importance of structured output formats and hybrid evaluation in delivering interpretable, reproducible benchmarks for precision-critical domains such as insurance claim processing, resale valuation, and fleet management. However, CarDamageEval is currently limited by its focus on two-dimensional imagery, reliance on English-language outputs, class imbalance across damage types, and potential sensitivity of semantic scoring to prompt design. Future extensions will address these limitations by incorporating multi-view and three-dimensional reasoning, expanding dataset coverage of rare and occluded damage cases, supporting multilingual and cost-aware evaluation, and refining semantic assessment for greater consistency. Collectively, these developments aim to establish CarDamageEval as a standardised and reusable benchmark not only for automotive applications but also for broader structured vision language tasks in industrial and healthcare domains.

Use of Generative AI (e.g., LLMs)

Generative AI tools were used solely for linguistic proofreading and grammatical review during the drafting of the manuscript. All technical content was developed independently by the authors, who assume full responsibility for the originality and accuracy of the submission.

References

1. Radford, A., et al.: Learning transferable visual models from natural language supervision. In: International Conference on Machine Learning, pp. 8748–8763. PMLR (2021)
2. Bai, S., et al.: Qwen2. 5-VL technical report. arXiv preprint: arXiv:2502.13923 (2025)
3. Grattafiori, A., et al.: The Llama 3 herd of models. arXiv preprint: arXiv:2407.21783 (2024)
4. Reddy, D.A., Shambharkar, S., Jyothsna, K., Kumar, V.M., Bhoyar, C.N., Somkunwar, R.K.: Automatic vehicle damage detection classification framework using fast and mask deep learning. In: 2022 Second International Conference on Computer Science, Engineering and Applications (ICCSEA), pp. 1–6 (2022)
5. Mishra, S., Kamal, D., Kumar K, S.: Vehicle damage identification using deep learning techniques. In: 2024 IEEE International Students' Conference on Electrical, Electronics and Computer Science (SCEECS), pp. 1–6 (2024)

6. González-Chávez, O., Ruiz, G., Moctezuma, D., Ramirez-delReal, T.: Are metrics measuring what they should? An evaluation of image captioning task metrics. Sign. Process.: Image Commun. **120**, 117071 (2024). https://www.sciencedirect.com/science/article/pii/S0923596523001534

7. ConfidentAI, DeepEval: the LLM evaluation framework (2025). https://github.com/confident-ai/deepeval. Accessed 24 Sep 2025

8. Rababaah, A.R.: Investigation of deep learning models for vehicle damage classification. In: 2023 10th International Conference on Signal Processing and Integrated Networks (SPIN), pp. 25–30. IEEE (2023)

9. Wang, X., Li, W., Wu, Z.: CarDD: a new dataset for vision-based car damage detection. IEEE Trans. Intell. Transp. Syst. (2023)

10. Huynh, N.T., Tran, N.N., Huynh, A.T., Hoang, V.-D., Nguyen, H.D.: VehiDE dataset: new dataset for automatic vehicle damage detection in car insurance. In: 2023 15th International Conference on Knowledge and Systems Engineering (KSE), pp. 1–6. IEEE (2023)

11. Chen, X., et al.: Microsoft coco captions: data collection and evaluation server. arXiv preprint: arXiv:1504.00325 (2015)

12. Antol, S., et al.: VQA: visual question answering. In: Proceedings of the IEEE International Conference on Computer Vision (ICCV) (2015)

13. Hudson, D.A., Manning, C.D.: GQA: a new dataset for real-world visual reasoning and compositional question answering. arXiv preprint: arXiv:1902.09506 (2019)

14. Goyal, Y., Khot, T., Summers-Stay, D., Batra, D., Parikh, D.: Making the V in VQA matter: elevating the role of image understanding in visual question answering. In: Proceedings of the IEEE Conference on Computer Vision and Pattern Recognition, pp. 6904–6913 (2017)

15. Chen, J., et al.: Revisiting referring expression comprehension evaluation in the era of large multimodal models. arXiv preprint: arXiv:2406.16866 (2024)

16. Hu, E.J., et al.: LoRA: low-rank adaptation of large language models. ICLR **1**(2), 3 (2022)

17. Redmon, J., Divvala, S., Girshick, R., Farhadi, A.: You only look once: unified, real-time object detection. In: Proceedings of the IEEE Conference on Computer Vision and Pattern Recognition, pp. 779–788 (2016)

18. Liu, H., Li, C., Li, Y., Lee, Y.J.: Improved baselines with visual instruction tuning. arXiv preprint: arXiv:2310.03744 (2023)

Regularizing StyleGAN with Inter-resolution Residual Pattern Consistency via a Laplacian Pyramid

Takahiro Koide[1] and Keisuke Kameyama[2]

[1] Degree Programs in Systems and Information Engineering and Graduate School of Science and Technology, University of Tsukuba, Tsukuba 305-8577, Japan
`koide@adapt.cs.tsukuba.ac.jp`
[2] Institute of Systems and Information Engineering, University of Tsukuba, Tsukuba 305-8577, Japan
`keisuke@cs.tsukuba.ac.jp`

Abstract. In image generation using StyleGAN2, which utilizes multiple layers corresponding to different resolutions, architectural flexibility allows higher-resolution layers to dominate the generation process. This may disrupt the functional hierarchy and lead to structural failures such as distorted faces or inconsistent layouts. We propose Residual Pattern Consistency (RPC), a lightweight regularization method that aligns each intermediate output with the corresponding frequency band from the Laplacian decomposition of the final output, encouraging resolution-specific generation without modifying the generator. We further introduce the Structural Consistency Score (SCS) to quantify the alignment between intermediate residuals and their expected roles. Experiments on the FFHQ, AFHQv2, and LSUN Church datasets show that RPC reduces Perceptual Path Length (PPL) and improves SCS, while incurring only a slight increase in Fréchet Inception Distance (FID). Qualitative results confirm that RPC notably reduces the frequency and severity of structural failures.

Keywords: Representation Learning · Generative Model · GAN · Regularization

1 Introduction

Generative Adversarial Networks (GANs) [5,11,15] have achieved remarkable success in image synthesis and editing, especially in generating high-resolution, photorealistic images with less computation. This success is due to the hierarchical nature of image generation; lower-resolution layers are responsible for global structure, while higher-resolution layers refine local textures and details. This assumption is reflected in both architectural designs [10,11,17] and downstream editing methods [6,19,21], which rely on hierarchical disentanglement to enable semantic control. Given this hierarchical foundation, the generated outputs are typically coherent and structurally plausible (Fig. 1 (left)). Despite this assumed hierarchy, we occasionally do observe structurally inconsistent images that violate spatial or semantic consistency (Fig. 1 (right)). This indicates that the assumed hierarchy is not always enforced during generation; this observation motivates our investigation into the causes and solutions to mitigate such failures.

Fig. 1. Random samples from the PPL distribution. **Left**: images in the lowest 10% (shortest PPL, smoother latent transitions). **Right**: images in the highest 10% (longest PPL, more likely to show structural artifacts). Red frames indicate images that appear to containing structural artifacts.

Despite the presumed hierarchical nature of GAN-based generation, the occurrence of structural failures raises the question of which underlying factors lead to such artifacts. Karras et al. [11] introduced Perceptual Path Length (PPL), a metric designed to quantify the smoothness of the latent space and its impact on editability. They later showed that higher PPL values were associated with more visual artifacts and structural inconsistencies (Fig. 1). Furthermore, follow-up work [9] has indicated that incorporating hierarchical structures into the generator can reduce PPL values and lead to more stable outputs. Taken together, these findings support the hypothesis that the lack of explicit hierarchical constraints in the GAN loss may contribute to structural inconsistencies. These observations motivate our approach; we augment the GAN training objective with explicit hierarchical constraints to enforce functional hierarchy and suppress structural artifacts.

Before introducing explicit hierarchical constraints into the GAN loss, we first investigate why a hierarchical structure fails to emerge under standard training. As an example of a hierarchical generator, StyleGAN2 [12] employs skip connections from multiple resolutions to the output, combining information from multiple scales in the final image. Visualizing the intermediate outputs at each resolution reveals that contributions concentrate in the higher-resolution stages, which can undermine the intended hierarchy (Fig. 2 (top)). We refer to these resolution-specific intermediate outputs, which are linearly combined to form the final image, as *residual patterns*, capturing how the image is progressively constructed across scales. This observation motivates our investigation into training objectives that explicitly reinforce resolution-specific roles to preserve the functional hierarchy of the generator.

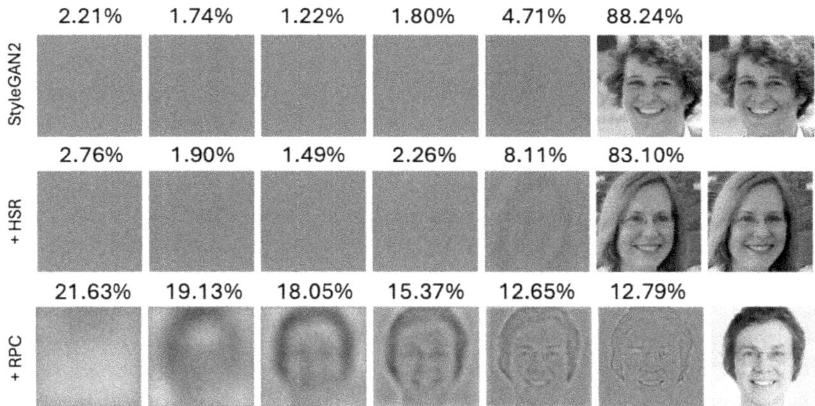

Fig. 2. Comparison of final output images and their residual patterns. **Top**: StyleGAN2 (baseline), **Middle**: HSR-based generator, **Bottom**: RPC-based generator (ours). From left to right, columns correspond to progressively higher resolutions, with the rightmost column showing the reconstructed final image obtained by summing all residuals. Percentages above each residual indicate its contribution, measured as the ratio of the average standard deviation of pixel values.

To address the underutilization of the hierarchy observed in current GANs, we propose a method that explicitly enforces the hierarchical structure by constraining the residual patterns. Our central idea is that lower-resolution layers should contribute to global structure, while higher-resolution layers should refine local details. To this end, we begin by analyzing how resolution-wise contributions should behave to support functional hierarchy. Based on these findings, we introduce a regularization term that penalizes disproportionate contributions from higher-resolution layers, encouraging each resolution level to specialize in its intended role within the generation process. This regularization promotes active contributions from lower-resolution layers and prevents higher-resolution layers from overriding them, thereby maintaining the intended role across resolutions.

Our contributions are summarized as follows:

- We propose a regularization method that enforces hierarchy in GANs by constraining the residual patterns at each resolution.
- The proposed method requires no architectural modifications or external supervision, and can be seamlessly integrated into existing GAN frameworks with minimal computational cost.
- Experimental results show improved structural consistency and reduced PPL across multiple datasets.

2   Related Work

2.1   Image Synthesis with a Hierarchical Process

Progressive Growing of GANs (PGGAN) [10] introduced a training strategy in which the output resolution increases gradually. This approach promotes a hierarchical generation process, where the lower-resolution layers model the global structure and higher-resolution layers add local details. The resulting architecture improves the visual quality, particularly in facial image synthesis. Notably, the hierarchical behavior is induced by the training procedure rather than being embedded in the model design.

StyleGAN [11] incorporated a style-based architecture, where latent codes are injected into each layer to modulate the features independently. This mechanism facilitates the layer-wise control over visual attributes, with the lower-resolution layers influencing the structural features and the higher-resolution layers affecting the fine-scale appearance. Subsequent methods such as StyleSpace [21] and InterFaceGAN [18] utilized this structure to enable attribute-specific manipulations across resolutions.

However, the generator does not include explicit constraints to maintain the functional hierarchy. As a result, resolution-specific roles may degrade during training, which can affect the consistency of the generated outputs.

2.2   Enforcing Hierarchical Structure via Regularization

To address the lack of explicit hierarchical control in GANs, several works have proposed methods to regularize or guide the behavior of individual layers during generation. For instance, some editing-based approaches [18, 19] exploit linear directions in the latent space that correspond to semantic factors, implicitly relying on the hierarchical nature of generators. These methods suggest the benefit of resolution-specific constraints, but they are designed primarily for post-hoc manipulation, rather than enforcing hierarchy during training.

More recent approaches such as HSR [9] introduce explicit structural supervision by incorporating pretrained vision models like DINO [3] to guide the generator toward semantically coherent outputs. While effective, these methods depend on external models trained on large-scale datasets such as ImageNet, raising concerns about feature leakage and limited generalizability to unseen domains. Additionally, supervision from pretrained embeddings may not align with the internal inductive biases of GANs. These drawbacks highlight the need for an internal mechanism that promotes hierarchical organization without relying on external supervision or architectural modifications.

2.3   Suppression of Structural Artifacts in GANs

Song et al. [20] investigate a phenomenon they term Feature Proliferation, where certain feature activations amplify through forward propagation in the generator, leading to characteristic artifacts in StyleGAN outputs. To mitigate these defects, they propose a post-processing feature rescaling method that identifies the "risky" features prone to proliferation and scales them down. This method is more fine-grained and less destructive than the traditional truncation trick [1], as it operates in the lower-level feature space rather than the latent space and preserves image diversity while improving the perceptual quality. However, since it is applied after image generation, it does not directly suppress the proliferation during training, and thus does not enforce structural consistency in the generation process itself.

2.4   Positioning of Our Method

In contrast to prior methods that either rely on implicit assumptions about generator structure or depend on external supervision, our approach directly addresses the breakdown of functional hierarchy in GANs. Specifically, we introduce a training-time regularization that penalizes over-reliance on higher-resolution layers, thereby encouraging each resolution level to fulfill its intended semantic role. Unlike methods such as HSR, our technique does not require pre-trained vision encoders or architectural modifications, making it broadly applicable to existing GAN frameworks. By promoting the resolution-specific roles from within the model itself, our method offers a lightweight yet effective mechanism for reinforcing the hierarchical image structure and improving the reliability of image synthesis.

3   Residual Pattern Consistency

3.1   Motivation: Structural Inconsistency and PPL

Modern GAN architectures, such as PGGAN and StyleGAN, adopt a coarse-to-fine generation process. This design implicitly assumes a functional hierarchy across resolution levels. However, in practice, the generated images often exhibit structural failures, including spatial distortions, duplicated patterns, or malformed features. These artifacts suggest that the expected division of responsibilities across layers does not always hold, raising the possibility that the hierarchical organization is not preserved during training.

Understanding the factors that cause structurally inconsistent images is essential for addressing this problem. Previous studies have reported that higher

PPL scores are associated with unstable generation and reduced structural consistency. In our experiments, we observed that the images in the highest 10% by PPL exhibit clear signs of structural failure, such as deformed faces or inconsistent backgrounds as shown in Fig. 1 (right). Empirical analysis [12] has shown that removing the progressive growing from the generator, which reduces its hierarchical structure, results in a substantial increase in PPL. Furthermore, HSR [9], which explicitly encourages hierarchical behavior during training, reports consistent improvements in PPL. This observation suggests a potential relationship: preserving functional hierarchy within the generator can mitigate such artifacts. However, to devise a principled approach for enforcing such hierarchy, we must first understand why it fails to emerge reliably in current architectures.

3.2 Problem Statement

To understand how the hierarchical structure may break down in modern GANs, we begin by formalizing the skip-connected generator architecture employed in StyleGAN2 (see Fig. 3). The generator produces the final image I_{final} by summing the residual patterns I_k at resolution levels $k \in \{2, \ldots, K\}$, upsampled to the final output size:

$$I_{\text{final}} = \sum_{k=2}^{K} \text{UP}^{K-k}(I_k),$$

where K is the total number of resolution layers in the generator and UP denotes a $2\times$ upsampling operator.

This formulation enables every resolution layer to contribute additively to the final output, providing flexibility for each layer to generate content at the full image scale. However, the generator places no constraints on what these layers should generate. As a result, the generator is free to distribute the contributions arbitrarily across layers, regardless of their intended roles.

Because the higher-resolution layers can represent both fine textures and coarse global structures, they can overwrite the outputs of the lower-resolution layers, potentially undermining the intended resolution-specific roles. Empirical visualizations of residual patterns (Fig. 2) support this; as training progresses, lower-resolution contributions become minimal, while higher-resolution layers increasingly take control of the final image. These observations suggest that, in the absence of explicit constraints, the generator may fail to realize a functionally hierarchical process, leading to structurally unstable outputs.

3.3 Desirable Residual Pattern

To enforce the functional hierarchy during generation, we first clarify the resolution-wise behavior that defines desirable residual patterns. In a well-structured generator, each resolution layer should contribute to the final image I_{final} in a way that reflects its spatial scale. To avoid redundancy and interference across scales, these contributions should be disjoint in the image space spanned by a basis of spatial-frequency components. In this case, each layer is responsible

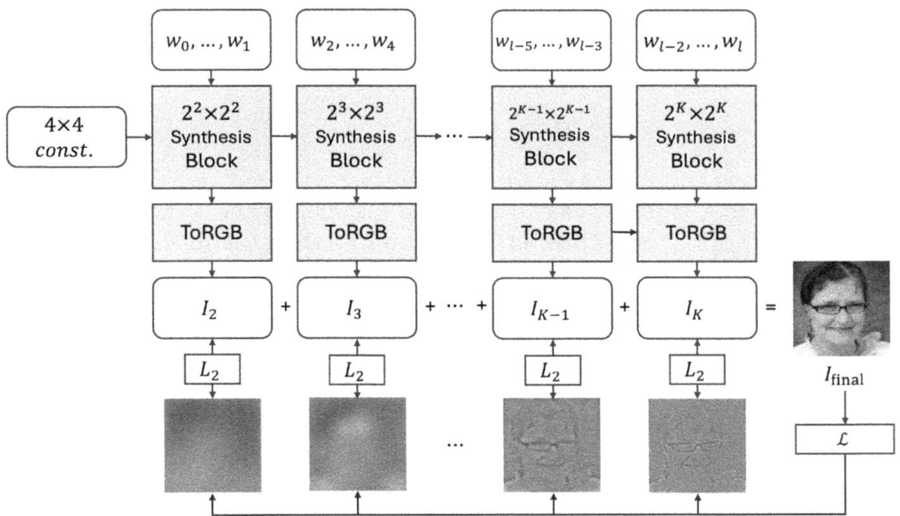

Fig. 3. Illustration of the StyleGAN2 generator architecture with the proposed RPC regularization. Each resolution includes a `ToRGB` layer contributing to the final output via skip connections. The Laplacian decomposition $\boxed{\mathcal{L}}$ and residual loss computation $\boxed{L_2}$ are applied at each resolution to align residual patterns with their corresponding frequency bands.

only for the frequency band corresponding to its resolution. This band separation forms a foundation for a functionally hierarchical generator, where each layer specializes in an appropriate scale of visual information.

To achieve this band separation, we adopt the Laplacian pyramid [2] as a practical decomposition method aligned with the generator's multiscale architecture. The Laplacian pyramid decomposes an image into a set of spatial frequency bands, with each band $\mathcal{L}_k(I_{\text{final}})$ corresponding to a particular resolution level k. In our context, these frequency bands serve as targets for the residual patterns $\{I_k\}$ produced at each layer of the generator. This decomposition is computationally efficient and naturally fits the skip-connected architecture of StyleGAN2, where residual patterns are summed after upsampling. It thus provides a scalable and interpretable reference for enforcing resolution-specific responsibilities during training.

3.4   Regularization and Training Algorithm

To enforce this hierarchical residual pattern into the generation behavior, we introduce a Laplacian-based regularization loss L_{rpc} (Residual Pattern Consistency; RPC). This loss penalizes discrepancies between each residual pattern I_k

and its corresponding Laplacian frequency band $\mathcal{L}_k(I_{\text{final}})$:

$$L_{\text{rpc}} = \frac{1}{K-1} \sum_{k=2}^{K} \|I_k - \mathcal{L}_k(I_{\text{final}})\|_2^2 .$$

This formulation encourages each resolution layer to specialize in its designated frequency range, resulting in well-structured residual patterns that reflect functional hierarchy. See Fig. 3 for an illustration of the generator structure and the location of the proposed regularizer. By penalizing the deviations from the desirable frequency-aligned decomposition, the regularizer promotes functional separation across layers and supports the emergence of a coarse-to-fine generation hierarchy.

To integrate the proposed RPC loss into the GAN training, we augment the standard objective with an additional regularization term:

$$L_{\text{total}} = L_{\text{GAN}} + \lambda_{\text{rpc}} L_{\text{rpc}},$$

where λ_{rpc} controls the strength of the frequency alignment constraint. Importantly, this modification requires no changes to the generator architecture. The computation of the Laplacian pyramid and corresponding residual patterns involves only simple upsampling and downsampling operations, resulting in a negligible computational overhead. This makes the RPC regularization broadly applicable and lightweight, facilitating an easy integration into existing StyleGAN-based frameworks.

Compatibility with Style Mixing Regularization StyleGAN models often employ style mixing regularization [11] during training to encourage disentangled representations across different layers. In this scheme, two latent codes are sampled and mapped to two style vectors $\boldsymbol{w}^{(A)}$ and $\boldsymbol{w}^{(B)}$ of length l, and their layer-wise style vectors are mixed at a randomly selected cutoff layer $c \in \{0, \ldots, l\}$. The mixed style vector is defined as $\boldsymbol{w}^{(\text{mix})} = (\boldsymbol{w}_0^{(A)}, \ldots, \boldsymbol{w}_c^{(A)}, \boldsymbol{w}_{c+1}^{(B)}, \ldots, \boldsymbol{w}_l^{(B)})$, and the resulting image, synthesized by the generator G, $I_{\text{mix}} = G(\boldsymbol{w}^{(\text{mix})})$, is included in the training batch.

However, when combining this style mixing regularization with our RPC, unintended leakage of style information may occur. Specifically, although styles above the cutoff originate from $\boldsymbol{w}^{(B)}$, the final image tends to contain features from both styles. The Laplacian pyramid computed from I_{mix} may reflect the higher-resolution style $\boldsymbol{w}^{(B)}$ even in low-frequency bands, thereby introducing a conflict with the residual pattern loss, which assumes frequency-localized responsibility.

To address this issue, we modify the application of RPC during style-mixed training iterations. Specifically, we apply the L_{rpc} loss only to layers at or above the resolution indexed by the cutoff layer c. This strategy ensures that the regularization is only imposed on layers whose style originates from a single latent code, avoiding "contamination" across layers and preserving compatibility with the objectives of style mixing.

4   Experiments

4.1   Experimental Setup

We evaluate our method on three widely used datasets for image synthesis: FFHQ [11], AFHQv2 [4], and LSUN Church [22]. All datasets are resized to a spatial resolution of 128×128 for consistency across experiments. Preprocessing involves mean and variance normalization, and horizontal flipping is applied as data augmentation for all datasets. For LSUN Church, we additionally apply center cropping after resizing.

In the experiments, we compare the following generative models:

- **StyleGAN2 (baseline)**: Standard architecture with skip-connected generator as described in Sect. 3, along with a residual discriminator. It uses Modulated Convolution, LeakyReLU activations, and equalized learning rate.
- **HSR** [9]: StyleGAN2 backbone with additional resolution-specific supervision. No architectural changes; supervision applied to intermediate representations.
- **RPC (ours)**: StyleGAN2 backbone with the proposed Residual Pattern Consistency (RPC) regularization term applied to residual patterns.

All models are trained using the Adam optimizer [13] with a learning rate of 0.002, $\beta_1 = 0$, and $\beta_2 = 0.99$. The total number of training iterations is 25M images for FFHQ, 10M images for AFHQv2, and 20M images for LSUN Church, with early stopping if the FID score shows no improvement.

4.2   Evaluation Metrics

We evaluate the methods using three metrics:

- Fréchet Inception Distance (FID) [7]: evaluates fidelity by measuring the distributional distance from real images. We use 50k real and 50k generated images for evaluation.
- Perceptual Path Length (PPL) [11]: measures the smoothness of latent space interpolations, indicating structural stability. We use 20k pairs of images for evaluation.
- Structural Consistency Score (SCS): our proposed metric to quantify how well the residual patterns contribute consistently to their designated frequency bands. 10k generated images are used for evaluation.

Structural Consistency Score. To directly evaluate the hierarchical consistency of the generation process, we introduce a new metric called the Structural Consistency Score. This metric quantifies how coherently each resolution contributes to the final image without being overwritten by subsequent higher-resolution layers.

Specifically, for each intermediate resolution level $2 \leq j < K$, we compute the following:

1. **Accumulated Image** ($I^{(j)}_{\text{accum}}$): An image obtained by summing up all residual patterns I_k from resolutions 2 to j, upsampled accordingly:

$$I^{(j)}_{\text{accum}} = \sum_{k=2}^{j} \text{UP}^{j-k}(I_k)$$

2. **Target Low-Frequency Component** ($\mathcal{G}_j(I_{\text{final}})$): From the final output I_{final}, the j-th level of the Gaussian pyramid [2] $\mathcal{G}_j(I_{\text{final}})$ is computed, representing the desirable low-frequency content at that scale.

The consistency score at resolution j, denoted by S_j, is defined as the peak signal-to-noise ratio (PSNR) between the accumulated image and the Gaussian target:

$$S_j = \text{PSNR}\left(I^{(j)}_{\text{accum}}, \mathcal{G}_j\left(I_{\text{final}}\right)\right)$$

A higher S_j indicates that the structure constructed at lower resolutions is preserved at that scale and not overwritten by higher-resolution layers. We report these scores across all intermediate layers to analyze and compare the internal generation strategies of different models.

4.3   Effectiveness of Enforcing Hierarchical Generation

Using the training configurations described in Sect. 4.1, we evaluate RPC against StyleGAN2 (baseline). The experiments are organized as follows:

- FFHQ: comparisons among StyleGAN2, HSR, and RPC.
- AFHQv2 and LSUN Church: comparisons between StyleGAN2 and RPC.

Quantitative comparisons of FID and PPL are reported in Tables 1 and 3, while SCS is reported in Table 2. Distributional changes in PPL are shown in Fig. 4, and qualitative examples are provided in Fig. 5.

Reduction in PPL. As discussed in Sect. 3, lower PPL indicates smoother image transitions under latent-space interpolation and correlates with structural consistency. In Tables 1 and 3, across all datasets, the proposed RPC regularizer consistently reduced PPL compared to StyleGAN2 and HSR. For example, on FFHQ, PPL dropped from 989.5 to 491.9 (a 50.3% reduction), while similar improvements were observed for AFHQv2 and LSUN Church. This result suggests that the proposed frequency-constrained generation enhances semantic stability along latent paths, supporting our goal of restoring hierarchical behavior.

Beyond the averages, in Fig. 4, RPC shifted the entire PPL distribution leftward and shortened its right tail. In practical terms, this means that outlier cases with severe structural instability become substantially rarer under RPC.

Table 1. Comparison of FID and PPL scores on the FFHQ dataset. Lower values indicate better fidelity (FID) and smoother latent space traversal (PPL).

Methods	FID($\downarrow$)	PPL($\downarrow$)
StyleGAN2	**10.04**	989.5
+HSR	12.10	764.6
+RPC ($\lambda_{\mathrm{rpc}} = 2.0$)	10.57	**491.9**

Table 2. Comparison of Structural Consistency Score on FFHQ dataset. Higher values indicate the consistency between the final image I_{final} and each residual pattern I_k.

Methods	4×4	8×8	16×16	32×32	64×64
StyleGAN2	7.51	6.49	5.88	5.59	5.66
+HSR	7.53	6.48	5.90	5.52	6.12
+RPC ($\lambda_{\mathrm{rpc}} = 2.0$)	**13.16**	**14.61**	**16.25**	**17.37**	**17.95**

Table 3. Comparison of FID and PPL on AFHQv2 and LSUN Church datasets.

Methods	AFHQv2		LSUN Church	
	FID($\downarrow$)	PPL ($\downarrow$)	FID ($\downarrow$)	PPL ($\downarrow$)
StyleGAN2	**11.17**	1161.6	**7.07**	1337.8
+RPC ($\lambda_{\mathrm{rpc}} = 2.0$)	11.60	**720.0**	8.03	**813.0**

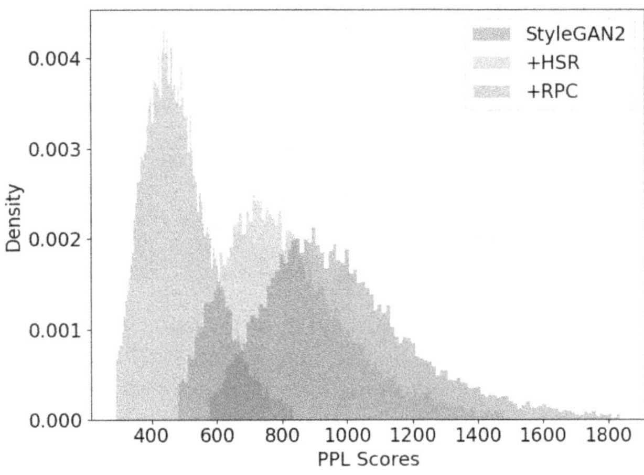

Fig. 4. Distributions of PPL for StyleGAN2, HSR, and the proposed RPC across generated image samples. Each bar chart represents the empirical distribution of PPL scores computed from random latent interpolations.

Improvement in Structural Consistency Score (SCS). While FID and PPL capture fidelity and latent smoothness, they do not directly quantify hierarchical consistency. In order to measure it, we use SCS. SCS quantifies how much each residual pattern contributes to the final image in a manner aligned with its frequency band, as defined by the Gaussian pyramid. A higher SCS value indicates that lower-resolution structures are preserved and not overwritten by higher-resolution layers.

On FFHQ, RPC increased SCS across all scales (Table 2), with especially large gains at early resolutions (e.g., from 7.51 to 13.16 at 4×4). This confirms that RPC encourages each layer to specialize in its intended frequency band, reducing redundancy and reinforcing a clean hierarchical decomposition.

Qualitative Reduction of Structural Failures. To visually validate structural consistency, we sample from the highest 10% of latent interpolations ranked by PPL. Figure 5 compares results from HSR and RPC under this protocol. While both methods occasionally exhibit structural failures, such as facial parts inconsistency or inconsistent backgrounds, the frequency and severity of such artifacts are reduced. Compared to HSR, RPC produces cleaner spatial arrangements and better-preserved textures, qualitatively supporting the improvement in structural consistency observed in quantitative metrics.

Fig. 5. Random samples from the highest 10% of the PPL distribution. **Left**: Generated images with HSR. **Right**: Generated images with RPC ($\lambda_{\mathrm{rpc}} = 2.0$). Red frames indicate images that appear to containing structural artifacts.

Additional Evidence from Residual Pattern. To further validate the effectiveness of RPC in promoting resolution-specific specialization, we visualize the residual pattern $\{I_k\}$ across multiple layers for StyleGAN2, HSR, and our method. As shown in Fig. 2, StyleGAN2 shows minimal contribution in the lower-resolution layers, with later layers dominating the final output. HSR partially alleviates this effect but retains considerable redundancy across scales. In contrast, RPC yields residual patterns that align closely with their expected frequency bands. Early layers synthesize low-frequency structure, while higher layers add fine details without overriding global content. This visual evidence is consistent with the improvements in Structural Consistency Score (SCS), supporting the claim that RPC effectively restores functional hierarchy.

Trade-off in Fidelity. While RPC improves structural consistency and reduces PPL by enforcing frequency-specific generation, it causes a slight increase in FID (e.g., FFHQ: 10.04←10.57 in Table 1). This trade-off can be attributed to the suppression of cross-band redundancy.

In StyleGAN2, features at each resolution are reused by subsequent layers, allowing higher-resolution blocks to refine both fine details and coarse structures. This enables a joint optimization across scales. In contrast, RPC constrains each layer to operate within its frequency band, limiting access to the low-frequency context in the higher-resolution layers. As a result, while structural roles become more specialized (higher SCS), the ability to co-optimize across resolutions is reduced and the representational flexibility of higher-resolution layers is constrained, slightly degrading distributional fidelity.

4.4   Ablation Study

We evaluate how the regularization strength λ_{rpc} affects the model's behavior by varying it from 0.1 to 10.0 on the FFHQ dataset. Results for FID, PPL, and Structural Consistency Score (SCS) are shown in Tables 4 and 5.

As λ_{rpc} increases, PPL generally decreases, indicating smoother latent interpolations and improved structural stability. For example, increasing λ_{rpc} from 0 to 5.0 reduces PPL from 989.5 to 365.0. Similarly, SCS improves across all resolutions, showing that hierarchical consistency is better enforced with stronger regularization.

However, higher λ_{rpc} also leads to a gradual rise in FID, likely because the strict frequency separation reduces the model's ability to reuse low-frequency information at later stages. This may negatively affect the sample fidelity or diversity.

Very small λ_{rpc} (e.g., 0.1) yielded limited gains, suggesting that minimal regularization is insufficient to enforce resolution-wise specialization.

Among all settings, $\lambda_{\mathrm{rpc}} = 2.0$ offered the best balance, roughly halving PPL, keeping FID close to baseline, and yielding consistent SCS improvements. This makes it a practical choice for general use.

In summary, RPC provides a tunable mechanism to balance structural consistency and sample quality. The strength of regularization can be adjusted depending on task priorities, such as robustness or realism.

Table 4. Comparison of the effects of λ_{rpc} on FID, PPL on FFHQ dataset.

λ_{rpc}	FID($\downarrow$)	PPL($\downarrow$)
0	10.04	989.5
0.1	12.48	648.2
1.0	11.50	475.5
2.0	**10.57**	491.9
5.0	13.60	**365.0**
10.0	12.67	386.5

Table 5. Comparison of the effects of λ_{rpc} on Structural Consistency Score on FFHQ dataset.

λ_{rpc}	4×4	8×8	16×16	32×32	64×64
0	7.51	6.49	5.88	5.59	5.66
0.1	7.94	6.92	7.10	9.05	13.27
1.0	12.23	13.16	14.51	15.96	17.52
2.0	13.16	14.61	16.25	17.37	17.95
5.0	14.17	16.24	17.93	18.38	18.10
10.0	**14.85**	**17.24**	**18.99**	**19.00**	**18.26**

5 Conclusion

In this work, we addressed the breakdown of functional hierarchy in GANs, where higher-resolution layers often dominate generation, undermining the intended coarse-to-fine structure. To resolve this, we proposed the Residual Pattern Consistency (RPC), a lightweight regularization method that encourages each resolution layer to specialize in its corresponding frequency band using Laplacian pyramid decomposition.

Experiments on FFHQ, AFHQv2, and LSUN Church datasets demonstrated that RPC significantly reduced PPL and improved our proposed Structural Consistency Score (SCS), indicating more stable and hierarchically organized generation. Although FID increased slightly due to reduced cross-resolution redundancy, this trade-off favors structural consistency and is especially beneficial for applications requiring semantic robustness.

RPC is architecture-agnostic, supervision-free, and can be easily integrated into existing StyleGAN frameworks, making it applicable to tasks such as semantic editing or domain adaptation. Moreover, the core idea of enforcing resolution-wise specialization may extend to diffusion models [8, 14, 16], suggesting future directions for improving structural consistency in generative models.

Disclosure of Interests. The authors have no competing interests to declare that are relevant to the content of this article.

References

1. Brock, A., Donahue, J., Simonyan, K.: Large scale GAN training for high fidelity natural image synthesis. In: ICLR (2019)
2. Burt, P.J.: The pyramid as a structure for efficient computation. In: Multiresolution Image Processing and Analysis, pp. 6–35. Springer (1984)
3. Caron, M., et al.: Emerging properties in self-supervised vision transformers. In: Proceedings of the IEEE/CVF International Conference on Computer Vision, pp. 9650–9660 (2021)

4. Choi, Y., Uh, Y., Yoo, J., Ha, J.W.: StarGAN v2: diverse image synthesis for multiple domains. In: Proceedings of the IEEE Conference on Computer Vision and Pattern Recognition (2020)
5. Goodfellow, I., et al.: Generative adversarial nets. In: NeurIPS, vol. 27 (2014)
6. Härkönen, E., Hertzmann, A., Lehtinen, J., Paris, S.: GANspace: discovering interpretable GAN controls. NeurIPS **33**, 9841–9850 (2020)
7. Heusel, M., Ramsauer, H., Unterthiner, T., Nessler, B., Hochreiter, S.: GANs trained by a two time-scale update rule converge to a local Nash equilibrium. In: NeurIPS (2017)
8. Ho, J., Jain, A., Abbeel, P.: Denoising diffusion probabilistic models. In: Advances in Neural Information Processing Systems, vol. 33, pp. 6840–6851 (2020)
9. Karmali, T., et al.: Hierarchical semantic regularization of latent spaces in Style-GANs. In: ECCV. Lecture Notes in Computer Science, vol. 13675, pp. 443–459. Springer Nature Switzerland (2022)
10. Karras, T.: Progressive growing of GANs for improved quality, stability, and variation. arXiv preprint: arXiv:1710.10196 (2017)
11. Karras, T., Laine, S., Aila, T.: A style-based generator architecture for generative adversarial networks. In: CVPR, pp. 4401–4410 (2019)
12. Karras, T., Laine, S., Aittala, M., Hellsten, J., Lehtinen, J., Aila, T.: Analyzing and improving the image quality of StyleGAN. In: CVPR, pp. 8110–8119 (2020)
13. Kingma, D.P., Ba, J.: Adam: a method for stochastic optimization. In: ICLR (2015)
14. Lu, Y., Zhang, M., Ma, A.J., Xie, X., Lai, J.: Coarse-to-fine latent diffusion for pose-guided person image synthesis. In: Proceedings of the IEEE/CVF Conference on Computer Vision and Pattern Recognition, pp. 6420–6429 (2024)
15. Radford, A., Metz, L., Chintala, S.: Unsupervised representation learning with deep convolutional generative adversarial networks. In: ICLR (2016)
16. Ryu, D., Ye, J.C.: Pyramidal denoising diffusion probabilistic models. arXiv preprint: arXiv:2208.01864 (2022)
17. Sauer, A., Karras, T., Laine, S., Geiger, A., Aila, T.: StyleGAN-T: unlocking the power of GANs for fast large-scale text-to-image synthesis. In: International Conference on Machine Learning, pp. 30105–30118. PMLR (2023)
18. Shen, Y., Yang, C., Tang, X., Zhou, B.: InterfaceGAN: interpreting the disentangled face representation learned by GANs. IEEE Trans. Pattern Anal. Mach. Intell. **44**(4), 2004–2018 (2020)
19. Shen, Y., Zhou, B.: Closed-form factorization of latent semantics in GANs. In: CVPR, pp. 1532–1540 (2021)
20. Song, S., Liang, Y., Wu, J., Lai, Y.K., Qin, Y.: Feature proliferation–the "cancer" in StyleGAN and its treatments. In: Proceedings of the IEEE/CVF International Conference on Computer Vision, pp. 2360–2370 (2023)
21. Wu, Z., Lischinski, D., Shechtman, E.: StyleSpace analysis: disentangled controls for StyleGAN image generation. In: CVPR, pp. 12863–12872 (2021)
22. Yu, F., Seff, A., Zhang, Y., Song, S., Funkhouser, T., Xiao, J.: LSUN: construction of a large-scale image dataset using deep learning with humans in the loop. arXiv preprint: arXiv:1506.03365 (2015)

Mixup and Local-FOMA Based Two-Phase Manifold Augmentation in Image Classification

Takumi Shirahama[1]([✉]) [iD] and Keisuke Kameyama[2] [iD]

[1] Degree Programs in Systems and Information Engineering,
Graduate School of Science and Technology, University of Tsukuba, Tsukuba, Japan
`shirahama@adapt.cs.tsukuba.ac.jp`
[2] Institute of Systems and Information Engineering, University of Tsukuba,
Tsukuba, Japan
`kameyama@cc.tsukuba.ac.jp`

Abstract. In deep learning-based image classification, it is necessary to achieve high generalization performance while suppressing overfitting in situations when training data are limited. Mixup, a data augmentation using linear interpolation in the input space, is effective in smoothing the decision boundary, but it can hinder the model's ability to learn fine-grained, class-specific details in the later training stages. On the other hand, augmentation by First-Order Manifold Augmentation (FOMA) utilizes local geometry but is designed for regression, making its direct application to classification difficult. We propose a two-phase augmentation method that changes its strategy according to the phase of training. In Phase 1, standard Mixup is used for global interpolation to smooth the class boundary. In Phase 2, we apply Local-FOMA, which is designed for classification, to estimate principal component directions based on local neighborhoods in the feature space and perform data augmentation while preserving the manifold structure. The weights of the loss function are linearly transitioned during the learning process, enabling a smooth switch from Mixup to Local-FOMA. On CIFAR-100 and CIFAR-10, our method outperformed strong Mixup variants. Under distribution shifts (CIFAR-C), it also improves robustness and was particularly effective on CIFAR-100-C, suggesting that the method is particularly advantageous in complex scenarios involving datasets with many classes. These results demonstrate that Mixup-FOMA provides an effective strategy to enhance both classification performance and robustness by gradually combining global generalization with local refinement.

Keywords: Image Classification · Data Augmentation · Curriculum Learning · Mixup · Manifold Learning

1 Introduction

Deep learning has made remarkable progress in image classification, with powerful models such as Wide-ResNet [18] and Vision Transformer (ViT) [4] achieving high accuracy. However, these models have a large number of learnable parameters and are prone to overfit when the available training data is limited. To mitigate this issue, data augmentation has become an essential technique for improving generalization performance and preventing overfitting [12,14].

Traditional data augmentation methods primarily rely on geometric transformations such as rotation, flipping, and cropping. While effective, these methods require domain-specific adjustments and may compromise important structural features [2]. In contrast, Mixup [19], a domain-independent method, generates synthetic samples by linearly interpolating the images and the label vectors. It promotes smoothing of decision boundaries and enables to learn the class relationships across the entire input space [2,15,19].

While Mixup significantly improves the generalization performance, Zou et al. [20] have pointed out its limitations. They demonstrated that while Mixup is effective in acquiring rare features that are difficult to capture in standard learning, its positive contributions are concentrated in the early stages of learning, and in the latter stages, the regularization effect may diminish, hindering the effective utilization and refinement of fine-grained, class-specific features in the later training stages. Additionally, since interpolation is performed even between samples with low semantic relevance, there is a risk of distorting the local decision boundaries in the later stages.

On the other hand, First-Order Manifold Augmentation (FOMA) [8] achieves augmentation based on the data manifold by perturbing along the tangent space direction of the data manifold, thereby preserving the geometric structure of the data manifold even after the augmented data are added. While FOMA has shown promising results in regression tasks, its direct application to classification tasks with discrete class labels is challenging.

In this study, we propose a two-phase hybrid learning method that integrates the complementary strengths of Mixup and FOMA. In the first phase, Mixup is used to obtain global generalization and promote smoothing of the decision boundary. In the second phase, we introduce Local-FOMA, which has been redesigned for classification, to estimate the principal component direction using local data in the feature space and generate extended samples along the direction of the neighboring data manifold. This two-phase framework is expected to improve classification accuracy and robustness under distribution shifts. In the first phase, global interpolation promotes broad generalization ability. In the second phase, utilizing local structures enhances fine-grained recognition performance.

2   Related Works

2.1   Geometric Data Augmentation

In image classification, geometric data augmentation, such as translation, rotation, scaling, flip, crop, and cutout [3], has long been a fundamental and widely used technique for reducing overfitting and improving generalization. Especially, modern deep networks often have far more trainable parameters than there are training examples, making these augmentations essential [12,16]. However, this approach has its limitations. In particular, choosing which transformations to apply and at what intensity requires domain expertise [5,14].

2.2 Mixup

Mixup [19] is a data augmentation method that forms new training examples by taking two samples $(\boldsymbol{x}_i, \boldsymbol{y}_i)$ and $(\boldsymbol{x}_j, \boldsymbol{y}_j)$ and linearly interpolating both their inputs and labels:

$$\tilde{\boldsymbol{x}} = \gamma\boldsymbol{x}_i + (1 - \gamma)\boldsymbol{x}_j, \quad \tilde{\boldsymbol{y}} = \gamma\boldsymbol{y}_i + (1 - \gamma)\boldsymbol{y}_j \tag{1}$$

where $\gamma \in [0, 1]$ is typically drawn from a Beta distribution $\gamma \sim Beta(\alpha, \alpha)$ ($\alpha > 0$). Figure 1 shows an example of Mixup applied to a dog image ($\gamma = 0.6$) and a bird image ($1 - \gamma = 0.4$). As shown in the figure, the input is not a simple geometric transformation, but rather a pixel-by-pixel linear superposition, and the labels are similarly linearly interpolated.

Fig. 1. An example of Mixup: a Dog and a Bird image are combined with $\gamma = 0.6$, yielding a mixed image and label (Dog 0.6 : Bird 0.4).

By blending inputs and targets in this way, Mixup encourages the model to learn smoother decision boundaries and become more robust, thereby helping to prevent overfitting. Moreover, as noted by Cao et al. (2024), Mixup operates on a global mixing scheme across the input space, which facilitates learning of the broader structural relationships between the classes [2]. This enables the model not only to generalize better on unseen data but also to capture the global data structure. Following the success of Mixup, many variants have been proposed, e.g., mixing in feature space, cut-and-paste strategies, or similarity- and regularization-based formulations [1,13,15,17]. In addition, Zou et al. have contributed to the theoretical understanding of Mixup [20]. They demonstrated that Mixup promotes the learning of not only common features but also rare features that are often overlooked in standard learning. Furthermore, they showed that the effect of Mixup mainly appears in the early stages of learning, and based on this property, early-stopped Mixup achieved performance equivalent to or better than normal Mixup [20].

2.3   First-Order Manifold Augmentation

First-Order Manifold Augmentation (FOMA) [8] is a domain-agnostic augmentation method originally proposed for regression. It operates on a mini-batch at an intermediate layer by constructing a joint feature-output matrix and perturbing directions that are orthogonal to the dominant components of that joint space. Specifically, at layer l, let $\boldsymbol{Z}_l \in \mathbb{R}^{b \times n_l}$ denote the batch features (where b is the mini-batch size and n_l is the number of feature dimensions at layer l) and $\boldsymbol{Y} \in \mathbb{R}^{b \times m}$ the corresponding continuous targets (where m is the output dimension). Form the joint matrix

$$\boldsymbol{A} = [\boldsymbol{Z}_l \,|\, \boldsymbol{Y}] \in \mathbb{R}^{b \times (n_l + m)}, \tag{2}$$

and compute the Singular Value Decomposition (SVD)

$$\boldsymbol{A} = \boldsymbol{U}\boldsymbol{S}\boldsymbol{V}^\top, \quad \boldsymbol{S} = \mathrm{diag}(\sigma_1, \dots, \sigma_r) \tag{3}$$

with $r = \min(b, n_l + m)$ and singular values $\sigma_1 \geq \cdots \geq \sigma_r \geq 0$. To obtain a tangent-plane perturbation, the top h singular directions are preserved and the others are scaled by a factor $\epsilon \sim Beta(\alpha, \alpha)$:

$$\boldsymbol{S}(\epsilon, h) = \mathrm{diag}\big(\sigma_1, \dots, \sigma_h, \ \epsilon\sigma_{h+1}, \dots, \epsilon\sigma_r\big). \tag{4}$$

The augmented joint matrix is reconstructed as

$$\boldsymbol{A}(\epsilon, h) = \boldsymbol{U}\,\boldsymbol{S}(\epsilon, h)\,\boldsymbol{V}^\top, \tag{5}$$

and then split back into features and outputs by the columns,

$$\boldsymbol{Z}_l(\epsilon, h) = \boldsymbol{A}(\epsilon, h)_{1:n_l} \in \mathbb{R}^{b \times n_l}, \qquad \boldsymbol{Y}(\epsilon, h) = \boldsymbol{A}(\epsilon, h)_{n_l+1:n_l+m} \in \mathbb{R}^{b \times m}. \tag{6}$$

Choosing h. The number of preserved directions h can be determined by a hyper-parameter $\rho \in [0, 1]$ that specifies the amount of signal to keep unchanged. Using the cumulative sum of singular values, h is set as the largest index whose cumulative fraction does not exceed ρ:

$$h = \arg\max_{h^*} \frac{\sum_{s=1}^{h^*} \sigma_s}{\sum_{s=1}^{r} \sigma_s} \leq \rho.$$

The top h singular values are left unchanged, and only the remaining directions ($s > h$) are scaled according to Eq. 4. These augmented examples are generated by preserving the dominant components of the data—patterns and features that explain most of its variation—while making slight modifications to less important details. Training with such examples enables the model to focus on these core structures and improves its ability to capture the essential patterns that define the data.

2.4　Objective of This Research

Following the introduction of Mixup [19], many variants pursue spatially consistent and semantically coherent mixing [1,6,15,17]. Some methods use similarity-aware pairing or feature-space interpolation, but most apply a *fixed* augmentation policy throughout the training and do not account for the *temporal* dynamics of training. Zou et al. [20] show that Mixup is most beneficial early in training and can hinder the learning of fine, local features later. In contrast, First-Order Manifold Augmentation (FOMA) [8] leverages local geometric structure by perturbing directions orthogonal to dominant components in the joint feature-output space, but it was designed for regression and there remain open challenges in applying it to classification tasks.

Motivated by these observations, our primary objective is to improve classification accuracy; achieve competitive robustness under distribution shift with a two-phase augmentation strategy that is time-aware and geometry-aware. In Phase 1, standard Mixup promotes global interpolation and smooth decision boundaries. In Phase 2, operating on a more mature feature space, we apply Local-FOMA, our reformulation of FOMA for classification. Local-FOMA uses batch-local neighborhoods and an SVD of the joint feature-label matrix to generate geometry-consistent feature perturbations and converts the label part into soft targets. A time-dependent coefficient gradually increases the contribution of Local-FOMA to ensure a smooth transition from Mixup and to refine local, class-discriminative structure.

We test the hypothesis that this curriculum, which applies global interpolation first and local geometry-aware perturbations second, improves accuracy over single-policy baselines and strengthens robustness on corrupted benchmarks such as CIFAR-100-C/CIFAR-10-C test set.

3　Proposed Method

Local-FOMA. FOMA is a data augmentation method that performs Singular Value Decomposition (SVD) on the joint space of features and labels and then perturbs the components orthogonal to the dominant directions to generate new data samples. It was originally designed for regression tasks assuming continuous outputs, which makes its direct use in classification non-trivial [8]. To apply it to classification, we propose a localized variant, Local-FOMA, that (1) leverages batch-local neighborhood structure in feature space and (2) converts the label-side vector into soft targets.

Given an input image $\boldsymbol{x}_i$ ($i = 1, ..., b$), let $\boldsymbol{z}_i = f(\boldsymbol{x}_i) \in \mathbb{R}^D$ denote the feature vector extracted from a selected intermediate layer. We then construct a batch-local neighborhood $\mathcal{N}_i$ of size k that includes the anchor $\boldsymbol{x}_i$ together with its $k-1$ nearest neighbors (Euclidean distance in feature space) among the samples in the current mini-batch. We stack the k feature vectors and their one-hot labels (for C classes) row-wise to form

$$\boldsymbol{Z}_i \in \mathbb{R}^{k \times D}, \qquad \boldsymbol{Y}_i \in \mathbb{R}^{k \times C}.$$

Let $p \in \{1, \dots, k\}$ denote the row index of the anchor within this neighborhood. We then build the joint matrix

$$A_i = \begin{bmatrix} Z_i \mid Y_i \end{bmatrix} \in \mathbb{R}^{k \times (D+C)}. \tag{7}$$

Before SVD, we simply apply a column-wise mean centering to A_i (subtract column means); after reconstruction. We compute the SVD on the centered matrix

$$\tilde{A}_i = U_i S_i V_i^\mathsf{T}, \tag{8}$$

and scale only the low-contribution singular components as in Eq. 4 to obtain $S_i(\epsilon, h)$. The augmented joint matrix is reconstructed by

$$\hat{A}_i(\epsilon, h) \ = \ U_i \, S_i(\epsilon, h) \, V_i^\mathsf{T}. \tag{9}$$

Finally, we extract the p-th row of $\hat{A}_i(\epsilon, h)$ and split it back into feature and label parts:

$$z_i^{\text{foma}} \in \mathbb{R}^D \text{ (first } D \text{ dimensions)}, \qquad y_i^{\text{foma}} \in \mathbb{R}^C \text{ (last } C \text{ dimensions)}.$$

After reconstruction, the label-side vector y_i^{foma} is a real-valued vector that is not a probability distribution (its entries can be negative and do not sum to 1). We therefore apply the softmax function to y_i^{foma}, treating it as unnormalized logits, to obtain a valid soft target distribution, as given in Eq. 10:

$$\hat{y}_i^{\text{foma}} = \text{softmax}\left(y_i^{\text{foma}}\right). \tag{10}$$

Thus, the augmented sample $(z_i^{\text{foma}}, \hat{y}_i^{\text{foma}})$ preserves local geometry through neighborhood based SVD while providing flexible multi-class supervision through soft targets, unlike Mixup-style global interpolation.

Loss Function for Learning with Local-FOMA. Preliminary experiments revealed that training with only the L_{foma} loss term leads to unstable learning and significantly reduced classification accuracy. This instability is mainly due to the soft labels generated by Local-FOMA, which reflect the local structure but often lack sufficient class-discriminative information to guide the learning effectively.

To address this issue, we combine the loss from clean features and labels with the Local-FOMA loss, as shown in the following formulation:

$$L_{\text{Local-FOMA}} = L_{\text{orig}} + \lambda_{\text{foma}} L_{\text{foma}} \tag{11}$$

$$= \text{CE}(g(z), y^{\text{true}}) + \lambda_{\text{foma}} \text{CE}(g(z^{\text{foma}}), \hat{y}^{\text{foma}}) \tag{12}$$

Here, $z = f(x)$ denotes the feature representation of the input x, z^{foma} is the locally perturbed feature vector generated by Local-FOMA, and $\hat{y}^{\text{foma}}$ is the corresponding soft label. The coefficient λ_{foma} is a weight that controls the contribution of the Local-FOMA term (see Sect. 3.2 for details). CE denotes the cross-entropy loss. This combined loss provides both strong supervision from ground-truth labels and regularization through local feature-space augmentation, thereby improving stability and performance.

Assuming that the mapping by the feature extractor is $f(\cdot)$, the mapping by the classifier is $g(\cdot)$, and $\boldsymbol{y}_i^{\text{true}}$ is the true label for image $\boldsymbol{x}_i$, the learning of the model by Local-FOMA can be represented in Fig. 2.

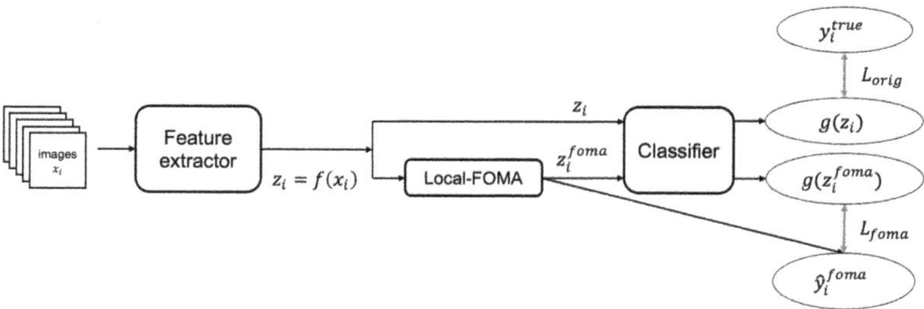

Fig. 2. Schematic of Local-FOMA learning. L_{orig}: the loss on the original clean data; L_{foma}: the loss on the extended data obtained by Local-FOMA. Minimizing the combined loss of L_{orig} and L_{foma} allows the model to gain local robustness.

3.1 Two-Phase Training with Mixup and Local-FOMA

In this research, we propose a two-phase hybrid learning method that uses Mixup to promote global generalization performance in the first phase of learning and then improves model robustness using Local-FOMA based on local structure in the subsequent learning phase for image classification tasks.

Phase 1: Early-Stage Global Regularization with Mixup. In the initial phase of training (i.e., the first T_{mixup} epochs), we employ the Mixup technique [19] to generate augmented samples by linearly interpolating between randomly paired training examples and their corresponding labels. This phase encourages the model to learn smooth decision boundaries and global structural relationships among classes. Previous research [20] has shown that Mixup is particularly effective in the early stages of training, helping models avoid overfitting to noise and capturing broad, shared features across the dataset.

Phase 2: Local Structure Learning with Local-FOMA. In the second phase (epoch $t > T_{\text{mixup}}$), the training gradually shifts its focus from Mixup-based global interpolation to Local-FOMA, which augments features along the local tangent space of the underlying manifold by exploiting neighborhood information. Instead of discarding Mixup entirely, we retain it as a complementary branch to stabilize optimization and preserve the global boundary-smoothing effect achieved in Phase 1.

Specifically, let T_{mixup} be the epoch at which Phase 2 begins and T_{total} the total number of epochs; the coefficient $\lambda_{\mathrm{foma}}(t)$ is then defined as

$$\lambda_{\mathrm{foma}}(t) = \min\left(1.0, \ \frac{t - T_{\mathrm{mixup}}}{(T_{\mathrm{total}} - T_{\mathrm{mixup}})/2}\right), \tag{13}$$

which linearly increases from 0 to 1 during the first half of Phase 2 and remains 1 thereafter. The Mixup branch is weighted by its complement:

$$\lambda_{\mathrm{mix}}(t) = 1 - \lambda_{\mathrm{foma}}(t). \tag{14}$$

This warm-up design stabilizes the training dynamics by allowing a smooth transition of the model's focus from smoothing global decision boundaries in Phase 1 to capturing fine local structures in Phase 2, while preserving and refining the feature representations acquired in the first phase.

Combined Loss Function. The overall training loss at epoch t is defined as

$$\mathcal{L}(t) = \begin{cases} \mathrm{CE}\big(g(\boldsymbol{z}), \boldsymbol{y}^{\mathrm{true}}\big) + \mathrm{CE}\big(g(\tilde{\boldsymbol{z}}), \tilde{\boldsymbol{y}}\big), & t \leq T_{\mathrm{mixup}}, \\ \mathrm{CE}\big(g(\boldsymbol{z}), \boldsymbol{y}^{\mathrm{true}}\big) + \lambda_{\mathrm{mix}}(t)\,\mathrm{CE}\big(g(\tilde{\boldsymbol{z}}), \tilde{\boldsymbol{y}}\big) + \lambda_{\mathrm{foma}}(t)\,L_{\mathrm{foma}}, & t > T_{\mathrm{mixup}}, \end{cases} \tag{15}$$

where $g(\cdot)$ is the classifier head, $\mathrm{CE}(\cdot)$ denotes the cross-entropy loss, $\tilde{\boldsymbol{z}}$ is the Mixup-augmented representation with soft label $\tilde{\boldsymbol{y}}$, and L_{foma} denotes the cross-entropy loss between the classifier prediction on the Local-FOMA perturbed representation $\boldsymbol{z}^{\mathrm{foma}}$ and its soft label $\hat{\boldsymbol{y}}^{\mathrm{foma}}$. The coefficients are given by Eqs. 13 and 14.

In Phase 1, Mixup learning encourages global decision boundary smoothing and helps capture rare features that are difficult to learn through standard training, thereby improving generalization while mitigating overfitting. In Phase 2, Local-FOMA performs local geometry-preserving feature-space augmentation, which enhances fine discriminative ability and strengthens robustness under distribution shifts. The coefficient $\lambda_{\mathrm{foma}}(t)$ follows Eq. 13, increasing linearly during the first half of Phase 2. This gradual adaptation ensures a smooth transition from global boundary smoothing to local structure learning, preventing the abrupt degradation of feature representations acquired during Phase 1.

3.2   Expected Benefits of the Proposed Method

The two-phase data augmentation method proposed in this research, which uses Mixup and Local-FOMA in stages, is expected to be effective in two main areas: improving classification accuracy and strengthening robustness against distribution shifts.

First, with respect to classification accuracy, Mixup [19] smooths the decision boundaries between classes and improves generalization performance, but it has been pointed out that it may hinder the learning of fine features in the latter half of training [20]. In this method, Mixup is applied in the early phase of learning to learn the global structure, and then Local-FOMA is applied in the latter phase to perform data augmentation utilizing the local structure in the feature space. This two-phase strategy is expected to effectively capture both inter-class and intra-class information, leading to high classification performance.

Next, regarding robustness under distribution shifts, in real-world environments, data containing noise or transformations different from those during training is likely to be input. Local-FOMA performs semantically consistent expansion based on neighborhood structures in the feature space, promoting the acquisition of representations robust to minor perturbations and environmental changes. As a result, the proposed method is expected to maintain stable performance even in situations with distribution shifts, such as CIFAR-C [7].

4   Experiments

4.1   Experimental Settings

Datasets. We used CIFAR-100 and CIFAR-10 datasets [10] in our experiments. Both datasets consist of 32×32 color images, where CIFAR-100 contains 100 classes grouped into 20 superclasses, and CIFAR-10 contains 10 classes. Each dataset comprises 50,000 training images and 10,000 test images.

Experiment 1. We trained on CIFAR-100 and CIFAR-10 datasets and evaluated on their original test sets. From the training images, 10,000 were set aside for validation.

Experiment 2. We trained on both CIFAR-100 and CIFAR-10 datasets, and evaluated under distribution shifts using CIFAR-100-C and CIFAR-10-C datasets [7]. CIFAR-100-C and CIFAR-10-C are corrupted versions of the respective test sets, each containing 15 types of common corruptions (e.g., blur, noise, weather effects, and digital artifacts) at 5 severity levels. In this study, we evaluate across all five severity levels and report the average performance over severities 1–5, resulting in a total of 750,000 corrupted images for evaluation in each dataset.

Experiment 3. In this experiment, we conduct a sensitivity analysis of the neighbor size k used in the proposed Mixup-FOMA method. The evaluation was performed on CIFAR-100 (clean) and CIFAR-100-C (corrupted) test sets to analyze the effect of varying k on both accuracy and robustness.

Compared Methods. To evaluate the effectiveness of the proposed method, we compared it with the following methods:

- Baseline: A model trained with only standard data augmentation (random cropping and horizontal flipping), without using Mixup or Local-FOMA.
- Mixup [19]: A global mixed data augmentation technique that performs linear interpolation in the input space.
- Manifold Mixup [15]: A mixing technique that performs linear interpolation in intermediate feature spaces.
- SK-Mixup [1]: A Mixup variant that adjusts the interpolation coefficient based on the distance between samples.
- RegMixup [13]: An extension of Mixup that treats mixed-sample training as an explicit regularization term. It combines the standard cross-entropy loss on clean samples with an additional Mixup loss on interpolated samples.
- CutMix [17]: A method that pastes patches between images while mixing the corresponding labels.
- Local-FOMA (Proposed): The Local-FOMA method used in this study, applied throughout the entire training process without Mixup.
- Mixup-FOMA (Proposed): A two-phase training method that applies Mixup in the early phase and Local-FOMA in the later phase.

Experimental Condition. We used Wide-ResNet-28-10 [18] as the classification model, following common practice for the CIFAR datasets. The optimizer used was Adam [9] with its default hyper-parameters. The total number of training epochs, T_{total} was set to 250 for CIFAR-10 and 400 for CIFAR-100 with batch size 128. All methods were trained with standard data augmentations, including random cropping and horizontal flipping. In the proposed method, the length of the Mixup phase (Phase 1), denoted as T_{mixup}, was set to 90% of the total number of training epochs for each dataset. This design allows Mixup to sufficiently smooth decision boundaries in the earlier part of training, while Local-FOMA, which relies on local neighborhood structures, is introduced in the later stage after feature representations have been well matured. For Mixup, the interpolation coefficient γ is sampled from $\text{Beta}(\alpha, \alpha)$ with $\alpha = 1.0$. For Local-FOMA, the number of neighbors k is set to 32 (the reason for this choice will be explained in Experiment 3), and the threshold for the explained variance in SVD is set to $\rho = 0.9$. This value was selected to retain the dominant geometric components while allowing sufficient perturbation in less informative directions. Local-FOMA was applied at the final feature-extraction layer. The singular-value scaling coefficient ϵ is sampled from $\text{Beta}(\alpha, \alpha)$ with $\alpha = 1.0$. Model performance was evaluated using Top-1 Accuracy, and we report the mean $\pm$ standard deviation over four independent runs with different random seeds.

4.2   Experiment 1: Classification Performance on Clean Data

The purpose of this experiment is to examine how much the proposed method (Mixup-FOMA) can improve classification performance compared to existing mixing methods and Local-FOMA alone on clean test sets from the CIFAR-10 and CIFAR-100 datasets.

Result. As shown in Table 1, the proposed Mixup-FOMA method achieved the highest accuracy on both datasets. On the CIFAR-100 dataset, it achieved $78.93 \pm 0.20\%$, which is $+1.04$ pp better than the next best method, RegMixup, and $+1.09$ to $+1.25$ pp better than other Mixup-based methods (Mixup, SK-Mixup). This represents a significant improvement of $+8.85$ pp over the baseline. On the CIFAR-10 dataset, it achieved $95.62 \pm 0.21\%$, which is 0.29 pp higher than RegMixup, 1.73 pp higher than Mixup, and 0.66 pp higher than SK-Mixup. Local-FOMA alone showed only a slight improvement of $+1.77$ pp for CIFAR-100 and $+0.27$ pp for CIFAR-10 compared to the baseline.

Table 1. Top-1 Accuracy (%) on CIFAR-100 and CIFAR-10. Best in **bold**, second best underlined. All results are mean $\pm$ standard deviation over four runs.

Method	CIFAR-100 Acc. (%)	CIFAR-10 Acc. (%)
Baseline	70.08 ± 0.95	92.01 ± 0.24
Mixup	77.84 ± 0.79	93.89 ± 0.23
Manifold Mixup	76.80 ± 0.52	94.05 ± 0.18
SK-Mixup	77.68 ± 0.41	94.96 ± 0.15
RegMixup	$\underline{77.89 \pm 0.39}$	$\underline{95.33 \pm 0.09}$
CutMix	75.41 ± 0.60	94.45 ± 0.21
Local-FOMA	71.85 ± 0.38	92.28 ± 0.24
Mixup-FOMA	$\mathbf{78.93 \pm 0.20}$	$\mathbf{95.62 \pm 0.21}$

Furthermore, to visually confirm the differences in feature representations achieved by the proposed method, one class was randomly selected from each of the 20 superclasses in the CIFAR-100 test set, and the resulting feature vectors were visualized using t-SNE [11] (Fig. 3). In Baseline, we can see that clusters significantly overlap between classes, and that the discriminant boundaries are unclear, especially between nearby superclasses. Mixup improved cluster separation for some classes, but overall overlap remained. In contrast, Mixup-FOMA formed clearer and more compact clusters, and the boundaries between different superclasses were also sharper. These results demonstrate that Mixup-FOMA not only improves numerical classification accuracy, but also significantly strengthens the discriminant nature of representations in feature space.

Discussion. The results of this experiment show that the proposed Mixup-FOMA method consistently outperforms existing Mixup-based approaches on both CIFAR-10 and CIFAR-100 datasets. Conventional methods such as Mixup, SK-Mixup, and RegMixup perform well by interpolating input samples, but

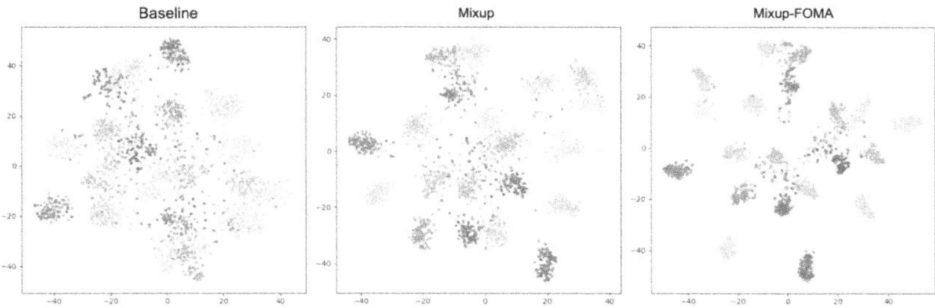

Fig. 3. t-SNE visualization of final-layer features from the CIFAR-100 test set. One class was randomly selected from each of the 20 superclasses. From left to right: Baseline, Mixup, and Mixup-FOMA (proposed).

their improvements tend to level off at a similar point. In contrast, Mixup-FOMA combines techniques in a stepwise manner to create a synergistic effect. By combining Mixup's global generalization with Local-FOMA's local refinement in sequence, Mixup-FOMA makes the most of both strengths.

It is also important to note that Local-FOMA alone brought only a small improvement over the baseline. This suggests that, without first learning the global data structure and applying it to a mature feature space, local perturbations by themselves are not very effective. Therefore, the strong performance of Mixup-FOMA shows that a staged combination of global and local strategies works better than using either one on its own.

In addition, the t-SNE visualization provides qualitative evidence supporting these numerical results. While the Baseline model shows substantial overlap between clusters, and Mixup only partially alleviates this issue, Mixup-FOMA leads to clearer and more compact cluster formation with sharper class boundaries. This indicates that Mixup-FOMA not only enhances classification accuracy but also improves the discriminative quality of learned feature representations.

4.3   Experiment 2: Evaluation of Robustness Using Corrupted Data

The purpose of this experiment is to evaluate the robustness of Mixup-FOMA under distributional shifts using CIFAR-10-C and CIFAR-100-C. By comparing it with baseline and existing augmentation methods, we investigate its ability to generalize to out-of-distribution settings and whether the integration of global and local feature-based augmentation contributes to robustness.

Result. The results are shown in Table 2. For the CIFAR-100-C test set, the proposed method Mixup-FOMA achieved the best result of $53.42 \pm 0.55\%$, outperforming the next best method, RegMixup, by $+2.14$ pp, and surpassing Mixup and SK-Mixup by $+2.23$ pp and $+3.03$ pp, respectively. Compared to the baseline, this represents an improvement of $+10.70$ pp. On the CIFAR-10-C test set, the proposed method also achieved the best result of $78.96 \pm 0.50\%$, outperforming SK-Mixup and RegMixup by $+0.37$ pp and $+0.44$ pp, respectively. It also improved upon Mixup by $+1.89$ pp and the baseline by $+7.94$ pp. Local-FOMA alone achieved $44.59 \pm 1.04\%$ on CIFAR-100-C and $70.96 \pm 1.14\%$ on CIFAR-10-C, confirming that its robustness improvement is limited when used alone.

Table 2. Top-1 Accuracy (%) on CIFAR-100-C and CIFAR-10-C. Best in **bold**, second best underlined. All results are mean $\pm$ standard deviation over four runs.

Method	CIFAR-100-C Acc. (%)	CIFAR-10-C Acc. (%)
Baseline	42.72 ± 0.89	71.02 ± 1.90
Mixup	51.19 ± 0.31	77.07 ± 0.44
Manifold Mixup	49.95 ± 0.65	71.68 ± 0.72
SK-Mixup	50.39 ± 0.58	$\underline{78.59 \pm 0.63}$
RegMixup	$\underline{51.28 \pm 1.29}$	78.52 ± 0.51
Local-FOMA	44.59 ± 1.04	70.96 ± 1.14
Mixup-FOMA	$\mathbf{53.42 \pm 0.55}$	$\mathbf{78.96 \pm 0.50}$

Discussion. The results demonstrate that the proposed Mixup-FOMA can improve robustness against severe corruptions, particularly on CIFAR-100-C. The method achieved the best performance among all compared techniques, surpassing RegMixup and other Mixup variants. This suggests that the two-phase strategy—combining global interpolation in the first phase with local geometry-based perturbations in the second—helps the model learn features resilient to distribution shifts in complex datasets.

On CIFAR-10-C, Mixup-FOMA also achieved the best performance, slightly surpassing SK-Mixup and RegMixup. Although the margin was smaller, the results confirm that integrating global and local strategies contributes to robustness. Importantly, Mixup-FOMA consistently outperformed standard Mixup and the baseline, demonstrating the effectiveness of the proposed framework.

Overall, these findings indicate that Mixup-FOMA maintains strong generalization under clean conditions and provides enhanced robustness under distributional shifts, with particularly clear benefits in complex datasets.

4.4 Experiment 3: Sensitivity Analysis of the Neighbor Size k in Mixup-FOMA

The purpose of this experiment is to verify the effect of the number of neighbors k in the proposed Mixup-FOMA method on model performance. Specifically, we aim to investigate the behavior when the value of k is changed in terms of both classification accuracy on clean data and robustness under distribution perturbation (CIFAR-100-C), and to clarify the appropriate setting range.

Result. Table 3 shows the performance of CIFAR-100 (clean) and CIFAR-100-C (corrupted) when varying the number of neighbors k in Mixup-FOMA. Overall, we observed a gradual improvement in both accuracy and robustness when k was increased from smaller values (10, 16, 20) to medium values (32, 64). The best results were obtained at $k = 32$, corresponding to improvements of $+0.51$ pp and $+0.81$ pp, respectively, compared to $k = 10$. At $k = 64$, the robustness was nearly identical but the clean accuracy slightly decreased.

Table 3. Ablation on the number of neighbors k in Mixup-FOMA. We report Top-1 accuracy (%) on the CIFAR-100 and CIFAR-100-C as mean $\pm$ standard deviation over four independent runs. Best in **bold**, second best underlined.

k	CIFAR-100 (%)	CIFAR-100-C (%)
10	78.42 ± 0.36	52.61 ± 0.46
16	78.57 ± 0.17	52.60 ± 0.42
20	78.61 ± 0.29	52.50 ± 0.26
32	**78.93 ± 0.20**	**53.42 ± 0.55**
<u>64</u>	78.82 ± 0.30	53.41 ± 0.35

Discussion. The experimental results showed that the selection of the number of neighbors k has a certain impact on the performance of Mixup-FOMA. When k is small (e.g., 10 or 16), the available neighborhood information is limited, leading to bias in local sample relationships, resulting in insufficient accuracy and robustness. On the other hand, when k is set too large, a more diverse neighborhood can be considered, but the boundaries between different classes are excessively smoothed, causing the improvement in clean accuracy to plateau and, in some cases, even decrease slightly. A value of $k = 32$, which is in the middle of the two, achieves a balance between utilizing neighborhood information appropriately and not blurring the boundaries between classes unnecessarily, achieving the best results in terms of both accuracy and robustness. These results suggest that, although Mixup-FOMA is not extremely sensitive to k, in practice, selecting $k \sim 32$ provides the most stable and high performance.

5   Conclusion

In this paper, we proposed a two-phase augmentation strategy, Mixup-FOMA, which combines the global boundary-smoothing effect of Mixup with the local data manifold perturbation of Local-FOMA. The key idea is to leverage Mixup in the early phase of training to improve generalization and prevent overfitting, while gradually transitioning to Local-FOMA in the later phase to refine class-discriminative features using local manifold structures.

Through experiments on CIFAR-10/100 datasets and their corrupted variants (CIFAR-10-C/100-C), we demonstrated that the proposed method achieves superior classification performance compared to strong Mixup-based baselines. Mixup-FOMA improved accuracy on clean test sets and showed robustness under distribution shifts, particularly CIFAR-100-C. This supports the effectiveness of combining global and local augmentation within a time-aware curriculum. A sensitivity analysis further revealed that the neighborhood size k in Local-FOMA has moderate influence, with $k = 32$ providing the best balance between accuracy and robustness. These findings suggest that Mixup-FOMA is especially beneficial for complex datasets with many classes, where both global interpolation and local feature refinement contribute to reliable learning.

For future work, promising directions include exploring adaptive scheduling strategies to dynamically adjust the transition between Mixup and Local-FOMA, and systematically investigating the optimal neighborhood size k across different datasets and tasks to better understand the role of local manifold structures.

In this study, we used Wide-ResNet as an example, but Mixup-FOMA is expected to apply to larger architectures such as ViTs, showing it is not limited to CNNs.

References

1. Bouniot, Q., Mozharovskyi, P., d'Alché Buc, F.: Tailoring mixup to data for calibration. In: The Thirteenth International Conference on Learning Representations (2025)
2. Cao, C., Zhou, F., Dai, Y., Wang, J., Zhang, K.: A survey of mix-based data augmentation: taxonomy, methods, applications, and explainability. ACM Comput. Surv. **57**(2), 1–38 (2024)
3. DeVries, T., Taylor, G.W.: Improved regularization of convolutional neural networks with cutout. arXiv preprint: arXiv:1708.04552 (2017)
4. Dosovitskiy, A., et al.: An image is worth 16x16 words: transformers for image recognition at scale. In: International Conference on Learning Representations (2021)
5. Goceri, E.: Medical image data augmentation: techniques, comparisons and interpretations. Artif. Intell. Rev. **56**(11), 12561–12605 (2023)
6. Guo, H., Mao, Y., Zhang, R.: MixUp as locally linear out-of-manifold regularization. In: Proceedings of the AAAI Conference on Artificial Intelligence, vol. 33, pp. 3714–3722 (2019)
7. Hendrycks, D., Dietterich, T.: Benchmarking neural network robustness to common corruptions and perturbations. In: International Conference on Learning Representations (ICLR) (2019)
8. Kaufman, I., Azencot, O.: First-order manifold data augmentation for regression learning. In: Proceedings of the 41st International Conference on Machine Learning (2024)
9. Kingma, D.P., Ba, J.: Adam: A method for stochastic optimization. In: 3rd International Conference on Learning Representations, ICLR 2015, San Diego, CA, USA, 7–9 May 2015, Conference Track Proceedings (2015)
10. Krizhevsky, A., Hinton, G.: Learning multiple layers of features from tiny images. Tech. rep., University of Toronto (2009)

11. Maaten, L.v.d., Hinton, G.: Visualizing data using t-SNE. J. Mach. Learn. Res. **9**(Nov), 2579–2605 (2008)
12. Perez, L., Wang, J.: The effectiveness of data augmentation in image classification using deep learning. arXiv preprint: arXiv:1712.04621 (2017)
13. Pinto, F., Yang, H., Lim, S.N., Torr, P.H., Dokania, P.K.: RegMixup: mixup as a regularizer can surprisingly improve accuracy & out-of-distribution robustness. In: Proceedings of the 36th International Conference on Neural Information Processing Systems (2022)
14. Shorten, C., Khoshgoftaar, T.M.: A survey on image data augmentation for deep learning. J. Big Data **6**(1), 1–48 (2019)
15. Verma, V., et al.: Manifold mixup: better representations by interpolating hidden states. In: International Conference on Machine Learning, pp. 6438–6447 (2019)
16. Yang, S., Xiao, W., Zhang, M., Guo, S., Zhao, J., Shen, F.: Image data augmentation for deep learning: a survey. arXiv preprint: arXiv:2204.08610 (2022)
17. Yun, S., Han, D., Oh, S.J., Chun, S., Choe, J., Yoo, Y.: CutMix: regularization strategy to train strong classifiers with localizable features. In: Proceedings of the IEEE/CVF International Conference on Computer Vision, pp. 6023–6032 (2019)
18. Zagoruyko, S., Komodakis, N.: Wide residual networks. In: BMVC (2016)
19. Zhang, H., Cisse, M., Dauphin, Y.N., Lopez-Paz, D.: mixup: beyond empirical risk minimization. In: International Conference on Learning Representations (2018)
20. Zou, D., Cao, Y., Li, Y., Gu, Q.: The benefits of mixup for feature learning. In: Proceedings of the 40th International Conference on Machine Learning (2023)

BARE: Boundary-Aware with Resolution Enhancement for Tree Crown Delineation

Attavit Wilaiwongsakul[1(✉)] ⓘ, Bin Liang[1] ⓘ, Wenfeng Jia[2] ⓘ, Bryan Zheng[1] ⓘ, and Fang Chen[1] ⓘ

[1] University of Technology Sydney, Sydney, Australia
`attavit.wilaiwongsakul@student.uts.edu.au`,
`{bin.liang,boyuan.zheng,fang.chen}@uts.edu.au`
[2] Charles Sturt University, Estella, Australia
`wjia@csu.edu.au`

Abstract. Accurate automated tree crown delineation (TCD) requires highly precise boundary segmentation, yet reduced-resolution architectures face limitations from decoder outputs at lower spatial resolutions. We propose BARE (Boundary-Aware with Resolution Enhancement), a simple architecture-preserving training strategy combining external full-resolution loss supervision with class weighting. BARE upsamples decoder output solely during training, maintaining inference efficiency while improving boundary precision. Through comprehensive evaluation on SegFormer, PSPNet, and SETR using the OAM-TCD dataset, we demonstrate that external full-resolution supervision universally benefits all tested architectures, achieving significant boundary quality improvements. We introduce B-IoU (Boundary-Intersection over Union) to TCD research, enabling rigorous boundary quality assessment. Our systematic evaluation reveals architecture-dependent optimization characteristics, providing actionable guidelines for practitioners seeking to enhance boundary precision in reduced-resolution segmentation architectures via training-only modifications. Code: https://github.com/attavit14203638/bare.

Keywords: Tree Crown Delineation · Boundary IoU · Vision Transformers

1 Introduction

Precise tree crown boundaries are essential for accurate forest monitoring, yet automated methods struggle to achieve the boundary precision required for critical applications such as biomass estimation, biodiversity assessment, and climate change mitigation planning [6,14]. While pixel-level accuracy metrics may suggest high performance, boundary imprecision—manifesting as edge blurring, crown merging, and fragmented delineations—undermines the reliability of downstream analyses in dense forest canopies where individual tree discrimination is paramount. Traditional computer vision approaches face particular

Q. V. Nguyen et al. (Eds.): AusDM 2025, CCIS 2765, pp. 302–315, 2026.
https://doi.org/10.1007/978-981-95-6786-7_20

challenges with high-resolution aerial imagery, where fine spatial details directly impact the quality of forest inventory and ecosystem monitoring applications.

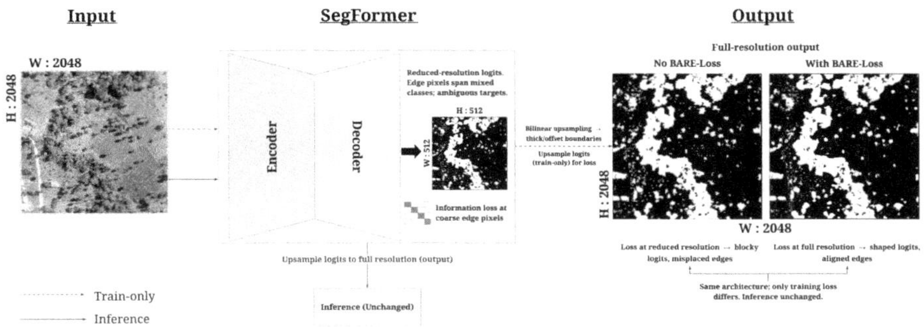

Fig. 1. Reduced-resolution decoder challenge.

The evolution to deep learning has revolutionized TCD, with Vision Transformer architectures demonstrating significant advantages through global context modeling [6–8]. However, several segmentation architectures including SegFormer [17], PSPNet [21], and SETR [22] share a fundamental bottleneck: their decoders produce predictions at reduced spatial resolution requiring upsampling, potentially compromising boundary quality. This architectural limitation manifests as a critical resolution mismatch where high-resolution input images (2048×2048 pixels) are processed through encoders that maintain spatial detail, but decoders output predictions at significantly reduced resolutions—SegFormer at 1/4, PSPNet at 1/8, and SETR at 1/16 of the original size. The subsequent bilinear upsampling to restore full resolution introduces boundary artifacts that severely impact crown separation accuracy in dense forest canopies where precise delineation between adjacent trees is paramount (Fig. 1). To address current evaluation limitations in boundary assessment, we introduce boundary IoU (B-IoU) [2] to TCD evaluation—to our knowledge, the first application of this boundary-focused metric in TCD research. Unlike standard IoU metrics that treat all pixels equally, B-IoU specifically evaluates segmentation quality within narrow boundary regions, revealing that full-resolution loss supervision during training produces markedly cleaner crown boundaries with reduced edge artifacts and improved crown separation compared to standard reduced-resolution training approaches (Fig. 2).

We propose **BARE** (Boundary-Aware with Resolution Enhancement): an architecture-preserving training strategy that combines external full-resolution loss supervision with class weighting to address data set imbalance. Rather than modifying architectural structures, our approach upsamples decoder output solely during training for loss computation, maintaining inference efficiency while improving boundary quality. Through systematic evaluation on the OAM-TCD dataset, we demonstrate that methodical optimization of training supervision

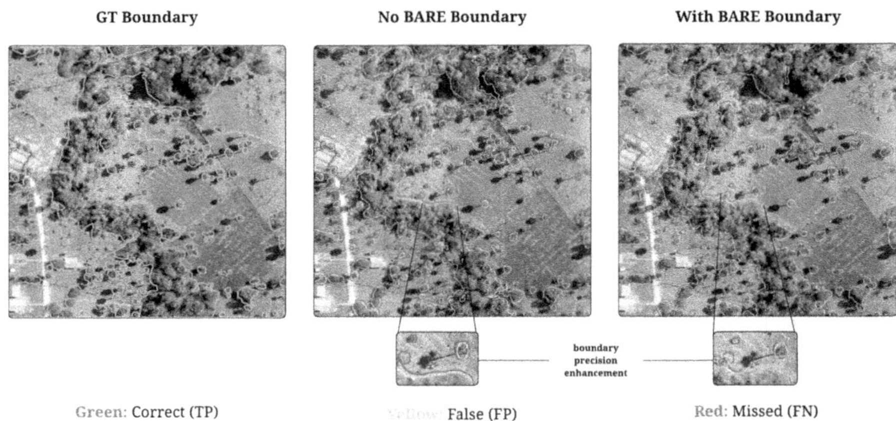

Fig. 2. B-IoU boundary precision evaluation.

can be more effective than architectural modifications for boundary-sensitive applications. Our contributions are threefold: First, we propose BARE as an architecture-preserving approach that improves boundary quality without structural changes, offering a practical solution for deployment scenarios requiring computational efficiency. Second, we introduce B-IoU to TCD research, establishing the first boundary-focused assessment approach in this domain. Third, we demonstrate that external full-resolution supervision universally benefits multiple reduced-resolution architectures (SegFormer, PSPNet, and SETR), while revealing that combining it with class weighting exhibits highly architecture-dependent effectiveness—only hierarchical transformers benefit from the complete approach, while pure-transformer and CNN-based architectures perform best with full-resolution supervision alone.

2   Related Work

We review the evolution toward Transformer-based TCD architectures, challenges in preserving spatial details, and existing approaches for high-resolution semantic segmentation and boundary refinement.

Deep learning for TCD has evolved from CNNs like U-Net [6,12] to self-attention models such as SegFormer [1,7,8,14]. Despite F1 scores exceeding ninty-five percent [8,16], precise boundary delineation remains challenging due to dense canopies [4,14,23], small instances [6], shadows [13], and overlapping crowns [20]. Advanced models often introduce significant computational costs [3,14], necessitating accuracy-efficiency trade-offs [9].

Semantic segmentation with Transformers in TCD features distinct architectural approaches. SETR [22] represents pure Vision Transformer design without hierarchical feature extraction, while SegFormer [17] employs hierarchical Transformer encoders with lightweight Multi-Layer Perceptron (MLP) decoders for multi-scale feature fusion. Both output logits at reduced resolution requiring upsampling, potentially losing fine-grained details crucial for TCD. This architectural diversity motivates evaluating training strategies that enhance boundary quality without architectural modifications [5].

High-resolution semantic segmentation in TCD addresses spatial detail recovery through learned upsampling, multi-scale fusion, and attention mechanisms [10,11,19], though increasing complexity. Alternative approaches include interpolation-based upsampling and boundary refinement modules [15, 18,20] at increased computational cost. Unlike dedicated boundary refinement modules adding inference overhead, BARE uses a training-only strategy to enhance boundary learning while preserving efficiency. Classical CNNs like PSP-Net [21] face similar challenges through pyramid pooling at reduced resolution. We investigate external upsampling strategies requiring careful optimization for meaningful boundary improvements.

3   Methodology

We present BARE, an architecture-preserving training strategy addressing the resolution bottleneck in segmentation models for TCD. BARE combines external full-resolution loss supervision with class weighting to improve boundary precision without architectural modifications.

3.1   BARE Framework Overview

BARE targets architectures producing decoder outputs at reduced resolution: SegFormer ($\frac{H}{4} \times \frac{W}{4}$), PSPNet ($\frac{H}{8} \times \frac{W}{8}$), and SETR ($\frac{H}{16} \times \frac{W}{16}$). Figure 3 illustrates our approach: rather than modifying architectures, BARE applies two complementary training strategies. Panel (a) shows standard training with loss computed at reduced resolution, while panel (b) demonstrates BARE with external upsampling for full-resolution supervision. The approach applies across different reduction factors while maintaining inference efficiency through training-only modifications. BARE combines (1) external full-resolution supervision upsampling logits solely during training, and (2) class weighting addressing dataset imbalance. While external full-resolution supervision universally benefits all tested architectures, combining both components shows architecture-dependent effectiveness—only hierarchical transformers (SegFormer) benefit from the complete approach, while pure-transformer (SETR) and CNN-based (PSPNet) architectures show negative interactions.

3.2   Reduced-Resolution Processing Across Architectures

Contemporary segmentation architectures produce decoder outputs at reduced resolution for computational efficiency. For input $I \in \mathbb{R}^{H \times W \times 3}$, these architectures generate logits $L_{reduced} \in \mathbb{R}^{\frac{H}{R} \times \frac{W}{R} \times N_{cls}}$ where R is the reduction factor and N_{cls} the number of classes. The factor varies by architecture: SegFormer uses

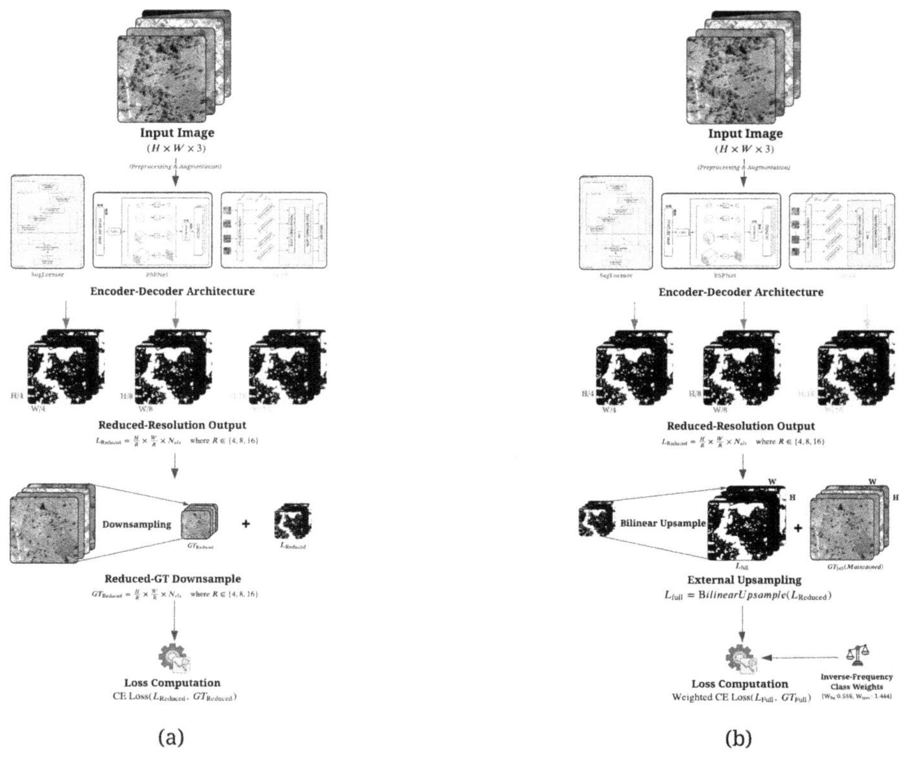

Fig. 3. BARE framework: (a) Standard vs (b) BARE training.

$R = 4$, PSPNet $R = 8$, and SETR $R = 16$. Standard training computes loss on $L_{reduced}$ or its upsampled version, potentially compromising boundary quality due to resolution mismatch with full-resolution ground truth masks.

3.3 External Full-Resolution Supervision Strategy

BARE's core component is external full-resolution supervision, bridging the resolution gap between decoder outputs and ground truth during training. As depicted in Fig. 3(b), this strategy externally upsamples reduced-resolution logits for loss computation while maintaining the architecture unchanged. For input size $H \times W$ and logits $L_{reduced} \in \mathbb{R}^{\frac{H}{R} \times \frac{W}{R} \times N_{cls}}$ (R: reduction factor; N_{cls}: number of classes), we bilinearly upsample to obtain full-resolution logits:

$$L_{full} = \text{BilinearUpsample}(L_{reduced}, \text{size} = (H, W))$$ (1)

Here, $L_{full} \in \mathbb{R}^{H \times W \times N_{cls}}$ matches ground truth dimensions. This upsampling is applied during training only, preserving inference efficiency. Predicted probabilities are computed via per-pixel softmax:

$$P_{i,c} = \frac{\exp(L_{full,i,c})}{\sum_{j=1}^{N_{cls}} \exp(L_{full,i,j})} \tag{2}$$

where i indexes pixels and c and j index classes.

3.4   Class Weighting Strategy

TCD datasets exhibit class imbalance, with tree crowns typically a minority class. BARE uses inverse-frequency class weights with unit-mean normalization. For a dataset with N_{ds} total pixels and n_c pixels in class c, the class frequency is $f_c = \frac{n_c}{N_{ds}}$ and the weight:

$$w_c = \frac{\frac{1}{f_c}}{\frac{1}{N_{cls}} \sum_{j=1}^{N_{cls}} \frac{1}{f_j}} \tag{3}$$

where N_{cls} is the number of classes (2 for binary TCD), j indexes classes, and f_j is the frequency of class j. This normalization keeps weights inversely proportional to frequency while ensuring unit mean. For OAM-TCD with 27.8% tree crown pixels ($f_{\text{tree}} \approx 0.278$, $f_{\text{background}} \approx 0.722$), this yields $w_{\text{background}} = 0.556$ and $w_{\text{tree}} = 1.444$.

3.5   Complete Loss Function Formulation

BARE employs Weighted Cross-Entropy (WCE) loss computed on full-resolution predictions. The complete loss formulation is:

$$\mathcal{L}_{\text{WCE}} = -\frac{1}{N} \sum_{i=1}^{N} \sum_{c=1}^{N_{cls}} w_c \cdot y_{i,c} \cdot \log(P_{i,c} + \epsilon) \tag{4}$$

where $N = H \times W$ is the total pixels, N_{cls} the number of classes (2 for binary TCD), w_c the class weight from Eq. 3, $y_{i,c} \in \{0,1\}$ the ground truth label for pixel i and class c, and $P_{i,c}$ the predicted probability after softmax. For numerical stability, we add $\epsilon = 1 \times 10^{-7}$ to prevent $\log(0)$ undefined values while having negligible effect on typical probabilities. This formulation ensures boundary pixels receive full-resolution supervision while addressing class imbalance.

4   Experiments

We present comprehensive experimental evaluation of BARE across multiple reduced-resolution architectures, beginning with dataset characteristics and experimental setup, followed by implementation details and results demonstrating BARE's effectiveness in improving boundary precision across different architectural paradigms.

4.1   Dataset and Experimental Setup

We use the OpenAerialMap Tree Crown Delineation (OAM-TCD) dataset [14], sourced from Hugging Face Hub as `restor/tcd`. The dataset provides global coverage across diverse forest environments spanning multiple continents and biomes, ideal for evaluating BARE's generalizability. Figure 4 shows the dataset encompasses temperate forests, tropical regions, and agricultural landscapes across North America, Europe, Africa, Asia, and Australia. The dataset features high-resolution aerial imagery (10 cm ground sampling distance) with 2048×2048 pixel images split into 4169/416/439 samples for training/validation/testing. The binary segmentation task exhibits class imbalance with tree crowns representing 27.8% of training pixels, motivating inverse-frequency class weighting [0.556, 1.444] for background and tree crown classes.

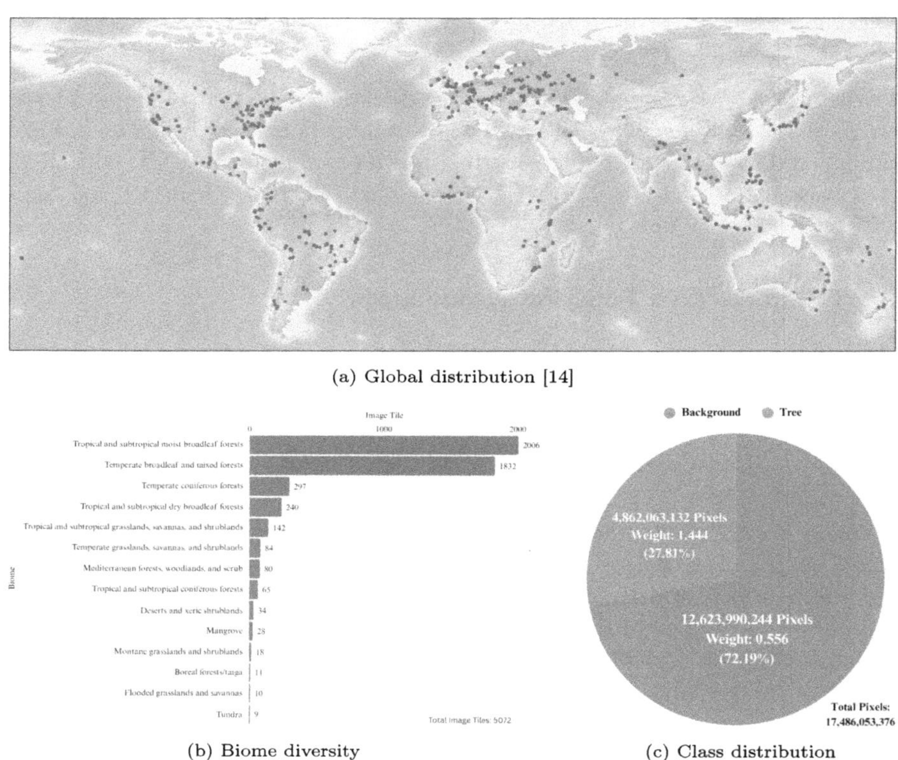

(a) Global distribution [14]

(b) Biome diversity   (c) Class distribution

Fig. 4. OAM-TCD dataset characteristics.

We apply comprehensive augmentation exclusively to training. Geometric transformations include random crops to 1024×1024 pixels (maintaining full-resolution supervision compatibility), horizontal/vertical flips, and rotations up to $\pm 180°$ preserving boundary integrity. Color augmentations include brightness/contrast/saturation/hue jittering (P=0.7), Gaussian blur (P=0.5), and atmospheric/shadow effects (P=0.2 each) simulating real-world conditions without compromising edge quality. This maintains spatial dimensions for external full-resolution loss while enhancing robustness.

4.2   Implementation Details and Model Configurations

We evaluate BARE across three architectures: SegFormer (MiT-B5), PSPNet (ResNet-50), and SETR (ViT-base), using pre-trained ImageNet weights when available. BARE applies external upsampling to decoder outputs during training: SegFormer logits at $\frac{H}{4} \times \frac{W}{4}$, PSPNet at $\frac{H}{8} \times \frac{W}{8}$, and SETR at $\frac{H}{16} \times \frac{W}{16}$ are bilinearly upsampled to $H \times W$ for loss computation while maintaining original inference paths.

All models use consistent optimization settings: AdamW (lr=1×10^{-5}, weight decay=0.01), cosine annealing with 10% warmup, and FP16 precision. Due to memory constraints, batch configurations differ while maintaining effective batch size 8: SegFormer uses per-device batch 1 (eval batch 2) with 8 gradient accumulation steps, while PSPNet and SETR use larger per-device batches with 1 accumulation step. Training spans 50 epochs on NVIDIA L4 GPUs using PyTorch 2.6.0 and Transformers 4.51.3. Class weights [0.556, 1.444] address imbalance when enabled.

For systematic evaluation, we implement four configurations per architecture: (1) Baseline (standard training with reduced-resolution loss), (2) Class Weighting only (CW), (3) Full-Resolution supervision only (Full-Res), and (4) BARE (our complete approach combining full-resolution supervision with class weighting).

4.3   Evaluation Metrics and Boundary Assessment

Precise boundary delineation is crucial for TCD. We employ B-IoU [2]—to our knowledge, its first application in TCD research. B-IoU evaluates segmentation within narrow bands around contours, providing direct boundary precision measurement complementing traditional pixel-wise metrics.

Boundary IoU Definition. Given ground truth mask G and prediction mask P, B-IoU computes IoU for mask pixels within distance d from the respective contours. Formally, B-IoU is defined as:

$$\text{B-IoU}(G, P) = \frac{|(G_d \cap G) \cap (P_d \cap P)|}{|(G_d \cap G) \cup (P_d \cap P)|} \tag{5}$$

where G_d and P_d are boundary regions—pixels within distance d from ground truth and prediction contours. Unlike standard IoU treating all pixels equally, B-IoU focuses on boundary regions, making it more sensitive to boundary quality—particularly valuable for TCD where precise crown separation is critical. We set d to 2% of image diagonal following established guidelines [2], corresponding to approximately 15 pixels for 2048×2048 images. This captures meaningful boundary errors while remaining robust to minor annotation ambiguities.

We also evaluate using standard segmentation metrics: IoU, F1-Score (Dice), Precision, Recall, and Pixel Accuracy. These provide complementary perspectives, with IoU measuring region overlap, F1-Score offering balanced precision-recall assessment, and Precision/Recall quantifying exactness and completeness.

4.4   Results and Analysis

Performance Analysis. Table 1 presents comprehensive evaluation across three architectures and training configurations, revealing architecture-dependent effectiveness of BARE strategies.

Table 1. Performance metrics across configurations and architectures.

Config	Architecture	IoU	F1	Prec	Rec	Acc	B-IoU
Baseline	SETR	0.881	0.937	**0.948**	0.926	**0.970**	0.643
Baseline	SegFormer	0.817	0.899	0.931	0.870	0.953	0.590
Baseline	PSPNet	0.743	0.853	0.847	0.859	0.928	0.415
CW	SETR	0.879	0.935	0.908	0.964	0.968	0.627
CW	SegFormer	0.845	0.916	0.884	0.950	0.959	0.606
CW	PSPNet	0.727	0.842	0.775	0.922	0.916	0.399
Full-Res	SETR	**0.882**	**0.938**	**0.948**	0.927	**0.970**	**0.644**
Full-Res	SegFormer	0.828	0.906	0.931	0.882	0.957	0.610
Full-Res	PSPNet	0.743	0.852	0.852	0.853	0.928	0.418
BARE	SETR	0.879	0.936	0.908	**0.965**	0.968	0.625
BARE	SegFormer	0.848	0.918	0.885	0.953	0.960	0.620
BARE	PSPNet	0.728	0.843	0.776	0.922	0.916	0.404

SegFormer shows strongest response to complete BARE, with B-IoU improving 5.1% from 0.590 to 0.620 and IoU from 0.817 to 0.848. Notably, SegFormer alone benefits from combining full-resolution supervision with class weighting. PSPNet reveals that full-resolution supervision alone (B-IoU: 0.418) outperforms BARE (B-IoU: 0.404), indicating class weighting creates negative interactions— reducing boundary quality 3.3% despite similar overall IoU. SETR exhibits similar behavior: full-resolution supervision alone achieves optimal performance (IoU: 0.882, B-IoU: 0.644), while adding class weighting reduces performance (IoU: 0.879, B-IoU: 0.625).

These results demonstrate that external full-resolution supervision universally benefits all architectures, but combining it with class weighting shows architecture-dependent behavior. PSPNet and SETR achieve optimal boundary quality with full-resolution supervision alone—class weighting degrades performance. Only SegFormer's hierarchical transformer design benefits from complete

BARE. This highlights the importance of empirical validation when combining training optimizations, as optimal configurations vary architecturally—pure-transformer and CNN-based models generally prefer full-resolution supervision alone, while hierarchical transformers benefit from the combined approach.

Qualitative Analysis. Figure 5 shows detailed visual comparisons demonstrating BARE's effectiveness across SegFormer, PSPNet, and SETR, with cleaner boundaries and reduced artifacts versus baselines. BARE variants show enhanced boundary definition across diverse landscapes: rural (Sample 1), dense forest (Sample 2), urban mixed (Sample 3), and agricultural (Sample 4). The analysis reveals systematic boundary coherence improvements, particularly in challenging scenarios with dense canopies and complex urban-forest interfaces.

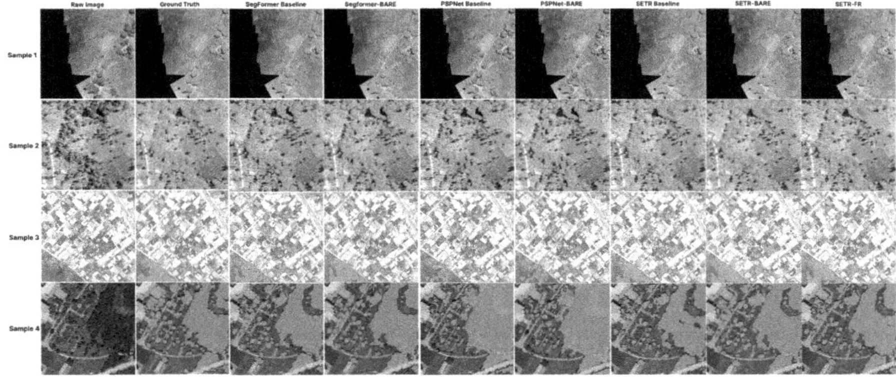

Fig. 5. Cross-architecture prediction comparison.

Figure 6 shows error patterns where green indicates correct predictions, yellow shows false positives, and red shows missed detections. BARE variants exhibit reduced false positive artifacts and improved boundary precision with balanced error patterns. The systematic reduction in artifacts demonstrates BARE's effectiveness across all architectures.

Figure 7 shows boundary-focused validation where green indicates correct boundaries. Progressive improvement from baseline to BARE variants is evident across all architectures, with BARE achieving superior boundary coherence, particularly in complex boundary regions.

Computational Efficiency Analysis. BARE maintains efficiency by applying modifications only during training. Inference times remain largely unchanged (SegFormer: 274 ms to 290 ms, PSPNet: 195 ms to 201 ms, SETR: 184 ms). Parameter counts stay constant (84.6M SegFormer, 49.0M PSPNet, 92.1M

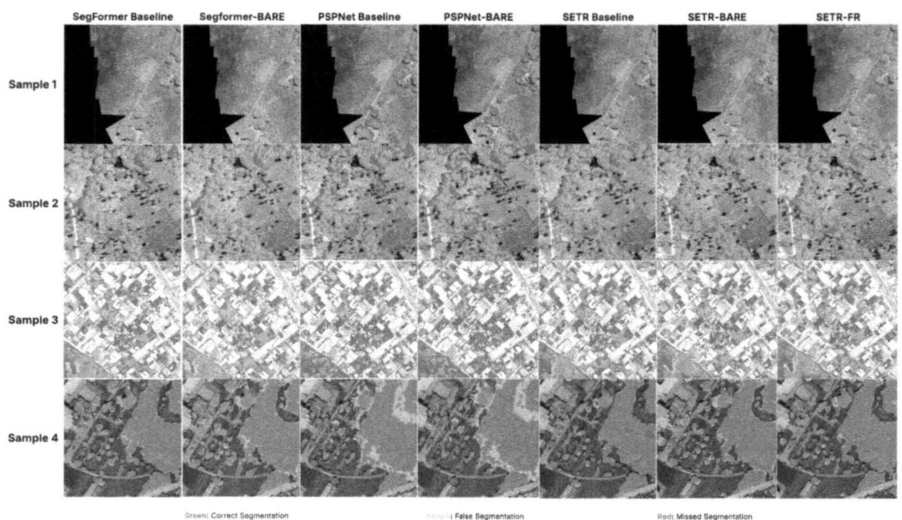

Fig. 6. Error pattern analysis across architectures.

SETR), as BARE adds no parameters. Memory increases are minimal (Seg-Former: 14.4 GB to 14.7 GB, PSPNet: 12.1 GB to 12.3 GB), making BARE practical for accuracy-constrained and efficiency-constrained deployments.

4.5 Discussion and Implications

Our evaluation reveals that external full-resolution supervision universally improves boundary precision across tested architectures through training-only modifications.

Architecture-Dependent Effectiveness. BARE's two components show different generalization characteristics. External full-resolution supervision universally benefits all architectures, with SETR Full-Res achieving highest performance (IoU: 0.882, B-IoU: 0.644) and PSPNet Full-Res showing strongest boundary improvement (B-IoU: 0.415 to 0.418). However, combining full-resolution supervision with class weighting shows architecture-dependent behavior: only SegFormer benefits from complete BARE, while PSPNet (B-IoU: 0.418 to 0.404) and SETR (B-IoU: 0.644 to 0.625) experience negative interactions. This suggests pure-transformer and CNN-based architectures achieve near-optimal precision-recall balance with full-resolution supervision alone, making class weighting counterproductive. Only hierarchical transformer design (SegFormer) benefits from the combined approach, likely due to its multi-scale feature fusion requiring explicit class balance guidance.

Practical Implications. These findings provide actionable guidance: (1) apply external full-resolution supervision broadly as it universally benefits all tested models, (2) use class weighting cautiously—only hierarchical transform-ers (SegFormer) benefit from combining it with full-resolution supervision, while

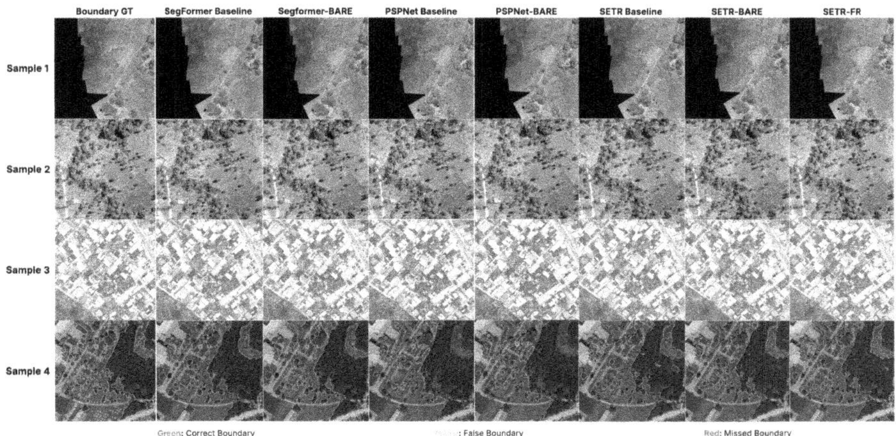

Fig. 7. Boundary-specific analysis results.

pure-transformer (SETR) and CNN-based (PSPNet) architectures perform best with full-resolution supervision alone, and (3) for pure-transformer and CNN-based architectures, avoid adding class weighting unless empirical validation shows benefits. The training-only nature makes empirical validation computationally feasible.

Limitations. Our evaluation uses only OAM-TCD; generalizability across remote sensing domains requires validation. The study examines three architectural paradigms; effectiveness on newer transformers (e.g., Swin) remains unexplored. We employ only bilinear upsampling; learned upsampling may yield different results. While maintaining effective batch size 8, memory constraints necessitated different gradient accumulation strategies, which could theoretically influence dynamics. Finally, deeper analysis of mechanisms driving architecture-specific responses would strengthen understanding.

Conclusion

We introduce BARE, an architecture-preserving training strategy combining external full-resolution supervision with class weighting to enhance boundary precision in tree crown delineation. Evaluation results reveal that external full-resolution supervision consistently benefits all tested architectures (SegFormer, PSPNet, SETR), achieving boundary IoU improvements, while class weighting shows architecture-dependent behavior—benefiting only hierarchical transformers (SegFormer) but degrading performance in pure-transformer (SETR) and CNN-based (PSPNet) architectures. The training-only nature maintains computational efficiency, enabling practitioners to enhance existing models without structural changes or inference costs.

We introduce the B-IoU evaluation to TCD research, establishing the first boundary-focused assessment in this domain. Future work should validate BARE across additional TCD datasets, explore learned upsampling strategies, and investigate mechanisms driving architecture-specific responses to inform refined guidelines for applying training optimizations across architectural paradigms.

Disclosure of Interests. The authors have no competing interests to declare that are relevant to the content of this article.

References

1. Al-Ruzouq, R., et al.: Spectral-Spatial transformer-based semantic segmentation for large-scale mapping of individual date palm trees using very high-resolution satellite data. Ecol. Ind. **163**, 112110 (2024). https://doi.org/10.1016/j.ecolind.2024.112110
2. Cheng, B., Girshick, R., Dollár, P., Berg, A.C., Kirillov, A.: Boundary IoU: Improving object-centric image segmentation evaluation. In: Proceedings of the IEEE/CVF conference on computer vision and pattern recognition. pp. 15334–15342. IEEE (2021). https://doi.org/10.1109/cvpr46437.2021.01508
3. Deng, G., Wu, Z., Xu, M., Wang, C., Wang, Z., Lu, Z.: Crisscross-global vision transformers model for very high resolution aerial image semantic segmentation. IEEE Trans. Geosci. Remote Sens. **61**, 1–19 (2023). https://doi.org/10.1109/tgrs.2023.3276172
4. Dersch, S., Schöttl, A., Krzystek, P., Heurich, M.: Towards complete tree crown delineation by instance segmentation with Mask R-CNN and DETR using UAV-based multispectral imagery and lidar data. ISPRS Open J. Photogram. Remote Sens. **8**, 100037 (2023). https://doi.org/10.1016/j.ophoto.2023.100037
5. Dosovitskiy, A., et al.: An image is worth 16x16 words: Transformers for image recognition at scale. arXiv preprint arXiv:2010.11929 (2020). https://doi.org/10.48550/arXiv.2010.11929
6. Freudenberg, M., Magdon, P., Nölke, N.: Individual tree crown delineation in high-resolution remote sensing images based on U-Net. Neural Comput. Appl. **34**(24), 22197–22207 (2022). https://doi.org/10.1007/s00521-022-07640-4
7. Georges Gomes, et al.: Urban Trees Mapping Using multi-scale rgb image and deep learning vision transformer-based. SSRN Electr. J. (2022). https://doi.org/10.2139/ssrn.4167085
8. Gibril, M.B.A., Shafri, H.Z.M., Al-Ruzouq, R., Shanableh, A., Nahas, F., Al Mansoori, S.: Large-scale date palm tree segmentation from multiscale Uav-based and aerial images using deep vision transformers. Drones **7**(2), 93 (2023). https://doi.org/10.3390/drones7020093
9. Gominski, D., Kariryaa, A., Brandt, M., Igel, C., Li, S., Mugabowindekwe, M., Fensholt, R.: Benchmarking Individual Tree Mapping with Sub-meter Imagery. arXiv.org (2023). https://doi.org/10.48550/arXiv.2311.07981
10. Liu, Y., Zhang, Y., Wang, Y., Mei, S.: Rethinking Transformers for Semantic Segmentation of Remote Sensing Images. IEEE Trans. Geosci. Remote Sens. **61**, 1–15 (2023). https://doi.org/10.1109/tgrs.2023.3302024
11. Ren, D., Li, F., Sun, H., Liu, L., Ren, S., Yu, M.: Local-enhanced multi-scale aggregation swin transformer for semantic segmentation of high-resolution remote sensing images. Int. J. Remote Sens. **45**(1), 101–120 (2023). https://doi.org/10.1080/01431161.2023.2292550

12. Ronneberger, O., Fischer, P., Brox, T.: U-Net: Convolutional Networks for Biomedical Image Segmentation. In: Navab, N., Hornegger, J., Wells, W.M., Frangi, A.F. (eds.) MICCAI 2015. LNCS, vol. 9351, pp. 234–241. Springer, Cham (2015). https://doi.org/10.1007/978-3-319-24574-4_28
13. Tao, Y., Wang, Z., Zhao, G.: Shadow-resilient tree crown detection in uav remote sensing images using deep learning. In: 2024 4th International Conference on Computer Science and Blockchain (CCSB). pp. 318–322. IEEE (2024). https://doi.org/10.1109/ccsb63463.2024.10735679
14. Veitch-Michaelis, J., et al.: Oam-TCD: A globally diverse dataset of high-resolution tree cover maps. arXiv.org (2024). https://doi.org/10.48550/arXiv.2407.11743
15. Wang, D., Chen, Y., Naz, B., Sun, L., Li, B.: Spatial-aware transformer (SAT): enhancing global modeling in transformer segmentation for remote sensing images. Remote Sens. **15**(14), 3607 (2023). https://doi.org/10.3390/rs15143607
16. Wang, Y., Yang, G., Lu, H.: Domain adaptive tree crown detection using high-resolution remote sensing images. J. Appl. Remote Sens.**16**(04) (2022). https://doi.org/10.1117/1.jrs.16.044505
17. Xie, E., Wang, W., Yu, Z., Anandkumar, A., Alvarez, J.M., Luo, P.: Segformer: simple and efficient design for semantic segmentation with transformers. Adv. Neural Inf. Process. Sys. **34**, 12077–12090 (2021). https://doi.org/10.48550/arXiv.2105.15203
18. Xu, Z., Zhang, W., Zhang, T., Yang, Z., Li, J.: Efficient Transformer for Remote Sensing Image Segmentation. Remote Sen. **13**(18), 3585 (2021). https://doi.org/10.3390/rs13183585
19. Zhang, C., Jiang, W., Zhang, Y., Wang, W., Zhao, Q., Wang, C.: Transformer and CNN Hybrid Deep Neural Network for Semantic Segmentation of Very-High-Resolution Remote Sensing Imagery. IEEE Trans. Geosci. Remote Sens. **60**, 1–20 (2022). https://doi.org/10.1109/tgrs.2022.3144894
20. Zhang, J., Shao, M., Wan, Y., Meng, L., Cao, X., Wang, S.: Boundary-aware spatial and frequency dual-domain transformer for remote sensing urban images segmentation. IEEE Trans. Geosci. Remote Sens. **62**, 1–18 (2024). https://doi.org/10.1109/tgrs.2024.3430081
21. Zhao, H., Shi, J., Qi, X., Wang, X., Jia, J.: Pyramid scene parsing network. In: Proceedings of the IEEE Conference on Computer Vision and Pattern Recognition. pp. 2881–2890 (2017). https://doi.org/10.48550/arXiv.1612.01105
22. Zheng, S., et al.: Rethinking semantic segmentation from a sequence-to-sequence perspective with transformers. In: Proceedings of the IEEE/CVF Conference on Computer Vision and Pattern Recognition. pp. 6881–6890 (2021). https://doi.org/10.48550/arXiv.2012.15840
23. Zhu, F., Chen, Z., Li, H., Shi, Q., Liu, X.: Cedanet: individual tree segmentation in dense orchard via context enhancement and density prior. IEEE J. Sel. Appl. Earth Obs. Remote Sens. **17**, 7040–7051 (2024). https://doi.org/10.1109/jstars.2024.3378167

Integrating Vision Transformers and Autoencoders for Interpretable Cancer Risk Assessment

Ahmad Hussein[1], Ali Anaissi[2,3] (iD), Mukesh Prasad[1] (iD), and Ali Braytee[1(✉)] (iD)

[1] School of Computer Science, University of Technology Sydney, Sydney, Australia
{ahmad.hussein,mukesh.prasad,ali.braytee}@uts.edu.au
[2] TD School, University of Technology Sydney, Sydney, Australia
ali.anaissi@uts.edu.au
[3] School of Computer Science, The University of Sydney, Sydney, Australia

Abstract. Cancer remains one of the leading causes of mortality worldwide, necessitating accurate diagnosis and prognosis. Whole Slide Imaging (WSI) has become an integral part of clinical workflows with advancements in digital pathology. While Vision Transformers (ViT) have been applied to WSI analysis, their lack of interpretability limits clinical adoption. In this paper, we propose PATH-X, a deep learning framework that integrates pretrained ViT with SHAP (Shapley Additive Explanations) to enhance model explainability for patient stratification and risk prediction using WSIs from The Cancer Genome Atlas (TCGA). A representative image slice is selected for each WSI, and numerical feature embeddings are extracted using Google's pre-trained ViT. These features are then compressed via an autoencoder and used for unsupervised clustering and classification tasks. Kaplan-Meier survival analysis evaluates risk stratification into two and three risk groups. SHAP is applied to identify key contributing features, which are mapped onto histopathological slices to provide spatial context. PATH-X was applied to three TCGA cancer types: kidney, glioma, and breast, selected for sufficient WSI sample sizes, and achieved robust stratification and strong classification performance across all cohorts, demonstrating the model's effectiveness.

Keywords: Whole Slide Images · Deep Learning · Vision Transformer · Autoencoder · Explainable AI · SHAP · Patient Stratification · Survival Analysis · Cancer Prognosis

1 Introduction

Whole-slide imaging (WSI) has revolutionized digital pathology by enabling the digitization of entire histology slides, facilitating remote consultations, and improving communication among healthcare providers, thereby making specialized care more accessible [9,10]. The integration of computational methods with WSIs has enhanced the analysis and interpretation of histopathological slides,

© The Author(s), under exclusive license to Springer Nature Singapore Pte Ltd. 2026
Q. V. Nguyen et al. (Eds.): AusDM 2025, CCIS 2765, pp. 316–330, 2026.
https://doi.org/10.1007/978-981-95-6786-7_21

contributing to more precise and data-driven clinical decisions. Machine learning (ML) techniques have played a crucial role in computational pathology, offering automated solutions for tumor segmentation, classification, and prognostic modeling [1,16]. Various deep learning (DL)-based frameworks have been developed to process histopathological images, extracting meaningful features for cancer diagnosis and outcome prediction [9,10]. Pathomics focuses on extracting high-dimensional features from WSIs and converting them into compact representations for clinical decision-making [6]. Studies integrating deep learning with WSI and pathomics have demonstrated improved tumor detection and classification, supporting AI-driven pathology applications [21]. Despite these advancements, several research gaps remain. Many existing deep learning models, particularly those leveraging convolutional neural networks (CNNs), lack interpretability, which hinders their clinical adoption. While some studies have incorporated WSIs into their frameworks, the extracted features may not fully capture the most relevant pathological information, and the absence of interpretability remains a significant limitation. To address these gaps, we propose PATH-X, a framework that integrates explainable AI (xAI) techniques with ViT-extracted pathomic features. Our study evaluates the effectiveness of pathomics alone in risk stratification across multiple cancer types.

The practical relevance of our proposed framework is evident, as it enables the early detection of specific risk groups, thereby supporting clinicians in making informed treatment decisions at various stages. Our approach enhances patient stratification within the context of precision oncology. Furthermore, incorporating SHAP improves model interpretability by highlighting key image features that contribute to risk assessment. PATH-X was validated on BRCA, GLIOMA, and KICH, demonstrating its effectiveness in patient stratification and disease classification tasks.

Our key contributions are as follows:

- We propose a WSI-driven patient stratification framework that integrates an autoencoder with Google's pre-trained Vision Transformer (ViT), trained on a large-scale image dataset, to generate compact and meaningful feature representations.
- We perform extensive experiments on three cancer types (BRCA, GLIOMA, and KICH), where sufficient WSIs are available, to demonstrate the effectiveness of PATH-X in pathomics-based analysis, achieving strong classification and stratification performance.

2   Related Work

Recent advances indicate that Vision Transformers (ViTs) are playing an increasingly important role in computational pathology. A comprehensive 2024 review [8] highlights that ViT architectures have begun to outperform traditional CNNs in various cancer diagnosis tasks, owing to their ability to model long-range dependencies in whole-slide images (WSIs). Unlike patch-level CNN approaches,

ViTs can capture global context across gigapixel pathology slides, which has driven their increasing adoption in digital pathology and cancer detection.

Various studies have explored deep learning-based pathomics for cancer diagnosis, prognosis, and treatment response prediction. Nibid et al. [14] applied a deep learning-based pathomics approach to predict response to chemoradiotherapy in stage III NSCLC patients using 35 digitized tissue slides. Five pretrained CNN models (AlexNet, VGG, MobileNet, GoogLeNet, ResNet) were evaluated, with GoogLeNet achieving the best performance, correctly classifying responders (8/12) and non-responders (10/11), demonstrating high specificity (TNR: 90.1) and sensitivity (TPR: 0.75). Similarly, Li et al. [13] used WSI images of H&E-stained histological specimens from ESCC patients receiving PD-1 inhibitors, employing a pre-trained ViT model for patch-level feature extraction and an RNN-based patient-level predictor to construct an ESCC-pathomics signature (ESCC-PS). Their model, trained on 486,188 image patches, achieved 84.5% accuracy, effectively stratifying patients into three risk groups based on progression-free survival (PFS). Another study by Kim et al. [11] utilized WSIs from 256 melanoma patients to predict BRAF mutations using a CNN-based pipeline. Their model identified tumor-rich areas (AUC = 0.96) and predicted BRAF mutation status (AUC = 0.71). In addition, a pathomics-based analysis quantified nuclear morphology, revealing BRAF-mutated nuclei as larger and rounder than wild-type nuclei. A combined model integrating clinical data, deep learning, and pathomics further improved predictive accuracy (AUC = 0.89). Furthermore, Williams et al. [9] reviewed a range of WSI-based deep learning approaches, including ensemble learning to address class imbalance, weakly supervised learning for large-scale image analysis, and patch-based aggregation strategies for efficient feature extraction.

ViTs have also been applied to specialized predictive tasks in pathology. For instance, DIOR-ViT [12] introduced a multi-task transformer capable of both standard classification and ordinal regression of tumor grades. By modeling the relative severity between samples, this approach better captured the inherent ordering of histologic grades. The DIOR-ViT model delivered accurate and reliable grading across three cancer datasets, outperforming conventional grading methods that treat grades as independent classes. Overall, the surge of ViT-driven WSI frameworks – from multi-class tumor classifiers to multi-task and multi-scale models – has significantly advanced the state of cancer diagnosis. These transformer-based approaches not only achieve state-of-the-art accuracy but also enhance interpretability through attention-based region highlighting or ordinal relationship learning, which is crucial for building clinician trust and enabling clinical adoption [19]. This growing body of work provides a strong foundation for integrating ViTs, alongside autoencoder-based representation learning [2,5] into interpretable cancer risk assessment systems, as pursued in our study.

3   Methodology

3.1   Proposed Framework

We propose PATH-X, a deep learning framework for pathology-driven cancer prognosis and classification, comprising three key components: Slicing Images, ViT and Autoencoder, followed by feature evaluation, as shown in Fig. 1. Additionally, GradientSHAP enhances interpretability by localizing the most influential features on histological slices. The framework extracts pathomics features from WSIs using Google's ViT model, generating high-dimensional representations. These features are then compressed via an autoencoder for efficiency, followed by clustering and classification for patient stratification.

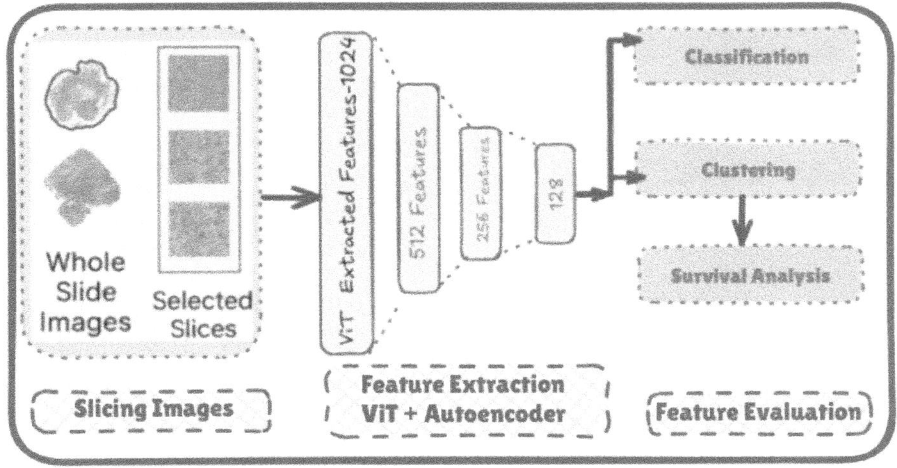

Fig. 1. Workflow of the PATH-X framework combining image slicing, Vision Transformer encoding, and autoencoder-based feature evaluation.

3.2   Slicing Images

Whole-slide images (WSIs) can range from tens of thousands to over 100,000 pixels per dimension and may reach sizes up to 10 GB [4]. The pre-trained Google Vision Transformer (ViT) model used in our study accepts input patches of size 224×224 pixels. Therefore, for each WSI, we select a representative patch based on a defined selection criterion to ensure compatibility with the model input requirements. Our method processes SVS files to extract 1024×1024 image slices. Our code runs on Google Colab's A100 GPU and utilizes GPU acceleration with cuCIM and CuPy libraries for efficient computation. The "best slice" for each sample is selected and saved using a composite scoring mechanism. The selection process integrates three key metrics: nuclei count, clarity, and blank space. Nuclei count is determined through color thresholding, where pixels within a defined range $[0, 0, 50]$ to $[120, 120, 200]$ are classified as nuclei. A binary mask is then applied to identify these pixels, and the number of connected components is computed using the `measure.label()` function. Clarity is assessed by computing the variance of the Laplacian, a mathematical measure of image sharpness. This involves calculating the gradients of the grayscale image along the x and y axes, summing their variances:

$$\text{clarity} = \text{Var}(\nabla_x) + \text{Var}(\nabla_y) \tag{1}$$

where ∇_x and ∇_y represent the gradients along the respective axes. The final composite score for each slice is computed as:

$$\text{score} = (\text{num\_nuclei} \times \text{clarity}) - \text{blank\_space} \tag{2}$$

Here, the number of detected nuclei (num_nuclei) is weighted by the clarity of the slice, while blank space is subtracted to penalize slices with large empty areas. This scoring mechanism ensures that only high-quality slices with well-preserved tissue structures are selected, facilitating accurate downstream analysis across various cancer types.

3.3   ViT and Autoencoder

To extract features from the resulting slices we use Google's Vision Transformer (ViT), introduced in the paper [7], It redefines deep learning for computer vision by leveraging transformer architectures, which were originally designed for natural language processing. Unlike CNNs, ViT processes images as a sequence of non overlapping patches. Specifically, an image $x \in \mathbb{R}^{H \times W \times C}$ is divided into N patches of size $P \times P$, where:

$$N = \frac{H \times W}{P^2} \tag{3}$$

Each patch is flattened into a vector and linearly projected into D-dimensional embeddings, forming an input sequence to the transformer:

$$z_0^i = W_p x_p^i + E_p \tag{4}$$

where W_p is the projection matrix, x_p^i represents a patch, and E_p is a learned positional embedding. These embeddings pass through multiple self-attention layers, following the Multi-Head Self-Attention (MHSA) mechanism:

$$\text{Attention}(Q, K, V) = \text{softmax}\left(\frac{QK^T}{\sqrt{d_k}}\right) V \tag{5}$$

where Q, K, V are query, key, and value matrices derived from input embeddings. ViT processes images efficiently, achieving state-of-the-art performance on large datasets like ImageNet and demonstrating superior scalability compared to convolutional architectures.

For feature extraction, we utilize the Vision Transformer (ViT) model (google/vit-large-patch16-224), processing images through a transformation pipeline that resizes, normalizes, and tokenizes them into non-overlapping 16×16 patches. The extracted features from the model's final hidden state are averaged and stored in a structured CSV file for downstream analysis.

Autoencoder: To transform high-dimensional embeddings into compact representations, we employed an autoencoder in our framework to denoise and refine the extracted features, enhancing their suitability for downstream analysis. Autoencoders have been widely applied in dimensionality reduction, anomaly detection, and feature extraction [3]. Our employed autoencoder utilizes an encoder, decoder, loss function, and optimizer to effectively compress high-dimensional image features into a lower-dimensional latent representation. The encoder function transforms the 1024-dimensional input features into a 128-dimensional latent representation using fully connected layers with ReLU activation:

$$z = f_{\text{encoder}}(x) = \sigma(W_3\sigma(W_2\sigma(W_1 x + b_1) + b_2) + b_3) \tag{6}$$

where W_1, W_2, W_3 are the weight matrices for each dense layer, and b_1, b_2, b_3 are the corresponding biases. The decoder function reconstructs the original feature space from the latent representation, employing a similar multi-layer transformation with a sigmoid activation function:

$$\hat{x} = f_{\text{decoder}}(z) = \sigma(W_3'\sigma(W_2'\sigma(W_1'z + b_1') + b_2') + b_3') \tag{7}$$

where W_1', W_2', W_3' and b_1', b_2', b_3' are the weights and biases of the decoder layers. The sigmoid activation function ensures that values remain between 0 and 1, enhancing numerical stability:

$$\sigma(x) = \frac{1}{1 + e^{-x}} \tag{8}$$

To optimize reconstruction quality, the model minimizes the Mean Absolute Error (MAE) loss, defined as:

$$\mathcal{L}_{\text{MAE}} = \frac{1}{n}\sum_{i=1}^{n} |x_i - \hat{x}_i| \tag{9}$$

where $n = 1024$ is the number of input features, and $x_i, \hat{x}_i$ represent the original and reconstructed feature values, respectively. The model is trained using the Adam optimizer with a learning rate of 0.001, updating parameters as follows:

$$\theta \leftarrow \theta - \eta\frac{\partial\mathcal{L}}{\partial\theta} \tag{10}$$

where $\eta = 0.001$ is the learning rate, $\mathcal{L}$ is the MAE loss, and θ represents the model parameters (weights and biases).

3.4 Feature Evaluation

Our feature evaluation involves clustering the cases into two and three risk groups using hierarchical clustering applied to the extracted latent features, followed by Kaplan-Meier survival analysis to assess differences in patient outcomes. Additionally, we apply several classification algorithms across each cancer type to evaluate predictive performance. GradientSHAP is employed to interpret the contribution of individual ViT-extracted image features to the learned representations. GradientSHAP is an explainability method implemented in Captum, a model interpretability library for PyTorch. It approximates SHAP values using integrated gradients by computing the expectation of gradients over multiple noisy input perturbations and baseline references [18]. Mathematically, it is computed as:

$$\phi_i = \mathbb{E}_{x_\alpha \sim P} \left[(x_i - x_i') \cdot \frac{\partial f(x_\alpha)}{\partial x_i} \right] \tag{11}$$

where x represents the input sample, while x' is a randomly chosen baseline (reference). An interpolated sample between x and x' is denoted as x_α, with P representing a probability distribution over these interpolations. The function $f(x)$ denotes the model output, and $\frac{\partial f(x_\alpha)}{\partial x_i}$ represents the gradient of the model output with respect to the feature i. This equation integrates gradients over random paths between the input and baseline while incorporating Gaussian noise for robustness. We applied GradientSHAP to identify the top 10 ViT-extracted features contributing to the encoded representations.

4 Experiments

For this study, clinical data for breast, glioma, and kidney cancer patients were sourced from the LinkedOmics portal, a publicly available resource providing multi-omics data for all 32 TCGA cancer types and 10 Clinical Proteomics Tumor Analysis Consortium (CPTAC) cancer cohorts [15]. Additionally, biospecimen diagnostic slides for TCGA cases were obtained from the GDC Data Portal using R. The datasets include 1,065 breast cancer cases, 880 glioma cases, and 891 kidney cancer cases. For each case, 1,024 numerical features were extracted using Google's ViT model. These high-dimensional features were then compressed into 128 latent representations using an autoencoder.

Hierarchical clustering was performed separately for each cancer type, with patients stratified into two and three clusters, and the results visualized using t-SNE. Clinical data were used to conduct Kaplan–Meier survival analysis, assessing the prognostic significance of the stratified risk groups. In addition, classification tasks were carried out to distinguish breast cancer cases based on tumor purity (low vs. high), and to differentiate between glioma and kidney cancer subtypes. These classifications were evaluated using multiple machine learning models, including logistic regression, KNN, SVM, MLP, and random forest. Model performance was assessed using accuracy, weighted F1 score, and macro F1 score as evaluation metrics.

Finally, GradientSHAP was applied to identify the top 10 ViT-extracted features contributing to the autoencoder's latent representations for each cancer type. These important features were then mapped onto the corresponding histopathological slices to enhance model interpretability.

5   Results

5.1   Glioma

Clustering and Survival Analysis. Fig. 2 presents the t-SNE plots, which illustrate well-separated clusters, demonstrating meaningful risk stratification based on features extracted using PATH-X. Hierarchical clustering was employed to stratify patients into two and three risk groups. The corresponding Kaplan-Meier survival curves further validate this stratification, showing significant survival differences across risk groups, global $p = 1.4 \times 10^{-48}$. The extremely low p-values, as reported in Table 1, highlight the robustness of our method in distinguishing patient outcomes.

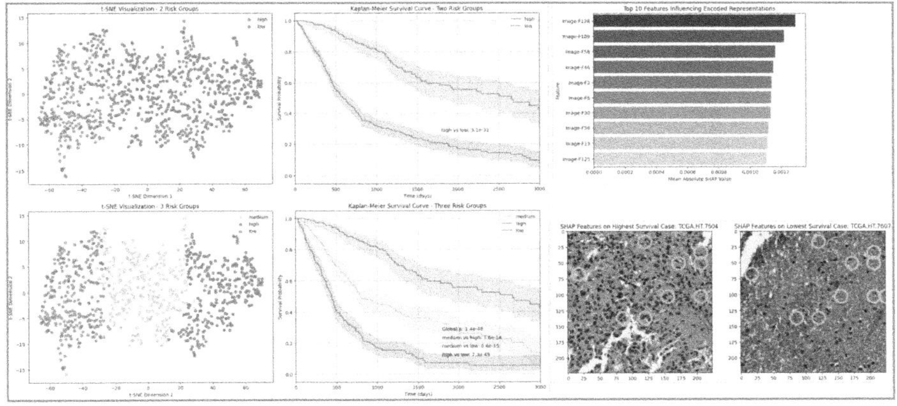

Fig. 2. Comparative visualizations for Glioma risk stratification. Left: t-SNE projection with Kaplan–Meier survival analysis illustrating patient subgroup separation. Right: Gradient SHAP analysis highlighting the top 10 features in selected samples.

Interpretation Using Gradient Shap. Fig. 2 illustrates the integration of GradientSHAP for selecting the most influential ViT-extracted features contributing to the encoded representations. It highlights the top 10 SHAP-selected features and visualizes their corresponding locations on histopathological slices of cases with the highest and lowest survival rates. This approach enables feature

Table 1. Pairwise and global log-rank p-values for glioma survival stratification into two or three risk groups.

Cancer Type	p-values
Glioma (2 Risk Groups)	3.1×10^{-31} (Low vs High)
Glioma (3 Risk Groups)	7.3×10^{-49} (Low vs High)
	6.6×10^{-15} (Low vs Medium)
	7.6×10^{-14} (Medium vs High)
	Global p: 1.4×10^{-48}

attribution, helping to understand which extracted features play a crucial role in downstream survival and clustering tasks. The small GradientSHAP values (0.0032 to 0.0042) are expected, as the inputs were min-max scaled to $[0, 1]$, and the autoencoder maps them to a 128-dimensional latent space without softmax or classification layers to amplify gradients. In this context, SHAP values tend to be small but still meaningful for identifying influential features.

Classification. As shown in Table 2, We trained five machine learning classifiers on a dataset of 879 glioma cases, each represented by 128 autoencoder-derived image features. The data were split into 70% training and 30% testing, with class labels balanced between 490 LGG and 389 GBM cases. Models were evaluated using accuracy, macro F1, and weighted F1 scores to assess classification performance. Based on the classification results, Random Forest achieved the best overall performance, with the highest accuracy (82.20%), macro F1 score (81.71%), and weighted F1 score (82.05%), indicating its superior ability to distinguish between GBM and LGG cases using autoencoder-derived image features.

Table 2. Classification performance on GBM vs LGG using image features

Method	Accuracy	Macro F1 Score	Weighted F1
Logistic Regression	0.7955	0.7924	0.7953
KNN	0.7917	0.7815	0.7868
SVM	0.7992	0.7958	0.7976
MLP	0.7841	0.7802	0.7855
Random Forest	**0.8220**	**0.8171**	**0.8205**

5.2 Kidney

Clustering and Survival Analysis. Figure 3 presents the t-SNE visualizations and Kaplan-Meier survival curves for kidney cancer risk stratification. The results show a clear separation between the two risk groups (low and high), indicating distinct patient clusters based on features extracted using PATH-X. The three-group stratification also demonstrates meaningful separation, although some overlap is observed between the medium- and high-risk groups. The Kaplan-Meier plots further confirm the clinical relevance of these stratifications, revealing significant survival differences across the risk groups. The pairwise and global p-values reported in Table 3 underscore the statistical robustness of PATH-X extracted features in differentiating patient outcomes.

Table 3. Pairwise and global log-rank p-values for kidney cancer survival analysis using two and three risk group stratification.

Cancer Type	p-values
Kidney (2 Risk Groups)	6.4×10^{-4} (Low vs High)
Kidney (3 Risk Groups)	2.7×10^{-3} (Low vs High)
	7.2×10^{-6} (Low vs Medium)
	5.2×10^{-2} (Medium vs High)
	Global p: 5.6×10^{-5}

Explainability with Gradient Shap: Visual inspection of the SHAP-highlighted regions shown in Fig. 3 reveals distinguishable patterns between high and low survival cases. The high-survival image appears more structurally uniform, while the low-survival case exhibits increased heterogeneity and denser cellular regions. These differences suggest that the ViT-derived features identified by GradientSHAP may capture relevant morphological cues correlated with patient outcomes in kidney cancer. If further validated, this type of interpretable AI-based analysis could help pathologists spot subtle features linked to patient outcomes and improve risk stratification in clinical practice.

Classification. To evaluate the discriminative power of ViT-derived features in distinguishing kidney cancer subtypes, we conducted a binary classification task comparing clear cell renal cell carcinoma (ccRCC) against other subtypes. For class balancing and interpretability, we grouped kidney papillary renal cell carcinoma and kidney chromophobe into a unified class labeled "other kidney subtypes." The dataset consisted of 891 samples in total, with 513 ccRCC and 378 other subtypes. Following preprocessing and scaling, the data was split into 70% training and 30% testing using stratified sampling. We trained five classifiers; Logistic Regression, KNN, SVM, MLP, and Random Forest, their performance were evaluated using accuracy, macro F1 score, and weighted F1 score. As showin in Table 4, the MLP classifier achieved the highest performance, with an accuracy of 70.90%, macro F1 score of 69.60%, and weighted F1 score of 70.50%, suggesting it effectively captured the latent differences between ccRCC and other subtypes.

Table 4. Classification performance on Kidney Clear Cell vs Other Subtypes using image features.

Method	Accuracy	Macro F1 Score	Weighted F1 Score
Logistic Regression	0.664	0.642	0.656
KNN	0.690	0.678	0.687
SVM	0.683	0.662	0.674
MLP	**0.709**	**0.696**	**0.705**
Random Forest	0.687	0.654	0.670

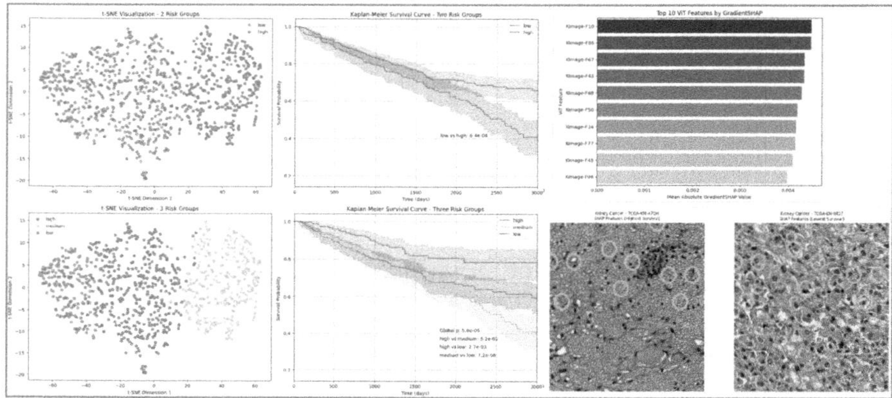

Fig. 3. Comparative visualizations for Kidney risk stratification. Left: t-SNE projection with Kaplan–Meier survival analysis illustrating patient subgroup separation. Right: Gradient SHAP analysis highlighting the top 10 features in selected samples.

5.3   Breast Cancer

Clustering and Survival Analysis: Fig. 4 illustrates the t-SNE visualizations and Kaplan-Meier survival curves for breast cancer risk stratification using learned latent features. The two-group stratification (low vs. high risk) shows a clear visual distinction between patient clusters, indicating that the extracted features effectively capture clinically relevant differences. The corresponding Kaplan-Meier plot reveals a significant survival divergence between the two groups, with a p-value of 2.5×10^{-05}, highlighting the prognostic value of the model-derived representations. Expanding this analysis to three risk groups (low, medium, and high) demonstrated a more nuanced stratification. Although the medium- and high-risk groups exhibit some spatial overlap in the t-SNE plot, the survival curves indicate statistically significant differences among all three groups. The global p-value 1.0×10^{-10}, along with strong pairwise p-values, see Table 5, further highlights the robustness of our approach in distinguishing survival outcomes.

Table 5. Pairwise and global p-values for breast cancer risk stratification.

Cancer Type	p-values
Breast (2 Risk Groups)	2.5×10^{-8} (Low vs High)
Breast (3 Risk Groups)	6.5×10^{-12} (Low vs High)
	4.5×10^{-3} (Low vs Medium)
	1.0×10^{-3} (Medium vs High)
	Global p: 1.0×10^{-10}

Explainability with GradientSHAP: Fig. 4 demonstrates the use of GradientSHAP to identify and visualize the most impactful ViT-derived features for breast cancer cases. The SHAP-identified patch locations are overlaid on images from two BRCA cases, one with the highest survival outcome and one with the lowest, highlighting spatially significant visual tokens that contributed most to the model's latent representations. This multi-view visualization enhances interpretability by linking abstract ViT features to tangible histopathological patterns, providing insight into how feature importance varies across different patient outcomes.

Table 6. Classification Performance on Breast Cancer Tumor Purity (Low vs High) Using Image Features

Method	Accuracy	Macro F1 Score	Weighted F1
Logistic Regression	0.615	0.536	0.603
KNN	0.599	0.519	0.588
SVM	**0.611**	**0.541**	**0.684**
MLP	0.606	0.534	0.598
Random Forest	0.621	0.502	0.588

Classification on Tumor Purity: We evaluated breast cancer tumor purity (low vs high) from image features on 1,054 cases (676 high, 378 low), using a 70/30 train–test split. Accuracy ranged 0.599–0.621, with Random Forest best at 0.621; SVM achieved the highest weighted F1= 0.684. Macro F1 scores were modest (0.502–0.541), indicating difficulty on the minority class and the effect of class imbalance. refer to Table 6 for results.

5.4   Comparison with State-of-the-Art Methods

Wessel et al. [20] propose an interpretable DINO–ViT pipeline for clear-cell renal cell carcinoma (ccRCC) that predicts overall survival (OS) and disease-specific survival (DSS). Self-supervised ViT features from TCGA WSIs are fed to

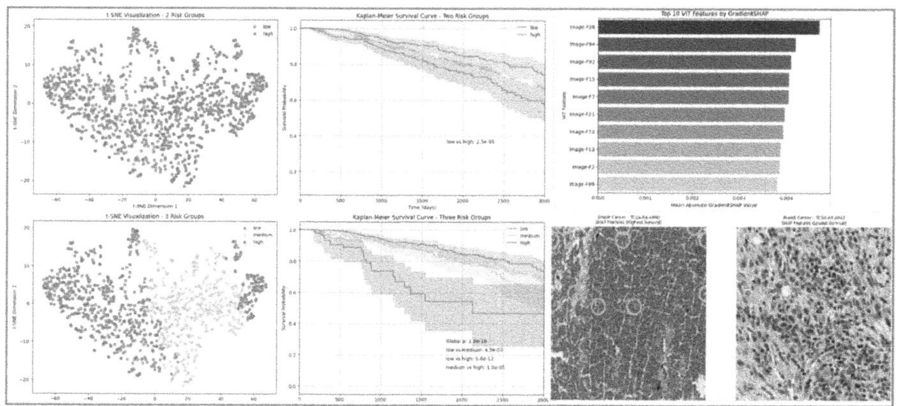

Fig. 4. Comparative visualizations for BRCA risk stratification. Left: t-SNE projection with Kaplan–Meier survival analysis illustrating patient subgroup separation. Right: Gradient SHAP analysis highlighting the top 10 features in selected samples.

Cox proportional-hazards models, which yielded significant Kaplan–Meier (KM) separation in univariable analyses and independent prognostic value (hazard ratios) in multivariable models. HVTSurv [17] introduces a hierarchical ViT with local (Manhattan), WSI-level (spatial shuffle), and patient-level (attention pooling) interactions, and reports higher concordance index (C-index) than weakly supervised WSI baselines across six TCGA cancers. In contrast, PATH-X compresses ViT tile embeddings with an autoencoder and performs unsupervised two and three group risk stratification, evaluated by KM curves and log-rank tests rather than supervised hazard modeling. Beyond WSI attention maps, PATH-X enhances interpretability by integrating GradientSHAP and mapping feature attributions onto selected slices.

6   Conclusion

We proposed a framework utilizing ViT-extracted pathomics features for cancer patient stratification. Our method successfully identified prognostically relevant features and achieved strong classification performance across multiple tasks. SHAP-based analysis enabled interpretable feature selection, while feature mapping localized critical visual tokens on histopathological slices. While the slicing algorithm used to extract informative regions from whole slide images is mathematically sound, clinical assessment remains essential to confirm the biological and diagnostic relevance of these selected slices. Additional challenges include the ViT model's fixed patching mechanism and the need for larger, more diverse datasets. Despite these limitations, our approach demonstrates the potential of explainable pathomics analysis in cancer prognosis. Future improvements will focus on multi-omics integration, attention-based model refinement, and expert validation to enhance the clinical applicability of our framework within precision oncology.

Acknowledgements. This research is supported by an Australian Government Research Training Program Scholarship.

– Generative AI tools were used only for grammar and language refinement.

References

1. Aerts, H.J., et al.: Decoding tumour phenotype by noninvasive imaging using a quantitative radiomics approach. Nat. Commun. **5**, 4006 (2014)
2. Anaissi, A., Zandavi, S.M., Suleiman, B., Naji, M., Braytee, A.: Multi-objective variational autoencoder: an application for smart infrastructure maintenance. Appl. Intell. **53**(10), 12047–12062 (2023)
3. Baldi, P.: Autoencoders, unsupervised learning, and deep architectures. Proc. ICML Workshop Unsupervised Transf. Learn. **27**, 37–50 (2012)
4. Bándi, P., Geessink, O., Manson, Q., Van Dijk, M., Balkenhol, M., Hermsen, M., et al.: From detection of individual metastases to classification of lymph node status at the patient level: The camelyon17 challenge. IEEE Trans. Med. Imaging **38**(2), 550–560 (2019). https://doi.org/10.1109/TMI.2018.2867350
5. Braytee, A., He, S., Tang, S., Sun, Y., Jiang, X., Yu, X., Khatri, I., Chaturvedi, K., Prasad, M., Anaissi, A.: Identification of cancer risk groups through multi-omics integration using autoencoder and tensor analysis. Sci. Rep. **14**, 11263 (2024). https://doi.org/10.1038/s41598-024-59670-8
6. Chen, R.J., et al.: Pathomic fusion: An integrated framework for fusing histopathology and genomic features for cancer diagnosis and prognosis. arXiv preprint arXiv:1912.08937 (2020)
7. Dosovitskiy, A., et al.: An image is worth 16x16 words: Transformers for image recognition at scale. In: International Conference on Learning Representations (ICLR). ICLR (2021)
8. Jiang, X., Wang, S., Zhang, Y.: Vision transformer promotes cancer diagnosis: a comprehensive review. Expert Sys. Appl.**252**, 124113 (2024). https://doi.org/10.1016/j.eswa.2024.124113
9. Jr, D.K.A.W., et al.: Digital pathology, deep learning, and cancer: a narrative review. Transl. Cancer Res. **13**(5), 2544–2560 (2024). https://doi.org/10.21037/tcr-23-964, https://dx.doi.org/10.21037/tcr-23-964
10. Khened, M., Kori, A., Rajkumar, H., Krishnamurthi, G., Srinivasan, B.: A generalized deep learning framework for whole-slide image segmentation and analysis. Sci. Rep. **11**, 11579 (2021). https://doi.org/10.1038/s41598-021-90444-8
11. Kim, R.H., et al.: Deep learning and pathomics analyses reveal cell nuclei as important features for mutation prediction of braf-mutated melanomas. J. Invest. Dermatol. **142**(6), 1650–1658.e6 (2022). https://doi.org/10.1016/j.jid.2021.09.034
12. Lee, J.C., Byeon, K., Song, B., Kim, K., Kwak, J.T.: Dior-vit: Differential ordinal learning vision transformer for cancer classification in pathology images. Med. Image Anal. **105**, 103708 (2025). https://doi.org/10.1016/j.media.2025.103708
13. Li, B., et al.: From pixels to patient care: deep learning-enabled pathomics signature offers precise outcome predictions for immunotherapy in esophageal squamous cell cancer. J. Transl. Med.**22**, 195 (2024). https://doi.org/10.1186/s12967-024-04997-z, https://doi.org/10.1186/s12967-024-04997-z

14. Nibid, L., et al.: Deep pathomics: A new image-based tool for predicting response to treatment in stage iii non-small cell lung cancer. PLOS ONE **18**(11), e0294259 (2023). https://doi.org/10.1371/journal.pone.0294259, https://doi.org/10.1371/journal.pone.0294259
15. Project, L.: Linkedomics: Analyzing multi-omics data within and across 32 cancer types (2023), https://www.linkedomics.org/login.php, retrieved Feb2023
16. Schmauch, B., et al.: A deep learning model to predict rna-seq expression of tumours from whole slide images. Nat. Commun. **11**(1), 3877 (2020)
17. Shao, Z., Chen, Y., Bian, H., Zhang, J., Liu, G., Zhang, Y.: Hvtsurv: Hierarchical vision transformer for patient-level survival prediction from whole slide image. In: Proceedings of the AAAI Conference on Artificial Intelligence. vol. 37, pp. 2209–2217 (2023). https://doi.org/10.1609/aaai.v37i2.25315
18. Team, C.: Gradientshap - captum. Available at https://captum.ai/api/gradient_shap.html (2024). Accessed Feb 2025
19. Tolkach, J., Schüffler, P.J.: Explainable artificial intelligence for pathology: hype or hope? Nat. Rev. Urol. **20**(1), 53–70 (2023)
20. Wessels, F., et al.: A self-supervised vision transformer to predict survival from histopathology in renal cell carcinoma. World J. Urol.**41**, 2233–2241 (2023). 29 Jun 2023
21. Wu, Y., Li, Y., Xiong, X., Liu, X., Lin, B., Xu, B.: Recent advances of pathomics in colorectal cancer diagnosis and prognosis. Frontiers Oncol. **13**, 1094869 (2023). https://doi.org/10.3389/fonc.2023.1094869

LightSkinNet: Lightweight CNN with Attention for Accurate, Mobile-Efficient Multiclass Skin Lesion Classification

F M Javed Mehedi Shamrat[1] , Xujuan Zhou[2]([✉]) , Mohd Yamani Idna Idris[1] , Raj Gururajan[2] , and Ka Ching Chan[2]

[1] Department of Computer System and Technology, Universiti Malaya, 50603 Kuala Lumpur, Malaysia
javedmehedicom@gmail.com, yamani@um.edu.my
[2] School of Business, University of Southern Queensland, Springfield Campus, Queensland, Australia
{Xujuan.zhou,Raj.Gururajan,KC.Chan}@unisq.edu.au

Abstract. Accurate multiclass classification of dermoscopic images is critical for early detection of malignant skin lesions, yet existing high-performing deep learning approaches often require computationally heavy backbones, limiting deployment on mobile or point-of-care devices. This study proposes LightSkinNet, a lightweight convolutional neural network (CNN) integrating Dilated Convolutional Enhancement (DCE) blocks, the Convolutional Block Attention Module (CBAM), and Efficient Channel Attention (ECA) to capture multi-scale features and refine salient lesion patterns with minimal computational cost. A hybrid class-balancing pipeline combining mild undersampling and oversampling was applied to the training set, followed by stratified splitting to preserve natural validation/test distributions, and targeted on-the-fly augmentation to enhance generalization. LightSkinNet was benchmarked on the HAM10000 dataset using five-fold stratified cross-validation against baseline models. The proposed model achieved a Top-1 accuracy of $90.27\% \pm 0.48$, Top-3 accuracy of $97.86\% \pm 0.22$, macro F1-score of $89.12\% \pm 0.53$, and Cohen's Kappa of 0.872 ± 0.006, outperforming all baselines with statistically significant improvements ($p < 0.05$) while maintaining a low parameter count. Gradient-Weighted Class Activation Mapping (Grad-CAM) showed that LightSkinNet focuses on key dermoscopic features, enhancing transparency in AI-assisted diagnosis. These results demonstrate that LightSkinNet delivers an effective accuracy–efficiency trade-off suitable for real-time clinical decision support on resource-constrained platforms.

Keywords: Dermoscopic image classification · Lightweight Convolutional Neural Network · Skin Lesion Analysis · Attention Mechanisms · Explainable Artificial Intelligence (XAI)

© The Author(s), under exclusive license to Springer Nature Singapore Pte Ltd. 2026
Q. V. Nguyen et al. (Eds.): AusDM 2025, CCIS 2765, pp. 331–345, 2026.
https://doi.org/10.1007/978-981-95-6786-7_22

1 Background Study

Skin cancer is among the most prevalent malignancies worldwide, with melanoma posing the highest mortality risk due to its aggressive progression and metastatic potential [1,2]. Early detection is critical, as survival rates drop significantly once melanoma metastasizes. Dermoscopy enables non-invasive visualization of skin lesions, improving diagnostic sensitivity compared to naked-eye examination, yet diagnostic accuracy remains highly dependent on clinical expertise and experience [3–5]. Automated computer-aided diagnosis (CAD) systems based on deep learning have emerged as promising tools to assist dermatologists, achieving performance comparable to experts in large-scale evaluations [6]. However, existing approaches often face challenges related to class imbalance, computational complexity, and reduced discriminative power for visually similar lesion types [7–9].

Early deep learning approaches for dermoscopic image analysis frequently relied on binary classification, most notably distinguishing melanoma from benign lesions, and achieved high accuracy in simplified setups [10]. While such settings are clinically valuable, they do not reflect real-world diagnostic complexity where multiple lesion categories coexist. Public benchmark datasets, such as ISIC 2016–2018 and HAM10000, have encouraged a shift toward multiclass classification involving up to seven distinct lesion types [11]. However, many high-performing models in this setting employ computationally heavy backbones (e.g., Inception-v4 [12], DenseNet-201 [13], ResNeXt-101 [14]) or ensemble strategies combining multiple CNNs [15], which hinder deployment on mobile or point-of-care devices. Recent baselines include ResNet variants, which leverage residual connections to facilitate deep architectures; EfficientNet families, which optimize depth–width–resolution scaling; and MobileNetV3, designed through neural architecture search with mobile efficiency in mind. On HAM10000, EfficientNet-B0 has demonstrated competitive accuracy [16], but with approximately 5.3M parameters, while MobileNetV3-Large (around 5.4M parameters) offers speed but can underperform on fine-grained lesion separation. Attention-enhanced approaches, such as the Convolutional Block Attention Module (CBAM) and Efficient Channel Attention (ECA), have improved feature selectivity in other domains but remain underexplored in lightweight multiclass dermoscopy contexts.

The multiclass dermoscopy setting presents several open challenges. Class imbalance, where common lesions such as melanocytic nevi (nv) are overrepresented and rare lesions such as dermatofibroma (df) are underrepresented, often biases models toward majority classes. Visual similarity between certain lesion types, such as melanoma and benign keratosis-like lesions, often results in cross-class misclassification. Hardware constraints in real-world deployments demand architectures with low parameter counts and fast inference without sacrificing accuracy. While several studies have incorporated data augmentation, transfer learning, and cost-sensitive loss functions to address imbalance, these approaches can increase computational overhead or fail to generalize to all minority classes. Moreover, few works combine multi-scale context aggregation with lightweight attention to achieve both efficiency and strong discriminative power in the seven-class setting.

This work makes the following key contributions:

- Introduces LightSkinNet, a 0.604 M-parameter CNN integrating Dilated Convolutional Enhancement (DCE), CBAM, and ECA to capture multi-scale features with high efficiency.
- Develops a hybrid class-balancing pipeline (undersampling + oversampling) for training data, combined with stratified splitting to maintain natural validation/test distributions, and targeted on-the-fly augmentation to enhance model generalization.
- Benchmarks LightSkinNet against EfficientNet-B0, MobileNetV3-Large, and ResNet50 on HAM10000, achieving the best Top-1/Top-3 accuracy, macro F1, and κ across five folds, and validating these improvements through statistical significance testing ($p < 0.05$).
- Integrates Grad-CAM for interpretable predictions, highlighting clinically relevant dermoscopic patterns.

2 Dataset Acquisition and Preprocessing Strategy

2.1 Dataset Description

The HAM10000 dataset [17] consists of 10,015 dermoscopic images, each belonging to one of seven histopathologically confirmed categories: melanocytic nevi (nv), melanoma (mel), benign keratosis-like lesions (bkl), basal cell carcinoma (bcc), actinic keratoses (akiec), vascular lesions (vasc), and dermatofibroma (df). Alongside the images, the dataset provides detailed metadata including lesion identifier, diagnosis label (dx), acquisition method, anatomical site, and patient demographics. This classification task is inherently imbalanced, with the largest class (nv) comprising 6,705 images and the smallest class (df) containing only 115 images. The class distribution is presented in Table 1. To mitigate imbalance during training, class weights were computed based on the training set distribution and applied in the loss function.

Table 1. Class distribution in the HAM10000 dataset.

Class Name	Label	Number of Images	Proportion (%)
Melanocytic nevi	nv	6,705	66.93
Melanoma	mel	1,113	11.11
Benign keratosis-like lesions	bkl	1,099	10.98
Basal cell carcinoma	bcc	514	5.13
Actinic keratoses	akiec	327	3.27
Vascular lesions	vasc	142	1.42
Dermatofibroma	df	115	1.15
Total	–	10,015	100.00

2.2 Data Preprocessing

The preprocessing pipeline ensured dataset integrity, standardized image formats, controlled class balancing, and augmentation for robustness. Metadata (CSV) was cross-referenced with the corresponding dermoscopic images, and corrupted or unreadable files were removed via image header verification.

Hybrid Class Balancing. The original HAM10000 distribution is highly imbalanced, with df, vasc, and akiec underrepresented relative to nv. SMOTE-family synthetic oversampling (e.g., Borderline-SMOTE, ADASYN) was considered but not adopted because pixel-space interpolation may distort dermoscopic morphology, so a morphology-preserving hybrid approach was preferred. To mitigate bias, a two-step hybrid strategy was applied toward a common target size $\bar{n}$ (the mean class count):

$$\bar{n} = \frac{1}{K} \sum_{c=1}^{K} n_c, \qquad K = 7, \tag{1}$$

consisting of (i) mild undersampling for classes with $n_c > \bar{n}$ and (ii) oversampling for classes with $n_c < \bar{n}$. Afterward, targeted augmentation (rotation, horizontal/vertical flips, zoom, brightness) was applied primarily to previously minority classes to introduce intra-class variability while preserving lesion morphology. This yielded an approximately even class distribution across all categories (Fig. 1).

Stratified Splitting. The balanced dataset was partitioned using five-fold stratified cross-validation. In each fold, class proportions are preserved. A stratified subset of the training split is held out for validation, and the remaining fold serves as the test split:

$$D = D_{\text{train}} \cup D_{\text{val}} \cup D_{\text{test}}, \quad D_{\text{train}} \cap D_{\text{val}} = \varnothing, \ D_{\text{train}} \cap D_{\text{test}} = \varnothing, \ D_{\text{val}} \cap D_{\text{test}} = \varnothing. \tag{2}$$

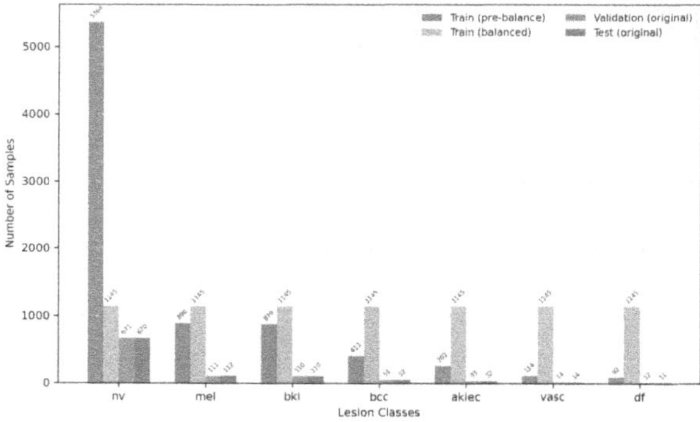

Fig. 1. Final dataset distribution after hybrid balancing.

Image Standardization. All images were resized to 224×224 red–green–blue (RGB). Pixel intensities were normalized to $[0,1]$:

$$I_{\text{norm}} = \frac{I_{\text{raw}}}{255}. \tag{3}$$

Lesion categories were label-encoded to $\{0, 1, 2, 3, 4, 5, 6\}$ for model compatibility.

On-the-Fly Data Augmentation. Augmentation was applied only to the training set during optimization:

$$\hat{x} = T(x), \quad T \in \mathcal{A}, \tag{4}$$

with $\mathcal{A}$ including controlled transformations (Table 2). Ranges preserve pigment-network structure and lesion borders; $\pm 25°/0.9 - 1.1/0.8 - 1.2$ reflect realistic scale and illumination variation in dermoscopy and align with common practice.

Table 2. On-the-fly data augmentation parameters applied to the training set.

Augmentation Type	Parameter Range
Rotation	$\pm 25°$
Horizontal Flip	Probability $= 0.5$
Vertical Flip	Probability $= 0.5$
Zoom	$[0.9, 1.1]$
Brightness Adjustment	$[0.8, 1.2]$
Width/Height Shift	$\pm 10\%$

3 Proposed LightSkinNet Architecture

LightSkinNet is a compact convolutional neural network (CNN) classifier designed for low-latency inference on ARM-based mobile devices while achieving accurate 7-class dermoscopic lesion recognition. The architecture follows a shallow–wide paradigm with early feature enhancement and lightweight attention mechanisms. It comprises: (i) an initial convolutional stem, (ii) two *DCE* (Depthwise–Conv–ECA) residual blocks, (iii) a downsampling stage with CBAM attention, (iv) a multi-branch context aggregation module, and (v) a global pooling classifier head. The total parameter count is 0.604M (trainable: 0.603M), satisfying the sub-1M mobile constraint. The overall computational workflow of the proposed LightSkinNet is illustrated in Fig. 2, covering dataset acquisition, preprocessing, architecture components, evaluation, and explainable AI visualization.

Given a feature map $X \in \mathbb{R}^{H \times W \times C}$, $\text{DWConv}_{3 \times 3}^{(d)}$ denotes a 3×3 depthwise convolution with dilation rate d, $\text{PWConv}_{1 \times 1}$ denotes a pointwise convolution, BN denotes batch normalization, and $\text{swish}(z) = z \cdot \sigma(z)$ denotes the swish activation. Throughout, GAP and GMP denote Global Average Pooling (GAP) and Global Max Pooling (GMP), respectively.

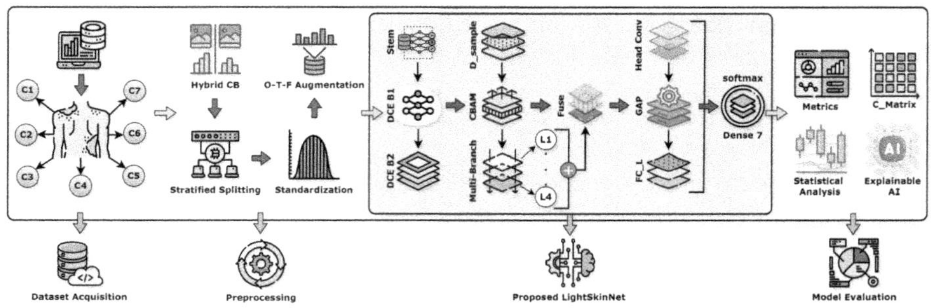

Fig. 2. Workflow of the proposed LightSkinNet framework for 7-class dermoscopic lesion classification, from dataset acquisition and preprocessing to model architecture, evaluation, and explainable AI.

3.1 ECA Attention

Efficient Channel Attention (ECA) is adopted as a lightweight channel reweighting mechanism on X. Let $g = \mathrm{GAP}(X) \in \mathbb{R}^C$ denote the channel descriptor obtained by Global Average Pooling (GAP). The 1D kernel size for channel interaction is selected adaptively:

$$t = \max\left(1, \left\lfloor \left| \tfrac{\log_2 C + b}{\gamma} \right| \right\rfloor \right), \quad k = \begin{cases} t, & t \text{ odd} \\ t+1, & t \text{ even} \end{cases} \tag{5}$$

with $(\gamma, b) = (2, 1)$ in our implementation. The attention vector is

$$\alpha = \sigma\big(\mathrm{Conv1D}(g; k, \mathrm{same})\big) \in \mathbb{R}^C, \qquad \mathrm{ECA}(X) = X \odot \alpha, \tag{6}$$

where $\odot$ denotes channel-wise multiplication broadcast to $H \times W$.

3.2 DCE Residual Block

A DCE block expands the receptive field via a dilated depthwise convolution, then applies pointwise mixing, BN, swish activation, and ECA attention. Residual skipping is used:

$$U = \mathrm{DWConv}_{3\times3}^{(d)}(X), \tag{7}$$
$$V = \mathrm{PWConv}_{1\times1}(U), \tag{8}$$
$$\tilde{V} = \mathrm{swish}(\mathrm{BN}(V)), \tag{9}$$
$$Y = \mathrm{ECA}(\tilde{V}), \tag{10}$$
$$\mathrm{DCE}(X) = \mathrm{Proj}(X) + Y, \tag{11}$$

where Proj is an identity mapping if channel dimensions match, otherwise a 1×1 projection with BN. In LightSkinNet, two DCE blocks are stacked with dilation rates $d \in \{2, 3\}$ and output width of 64 channels.

3.3 CBAM Attention Stage

After downsampling to 128 channels using a stride-2 3×3 convolution, CBAM refines salient channels and spatial locations. Given input X, channel attention is computed as:

$$\mathbf{c}_{\text{avg}} = \text{GAP}(X), \quad \mathbf{c}_{\text{max}} = \text{GMP}(X), \tag{12}$$

$$\mathbf{m}_c = \sigma\Big(W_2\,\delta(W_1\,\mathbf{c}_{\text{avg}}) + W_2\,\delta(W_1\,\mathbf{c}_{\text{max}})\Big), \tag{13}$$

followed by spatial attention:

$$S_{\text{avg}} = \text{Avg}_{\text{ch}}(X \odot \mathbf{m}_c), \quad S_{\text{max}} = \text{Max}_{\text{ch}}(X \odot \mathbf{m}_c), \tag{14}$$

$$M_s = \sigma\big(\text{Conv}_{7\times7}([S_{\text{avg}}, S_{\text{max}}])\big), \quad Y = (X \odot \mathbf{m}_c) \odot M_s. \tag{15}$$

3.4 Multi-branch Context Module

To aggregate multi-scale context efficiently, CBAM output Y is split into four parallel paths:

$$P_1 = \text{Conv}_{3\times3}^{\text{dil}=4}(Y) \rightarrow 64 \text{ ch}, \tag{16}$$

$$P_2 = \text{PWConv}_{1\times1}(\text{DWConv}_{3\times3}(Y)) \rightarrow 64, \tag{17}$$

$$P_3 = \text{Conv}_{5\times5}(Y) \rightarrow 64, \tag{18}$$

$$P_4 = Y \odot \sigma\big(\text{FC}_C(\delta(\text{FC}_{C/16}(\text{GAP}(Y))))\big), \tag{19}$$

where P_4 is a Squeeze-and-Excitation (SE) style gating that preserves original channels ($C = 128$). All paths are concatenated to form 320 channels and compressed:

$$Z = \text{swish}\big(\text{Conv}_{1\times1}([P_1, P_2, P_3, P_4])\big) \rightarrow 128 \text{ ch}. \tag{20}$$

3.5 Classifier Head

The final head applies Conv3×3 (128 ch) + BN + swish, dropout (0.3), global average pooling, and two Fully Connected (FC) layers with dropout (0.4) before the 7-way softmax:

$$h = \text{GAP}(Z), \tag{21}$$

$$o = \text{softmax}\big(W_3\,\delta(W_2\,\delta(W_1 h)) + b\big), \quad o \in \mathbb{R}^7. \tag{22}$$

Training Objective and Optimization. Let $x \in \mathbb{R}^{224\times224\times3}$ and $y \in \{1, \ldots, 7\}$ be an input image and its ground-truth class. The network outputs $\hat{\mathbf{p}} = \text{softmax}(\mathbf{z}) \in [0, 1]^7$.

Table 3. Layer-wise configuration of LightSkinNet.

Stage	Operation	Output size
Input	RGB image	$224 \times 224 \times 3$
Stem	Conv3×3-BN-swish; Conv3×3-BN-swish; MaxPool2×2	$112 \times 112 \times 32$
DCE×2	(DW dil. 2 → PW 64 → BN-swish-ECA) + skip; (DW dil. 3 *ldots*)	$112 \times 112 \times 64$
Downsample	Conv3×3, stride 2, BN-swish	$56 \times 56 \times 128$
Attention	CBAM	$56 \times 56 \times 128$
Multi-branch	P_1: dilated 3×3 (r=4) 64; P_2: DW3×3+PW 64; P_3: 5×5 64; P_4: SE-gated 128	concat $56 \times 56 \times 320$
Fuse	Conv1×1 → 128 ch (swish)	$56 \times 56 \times 128$
Head	Conv3×3-BN-swish; Dropout 0.3; GAP; Dense 128 (swish) → Dropout 0.4 → Dense 64 (swish)	–
Output	Dense 7 + softmax	7 classes

Class-Weighted Cross-Entropy: The weighted loss is:

$$\mathcal{L}_{\mathrm{wCE}}(y, \hat{\mathbf{p}}) = -w_y \log \hat{p}_y, \tag{23}$$

with class weights:

$$w_c = \frac{N}{7\, n_c}, \quad c \in \{1, \dots, 7\}. \tag{24}$$

Weights are normalized so that $\frac{1}{7} \sum_{c=1}^{7} w_c = 1$.

Top-k Accuracy. We report Top-1 and Top-3 accuracy:

$$\text{Top-}k = \frac{1}{|\mathcal{D}|} \sum_{(x,y) \in \mathcal{D}} \mathbf{1}[y \in \text{Top-}k(\hat{\mathbf{p}}(x))], \quad k \in \{1, 3\}. \tag{25}$$

Optimization Protocol: The model was trained using the Adam optimizer with an initial learning rate of 1×10^{-4} ($\beta_1 = 0.9$, $\beta_2 = 0.999$) and a batch size of 32 for up to 100 epochs. A cosine annealing learning rate scheduler was employed to adjust the learning rate dynamically. Early stopping with a patience of 15 epochs was applied based on validation loss to mitigate overfitting. Mixed precision training was adopted to accelerate convergence and reduce memory consumption.

Hyper-Parameter Selection: The training hyper–parameters were determined through a combination of preliminary tuning and guidance from prior studies on skin lesion classification. A random search was performed over learning rates $[10^{-5}, 10^{-2}]$, batch sizes $\{16, 32, 64\}$, and dropout rates $[0.2, 0.5]$. The configuration that provided the most stable convergence across five folds was the Adam optimizer with an initial learning rate of 1×10^{-4}, batch size of 32, dropout rates of $0.3/0.4$, and early stopping with patience 15. This setup was selected to ensure reproducibility across folds and to allow a fair comparison with widely adopted baselines such as ResNet50, EfficientNet-B0, and MobileNetV3-Large.

Regularization: Dropout (rates 0.3 and 0.4) is applied in the head. On-the-fly data augmentation is used for training only; validation and test sets remain unaltered. The detailed computational workflow of LightSkinNet is presented in Algorithm 1.

Algorithm 1: Computational Workflow of the Proposed LightSkinNet Architecture

---

Require: Input $x \in \mathbb{R}^{224 \times 224 \times 3}$; parameters θ
Ensure: Probabilities $\hat{\mathbf{p}} \in \mathbb{R}^{7}$; prediction $\hat{y} = \arg\max_c \hat{p}_c$
1: **procedure** LightSkinNet$(x; \theta)$
2:
$$z_0 \leftarrow \text{Conv}_{3\times3}\text{–BN–swish}(\text{Conv}_{3\times3}\text{–BN–swish}(x)); \quad z_0 \leftarrow \text{MaxPool}_{2\times2}(z_0)$$
3:    $z_1 \leftarrow \text{DCE}(z_0,\ \text{dil=2, out=64}); \quad z_2 \leftarrow \text{DCE}(z_1,\ \text{dil=3, out=64})$
4:    $z_3 \leftarrow \text{Conv}_{3\times3}-\text{BN}-\text{swish}(\text{Conv}_{3\times3}^{\text{stride=2, out=128}}(z_2))$
5:    $z_4 \leftarrow \text{CBAM}(z_3)$
6:    $Z \leftarrow \text{MultiBranch}(z_4)$
7:    $Z \leftarrow \text{Conv}_{3\times3}-\text{BN}-\text{swish}(Z); \quad h \leftarrow \text{GAP}(Z)$
8:    $h \leftarrow \text{Dense}(128, \text{swish}) \rightarrow \text{Dropout}(0.4) \rightarrow \text{Dense}(64, \text{swish})$
9:    $\hat{\mathbf{p}} \leftarrow \text{softmax}(Wh + b); \quad \hat{y} \leftarrow \arg\max_c \hat{p}_c$
10:    **return** $\hat{\mathbf{p}},\ \hat{y}$
11: **end procedure**

12: **procedure** DCE$(X;\ \text{dil, out}=F)$
13:    $U \leftarrow \text{DWConv}_{3\times3}^{(\text{dil})}(X); \quad V \leftarrow \text{PWConv}_{1\times1}^{\text{out}=F}(U)$
14:    $\tilde{V} \leftarrow \text{BN}(V); \quad \tilde{V} \leftarrow \text{swish}(\tilde{V})$
15:    $g \leftarrow \text{GAP}(\tilde{V}); \quad \alpha \leftarrow \sigma(\text{Conv1D}(g;\ k, \text{same}))$
16:    $E \leftarrow \tilde{V} \odot \alpha$
17:    $S \leftarrow \begin{cases} X, & C{=}F \\ \text{BN}(\text{PWConv}_{1\times1}^{\text{out}=F}(X)), & C \neq F \end{cases}$
18:    **return** $S + E$
19: **end procedure**

20: **procedure** CBAM(X)
21:    $\mathbf{m}_c \leftarrow \sigma(W_2\delta(W_1\,\text{GAP}(X)) + W_2\delta(W_1\text{GMP}(X))); \quad X' \leftarrow X \odot \mathbf{m}_c$
22:    $M_s \leftarrow \sigma(\text{Conv}_{7\times7}([\text{Avg}_{ch}(X'),\ \text{Max}_{ch}(X')]))$
23:    **return** $X' \odot M_s$
24: **end procedure**

25: **procedure** MultiBranch(Y)
26:    $P_1 \leftarrow \text{Conv}_{3\times3}^{\text{dil}=4}(Y) \rightarrow 64; \quad P_2 \leftarrow \text{PWConv}_{1\times1}(\text{DWConv}_{3\times3}(Y)) \rightarrow 64$
27:    $P_3 \leftarrow \text{Conv}_{5\times5}(Y) \rightarrow 64; \quad P_4 \leftarrow Y \odot \sigma(\text{FC}_C(\delta(\text{FC}_{C/16}(\text{GAP}(Y)))))$
28:    $Z \leftarrow \text{Conv}_{1\times1}([P_1, P_2, P_3, P_4]) \rightarrow 128;$
29:    $Z' \leftarrow \text{swish}(Z)$
30:    **return** Z'
31: **end procedure**

---

4   Results Analysis and Interpretation

4.1   Environmental Setup

All experiments were conducted on a desktop workstation equipped with an AMD Ryzen 7 3800X eight-core processor (3.90 GHz), 32 GB of RAM, and a 64-bit Windows 10 operating system. Model development and training were carried out using Python 3.10 with TensorFlow 2.15 and the Keras backend. GPU acceleration was provided by an NVIDIA GeForce RTX 4060 with 8 GB of VRAM, enabling efficient parallelized computation for deep learning model training.

To ensure reproducibility, fixed random seeds were applied across NumPy, TensorFlow, and Python's random module. The LightSkinNet model was trained for 100 epochs with a batch size of 32 using the Adam optimizer (initial learning rate of 1×10^{-4}) in conjunction with a cosine annealing learning rate scheduler. Early stopping with a patience of 15 epochs was employed based on validation loss to mitigate overfitting. In addition, real-time data augmentation was applied to the training set in each epoch to improve robustness and generalization capability.

4.2   Evaluation Metrics

Performance was assessed using Top-1 and Top-3 accuracy, class-wise precision, recall, and F1-score (macro- and micro-averaged), along with Cohen's Kappa (κ) to measure agreement beyond chance. Discriminative ability was evaluated via per-class and macro AUC-ROC and PR curves. A confusion matrix was used to visualize misclassification patterns. All results were reported as mean $\pm$ standard deviation over five stratified folds, with $p < 0.05$ considered statistically significant when comparing models. Qualitative analysis included Grad-CAM visualizations for interpretability.

4.3   Overall Classification Performance

Table 4 summarizes the average performance of the proposed LightSkinNet model compared to three baseline architectures: EfficientNet-B0, MobileNetV3-Large, and ResNet50. Results are reported as mean $\pm$ standard deviation over five stratified folds. LightSkinNet achieved a mean Top-1 accuracy of 90.27% $\pm$ 0.48 and a Top-3 accuracy of 97.86% $\pm$ 0.22, outperforming all baselines by a statistically significant margin ($p < 0.05$). The model also recorded the highest macro F1-score (89.12% $\pm$ 0.53) and Cohen's Kappa of 0.872 $\pm$ 0.006, indicating strong agreement with expert annotations.

Table 4. Overall classification performance comparison across models (mean $\pm$ std over 5 folds). Bold values indicate the best performance.

Metric	EfficientNet-B0	MobileNetV3-Large	ResNet50	LightSkinNet (Proposed)
Top-1 Acc. (%)	86.42 $\pm$ 0.61	84.97 $\pm$ 0.69	87.15 $\pm$ 0.57	**90.27 $\pm$ 0.48**
Top-3 Acc. (%)	95.18 $\pm$ 0.35	94.75 $\pm$ 0.40	95.94 $\pm$ 0.29	**97.86 $\pm$ 0.22**
Macro Precision (%)	85.33 $\pm$ 0.58	83.21 $\pm$ 0.65	86.12 $\pm$ 0.54	**89.45 $\pm$ 0.51**
Macro Recall (%)	84.91 $\pm$ 0.64	82.85 $\pm$ 0.70	85.88 $\pm$ 0.59	**88.81 $\pm$ 0.49**
Macro F1 (%)	85.11 $\pm$ 0.59	83.02 $\pm$ 0.67	85.98 $\pm$ 0.55	**89.12 $\pm$ 0.53**
Kappa	0.824 $\pm$ 0.007	0.809 $\pm$ 0.009	0.836 $\pm$ 0.008	**0.872 $\pm$ 0.006**

These results demonstrate that the combination of dilated convolutional enhancement (DCE), CBAM attention, and the multi-branch feature extraction module significantly improves classification accuracy while maintaining high computational efficiency. The proposed model achieved these improvements with only 0.604M trainable parameters, making it suitable for deployment on mobile or edge devices.

4.4 Per-class Performance Analysis

Table 5 presents the per-class classification performance of the proposed Light-SkinNet model on the held-out test set. The model achieved the highest F1-score for the vascular lesions (vasc) class (96.00%), followed by melanocytic nevi (nv) at 94.00%. Relatively lower performance was observed for actinic keratoses (akiec) and dermatofibroma (df) classes, primarily due to their limited representation in the training data. Macro-averaged precision, recall, and F1-score were 89.43%, 89.14%, and 89.21%, respectively, indicating balanced performance across classes despite the dataset imbalance. The micro-averaged F1-score of 90.27% is consistent with the overall Top-1 accuracy reported in Table 4.

Table 5. Class-wise performance of LightSkinNet on the held-out test set. Metrics are reported as percentages; Support denotes the number of test images per class.

Class (Label)	Precision (%)	Recall (%)	F1-score (%)	Support
Melanocytic nevi (nv)	95.00	93.00	94.00	670
Melanoma (mel)	88.00	89.00	88.50	112
Benign keratosis (bkl)	90.00	91.00	90.50	110
Basal cell carcinoma (bcc)	86.00	87.00	86.00	52
Actinic keratoses (akiec)	84.00	82.00	83.00	32
Vascular lesions (vasc)	97.00	95.00	96.00	14
Dermatofibroma (df)	86.00	87.00	86.50	11
Macro avg	89.43	89.14	89.21	—
Micro avg	—	—	90.27	1001

4.5   Confusion Matrix Analysis

Figure 3 presents the confusion matrices for all evaluated models, illustrating class-level prediction patterns. The proposed LightSkinNet demonstrated the highest concentration of predictions along the main diagonal, indicating strong true positive rates across all lesion categories. Notably, vascular lesions (vasc) and dermatofibroma (df) achieved 95.0% and 87.0% accuracy, respectively, despite their limited representation in the dataset.

Misclassifications were primarily observed between clinically similar lesion types. For instance, melanoma (mel) was occasionally predicted as benign kerato-sis (bkl) or melanocytic nevi (nv) due to overlapping dermoscopic patterns such as irregular pigmentation and asymmetry. Similarly, actinic keratoses (akiec) were sometimes confused with basal cell carcinoma (bcc), reflecting similarities in surface texture and pigmentation.

Compared with the baseline models (EfficientNet-B0, MobileNetV3-Large, and ResNet50), LightSkinNet consistently reduced cross-class errors. For instance, the misclassification of mel as nv decreased from 2.8% in ResNet50 to 2.2%, while bkl as nv reduced from 6.1% to 4.9%. These findings confirm that LightSkinNet is effective in distinguishing visually similar lesion categories while maintaining high performance for underrepresented classes.

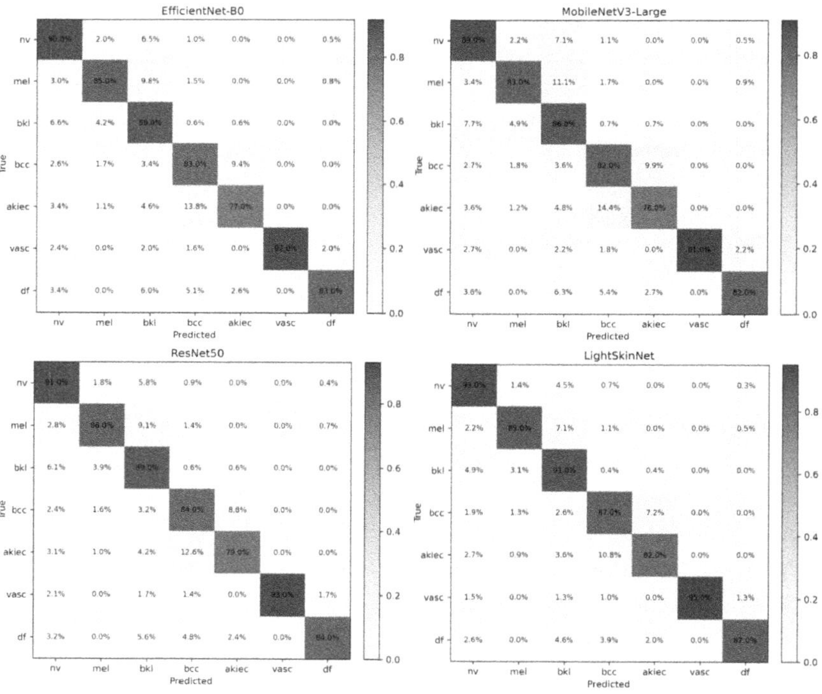

Fig. 3. Confusion matrices for the evaluated models: EfficientNet-B0, MobileNetV3-Large, ResNet50, and the proposed LightSkinNet.

4.6   Statistical Analysis

To assess whether the performance improvements of LightSkinNet over baseline models were statistically significant, a paired t-test was conducted on the per-fold Top-1 accuracy values obtained from the five stratified folds. This test was chosen because the same fold partitions were used for all models, allowing for direct paired comparisons. All p-values were two-tailed, with $p < 0.05$ considered statistically significant.

Table 6 summarizes the p-values for pairwise comparisons between Light-SkinNet and each baseline. Results indicate that the proposed model achieved statistically significant improvements in Top-1 accuracy over all baselines, with p-values ranging from 0.0015 to 0.0042. Similar trends were observed for macro F1-score and Cohen's Kappa, confirming that the observed performance gains are unlikely to have occurred by chance.

Table 6. Statistical significance of Top-1 accuracy improvements for LightSkinNet compared to baseline models, based on paired t-tests over five folds.

Comparison	p-value	Significant ($p < 0.05$)
LightSkinNet vs EfficientNet-B0	0.0031	Yes
LightSkinNet vs MobileNetV3-Large	0.0015	Yes
LightSkinNet vs ResNet50	0.0042	Yes

These findings support the conclusion that LightSkinNet's accuracy improvements over baseline models are both statistically significant and practically meaningful.

4.7   Grad-CAM-Based Model Interpretability

To further investigate the interpretability of the proposed LightSkinNet model, Gradient-weighted Class Activation Mapping (Grad-CAM) was applied to visualize class-discriminative regions within dermoscopic images. Figure 4 illustrates representative examples from the test set, where the first column (a) shows the original lesion images, the second column (b) presents the Grad-CAM heatmaps, and the third column (c) overlays the heatmaps onto the original images. The highlighted regions (red/yellow) correspond to areas that contributed most to the classification decision, whereas blue regions indicate low activation. It can be observed that LightSkinNet consistently attends to lesion-specific features such as irregular borders, color variegation, and localized pigmentation, which are clinically relevant cues in dermatological diagnosis. These visual explanations confirm that the model's decision-making process is aligned with dermatological assessment principles, thereby enhancing its potential clinical trustworthiness.

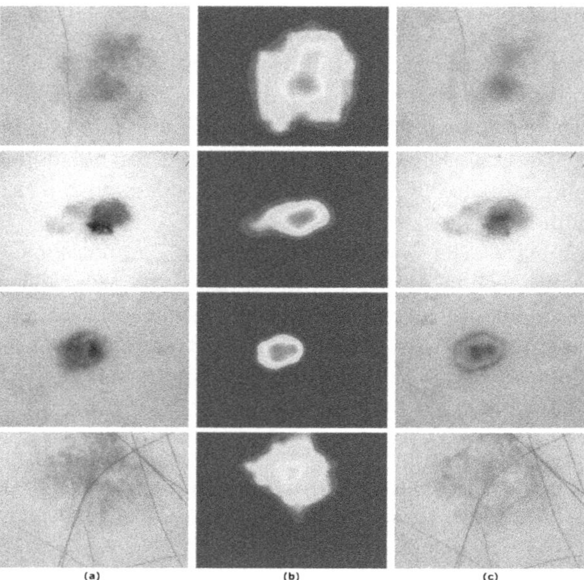

Fig. 4. Grad-CAM visualizations for representative test images using LightSkinNet. Column (a) shows the original dermoscopic images, column (b) displays the Grad-CAM heatmaps highlighting class-discriminative regions, and column (c) presents the overlay of heatmaps on the original images.

5   Conclusion

LightSkinNet demonstrates that lightweight deep learning architectures, when carefully designed with multi-scale feature extraction and attention mechanisms, can deliver high performance in multiclass dermoscopic classification while remaining computationally efficient for mobile or point-of-care deployment. The proposed combination of DCE, CBAM, and ECA modules allowed the model to maintain strong discriminative capability even for underrepresented lesion types, with statistical analysis confirming that observed gains over baseline models are unlikely due to chance. Grad-CAM visualizations further indicated that the network consistently attends to clinically meaningful dermoscopic structures, reinforcing its potential for trustworthy AI-assisted diagnosis. Nevertheless, the present work is limited to evaluation on the HAM10000 dataset, which may not fully capture variations in imaging conditions, device types, and patient demographics encountered in clinical practice. Future research will involve validating the model on multi-center, multi-device datasets, optimizing it for real-time on-device inference, and exploring integration with dermatological decision-support systems to assess its practical impact on early detection and treatment planning.

References

1. Arnold, M., et al.: Global burden of cutaneous melanoma in 2020 and projections to 2040. JAMA Dermatol. **158**(5), 495–503 (2022)
2. Lopes, J., Rodrigues, C.M., Gaspar, M.M., Reis, C.P.: Melanoma management: from epidemiology to treatment and latest advances. Cancers **14**(19), 4652 (2022)
3. Malvehy, J., Pellacani, G.: Dermoscopy, confocal microscopy and other non-invasive tools for the diagnosis of non-melanoma skin cancers and other skin conditions. Acta Derm. Venereol. **97**, 22–30 (2017)
4. Vestergaard, M., Macaskill, P., Holt, P., Menzies, S.: Dermoscopy compared with naked eye examination for the diagnosis of primary melanoma: a meta-analysis of studies performed in a clinical setting. Br. J. Dermatol. **159**(3), 669–676 (2008)
5. Bono, A., et al.: Melanoma detection: a prospective study comparing diagnosis with the naked eye, dermatoscopy and telespectrophotometry. Dermatology **205**(4), 362–366 (2002)
6. Asaduzzaman, A., Thompson, C.C., Sibai, F.N., Uddin, M.J.: Application of ensemble learning models in computer-aided diagnosis of skin diseases. Neural Comput. Appl. 1–17 (2025)
7. Yao, P., et al.: Single model deep learning on imbalanced small datasets for skin lesion classification. IEEE Trans. Med. Imaging **41**(5), 1242–1254 (2021)
8. Patil, A., Mehto, A., Nalband, S.: Enhancing skin lesion diagnosis with data augmentation techniques: a review of the state-of-the-art. Multimedia Tools Appl. 1–40 (2024)
9. Barata, C., Celebi, M.E., Marques, J.S.: Explainable skin lesion diagnosis using taxonomies. Pattern Recogn. **110**, 107413 (2021)
10. Naseri, H., Safaei, A.A.: Diagnosis and prognosis of melanoma from dermoscopy images using machine learning and deep learning: a systematic literature review. BMC Cancer **25**(1), 75 (2025)
11. Riaz, S., Naeem, A., Malik, H., Naqvi, R.A., Loh, W.-K.: Federated and transfer learning methods for the classification of melanoma and nonmelanoma skin cancers: a prospective study. Sensors **23**(20), 8457 (2023)
12. Emara, T., Afify, H.M., Ismail, F.H., Hassanien, A.E.: A modified inception-v4 for imbalanced skin cancer classification dataset. In: 2019 14th International Conference on Computer Engineering and Systems (ICCES), pp. 28–33. IEEE (2019)
13. Adegun, A.A., Viriri, S.: FCN-based DenseNet framework for automated detection and classification of skin lesions in dermoscopy images. IEEE Access **8**, 150377–150396 (2020)
14. Gu, M., Li, X., Chen, S.: Transfer learning-based intelligent diagnosis for skin lesion in cloud-edge computing networks. In: 2021 13th International Conference on Wireless Communications and Signal Processing (WCSP), pp. 1–6. IEEE (2021)
15. Amiri, S.A., Nasrolahzadeh, M., Mohammadpoory, Z., Kordkheili, A.H.Z.: Skin lesion classification via ensemble method on deep learning. Multimedia Tools Appl. **84**(18), 19379–19397 (2025)
16. Ali, K., Shaikh, Z.A., Khan, A.A., Laghari, A.A.: Multiclass skin cancer classification using EfficientNets-a first step towards preventing skin cancer. Neuroscience Inform. **2**(4), 100034 (2022)
17. Tschandl, P., Rosendahl, C., Kittler, H.: The HAM10000 dataset, a large collection of multi-source dermatoscopic images of common pigmented skin lesions. Sci. Data **5**(1), 1–9 (2018)

A DenseNet-YOLOv8 Fusion Model
for Intelligent Parasite Egg Detection
and Classification

Muhammad Bilal Zia, Xujuan Zhou[✉], Raj Gururajan, and Ka Ching Chan

School of Business, University of Southern Queensland, UniSQ Springfield Education City, 37 Sinnathamby Blvd, Springfield, QLD 4300, Australia
{MuhammadBilal.Zia,Xujuan.Zhou,Raj.Gururajan,KC.Chan}@unisq.edu.au

Abstract. Precise categorization of sheep parasite eggs is essential for improved veterinary diagnostics, automated monitoring, early disease detection, and efficient farm management in modern agriculture. However, most existing deep learning models struggle to achieve both rich feature extraction and fast inference, often excelling at one while sacrificing the other. To address this limitation, we propose a DenseNet–YOLOv8 hybrid model that combines the dense connectivity and strong feature propagation of DenseNet with the anchor-free detection and high-speed inference of YOLOv8. The framework was trained and evaluated on a sheep parasite egg dataset under challenging conditions such as occlusion, illumination variation, and background clutter. Experimental results show that the hybrid model outperforms the individual DenseNet and YOLOv8 baselines across all tested metrics, achieving 96.8% accuracy, 95.7% precision, 96.3% recall, and a mAP@0.5 of 97.1%, while sustaining an inference speed of 82 FPS. Unlike previous approaches that were limited by slower inference or weaker feature representation, our model provides a balanced solution that is accurate and efficient, offering strong potential for integration into laboratory diagnostics and intelligent livestock health management systems.

Keywords: Sheep Dataset · Hybrid Deep Learning · DenseNet · YOLOv8

1 Introduction

One major breakthrough in parasite egg detection has been the use of computer vision technology. This advancement has had a significant impact on agriculture, food security, and smart farm management [1,2]. Sheep as one of the most common animals on the farm have the great importance in the production of meat, milk, and wool [3]. Not only precisely through sheep classification is the foundation of automated livestock monitoring, but also it facilitates the increase of the capability of contagion identification, the selection of good breeding animals, and behavioral studies. Traditional manual classification methods are very

Q. V. Nguyen et al. (Eds.): AusDM 2025, CCIS 2765, pp. 346–358, 2026.
https://doi.org/10.1007/978-981-95-6786-7_23

time-consuming, error-prone, and require a lot of labor, indicating the need for intelligent and automated solutions [4].

Sheep farming is particularly vulnerable to parasitic infections caused by species such as *Haemonchus contortus* and *Teladorsagia*, which lead to severe economic losses through reduced weight gain, milk production, and quality of wool. Therefore, early and accurate identification of these parasites is crucial for maintaining animal health and farm productivity. We specifically focused on these two species due to their prevalence in sheep production systems world-wide and their close morphological resemblance, which makes manual diagnosis challenging.

Visual recognition tasks were completely transformed by deep learning, with convolutional neural networks (CNN) as the main image classification unit [5]. One such network is DenseNet, which has brought into focus the advantages of its dense connectivity pattern as it overcomes the problem of gradient disappearance and, at the same time [6], facilitates the reuse of features by different layers in a network. So this type of architecture is really good at extracting the necessary details to tell apart the sheep despite the changes in the environment. On the flip side, the advent of object detection models, especially the YOLO (You Only Look Once) family of algorithms, marks the milestone of their development as the quickest and most accurate detectors in the history of computer vision research [7]. The most recent version YOLOv8, comes with features such as anchor-free detection, live inference, and high precision across a wide variety of categories of objects.

Sheep egg classification is one of the most important procedures in veterinary parasitology, which is mainly performed for the correct identification of parasite eggs together with Haemonchus contortus and Teladorsagia species in the fecal sample [8,9]. Appropriate detection and classification contribute to the diagnosis of the disease at the right time, as well as the scheduling of treatment for the animals, thus minimizing the losses caused by sheep farming [10,11]. The standard microscopic method is not only time-consuming but also susceptible to mistakes; however, deep learning techniques such as CNNs and different YOLO models make the process more efficient and accurate. The combination of a feature-rich backbone such as DenseNet and an efficient detector such as YOLOv8 can result in a reliable performance of sheep egg classification even in the presence of various image abnormalities, leading to the implementation of smarter and more sustainable livestock health management. Beyond parasitology, hybrid deep learning frameworks have also proven valuable in other biomedical domains, such as CNN–BiLSTM models for pain intensity estimation from facial images [23], and nested ensemble classifiers for breast cancer diagnosis [24] Although CNN-based models have achieved strong results in parasite egg classification in controlled datasets, their performance often drops when facing variations such as occlusion, uneven illumination, or cluttered microscopic backgrounds. Recent work introduced BIAE-Net, which enhances parasite egg detection by combining border injection with transformer modules to improve feature extraction and localization accuracy [22]. On the other hand, YOLOv8 excels at fast object detection but may overlook fine-grained inter-class differences, such as between morphologically similar parasite eggs. In medical imaging, the combination of detection and classification in unified pipelines has been effective, for example, with two-step deep networks for lung nodule recognition [25] and multi-deep models with dual CNN and multitask learning to handle intra-class variations [26]. However, there is limited research exploring hybrid frameworks that fuse a feature-rich backbone with a fast detector specifically for veterinary parasitology.

Unlike prior YOLO-based adaptations for livestock monitoring, which primarily target macroscopic detection tasks such as sheep behavior, livestock identification, or pain recognition, our work directly addresses microscopic parasitology. To our knowledge, this is the first study to fuse DenseNet with YOLOv8 for veterinary parasitology, achieving a balance between fine-grained feature representation and real-time inference speed, a gap not covered in previous works.

To address this gap, we propose a DenseNet–YOLOv8 hybrid model that leverages DenseNet for deep feature extraction and YOLOv8 for efficient detection and classification, thus combining precision with inference efficiency.

The major contributions of this paper can be outlined as follows:

- We propose a hybrid DenseNet–YOLOv8 model that leverages DenseNet's rich feature representation together with YOLOv8's efficient detection, allowing accurate and efficient parasite egg detection and classification.
- Through experiments on a sheep parasite egg dataset, we show that the hybrid consistently outperforms the standalone DenseNet and YOLOv8 baselines across multiple evaluation metrics.
- We discuss the practical relevance of the proposed framework for laboratory diagnostics and livestock health monitoring, highlighting its potential for integration into real veterinary workflows.

The remainder of this paper is structured as follows. Section 2 reviews related work on deep learning models for livestock classification and object detection. Section 3 presents the proposed hybrid DenseNet YOLOv8 methodology in detail. Section 4 describes the experimental setup, including dataset preparation, training strategies, and evaluation metrics and experimental results. Finally, Sect. 5 concludes the study and outlines future research directions.

2   Literature Review

Fang et al. [12] Come up with CCS YOLOv8, a modified version of YOLOv8 that is upgraded with attention modules (CBAM), feature reassembly (CARAFE), and a new small-object detection layer. The authors report that when the method was applied to intricate grassy scenes for monitoring the livestock, the 84.4% mAP@0.5, 82.2% recall, and 83.1% F1-score were obtained, which were considerably better than those of the baseline YOLOv8. Guo et al. [13] Introduce FESS YOLOv8n, a minimalistic model for recognizing sheep motions such as feeding, standing, and lying down. By incorporating EMA attention, C2f Faster,

Table 1. YOLO-based studies Comparison.

Model	Domain & Innovation	Key Results
CCS-YOLOv8 (Fang et al.)	Livestock detection with attention and small-target focus	mAP@0.5: 84.4%; Recall: 82.2%
FESS-YOLOv8n (Behavior)	Sheep activity detection, lightweight architecture	mAP@0.5: 91.4%; Small model size
SMEA-YOLOv8n (Facial expressions)	Sheep pain expression recognition	mAP@0.5: 92.5%; Recall: 91%; Precision: 86%
Hybrid (YOLO + DenseNet + Bi-LSTM)	Medical image classification	Demonstrates hybrid benefits in feature–temporal fusion
DenseNet-YOLO-v3 (Xu et al.)	Remote sensing multi-scale detection	mAP boost to 88.73%; Small-object detection improved
YOLO v1–v10 Review	Agricultural YOLO applications survey	Contextualizes YOLO evolution in farming
Sheep flocks & predator dataset (Yang et al.)	Dataset contribution for livestock classification	Supports realistic evaluation platforms

and SCDown components into YOLOv8n, they reported 91.4% mAP@0.5 with a model size of just 5.13 MB.

Yu et al. [14] have come up with an SMEA-YOLOv8n innovative system to identify the facial expressions of sheep (which are indicative of pain). The use of attention modules and modified loss/feature pooling methods has led to the model achieving an mAP@0.5 of 92.5%, a recall of 91%, a precision of 86%, and an F1-score of 88%, representing substantial improvements over the performance of baseline models. Karacı et al. [15] merged YOLO detection, DenseNet feature extraction, and Bi-LSTM sequence modeling to identify brain tumors. They provided a proof of concept that such a combination of detection, deep-feature extraction, and context modeling can be successfully applied to the medical domain.

Xu et al. [16] combined DenseNet with the YOLO v3 backbone and extended the detection range by adding a fourth detection scale to enhance remote sensing object detection. The mAP increased from 77.10% to 88.73% on both RSOD and UCS-AOD datasets, with a more significant improvement in small-object detection of 12.1%, accompanied by only a slight decrease in speed. Hasan Alp et al. [17] In a non-destructive fashion, the researchers apply convolutional neural networks (CNNs) to determine the freshness of eggs. For this purpose, 30 eggs were photographed each day for 29 consecutive days, thereby creating a dataset comprising the pictures. A deep learning model was developed to identify the age of the eggs (i.e., the number of days the eggs had been stored) and achieved 91.78% accuracy in classification as a whole. The evidence presented here essentially inaugurates the concept of quick image-based evaluations, eliminating the need for manual preprocessing, and also highlights the potential of automated classification systems.

This work [18] machine vision system has been advanced to identify egg quality in a caged environment. This system is equipped with a robotic patrol system integrated with fixed video streams for monitoring purposes. The primary detection model is a tailor-made YOLOv8 small backbone, further enhanced with Shuffle Attention mechanisms, and supports real-time applications on a Jetson AGX Orin. The model has raised its capability for various metrics: Precision 94.0% (+2.4%), Recall 92.8% (+4.6%), AP50–95 91.5% (+3%), and F1 score 93.4% (+3.9%), as well as for a high inference speed (92 FPS). As shown in Table 1 while existing YOLO-based variants achieve strong performance in livestock monitoring and behavior analysis, there remains a gap in applying such models to microscopic parasite egg detection with both accuracy and efficiency

To better contextualize our work within the existing research, the main characteristics and results of related studies are summarized in Table 1. Recent YOLO-based variants have been successfully applied to livestock detection, behavior recognition, and even medical image analysis, often incorporating attention mechanisms, lightweight designs, or hybrid feature extractors. While these models demonstrate strong performance in their respective domains, most either prioritize accuracy at the expense of inference speed or focus on livestock behavior rather than microscopic parasite egg recognition. This highlights the remaining gap in achieving both rich feature extraction and efficient inference for parasite egg detection, which our proposed DenseNet-YOLOv8 hybrid framework aims to address.

3   Methodology

The proposed framework represents a combination of deep learning architectures that creatively blend DenseNet as a backbone feature extractor with YOLOv8 as the detection and classification head for sheep dataset classification. he main idea is to combine DenseNet's detailed feature learning ability with YOLOv8's fast detection. This integration helps to overcome the limitations of using each model in isolation, as illustrated in Fig. 1.

3.1   Model Architecture

DenseNet Backbone. The new hybrid model utilizes DenseNet (Densely Connected Convolutional Networks) as the primary feature extractor to implement the system's backbone. Typically, CNNs are designed such that each layer receives input from and sends output to the next layer. DenseNet, however, introduces the concept of dense connectivity, where every layer receives input from all earlier layers and provides its feature maps to all subsequent layers. In this way, it not only ensures maximum feature reuse but also alleviates the vanishing gradient problem, allowing the use of both low-level and high-level features within a single network. In the case of sheep dataset classification, this attribute is extremely advantageous, as minor variations in wool texture, shape, and facial features need to be accurately represented by the features. DenseNet aims to reduce the number of parameters by eliminating the recurring feature maps; as a result, it is more efficient than other deep CNNs of the same depth. The feature maps produced by DenseNet represent images in a multi-stage format and are then combined with the YOLOv8 detection head to provide improved classification and localization. In this way, the backbone instills the hybrid model with the necessary support for capturing the contextual, discriminative, and transferable visual features that are crucial for accurately recognizing the sheep.

YOLOv8 Head. The detection and classification module of the hybrid framework is the one YOLOv8, the latest and most cutting-edge member of the YOLO (You Only Look Once) family of models, has been put in charge of. Compared

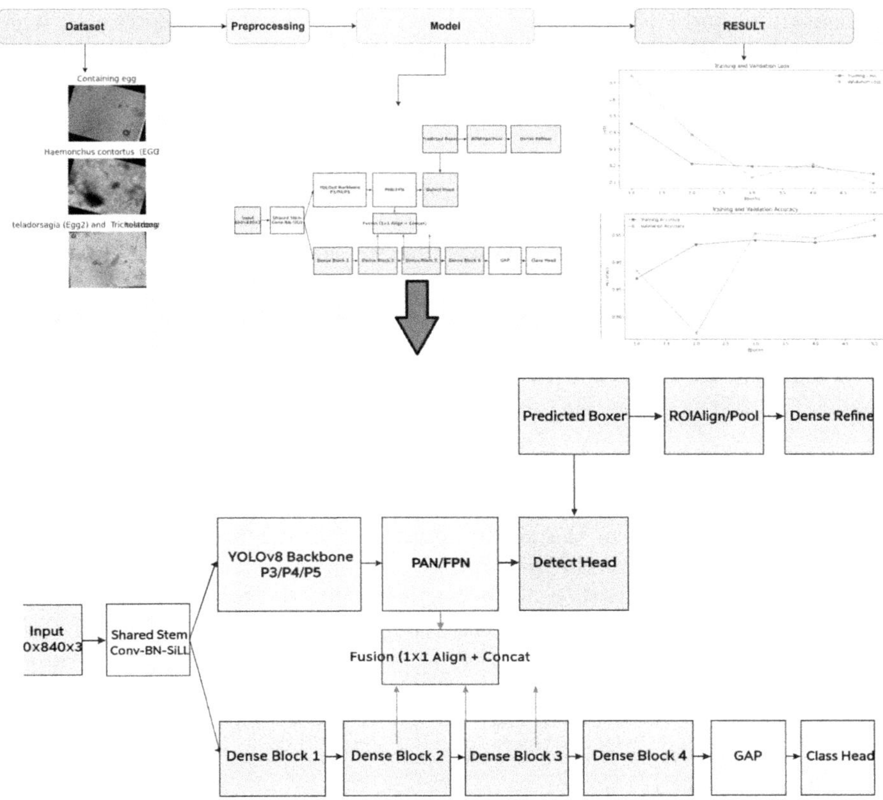

Fig. 1. The proposed Hybrid-framework for detection and classification

to the previous versions, YOLOv8 takes advantage of a completely anchor-free detection method with separate classification and regression heads, thereby substantially improving the detection precision while simultaneously lowering the load on the computer. YOLOv8 head fuses feature maps from the DenseNet backbone and, through its Neck (Path Aggregation Network and Feature Pyramid Network), carries out the operation, thus enabling multi-scale feature representation. Hence, the model can not only find the sheep of different sizes but also locate those which are partially covered. The detection head is the stage of the performance where the candidates for the bounding box coordinates, the objectness scores, and class probabilities are presented in one single forward pass, thus making it highly efficient for real-time applications. For example, with a sheep dataset, this design not only allows the computer to accurately find but also to assign the labels to the multiple sheep in a single image, no matter the severity of the farm environment or the background's intricacy. Moreover, YOLOv8 incorporates state-of-the-art training procedures such as label smoothing, mosaic augmentation, and CIoU loss that are used for bounding box regression, which in turn makes a big fraction of the overall performance.

Fusion Strategy. The fusion method combines the feature maps extracted from DenseNet with the detection layers of YOLOv8 to provide the best possible performance. Firstly, DenseNet outputs are converted via convolutional transformations for compatibility with YOLOv8's feature pyramid. The features thus processed are then integrated at different scales in the FPN/PAN modules, which allows the use of the detailed contextual information from DenseNet along with the fast detection capability of YOLOv8. Such a combination results in the improvement of the recognition of small objects, the decrease of the number of misclassifications, and the maintenance of the accuracy and speed of the classification of the sheep dataset, which can be confirmed in Fig. 1.

3.2   Experimental Setup

To assess the capability of the newly designed hybrid DenseNet–YOLOv8 system to identify sheep classification dataset, the authors ran some experiments following the scientific method. This part is about the dataset, preprocessing, training setup, and evaluation metrics. The experiments were designed to systematically vary the parameters of the classification framework and measure its performance on the sheep dataset. Details on the dataset, preprocessing, training configuration, and evaluation metrics are given in the following paragraphs.

3.3   Dataset Description

The dataset used in this study is one that we collected and prepared ourselves, specifically for this research. It consists of microscopic images of sheep parasite eggs gathered from fecal samples. To capture as much variation as possible, the images were taken under different conditions, including changes in lighting, background, and levels of occlusion. All samples were imaged using a UC50 digital camera mounted on a Nikon Eclipse 80i microscope at magnifications of $10\times$ and $20\times$. The data set focuses primarily on two species of veterinary importance, Haemonchus contortus and Teladorsagia, both of which are notoriously difficult to separate due to their overlapping morphological features. The dataset was divided into training (70%), validation (15%), and testing (15%) sets to ensure a fair model evaluation and also the sample of dataset containing Haemonchus contortus as egg1 and Teladorsagia egg22 is shown in Fig. 2. An assortment of data was used to depict various conditions of neighboring farms to facilitate a good generalization of the model.

In total, the dataset consisted of 1,344 images, with 980 samples of *Haemonchus contortus* and 364 samples of *Teladorsagia*. After applying data augmentation, the dataset was expanded to approximately 2312 images. Augmentation was designed to mimic real-world imaging variability: random flips and rotations accounted for variations in microscope orientation, brightness and

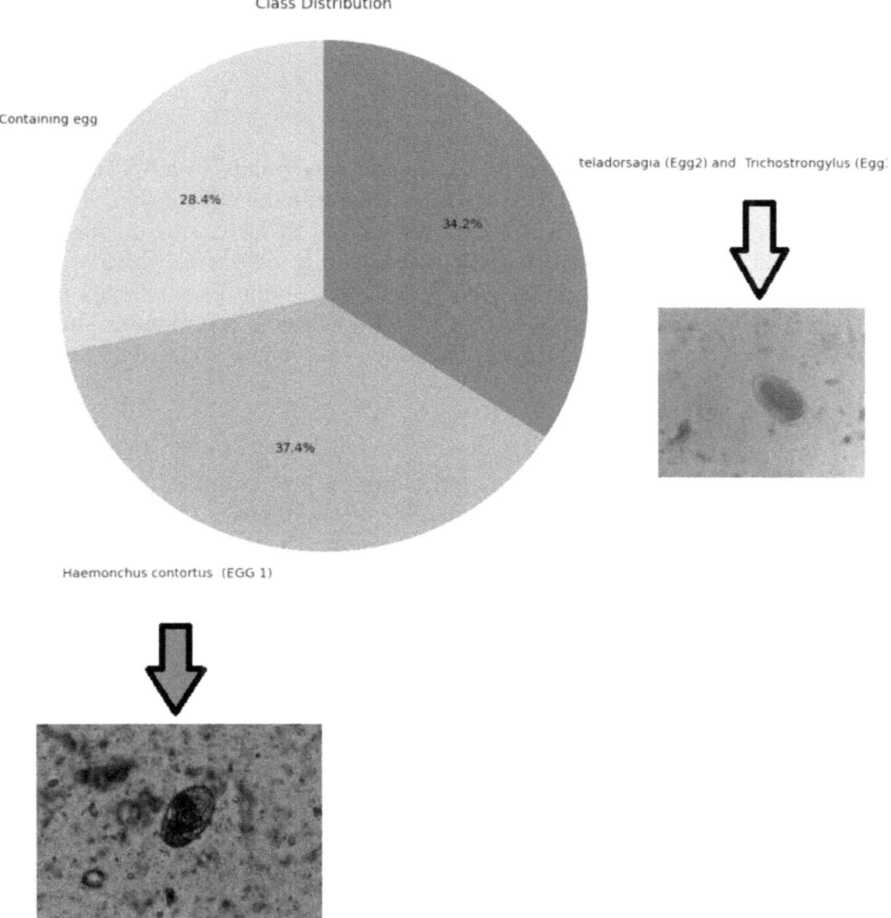

Fig. 2. Class distribution of the sheep egg dataset

contrast adjustments simulated changes in illumination, and occlusion-based transformations improved robustness against debris commonly present in fecal samples.

3.4   Data Preprocessing

Data processing before model training is necessary. To match the YOLOv8 input size, all images in the sheep dataset were resized to 640×640 pixels. Pixel values were then normalized to the [0, 1] range for stable convergence [19]. The model was also exposed to augmented data to enhance its resilience through several transformations, including random flips, rotations, contrast changes, and brightness variations. There was a change in the representation format of the bounding boxes and labels, from which the annotation structure compatible with YOLOv8 was derived, to facilitate easy integration into the training pipeline.

3.5   Training Configuration

The hybrid DenseNet–YOLOv8 model was developed on an NVIDIA GPU (RTX 3090, 24 GB VRAM) with CUDA, which accelerates the calculation. The GPU was used for 100 epochs with a batch size of 32. SGD with a momentum of 0.9 was used to optimize, starting the learning rate at 0.001, which was then modulated by the cosine decay schedule. A new training loss, which combines cross-entropy for classification and CIoU for bounding box regression, has been introduced to enhance both detection accuracy and localization precision.

3.6   Evaluation Metrics

The model's performance was evaluated using various standard metrics in the field. Accuracy (ACC) was the main feature of the metrics that measured the overall correctness of predictions, while Precision (P) and Recall (R) specified the model's ability to identify sheep instances correctly and, at the same time, obtain all relevant detections [20]. The F1-score, being the harmonic mean of precision and recall, provides a balanced measure of classification performance. Additionally, for the detection quality, Mean Average Precision (mAP@0.5 and mAP@0.5:0.95) was employed to combine the evaluation of localization and classification accuracy at different thresholds.

4   Results and Discussion

The proposed DenseNet–YOLOv8 hybrid framework demonstrates clear advantages over baseline architectures when evaluated on the sheep dataset. Quantitative metrics highlight its robustness: the model achieved an overall accuracy of **96.8%**, with precision of **95.7%**, recall of **96.3%**, and an F1-score of **96.0%**.

In detection benchmarks, the hybrid model reached a **mAP@0.5 of 97.1%** and a **mAP@0.5:0.95 of 94.5%**, outperforming both standalone DenseNet and YOLOv8. Importantly, it maintained a real-time inference speed of **82 FPS**, which underlines its practical feasibility for continuous farm monitoring an essential factor when deploying AI tools in real-world livestock environments. While these results were obtained using static microscopic images, the measured efficiency highlights the model's potential to process live microscope video streams without delay.

The confusion matrix in Fig. 3 illustrates the model's ability to differentiate between the three categories: *Haemonchus contortus*, *Teladorsagia*, and Background. The framework consistently recognized *Teladorsagia* with very few misclassifications, reinforcing its feature discrimination ability and robustness.

Fig. 3. Confusion Matrix of the Hybrid DenseNet-YOLOv8 Model.

4.1 Training Performance

The learning curves, shown in Fig. 4 confirm the hybrid framework's stability and generalization ability. Both training and validation loss steadily decreased, while accuracy consistently improved across epochs, surpassing **95%** without evidence of severe overfitting.

The training curves were monitored across all 50 epochs. For clarity, Fig. 4 currently shows the first 5 epochs; however, the same trends persisted throughout the training, with steady decreases in validation loss and stable convergence after approximately 60 epochs.

4.2 Discussion and Practical Implications

The findings confirm that integrating DenseNet's deep feature extraction with YOLOv8's efficient detection pipeline leads to a balanced system that is both highly accurate and computationally efficient. In practice, such a system can automatically localize and classify parasitic eggs in real time, where bounding boxes highlight the detected regions and confidence scores provide reliable classification certainty.

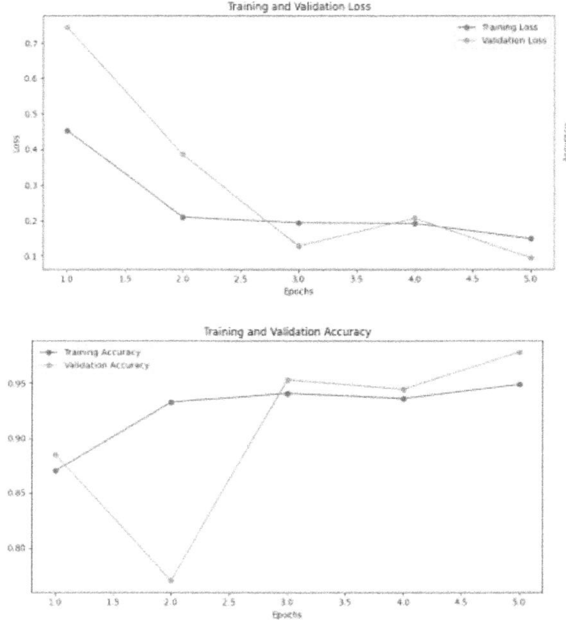

Fig. 4. Training and Validation Loss and Accuracy Curves.

This real-time capability (82 FPS) is not only a technical benchmark but also a practical enabler: it ensures that veterinary staff and farm technicians can rely on immediate, on site feedback during routine monitoring. Automated high-accuracy detection reduces diagnostic time, minimizes human error, and supports the broader goal of precision livestock farming.

5   Conclusion and Future Work

This work presented a DenseNet–YOLOv8 hybrid framework for parasite egg detection and classification in sheep. By combining DenseNet's fine-grained feature learning with YOLOv8's rapid detection, the model achieved 96.8% accuracy, 95.7% precision, 96.3% recall, and 97.1% mAP@0.5, while sustaining an inference speed of 82 FPS. Although this was measured in static microscope images, the efficiency strongly suggests that the model could also process live video streams from microscope camera without noticeable delay, making it a practical step toward real-time diagnostic use in veterinary settings.

Despite promising results, this study has several limitations. The dataset covers only two parasite species, which restricts generalization to broader parasitological contexts. The hybrid framework is more computationally intensive than a YOLOv8 model standalone, which may limit deployment on low-resource devices. Because the data were collected under controlled laboratory conditions, there is also a risk of overfitting when applying the model to more diverse field samples. Finally, reliance on microscope image quality could affect performance in real veterinary settings. As future work, we plan to explore model compression techniques such as pruning and quantization, expand testing to larger cross-breed datasets, and incorporate temporal modeling to capture behavioural patterns and support automated livestock health monitoring.

References

1. Giri, K.J., Wani, B.A., Raja, D.N., Mehraj, F.: Role of machine learning and computer vision in the agri-food industry. In: Artificial Intelligence in the Food Industry, pp. 56–83. CRC Press (2025)
2. Arshad, M.F., et al.: The groundbreaking impact of digitalization and artificial intelligence in sheep farming. Res. Vet. Sci. **170**, 105197 (2024)
3. Mazinani, M., Rude, B.: Population, world production and quality of sheep and goat products. Am. J. Anim. Vet. Sci. **15**(4), 291–299 (2020)
4. Jwade, S.A., Guzzomi, A., Mian, A.: On farm automatic sheep breed classification using deep learning. Comput. Electron. Agric. **167**, 105055 (2019)
5. Traore, B.B., Kamsu-Foguem, B., Tangara, F.: Deep convolution neural network for image recognition. Eco. Inf. **48**, 257–268 (2018)
6. Huang, G., Liu, Z., Pleiss, G., Van Der Maaten, L., Weinberger, K.Q.: Convolutional networks with dense connectivity. IEEE Trans. Pattern Anal. Mach. Intell. **44**(12), 8704–8716 (2019)
7. Diwan, T., Anirudh, G., Tembhurne, J.V.: Object detection using YOLO: challenges, architectural successors, datasets and applications. Multimedia Tools Appl. **82**(6), 9243–9275 (2023)
8. Taylor, M.: Parasitological examinations in sheep health management. Small Rumin. Res. **92**(1–3), 120–125 (2010)
9. Knoll, S., et al.: Practical guide for microscopic identification of infectious gastrointestinal nematode larvae in sheep from Sardinia, Italy, backed by molecular analysis. Parasites Vect. **14**(1), 505 (2021)
10. Hindson, J., Winter, A.: Manual of Sheep Diseases. John Wiley Sons, Hoboken (2008)
11. Saminathan, M., Rana, R., Ramakrishnan, M.A., Karthik, K., Malik, Y.S., Dhama, K.: Prevalence, diagnosis, management and control of important diseases of ruminants with special reference to Indian scenario. J. Exp. Biol. **4**, 3S (2016)
12. Fang, C., et al.: Enhancing livestock detection: an efficient model based on YOLOv8. Appl. Sci. **14**(11), 4809 (2024)
13. Guo, X., et al.: A sheep behavior recognition approach based on improved FESS-YOLOv8n neural network. Animals **15**(6), 893 (2025)
14. Yu, W., Yang, X., Liu, Y., Xuan, C., Xie, R., Wang, C.: SMEA-YOLOv8n: a sheep facial expression recognition method based on an improved YOLOv8n model. Anim. Open Access J. MDPI **14**(23), 3415 (2024)
15. Karacı, A., Akyol, K.: YoDenBi-NET: YOLO+ DenseNet+ Bi-LSTM-based hybrid deep learning model for brain tumor classification. Neural Comput. Appl. **35**(17), 12583–12598 (2023)
16. Xu, D., Wu, Y.: Improved YOLO-V3 with DenseNet for multi-scale remote sensing target detection. Sensors **20**(15), 4276 (2020)

17. Sahin, H.A., Onder, H.: Use of deep learning to determine the freshness of egg. J. Inst. Sci. Technol. **14**(1), 493–500 (2024)
18. Wu, Z., Zhang, H., Fang, C.: Research on machine vision online monitoring system for egg production and quality in cage environment. Poult. Sci. **104**(1), 104552 (2025)
19. Sant'Ana, D.A., et al.: Computer vision system for superpixel classification and segmentation of sheep. Eco. Inf. **68**, 101551 (2022)
20. Foody, G.M.: Challenges in the real world use of classification accuracy metrics: from recall and precision to the Matthews correlation coefficient. PLoS ONE **18**(10), e0291908 (2023)
21. LNCS Homepage. http://www.springer.com/lncs. Accessed 25 Oct 2023
22. Zia, M.B.: Adaptive enhancement network with border injection for parasitic eggs detection. IEEE Access **11**, 1–12 (2023)
23. Bargshady, G., Zhou, X., Deo, R.C., Soar, J., Whittaker, F., Wang, H.: Enhanced deep learning algorithm development to detect pain intensity from facial expression images. Expert Syst. Appl. **149**, 113305 (2020)
24. Abdar, M., et al.: A new nested ensemble technique for automated diagnosis of breast cancer. Pattern Recogn. Lett. **132**, 123–131 (2020)
25. Zia, M.B., Juan, Z.J., Xiao, N.: Detection and classification of lung nodule in diagnostic CT: a TsDN method based on improved 3D-Faster R-CNN and multi-scale multi-crop CNN. Int. J. Hybrid Inf. Technol. **13**(2), 45–56 (2020)
26. Zia, M.B., Juan, Z.J., Xiao, N., Wang, J., Khan, A., Zhou, X.: Classification of malignant and benign lung nodule and prediction of image label class using multi-deep model. Int. J. Adv. Comput. Sci. Appl. **11**(3), 35–45 (2020)

Application Track – Health and Social Good

An AI-Driven Framework for Real-Time Reporting and Identification of Lost Cats

Dezhou Zhang, Xiaodan Dong(✉), and Weidong Huang

University of Technology Sydney, Ultimo, Australia
`{Dezhou.Zhang,Xiaodan.Dong,Weidong.Huang}@uts.edu.au`

Abstract. Veterinary clinics and animal shelters frequently receive inquiries about lost pets, yet despite microchipping and other identification methods, many lost cats remain unclaimed. This causes emotional distress for owners and burdens animal care services. However, very limited previous study has investigated the effectiveness of integrating computer vision with deep learning to support the public in reporting lost animals. This paper proposes an AI-driven framework that allows users to identify, and report lost cats simply by capturing a photo with their mobile phones. The framework integrates breed classification through a three-stage deep neural network, along with tag detection and text extraction. Empirical results show that the system can accurately capture visual features, including breed and tag information, improving the timeliness and effectiveness of lost pet reporting. The approach can also be extended to other animals.

Keywords: Artificial Intelligence · Lost Pet · Computer Vision · Deep Learning

1 Introduction

With the growing number of lost pets in society, pet owners are often desperate to locate their beloved animals, while animal welfare organizations are seeing an increasing influx of lost and distressed pets. Statistics from the Royal Society for the Prevention of Cruelty to Animals (RSPCA), an animal welfare organization, indicate that more than 87,000 animals are admitted to shelters in Australia each year [15]. In Australia, estimated 3% of pets are homeless, including around 110,000 cats and 38,000 dogs. Australian shelters currently house approximately 62,000 dogs and nearly double that number of cats, at 117,000 [1,13]. This discrepancy is largely due to the more independent nature of cats, which are often permitted to roam freely outdoors, thereby increasing their risk of becoming lost. Hence, cat owners are generally more likely to provide identification tags compared to dog owners.

Recent advances in artificial intelligence (AI) and computer vision have led to the development of solutions aimed at reuniting pets with their owners. Examples include the Petco Love database [14] in the United States and the Pet Biometric

Q. V. Nguyen et al. (Eds.): AusDM 2025, CCIS 2765, pp. 361–370, 2026.
https://doi.org/10.1007/978-981-95-6786-7_24

Identification system [4], introduced recently. Zhang et al. [18] review applications of artificial intelligence technologies aimed at improving animal welfare, with a focus on current uses and recent research developments. Many applications have employed deep learning models for animal breed identification [11].

However, these methods typically rely on the animals being found and sheltered at a specific location, such as an animal welfare organization or shelter. Unfortunately, many lost cats never make it to these facilities, remaining on the streets and at risk of harm. Despite advances in artificial intelligence, there is little existing literature on the effectiveness of leveraging AI to enable the general public to report and identify lost animals. To address this gap, this paper proposes a new framework that enables the public to identify and report lost cats simply by taking a photo upon encountering a stray. The system aims to connect these reports to animal welfare organizations or social media networks, allowing owners to locate their lost pets in a timely manner and reducing the suffering of animals left without shelter for extended periods.

When reporting a lost cat, it is crucial for the public to accurately identify the cat's name and breed. With the increasing popularity of cat ownership and many recognized feline breeds, identifying a cat breed from an image remains a practical but challenging task. Adding to this problem is the difficulty in reuniting lost cats with their owners, especially when identification tags are missing or unreadable.

This project presents an intelligent, image-based classification and recognition framework to address these concerns. The system allows users to upload images of cats to automatically predict their breed, detect whether they are wearing a tag, and extract the tag's text if present. It combines three computer vision components: a ResNet50-based breed classifier, a YOLOv8-based tag detector, and an OCR pipeline using EasyOCR.

The primary objectives of this paper are to:

- Investigate the effectiveness of the proposed AI-driven framework in enabling the general public to efficiently report lost pets in real time.
- Assess the capability of the framework to accurately identify and facilitate the recovery of lost cats through AI enhanced image and tag-based text recognition, thereby contributing to pet owner support and broader animal welfare initiatives.

To ensure cost-effective accessibility, the solution will be deployed as a lightweight Streamlit web app, run locally with GPU support, and shared to the public via Ngrok. This eliminates the need for Docker container hosting or dedicated GPU servers, but requires prior coordination to ensure the app and public link are active during usage. To our knowledge, this is the first AI-driven real-time reporting system designed specifically for lost cats.

2   Related Work

Artificial intelligence (AI) has been increasingly applied across a wide array of domains, including healthcare, the pet industry, and animal welfare. Hamadani et al. [7] provided an overview of artificial neural networks (ANN) and their applications in animal sciences. Arshad et al. [2] reviewed the emerging applications of AI in the pet industry. Dave et al. [5] employed a deep learning model to track wild animals in the forest. Following this, several pet databases have become available, including the Petco Love database in the United States [14] and the recently introduced Pet Biometric Identification system [4], making AI-based breed identification more feasible and accurate.

One notable application is in animal species identification, where convolutional neural networks (CNNs) have shown strong performance in species classification tasks [9]. Liu t et al. [12] showed that machine learning could be an effective tool for breed identification. Kenneth et al. [11] employed deep neural networks to identify individual dogs. Their approach combined visual data with additional information, such as gender and breed, to enhance identification performance. The study found that performing dog breed classification prior to individual identification effectively reduced the search space and resulted in a 6.2% increase in accuracy [11]. However, their research did not incorporate the names of the animals, which is crucial for reuniting pets with their owners, nor did it explore the use of this information for reporting lost animals.

We extend the existing breed identification approach by integrating an efficient three-stage modeling framework with an additional tag recognition module. This enhancement enables the system to first detect whether the cat is wearing a tag and, if so, to extract the name from it. In doing so, we facilitate the reporting of lost animals by the general public, significantly improving both the efficiency and the timeliness of the identification process. In real-world scenarios, even a passerby who spots a lost cat can contribute by taking a photo and reporting the visible name and breed, enabling the owner to quickly locate and recover the animal. The proposed method can be extended to facilitate the identification and reporting of other lost pets, as long as a tag is present.

Beyond technical contributions, our approach has a meaningful social impact. Importantly, it empowers members of the general public to assist in this process: if someone spots a lost cat, they can simply take a photo and submit it along with visible tag information. This crowdsourced reporting capability greatly increases the chances of locating and recovering lost animals in a timely manner. The proposed framework reduces emotional distress for pet owners, lowers the burden on animal welfare organizations and shelters, and helps prevent unnecessary suffering of animals exposed to risks in the wild.

3   Methodology

This section presents the data collected for the study, describes the proposed modeling framework for facilitating the identification and reporting of lost cats, and provides a detailed account of the methodology implemented at each stage, including breed classification, tag detection, and tag text extraction.

3.1 The Data

We utilized the CatBreedsRefined-7k dataset [6] from Kaggle, which contains 7,000 images spanning 20 popular cat breeds. The dataset was split into 80% for training, 10% for validation, and 10% for testing. The validation set was used during training to evaluate model performance after each epoch and to enable early stopping. The test set was reserved for final evaluation and was used only once after training was complete to report the model's performance.

To construct a dataset of cats wearing tags, we manually annotated using the Roboflow platform [16]. We include 100 images of cats that were mainly collected from google images, specifically curated for the task of cat collar object detection. The single defined class is "cat collar", with precise bounding boxes drawn around each collar. Every image in the dataset is labeled, and no unannotated samples are included. To ensure balanced evaluation, the dataset is split into training , validation, and testing sets, following a 7:1.5:1.5 ratio. All images were resized to 640×640 pixels, and optional augmentations. The dataset is exported in YOLOv8-compatible format, including .txt annotation files and a data.yaml configuration file. To simulate real-world variability, 50% of the images were annotated with visible tags, while the remaining 50% were intentionally left untagged.

3.2 Modelling Framework

To develop an effective and lightweight system for cat classification and tag recognition, we employed a three-stage training strategy for the ResNet50 classifier [8], followed by the implementation of object detection and optical character recognition (OCR) modules. The proposed framework is illustrated in Fig. 1. It begins when a member of the general public spots a lost cat wandering outdoors or on the street. To help both the cat and its owner, the individual can simply take a photo using a mobile phone to report the sighting, making the reporting process accessible and user-friendly. The photo, along with the location data, is then processed by our proposed three-stage model and object detection framework. Subsequently, relevant information, such as the cat's breed, color, name, and location, is transmitted in real time to facilitate matching with the owner. This approach aims to accelerate the recovery process, support cat owners, minimize the duration cats remain lost, and alleviate the burden on social animal welfare services.

3.3 ResNet50 Breed Classifier

To identify cat breeds, we adopted the ResNet-50 breed classifier [19] and enhanced it through multi-phase fine-tuning. ResNet50 is a deep neural network architecture comprising 50 weighted layers [8]. A pretrained ResNet50 was used as the backbone, which is a pretrained feature extractor. To take advantage of a large pre-trained model (ResNet50) on our new dataset, we adopted a three-stage transfer learning strategy, frozen, partially unfrozen, and fully unfrozen,

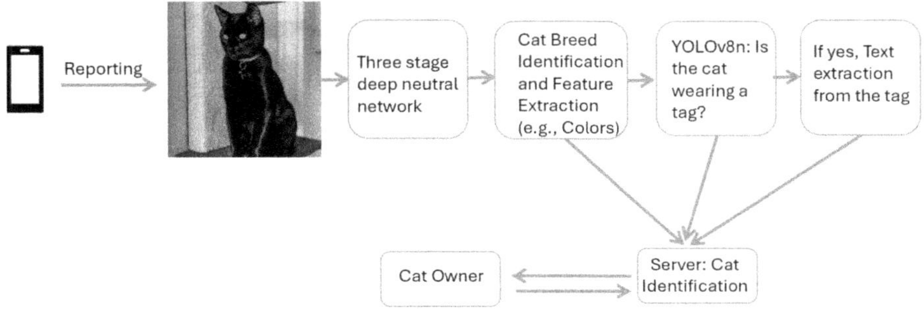

Fig. 1. Framework Architecture

which is an efficient and widely used approach for fine-tuning deep models. The following training stages were adopted with early stop trigger:

- Stage 1: Fully Frozen Backbone. Only the custom classification head was trained; all ResNet layers were frozen. During this stage, the final classification layer, which is automatically initialized, learns to map high-level features extracted from the frozen backbone to the target labels.
- Stage 2: Unfreeze the Last Two Stages for Layer3 and Layer4. The deeper layers of ResNet (Layer3 and Layer4) were unfrozen to enable gradual fine-tuning. These layers capture high-level semantic features such as object parts. Fine-tuning them allows the model to better adapt to the dataset while keeping training time and GPU usage moderate.
- Stage 3: Fully Unfrozen. All ResNet layers were unfrozen to allow full end-to-end training. This enables the model to fully adapt its feature representations to the target domain, potentially maximizing performance—albeit with increased computational cost and longer training time.

3.4   YOLOv8n for Tag Detection

To detect whether the lost cat is wearing a tag, we use YOLOv8n [17]. It is a compact and efficient variant of the YOLO (You Only Look Once) object detection models, developed by Ultralytics. We train the model using a small, manually annotated dataset. Images were resized to 640×640, and we used transfer learning with pretrained weights for fast convergence.

3.5   EasyOCR for Tag Text Extraction

EasyOCR [10] is an open source Optical Character Recognition (OCR) library that leverages deep learning techniques. It is widely used because of its lightweight architecture, high speed, and native support for multilingual text recognition. When a tag is detected on an animal, EasyOCR is employed to extract the text to enable accurate identification of the animal's name and to aid in the identification process. To simplify deployment, we used EasyOCR directly on the YOLO-predicted bounding boxes to extract readable characters from collar tags.

4 Results

The proposed framework comprises two distinct models: a deep neural network with a three-stage tuning strategy for cat breed classification, and a YOLO-based model for tag detection and text extraction. The corresponding results are summarized below.

4.1 Cat Breed Classification Results

For cat breed classification model, Fig. 2 illustrates the training and validation accuracy and loss curves over different epochs. The model demonstrates stable convergence without signs of overfitting, with early stopping triggered at epoch 32. Initially, the training accuracy is lower than the validation accuracy due to the use of data augmentation during training. Techniques such as random cropping and flipping were applied to introduce variability and enhance generalization. These transformations intentionally make the training process more challenging, helping the model learn robust features.

In contrast, the validation set was evaluated on unaltered and consistent data without augmentation. This allows for an accurate and reproducible assessment of the generalization performance of the model. As shown in Table 1, the test accuracy, F1 score, and mean Average Precision (mAP) all exceed 80%, demonstrating satisfactory model performance in identifying lost cats from photos.

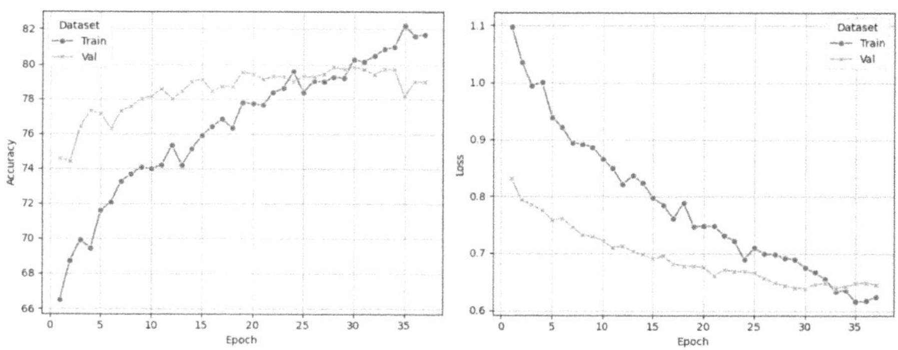

Fig. 2. Training and validation accuracy (left) and training and validation loss (right).

Table 1. Training, validation, and testing performance metrics across different stages

Stage	Train Acc	Train F1	Train mAP	Val Acc	Val F1	Val mAP	Test Acc	Test F1	Test mAP
S1: Frozen Backbone	0.7154	0.7112	0.0521	0.7214	0.7176	0.8039	0.7329	0.7309	0.8111
S2: Unfreeze 3&4	0.8363	0.8346	0.0519	0.7986	0.7980	0.8538	0.7886	0.7863	0.8658
S3: Fully Unfrozen	0.8675	0.8668	0.0505	0.7943	0.7949	0.8572	0.8143	0.8141	0.8733

4.2   Tag Identification Results

In the second detection stage, we use YOLOv8n to determine the presence of
a tag in a given photo. The validation results, presented in Fig. 3 and detailed
in Table 2, demonstrate satisfactory performance. The test accuracy rate of 93%
indicates a reasonable level of identification accuracy.

Fig. 3. Validation Results: Detecting Whether a Cat Is Wearing a Tag.

Table 2. Model Performance Metrics

Tag Identification Model	Accuracy Rate	F1 Score
Train	0.9714	0.9655
Validation	0.7333	0.6000
Test	**0.9333**	**0.9091**
Overall	0.9100	0.8941

5 Discussion and Future Improvements

The proposed framework is tested on the collected dataset from different sources. The ResNet50 backbone served as a reasonable foundation for breed classification, but its generalization on the test set was limited. This issue stems largely from the nature of the dataset. The most important factor is class imbalance Domestic Short Hair has over 53,000 images, while rarer breeds like York Chocolate have only one. This disparity leads the model to overfit common classes while underperforming on rare ones. Furthermore, certain breed categories contained mixed or unclear labels, possibly due to visual similarities or inconsistent labeling during data collection. These noisy labels hinder the model's ability to learn clear distinctions between breeds.

To further enhance the performance of the proposed framework, we plan to collaborate with animal welfare organizations, such as the RSPCA, to collect a larger and more diverse dataset. This will help improve the model's generalizability across different cat breeds, environments, and image qualities. We also aim to implement this solution as a mobile reporting application, enabling members of the public to easily report sightings of lost cats. As more images are collected through public contributions, the model's accuracy will continue to improve. Furthermore, denoising techniques can be applied as we gather more real-world data to improve image quality and model robustness. In addition to detecting tags and breed, the system can extract valuable information such as fur color, patterns, and location metadata. By integrating comprehensive data with immediate reporting functionality, it facilitates more rapid and effective reunification, thereby decreasing the dependence on animal welfare organizations. This system represents the first real-time reporting framework which empowers the general public to report missing cats.

As future work, the advancement of Time-Space Transformer [3] models could enable the use of video data to enhance identification of lost pets. Such models could capture distinctive physical traits, such as missing limbs, unique fur patterns, tail shapes, ear deformities, or visible scars, which would improve the accuracy and reliability of pet identification and reporting.

6 Conclusion

This project investigates the effectiveness of a new AI framework that allows the general public to report missing cats in a timely and user-friendly manner. The system integrates image classification, object detection, and optical character recognition (OCR) to facilitate accurate cat breed identification and tag recognition, thereby effectively extracting valuable information for cat identification. Empirical results demonstrate that the proposed framework is highly accurate in identifying lost cats.

While the model achieves high accuracy under controlled conditions, challenges such as data imbalance, label noise, and limited generalizability persist. However, as more people use the system, additional data will become available to further fine-tune the model. To our knowledge, this is the first AI driven, real-time reporting system designed specifically for lost cats. It provides a solid foundation for real-world applications in lost pet reporting and animal welfare support.

References

1. Animal Friendly Life: Global research reveals how many stray animals are on the street (2024). https://animalfriendlylife.com.au/news/global-research-reveals-how-many-stray-animals-are-on-the-streets/. Accessed 31 July 2024
2. Arshad, M., Ahmed, F., Nonnis, F., Tamponi, C., Scala, A., Varcasia, A.: Artificial intelligence and companion animals: Perspectives on digital healthcare for dogs, cats, and pet ownership. Res. Vet. Sci. **193**, 105776 (2025). https://doi.org/10.1016/j.rvsc.2025.105776. Accessed 24 Aug 2025
3. Bertasius, G., Wang, H., Torresani, L.: Is space-time attention all you need for video understanding? arXiv preprint arXiv:2102.05095 (2021)
4. Boulaouane, Y., Garifulla, M., Lim, J., Pak, D., Lim, J.: Advancing pet biometric identification: a state-of-the-art unified framework for dogs and cats. IEEE Access **12**, 188359–188372 (2024). https://doi.org/10.1109/ACCESS.2024.3516130
5. Dave, B., Mori, M., Bathani, A., Goel, P.: Wild animal detection using yolov8. In: 3rd International Conference on Evolutionary Computing and Mobile Sustainable Networks (ICECMSN 2023), vol. 230, pp. 100–111 (2023). https://doi.org/10.1016/j.procs.2023.12.065
6. Doctrinek: Catbreedsrefined-7k dataset (2023). https://www.kaggle.com/datasets/doctrinek/catbreedsrefined-7k. Accessed 5 July 2025
7. Hamadani, A., Ganai, N., Bashir, J.: Artificial neural networks for data mining in animal sciences. Bull. Natl. Res. Centre **47**, 68 (2023). https://doi.org/10.1186/s42269-023-01042-9
8. He, K., Zhang, X., Ren, S., Sun, J.: Deep residual learning for image recognition. In: 2016 IEEE Conference on Computer Vision and Pattern Recognition (CVPR), pp. 770–778 (2016). https://doi.org/10.1109/CVPR.2016.90
9. Huang, G., Liu, Z., Van Der Maaten, L., Weinberger, K.: Densely connected convolutional networks. In: Proceedings of the IEEE Conference on Computer Vision and Pattern Recognition (CVPR), Honolulu, HI, USA, 21–26 July 2017, pp. 2261–2269 (2017). https://doi.org/10.1109/CVPR.2017.243
10. Jaided AI: Easyocr: Ready-to-use ocr with 80+ supported languages (2020). https://github.com/JaidedAI/EasyOCR. Accessed 2 July 2025
11. Lai, K., Tu, X., Yanushkevich, S.: Dog identification using soft biometrics and neural networks. In: Proceedings of the 2019 International Joint Conference on Neural Networks (IJCNN), Budapest, Hungary, 14–19 July 2019, pp. 1–8 (2019). https://doi.org/10.1109/IJCNN.2019.8852322. Accessed 15 July 2025

12. Liu, R., et al.: Evaluation of six machine learning classification algorithms in pig breed identification using snps array data. Anim. Genet. (2022). https://doi.org/10.1111/age.13279
13. Mars Petcare or Affiliates: State of pet homelessness project: a global initiative for understanding pet homelessness. https://stateofpethomelessness.com. Accessed 31 May 2025
14. Petco Love: Petco love lost. https://petcolove.org/lost/about/. Accessed 2 July 2025
15. Petnow Inc.: Biometrics for dogs and cats: Petnow app launches in Australia and new zealand. News release (2024). https://prnmedia.prnewswire.com/news-releases/biometrics-for-dogs-and-cats-petnow-app-launches-in-australia-and-new-zealand-302205016.html. Accessed 31 May 2025
16. Roboflow: Cat tag dataset (2025). https://universe.roboflow.com/cat-tag/cat-tag. Accessed 1 July 2025
17. Ultralytics: Yolov8 - yolo by ultralytics (2023). https://github.com/ultralytics/ultralytics. Accessed 2 July 2025
18. Zhang, L., et al.: Advancements in artificial intelligence technology for improving animal welfare: current applications and research progress. Anim. Res. One Health **2**(1), 93–109 (2024). https://doi.org/10.1002/aro2.44
19. Zhang, X., Yang, L., Sinnott, R.: A mobile application for cat detection and breed recognition based on deep learning. In: 2019 IEEE 1st International Workshop on Artificial Intelligence for Mobile (AI4Mobile), pp. 7–12 (2019). https://doi.org/10.1109/AI4Mobile.2019.8672684

Benchmarking Preprocessing and Integration Methods in Single-Cell Genomics

Ali Anaissi[1,2]($\boxtimes$), Seid Miad Zandavi[2,3], Weidong Huang[1], Junaid Akram[2], Basem Suleiman[4], Ali Braytee[1], and Jie Hua[5]

[1] University of Technology Sydney, Ultimo, Australia
`{weidong.huang,ali.braytee}@uts.edu.au`
[2] University of Sydney, Sydney, Australia
`{ali.anaissi,Junaid.Akram}@uts.edu.au, szandavi@broadinstitute.org`
[3] Broad Institute, Cambridge, USA
[4] University of New South Wales, Kensington, Australia
`b.suleiman@unsw.edu.au`
[5] Shaoyang University, Shaoyang, China
`steven.hua@mq.edu.au`

Abstract. Single-cell data analysis has the potential to revolutionize personalized medicine by characterizing disease-associated molecular changes at the single-cell level. Advanced single-cell multimodal assays can now simultaneously measure various molecules (e.g., DNA, RNA, Protein) across hundreds of thousands of individual cells, providing a comprehensive molecular readout. A significant analytical challenge is integrating single-cell measurements across different modalities. Various methods have been developed to address this challenge, but there has been no systematic evaluation of these techniques with different preprocessing strategies. This study examines a general pipeline for single-cell data analysis, which includes normalization, data integration, and dimensionality reduction. The performance of different algorithm combinations often depends on the dataset sizes and characteristics. We evaluate six datasets across diverse modalities, tissues, and organisms using three metrics: Silhouette Coefficient Score, Adjusted Rand Index, and Calinski-Harabasz Index. Our experiments involve combinations of seven normalization methods, four dimensional reduction methods, and five integration methods. The results show that Seurat and Harmony excel in data integration, with Harmony being more time-efficient, especially for large datasets. UMAP is the most compatible dimensionality reduction method with the integration techniques, and the choice of normalization method varies depending on the integration method used.

1 Introduction

Technological advances have significantly increased our ability to generate high-throughput single-cell gene expression data [17]. However, single-cell data often originates from multiple experiments with variations in capturing time, personnel, reagents, equipment, and technology platforms, leading to large variations that can confound biological variations during data integration. scRNA-seq integration [2,9,26] addresses two main issues: generating cell-type feature clusters and determining whether clusters represent actual cell types or result from biological or technological variations, such as specific batch effects or high mitochondrial content. Despite its potential, scRNA-seq integration faces risks, including low-quality cluster identification due to meaningless variations and biased clustering from improper arrangement of similar cell types.

A popular strategy introduced by Haghverdi et al. [4] identifies cell mappings between datasets and reconstructs the data in a shared space by finding mutual nearest neighbors (MNNs) [4,17]. This method, while effective in generating a normalized gene expression matrix suitable for downstream analysis, is computationally intensive. To address this, the fastMNN algorithm applies the MNN technique in a PCA-computed subspace, improving performance and accuracy [8]. Similarly, Scanorama searches for MNNs in dimensionally reduced regions for batch integration [6].

scRNA-seq integration analysis typically involves four modules: data normalization, dimensionality reduction, data integration, and result visualization. Numerous algorithms are available for each module, creating a vast number of possible combinations that need evaluation to determine optimal performance. The performance of these combinations depends heavily on dataset size and type, posing a challenge in identifying the best algorithm and parameter settings. This challenge requires significant computational resources, time, and expertise.

This paper addresses this challenge by introducing an empirical evaluation framework to help scientists evaluate scRNA-seq algorithms and choose the best combinations for their datasets. We investigate optimal clustering model combinations for different types of datasets using various evaluation methods. The framework is divided into three parts: data normalization, dimensionality reduction, and data integration. For normalization, we investigate seven core methods: Log Normalization, Counts Per Million (CPM), SCTransform, TF-IDF, Linnorm, Scran, and TTM [18,31,32]. For dimensionality reduction, we evaluate PCA, UMAP, t-SNE, and PHATE. For data integration, we assess Seurat, Harmony, FastMNN [4,17], ComBat [7], and Scanorama [6]. We use three evaluation metrics—Silhouette Coefficient Score, Adjusted Rand Index, and Calinski-Harabasz Index—to examine clustering performance and time efficiency.

Our study selects the best models based on evaluation results for each dataset, analyzing reasons for different combinations' performance. We also provide insights into the rules of method selection for different dataset types and sizes, offering data support for future model selection.

The major contributions of our work are as follows:

1. We propose an empirical framework systematically assessing various computational strategies for scRNA-seq data integration. This framework includes seven normalization methods, four dimensionality reduction techniques, and five integration methods, providing a holistic approach to scRNA-seq data analysis.

2. Utilizing robust evaluation metrics—Silhouette Coefficient, Adjusted Rand Index, and Calinski-Harabasz Index—we analyze 140 combinations of the methods. This evaluation elucidates performance efficiency and scalability, offering critical insights into their applicability in clustering cell types and aligning datasets from varied sources.
3. Our comparative analysis identifies the most effective combinations of normalization, dimensionality reduction, and integration methods for scRNA-seq data. This provides a strategic roadmap for researchers, facilitating high-fidelity integration of heterogeneous single-cell datasets and enhancing biological insights.

2 Related Work

Single-cell RNA sequencing (scRNA-seq) has transformed the discovery and characterization of cellular phenotypes, aiding in the identification of biomarkers within the biomedical field [30]. The foundational principle of scRNA-seq involves measuring gene expression distributions across cell populations, as described by Tang et al. [24]. Since 2014, advancements have significantly reduced sequencing costs and enhanced protocols, broadening its application. scRNA-seq has been pivotal in profiling the molecular regulation of T lymphocytes, leading to new insights into molecular determinants [1]. The Human Cell Atlas (HCA) Global Alliance uses this technology to create a reference map of human tissues, promising advancements in understanding aging, disease, and potential treatments. Future applications extend to cell-based models, cell therapies, and regenerative medicine.

However, scRNA-seq data presents challenges, notably the batch effect, arising from variations in data collection and processing, which can hinder data integration and interpretation. Seurat is widely used for mitigating batch effects and integrating various single-cell data types. It employs Canonical Correlation Analysis (CCA) and anchoring techniques to address gene expression discrepancies through weighted-nearest neighbor analysis [5]. Despite its utility, Seurat's performance can decline with a high number of batches, particularly when dealing with non-highly variable genes [13]. To address this, Lakkis et al. introduced CarDEC, a deep learning model enhancing scRNA-seq data by increasing information content while denoising. Peng et al. [21] proposed the cFIT method, an unsupervised approach that integrates data from multiple sources with fewer restrictions, improving batch effect correction.

Normalization is crucial for reducing batch effects while preserving biological variation [3]. Techniques like TMM have shown success but can over-correct, prompting recommendations for methods like Linnorm and SCnorm, specifically designed for scRNA-seq [18]. scRNA-seq data, characterized by high dimensionality, sparsity, and noise, often requires dimensionality reduction to transform it into a lower-dimensional space while preserving meaningful properties. Methods such as PCA, UMAP, t-SNE, and deep count autoencoder (DCA) each have strengths and weaknesses, with UMAP preserving global structures but potentially introducing noise [27]. Visualization methods like UMAP and t-SNE are

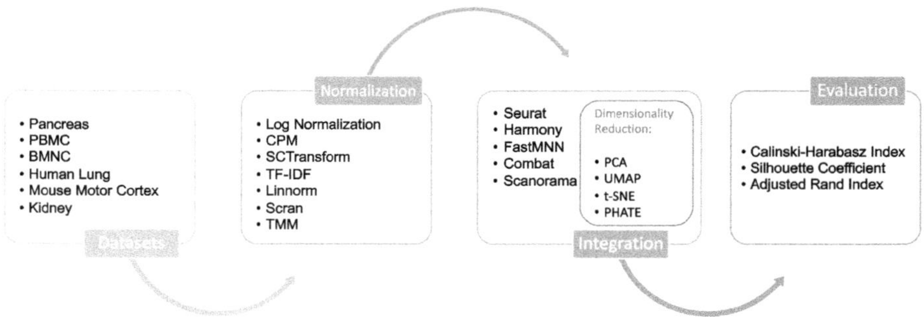

Fig. 1. Process Flow of the methods.

intuitive for evaluating integration effectiveness but need quantitative metrics like local inverse Simpson's index, average silhouette width, and adjusted rand index for rigorous assessment [15].

3   Methodology

We propose a comprehensive framework for the integration of scRNA-seq data, consisting of multiple stages: data preprocessing, dimensionality reduction, data integration, and evaluation of clustering performance. Each stage employs various established methods to ensure robust and accurate results.

Initially, data preprocessing involves normalization using several methods, including Log-Normalization, Counts Per Million (CPM), SCTransform, Term Frequency-Inverse Document Frequency (TF-IDF), Linnorm, Scran, and the Trimmed Mean of M-values (TMM). Following normalization, dimensionality reduction techniques such as Principal Component Analysis (PCA), Uniform Manifold Approximation and Projection (UMAP), t-Distributed Stochastic Neighbor Embedding (t-SNE), and Potential of Heat-diffusion for Affinity-based Transition Embedding (PHATE) are employed to transform high-dimensional data into a lower-dimensional space, facilitating visual inspection and further analysis.

Given the varied performance of dimensionality reduction methods in separating biological clusters and detecting rare cell populations, we systematically assess their effectiveness in conjunction with different scRNA-seq integration methods.

Next, we integrate the processed data and cluster cells using popular methods including Seurat, Harmony, Fast Mutual Nearest Neighbors (FastMNN), Combat, and Scanorama. The results are visualized using DimPlots, and evaluated using the Silhouette Coefficient Score, Calinski-Harabasz Index, and Adjusted Rand Index to measure clustering performance. Figure 1 illustrates our proposed framework.

3.1 Normalization Methods

We have chosen the following normalization methods to investigate as part of our framework:

- **Log Normalization**: This method uses the log function to scale larger values to a smaller interval, improving model accuracy by reducing the impact of large numerical weights.
- **Counts Per Million (CPM)**: CPM involves dividing the count columns by their total fragments and scaling by millions, followed by a log transformation. This method is used by Stuart et al. [23] for scaling and filtering scATAC-seq gene matrices before dimensionality reduction.
- **SCTransform**: An algorithm for normalization and variance stabilization, SCTransform uses a regularized negative binomial model, constructing a generalized linear model for each gene with sequencing depth as the explanatory variable and UMI counts as the response variable [3].
- **TF-IDF**: A method standard in text analysis, TF-IDF analyzes the importance of genes (words) in cells (documents) by their frequency and inverse document frequency [20].
- **Linnorm**: This normalization method uses a linear model and normality to perform accurate statistics and analysis on scRNA-seq datasets, using strictly selected homologous genes as a reference [29].
- **Scran**: An R package for RNA-seq data analysis, Scran's computeSumFactors method normalizes cell-specific biases by deconvolution [12].
- **Trimmed Mean of M-values (TMM)**: TMM uses weighted trimmed mean of log expression ratios to estimate RNA production, normalizing the data by calculating the M and A values, which represent log expression ratios and average expression levels, respectively.

3.2 Dimensionality Reduction Methods

The following dimensionality reduction methods are investigated as part of our proposed framework:

- **Principal Component Analysis (PCA)**: A linear dimensionality reduction method, PCA transforms correlated variables into a small number of uncorrelated principal components [16].
- **Uniform Manifold Approximation and Projection (UMAP)**: A nonlinear dimensionality reduction technique that preserves more of the global structure of the data compared to other methods, offering excellent runtime performance [22].
- **t-SNE**: This method converts high-dimensional data into a lower-dimensional space while maintaining the probability distribution of the data points before and after the reduction. t-SNE uses a t-distribution in the lower-dimensional space to improve separation between clusters [10].
- **PHATE**: A visualization method for high-dimensional data, PHATE retains the global structure of the data and shows the information-geometric distance between data points. It is robust to noise and scalable to large datasets [19].

3.3   Integration Methods

Integration methods are essential for removing unwanted technical variation while preserving valid biological variation. We employ the following integration methods for batch correction and data integration:

- **Seurat**: An R package designed for single-cell transcriptome sequencing and analysis, Seurat integrates various types of single-cell data and analyzes heterogeneity from single-cell transcriptomic measurements.
- **Harmony**: An efficient algorithm for integrating large single-cell datasets, Harmony starts by clustering cells in a low-dimensional embedding space and iteratively refines these clusters based on a metric that penalizes inappropriate cluster compositions [11].
- **FastMNN**: This method corrects batch effects using a modified mutual nearest neighbors (MNN) approach, identifying MNN pairs after dimensionality reduction and correcting batch effects accordingly [33].
- **ComBat**: An empirical Bayesian framework, ComBat corrects batch effects by standardizing data, estimating batch effect parameters, and adjusting data based on these estimates [7].
- **Scanorama**: This method integrates single-cell datasets from different technologies using panoramic batch correction and integration. It employs SVD for dimensionality reduction and constructs a nearest neighbor graph for integration [6].

3.4   Data Analysis

The Wilcoxon Rank-Sum Test is employed for data analysis. This non-parametric test compares the distribution of two independent samples to determine if they come from the same distribution. After calculating the Silhouette Coefficient, Calinski-Harabasz Index, and Adjusted Rand Index scores for each dataset, we rank the methods and apply the Wilcoxon Rank-Sum Test to identify the best normalization, dimensionality reduction, and integration approaches [14].

By following this comprehensive methodology, we aim to systematically assess the performance of various normalization, dimensionality reduction, and integration methods in scRNA-seq data analysis, ensuring robust and accurate results across different datasets.

4   Experiments and Results

This section discusses the model performance evaluation along with the time efficiency evaluation on five different datasets.

4.1   Datasets

We conducted experiments on five RNA gene sequences datasets which are described as follows:

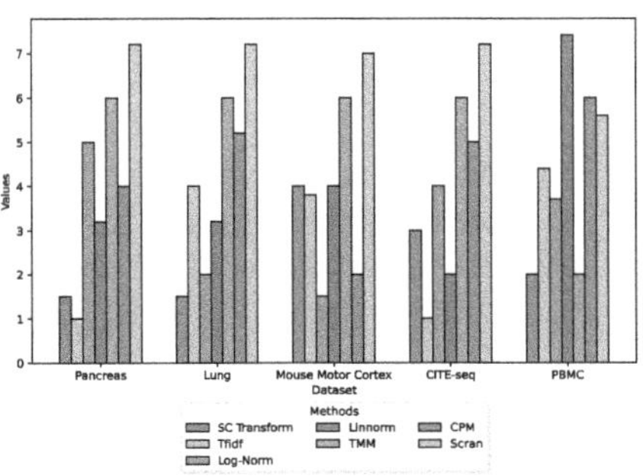

(a) Wilcoxon Rank-Sum Test of Normalization Methods.

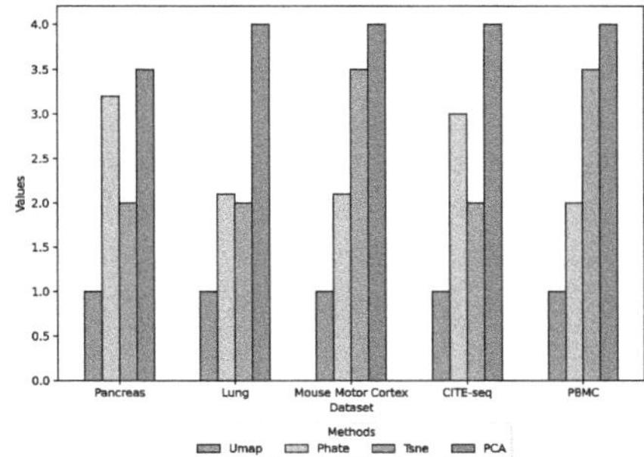

(b) Wilcoxon Rank-Sum Test of Dimension Reduction Methods.

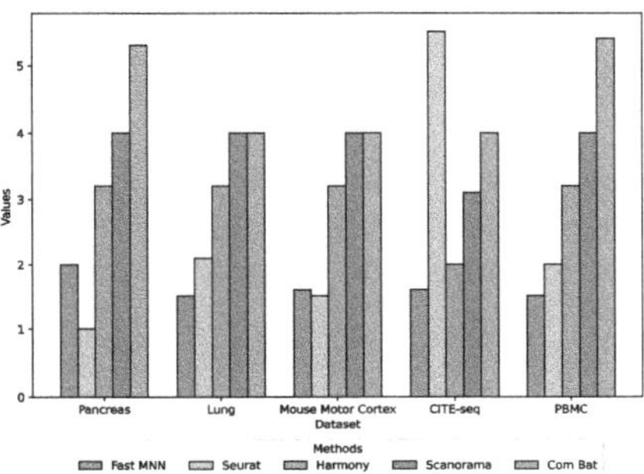

(c) Wilcoxon Rank-Sum Test of Integration Methods.

Fig. 2. Wilcoxon Rank-Sum Tests for Various Methods.

Table 1. Summary of Evaluation Results

Method	CH Sum	Silh. Sum	ARI Sum	CH Rank	Silh. Rank	ARI Rank	Rank Sum	Final Rank
Summary Normalisation Results								
Log-Norm	995503.848	32.3981	63.158689	5	7	7	19	1
CPM	904642.351	28.8685	58.4329301	3	3	3	9	5
SCTransform	1130895.78	31.8076	60.8940582	7	6	6	19	1
TFIDF	1004035.56	31.4413	60.7435297	6	5	5	16	3
Linnorm	935391.89	29.7849	59.721815	4	4	4	12	4
SCRAN	743686.811	9.28917	40.5276838	1	1	1	3	7
TMM	813891.279	27.62767	58.2278707	2	2	2	6	6
Summary Dimension Reduction Results								
PCA	203086.58	26.3973	83.709358	1	1	1	3	4
UMAP	2694239.70	67.8274	112.5972185	4	4	4	12	1
TSNE	1205768.39	56.3184	101.296	2	3	3	8	2
PHATE	2380788.68	39.88214	94.727	3	2	2	7	3
Summary Integration Results								
Seurat	1144219.94	47.57594	85.112	2	4	4	10	2
Harmony	1525235.92	39.9727	82.208	4	3	3	10	2
FastMNN	1538792.95	50.8561	96.145	5	5	5	15	1
Combat	846507.87	21.849	59.1045765	1	1	1	3	5
Scanorama	1251230.84	30.9551	79.137	3	2	2	7	4

- **Pancreas** dataset is a combination of different pancreas scRNA-seq datasets from eight studies using five different techniques. It is integrated into a single cell using Seurat [23].
- **Peripheral blood mononuclear cell (PBMC)** dataset is generated based on the eight volunteers enrolled in an HIV vaccine trial. It takes three-time point samples at days 0, 3, and 7 following vaccination to form 24 samples, which is processed by using the CITE-seq technique to produce RNA and ADT.
- **CITE-seq** dataset contains 30,672 samples of human bone marrow mononuclear cells (BMNC) and 25 antibodies, which were derived from eight individual donors. BMNC dataset, generated by the Human Cell Atlas, gains two assays, RNA and antibody-derived tags (ADT).
- **Human Lung cells** dataset [25] contains 58 molecular cell types from 65,662 human lung and blood cells, including bronchi, bronchiole, alveoli and circulating blood.
- **Mouse motor cortex** dataset is referenced from Yao et al. [28] which analyzes adult mouse isocortex and hippocampal formation to gain transcriptomic and epigenomic atlas from 12 individual mice.

4.2 Method Performance Evaluation

As a result of the evaluation performance of each dataset, a variety of different method combinations were determined to be the most effective. To combine the rankings across all metrics, we used the Wilcoxon Rank-Sum approach to rank methods based on each of the CH, SC, and ARI metrics. A lower rank-sum score indicates better performance when it comes to calculating the height of ridgelines across different datasets. Methods are ranked from top to bottom based on the sum of their rank scores for the six data sets, with the top-performing methods appearing at the top. Additionally, the datasets on the x-axis are sorted in ascending order of the size of the dataset, which is calculated by features across samples.

In terms of the normalization method (Fig. 2a), SCTransform, Log Normalization and TF-IDF come out as the top three methods with the most remarkable overall performance. These methods were ranked among the top three in four datasets, including Pancreas, PBMC, CITE-seq, and Mouse Motor Cortex. SCTransform produced the highest quality normalization results for Pancreas and CITE-seq, while poor results were obtained for Mouse Motor Cortex. Log Normalization ranked within the top four in all datasets except for Lung. Also, TF-IDF scored highest in Mouse Motor Cortex and lung, and best three in PBMC. TF-IDF and SCTransform performed well when dealing with small datasets, while SCTransform was also able to run the larger dataset successfully. Figure 2a shows that log normalization performed better for large datasets as a decreasing tendency.

Typically, batch integration is evaluated visually by examining t-SNE or UMAP plots, whilst our experiments also use PHATE plots. Figure 2b depicts

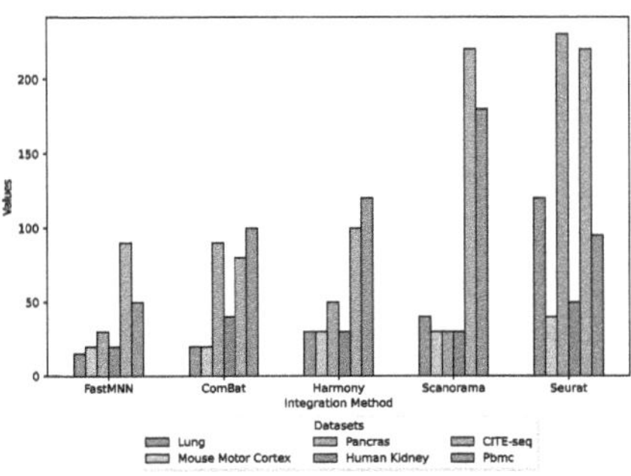

(a) Comparison of Integration Methods Across Datasets.

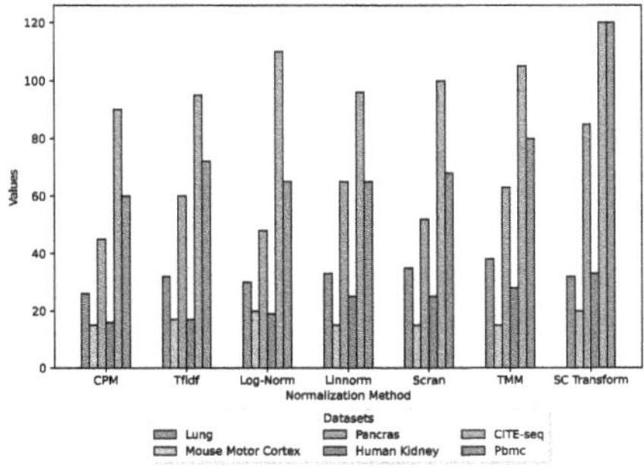

(b) Comparison of Normalization Methods Across Datasets.

Fig. 3. Rank of Running Efficiency.

UMAP's significant superiority over other methods of dimension reduction. UMAP consistently ranks first across all data sets without limiting the size of the data set, which proves that UMAP has a beneficial effect on the dimensionality reduction process of scRNA-seq integration. PHATE comes in second place in the overall results of evaluation metrics, and its evaluation performance is significant across most datasets. TSNE and PCA are the most under-performing methods for dimensionality reduction notably PCA is the least effective across all five datasets.

The assessment metrics for evaluating the integration methods relating to the different datasets are provided in Sect. 4 which outlines in detail the ranking of each method. As shown in Fig. 2c, the computed rank sum ranked FastMNN as the top method, with Seurat and Harmony ranked second (See Table 1). FastMNN produces the best results on mouse motor, pancreas, and PBMC datasets, but it does poorly on CITE-seq. Scanorama method was the least effective compared to other methods. It can be concluded that the FastMNN method is suitable for handling datasets of any size. Generally, Seurat performs better with smaller datasets, whereas Harmony performs well with large and small datasets.

4.3   Time Efficiency Evaluation

Computational time is another important factors to evaluate the pros and cons of a model. Figure 3 shows the comparative analysis of integration algorithms and standardization methods in terms of running efficiency. Because the file size of different datasets, operating environment and hardware equipment conditions have significant differences. To avoid interference, only the differences of methods between the same datasets are compared. For the data integration method, although the performance of Seurat is the best for clustering, the Seurat method takes a long time. Overall, the least time-consuming method is FastMNN, and the clustering performance is also relatively good, which means this method is more ideal.

For the data normalization method, there is no big difference between the different methods, but the SCTransform method takes a long time. However, the longer time can be accepted because of its excellent performance. Besides, Linnorm and SCTransform have almost the same good performance in the analysis of the model clustering performance, however in terms of efficiency, Linnorm has a more tremendous advantage, so Linnorm is better as a standardized method. In a nutshell, the optimal model for different data sets needs to be comprehensively determined. The above discussion can only be used as a reference.

5   Conclusion

We present a comparative analysis to evaluate the performance of workflows composed of different pre-processing methods and integration methods on six datasets. It can be seen from the result that it is necessary to choose different workflows according to the size and other characteristics of different datasets. In addition, using the subset of it for large datasets can greatly improve the efficiency of comparing different integration methods. We conduct experiments based combinations of seven normalization methods, four dimensional reduction methods, and five integration methods. Our results demonstrated that for the data integration module, the clustering performance of Seurat and Harmony are more prominent, but the time efficiency of Harmony was better. At the same time, the performance of Seurat for small data sets is superior. For the dimensionality reduction module, the UMAP method shows promising results in compatibility with the integration methods. Due to its significantly shorter computational time, FastMNN is recommended as the first method to try, with the other methods as viable alternatives.

References

1. Aldridge, S., Teichmann, S.A.: Single cell transcriptomics comes of age. Nat. Commun. **11**(1), 4307 (2020)
2. Alyassine, W., Raju, A.S., Braytee, A., Anaissi, A., Naji, M.: An efficient and reliable scrna-seq data imputation method using variational autoencoders. In: The International Conference on Innovations in Computing Research, pp. 84–97. Springer, Heidelberg (2024). https://doi.org/10.1007/978-3-031-65522-7_8
3. Hafemeister, C., Satija, R.: Normalization and variance stabilization of single-cell rna-seq data using regularized negative binomial regression. Genome Biol. **20**(1), 296 (2019)
4. Haghverdi, L., Lun, A.T., Morgan, M.D., Marioni, J.C.: Batch effects in single-cell rna-sequencing data are corrected by matching mutual nearest neighbors. Nat. Biotechnol. **36**(5), 421–427 (2018)
5. Hao, Y., et al.: Integrated analysis of multimodal single-cell data. Cell **184**(13), 3573–3587 (2021)
6. Hie, B., Bryson, B., Berger, B.: Efficient integration of heterogeneous single-cell transcriptomes using scanorama. Nat. Biotechnol. **37**(6), 685–691 (2019)

7. Johnson, W.E., Li, C., Rabinovic, A.: Adjusting batch effects in microarray expression data using empirical bayes methods. Biostatistics **8**(1), 118–127 (2007)
8. Jolliffe, I.T., Cadima, J.: Principal component analysis: a review and recent developments. Phil. Trans. R. Soc. A: Math. Phys. Eng. Sci. **374**(2065), 20150202 (2016)
9. Kharchenko, P.V.: The triumphs and limitations of computational methods for scrna-seq. Nat. Methods **18**(7), 723–732 (2021)
10. Kobak, D., Berens, P.: The art of using t-sne for single-cell transcriptomics. Nat. Commun. **10**(1), 5416 (2019)
11. Korsunsky, I., et al.: Fast, sensitive and accurate integration of single-cell data with harmony. Nat. Methods **16**(12), 1289–1296 (2019)
12. L. Lun, A.T., Bach, K., Marioni, J.C.: Pooling across cells to normalize single-cell rna sequencing data with many zero counts. Genome Biol. **17**, 1–14 (2016)
13. Lakkis, J., et al.: A joint deep learning model enables simultaneous batch effect correction, denoising, and clustering in single-cell transcriptomics. Genome Res. **31**(10), 1753–1766 (2021)
14. LaMorte, W.W.: Mann whitney u test (wilcoxon rank sum test). Boston University School of Public Health (2017)
15. Lazar, C., et al.: Batch effect removal methods for microarray gene expression data integration: a survey. Brief. Bioinf. **14**(4), 469–490 (2013)
16. Liu, Z.: Visualizing single-cell rna-seq data with semisupervised principal component analysis. Int. J. Mol. Sci. **21**(16), 5797 (2020)
17. Lun, A.: Further mnn algorithm development. GitHub repository (2019)
18. Lytal, N., Ran, D., An, L.: Normalization methods on single-cell rna-seq data: an empirical survey. Front. Genet. **11**, 501166 (2020)
19. Moon, K.R., et al.: Visualizing structure and transitions in high-dimensional biological data. Nat. Biotechnol. **37**(12), 1482–1492 (2019)
20. Moussa, M., Măndoiu, I.I.: Single cell rna-seq data clustering using tf-idf based methods. BMC Genom. **19**, 31–45 (2018)
21. Peng, M., Li, Y., Wamsley, B., Wei, Y., Roeder, K.: Integration and transfer learning of single-cell transcriptomes via cfit. Proc. Natl. Acad. Sci. **118**(10), e2024383118 (2021)
22. Schofield, F., Lensen, A.: Using genetic programming to find functional mappings for umap embeddings. In: 2021 IEEE Congress on Evolutionary Computation (CEC), pp. 704–711. IEEE (2021)
23. Stuart, T., et al.: Comprehensive integration of single-cell data. Cell **177**(7), 1888–1902 (2019)
24. Tang, F., et al.: mrna-seq whole-transcriptome analysis of a single cell. Nat. Methods **6**(5), 377–382 (2009)
25. Vieira Braga, F.A., et al.: A cellular census of human lungs identifies novel cell states in health and in asthma. Nat. Med. **25**(7), 1153–1163 (2019)
26. Wu, Y., Ji, Y., Lee, S., Akram, J., Braytee, A., Anaissi, A.: Simplified swarm learning framework for robust and scalable diagnostic services in cancer histopathology. In: International Conference on Computational Science, pp. 225–232. Springer, Heidelberg (2025). https://doi.org/10.1007/978-3-031-97635-3_27
27. Xiang, R., Wang, W., Yang, L., Wang, S., Xu, C., Chen, X.: A comparison for dimensionality reduction methods of single-cell rna-seq data. Front. Genet. **12**, 646936 (2021)
28. Yao, Z., et al.: A transcriptomic and epigenomic cell atlas of the mouse primary motor cortex. Nature **598**(7879), 103–110 (2021)

29. Yip, S.H., Wang, P., Kocher, J.P.A., Sham, P.C., Wang, J.: Linnorm: improved statistical analysis for single cell rna-seq expression data. Nucleic Acids Res. **45**(22), e179–e179 (2017)
30. Yu, B., Li, L., Zhang, J., Wang, X., Zeng, Y.: Single-Cell Sequencing and Methylation. Springer, Heidelberg (2020)
31. Zandavi, S.M., et al.: Disentangling single-cell omics representation with a power spectral density-based feature extraction. Nucleic Acids Res. **50**(10), 5482–5492 (2022)
32. Zandavi, S.M., Liu, D., Chung, V., Anaissi, A., Vafaee, F.: Fotomics: Fourier transform-based omics imagification for deep learning-based cell-identity mapping using single-cell omics profiles. Artif. Intell. Rev. **56**(7), 7263–7278 (2023)
33. Zhang, F., Wu, Y., Tian, W.: A novel approach to remove the batch effect of single-cell data. Cell Disc. **5**(1), 46 (2019)

Towards Automated Differential Diagnosis of Skin Diseases Using Deep Learning and Imbalance-Aware Strategies

Ali Anaissi[1,2]([envelope]), Ali Braytee[1], Weidong Huang[1], Junaid Akram[2], Alaa Farhat[1], and Jie Hua[3]

[1] University of Technology Sydney, Ultimo, Australia
{ali.anaissi,ali.braytee,weidong.huang}@uts.edu.au,
aaf261@student.bau.edu.lb
[2] University of Sydney, Camperdown, Australia
Junaid.Akram@uts.edu.au
[3] Shaoyang University, Shaoyang, China
steven.hua@mq.edu.au

Abstract. As dermatological conditions become increasingly common and the availability of dermatologists remains limited, there is a growing need for intelligent tools to support both patients and clinicians in the timely and accurate diagnosis of skin diseases. In this project, we developed a deep learning-based model for the classification and diagnosis of skin conditions. By leveraging pretraining on publicly available skin disease image datasets, our model was able to effectively extract visual features and accurately classify various dermatological cases.

Throughout the project, we refined the model architecture, optimized data preprocessing workflows, and applied targeted data augmentation techniques to improve overall performance. The final model, based on the Swin Transformer, achieved a prediction accuracy of **87.71%** across eight skin lesion classes on the ISIC2019 dataset. These results demonstrate the model's potential as a reliable diagnostic support tool for clinicians and a self-assessment aid for patients.

Keywords: Skin Disease Classification · Swin Transformer · BatchFormer · Focal Loss · ReduceLROnPlateau

1 Introduction

Skin diseases, particularly skin cancer, have become a significant global health concern. Early detection and accurate diagnosis are critical for effective treatment and positive patient outcomes. However, access to dermatologists remains limited worldwide, especially in regions where skin disease diagnosis is primarily handled by general practitioners [3,4,12,21]. Due to time constraints, limited resources, and a lack of specialized training, diagnostic accuracy by general practitioners can be suboptimal.

As a result, the use of advanced technologies, especially deep learning, has emerged as a promising solution to support both patients and clinicians in making faster and more accurate diagnostic decisions. In this context, the goal of our project is to apply deep learning techniques to improve skin disease classification, particularly under constraints such as limited image data and class imbalance. By incorporating domain knowledge with state-of-the-art deep learning methods, we aim to overcome current limitations and enhance diagnostic accuracy, ultimately contributing to better healthcare outcomes.

In this project, we used the widely adopted ISIC-2019 dermatology dataset and explored innovations in both model architecture and data augmentation. For deep learning models, we implemented the Swin Transformer, traditional CNNs, and ensemble CNN models. On the data augmentation side, we applied techniques such as SAM, AutoAugment, elastic deformation, and Fourier transforms to improve the model's ability to extract meaningful features from images.

Our best performance was achieved using the Swin Transformer in combination with these data enhancement techniques and optimized loss functions, resulting in an accuracy of 87.71% on the classification task. This result demonstrates significant potential for supporting physicians in clinical settings by improving diagnostic precision and assisting in early detection of skin diseases.

2   Related Work

Dermatological classification has become a key area of research due to the increasing need for accurate and efficient skin disease diagnosis [17,21]. In recent years, several large datasets have been established to support the development of deep learning models for this purpose, with the ISIC series, SD-198, and HAM10000 being the most widely used. Among these, the ISIC-2019 dataset has gained particular prominence, validated by numerous studies for its robust and comprehensive data. For this research, the ISIC-2019 dataset has been selected as the target dataset, given its broad adoption and reliability in the field.

Over the past few years, a variety of deep learning models have been applied to dermatological image classification, with convolutional neural networks (CNNs), Transformer-based models, and multi-model integration approaches leading the way. CNNs, such as ResNet, DenseNet, and MobileNetV2, have consistently shown strong performance in image classification tasks. For instance, Rodrigues et al. [15] introduced a skin lesion classification method combining transfer learning and Internet of Things (IoT) systems, leveraging CNNs to improve diagnostic accuracy. Similarly, Gessert et al. [7] applied transfer learning to pre-trained CNN models on the ISIC series, achieving an impressive AUC of 97.9%. ResNet models have also been tested extensively, with ResNet152 demonstrating the best performance in studies by Tan et al. [16] and Mishra et al. [14].

In more recent advancements, Transformer models have gained significant attention in dermatological classification. Liu et al. [13] introduced the Swin Transformer, which employs a hierarchical structure with shifted windows to improve computational efficiency and capture a broader context. This method has shown substantial promise in enhancing the performance of dermatological image classification tasks. Zhou et al. [20] further advanced this by integrating the Vision Transformer (ViT) backbone with specialized modules to improve diagnostic accuracy, achieving a notable 74.5% accuracy in clinical skin image diagnoses. Additionally, multimodal approaches that combine macro and micro images, as explored by Zhang et al. [19], have also proven to enhance diagnostic efficiency.

In addition to well-established models, other mainstream approaches have contributed to dermatological classification. Srinivasu et al. [1] integrated MobileNetV2 with a long short-term memory (LSTM) structure to improve model design. This combined approach achieved an impressive accuracy of 87.17%, showcasing excellent performance in skin lesion classification. Similarly, Groh et al. [8] developed a pre-trained CVGG model tailored to assist physicians by automating parts of the classification system, thereby enhancing the decision-making process and solidifying the role of deep learning in clinical dermatological diagnoses.

Beyond individual model approaches, multi-model integration has become increasingly popular. For example, Alshahrani et al. [2] combined the features of DenseNet121, MobileNet, and VGG19 into a high-dimensional eigenmatrix, achieving an accuracy of 88.79% after applying dimension reduction through t-SNE. Similarly, Gessert et al. [7] used a combination of EfficientNet, SENet, and ResNeXt models with soft voting to further improve classification accuracy and robustness.

One of the ongoing challenges in dermatological classification is the issue of class imbalance in datasets, which can hinder model performance. Yao et al. [18] addressed this by developing the MWNL (Modified Weighted Negative Loss) method, which assigns higher loss weights to underrepresented categories, thereby improving the model's attention to difficult or rare classes. This approach significantly reduces the impact of class imbalance, providing a more balanced and accurate model for skin disease diagnosis.

Overall, the integration of state-of-the-art deep learning models, large datasets, and innovative approaches like multi-model integration and class imbalance mitigation has propelled the field of dermatological classification forward. Continued advancements in these areas will be critical for improving diagnostic accuracy and accessibility in skin disease detection.

3   Proposed Framework

To address the challenges of imbalanced skin lesion datasets and enhance classification performance, we propose a hybrid framework built upon the Swin Transformer architecture pretrained on ImageNet-1K. While this pretrained backbone provides a strong starting point, domain-specific adaptation is necessary to improve generalizability across underrepresented disease categories.

To that end, we implement a series of architectural modifications and integrate multiple techniques—including BatchFormer, Focal Loss, and a dynamic learning rate scheduler (ReduceLROnPlateau)—to boost performance, particularly for tail classes. The model follows a four-stage hierarchical design, where each stage progressively reduces spatial resolution while increasing feature dimensionality, enabling extraction of increasingly abstract semantic features. The overall structure is illustrated in Fig. 1.

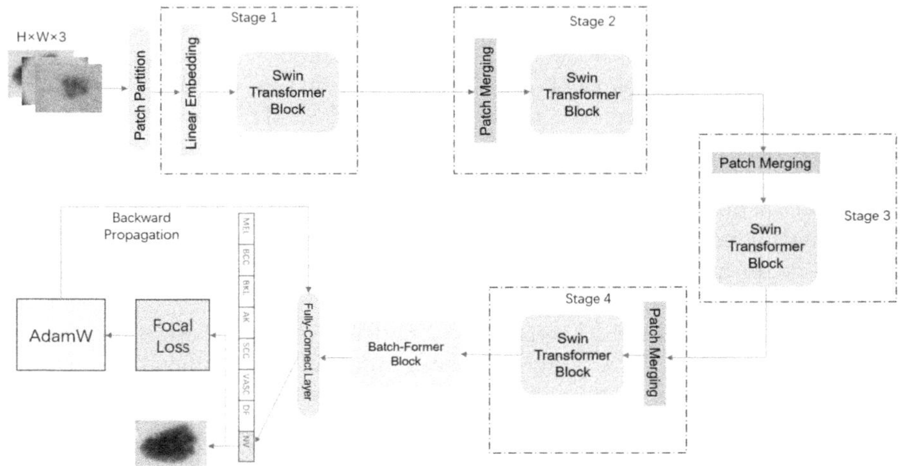

Fig. 1. Overview of the proposed model structure based on Swin-Transformer.

3.1 Swin Transformer

The Swin Transformer, proposed by Ze Liu et al. [13], is a hierarchical vision transformer that improves computational efficiency through localized self-attention mechanisms. Unlike traditional Vision Transformers (ViTs) that operate over global attention and are computationally expensive, the Swin Transformer introduces two novel mechanisms—Shifted Windows (SW-MSA) and Patch Merging—to process images in a hierarchical and efficient manner.

Built upon the success of Transformers in NLP, Swin Transformer combines the benefits of self-attention with the hierarchical representation capabilities of CNNs. This makes it particularly suitable for tasks involving low-resolution and variable-sized medical images. Each Swin Transformer Block consists of two core components: Window-based Multi-head Self-Attention (W-MSA) and Shifted Window-based Multi-head Self-Attention (SW-MSA).

W-MSA restricts attention computation within non-overlapping windows, while SW-MSA allows information exchange across windows by shifting the partitioned windows. This combination enables both local and global feature learning, improving the model's ability to capture complex spatial relationships. The full process of transitioning from W-MSA to SW-MSA is illustrated in Fig. 2.

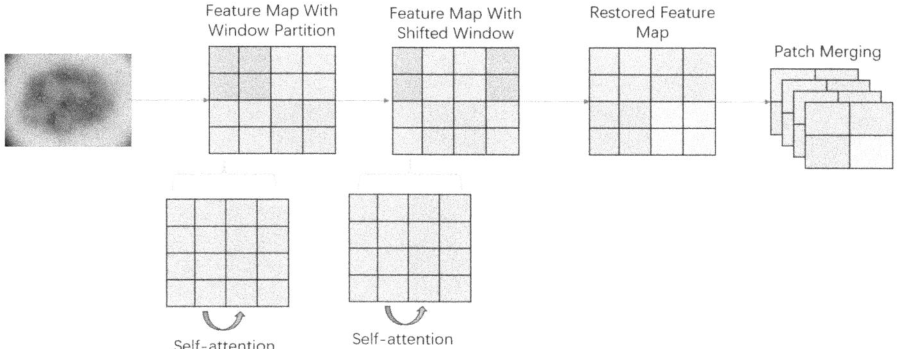

Fig. 2. W-MSA to SW-MSA Attention Process in the Swin-Transformer Block.

Patch merging is employed at each stage to reduce spatial resolution while increasing feature depth. For an input of size (H, W, D), patch merging reduces the spatial dimensions to $(H/2, W/2)$ and quadruples the feature depth to $4D$, enabling the model to extract more complex features at lower resolutions while minimizing computational cost.

This hierarchical structure, combined with localized attention and channel dimension expansion, equips the model with strong representational capacity suitable for medical imaging tasks, including skin disease classification.

3.2 BatchFormer

To address the issue of class imbalance—particularly the under-representation of rare diseases—we incorporate BatchFormer [9], a module designed to enhance representation learning for tail classes by leveraging inter-sample relationships within a mini-batch.

BatchFormer applies a Transformer encoder across the batch dimension, encoding not only individual sample features but also the similarity between samples. The architecture of the BatchFormer block is illustrated in Fig. 3.

This structure enables cross-sample gradient propagation, wherein gradients for each sample are influenced by those of others in the batch. As illustrated in Fig. 4, this mechanism allows knowledge transfer from head classes to tail classes, acting as a form of virtual data augmentation for underrepresented categories.

By aligning feature distributions between classes, BatchFormer helps mitigate bias toward majority classes and improves the model's ability to generalize to minority classes [5, 9].

This mechanism acts as a form of virtual data augmentation, enabling minority classes to benefit from rich gradients produced by majority samples, thereby improving classification performance across underrepresented classes.

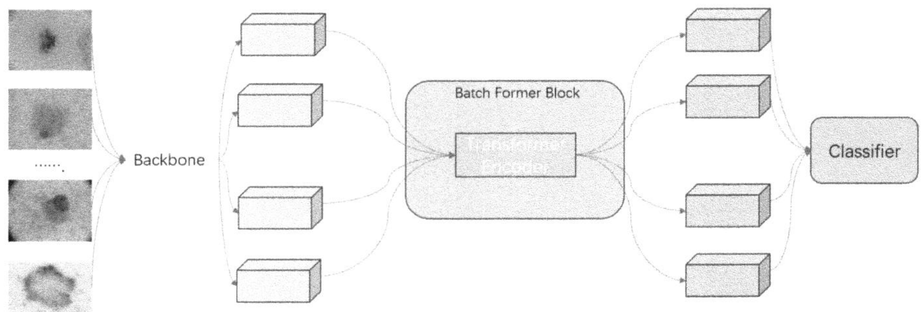

Fig. 3. BatchFormer Block Architecture.

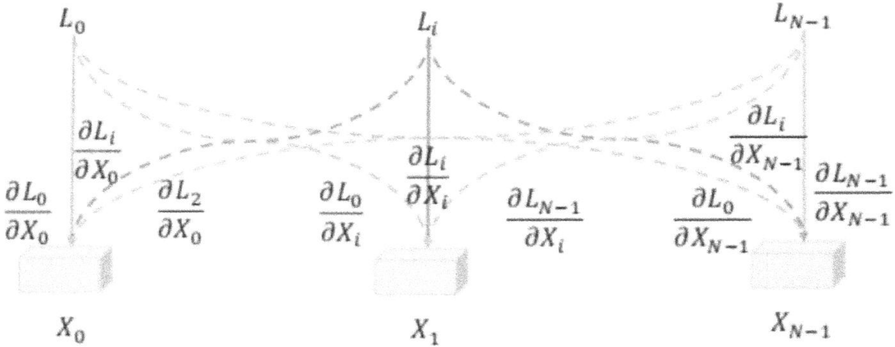

Fig. 4. Cross-sample Gradient Propagation in BatchFormer.

3.3   Focal Loss for Imbalanced Learning

To further address the challenge of class imbalance in our dataset, we incorporate the Focal Loss function, originally proposed by Lin et al. [11]. Unlike standard cross-entropy loss, Focal Loss dynamically scales the loss contribution from well-classified examples and focuses learning on hard misclassified instances. This is particularly beneficial in scenarios like medical image classification, where minority classes are often underrepresented and easily misclassified.

The Focal Loss introduces a modulating factor to the standard cross-entropy loss:

$$\mathcal{L}_{\text{focal}} = -\alpha_t (1 - p_t)^\gamma \log(p_t) \tag{1}$$

where:

- p_t is the model's estimated probability for the true class,
- α_t is a weighting factor for class t that balances positive and negative examples,
- γ is the focusing parameter that reduces the relative loss for well-classified examples.

When an example is correctly classified with high confidence (i.e., p_t is large), the term $(1 - p_t)^\gamma$ becomes small, thereby reducing the loss contribution from that example. Conversely, for misclassified or low-confidence examples, the loss remains high, forcing the model to focus on those harder cases during training.

In our application, Focal Loss enables the model to prioritize learning from underrepresented and difficult examples, which helps improve the classification of minority classes while mitigating overfitting to the majority class. Unlike traditional sampling strategies or cost-sensitive learning, Focal Loss reshapes the loss landscape directly, making it computationally efficient and straightforward to integrate into modern deep learning frameworks.

3.4   Learning Rate Scheduling with ReduceLROnPlateau

To optimize training stability and convergence, we employ the ReduceLROn-Plateau scheduler, which dynamically adjusts the learning rate based on the validation loss plateau. This scheduler reduces the learning rate when performance stagnates, allowing the model to fine-tune its parameters in later stages of training.

In combination with Focal Loss, this approach further enhances learning dynamics. Focal Loss encourages the model to focus on difficult samples, while ReduceLROnPlateau ensures that the learning rate is appropriately scaled to allow for fine-grained updates. Together, they help prevent overfitting, improve convergence speed, and boost classification performance on imbalanced datasets.

4   Experiments

4.1   Dataset Description and Preparation

In this study, we use the ISIC2019 dataset, a publicly available benchmark provided by the International Skin Imaging Collaboration (ISIC). The data set comprises thousands of dermoscopic images labeled in various categories of skin lesion and is accessible through the official ISIC archive.

To gain insight into the dataset and identify potential classification challenges, we performed a preliminary statistical analysis of class distributions. Using `NumPy`, we computed the number of samples in each category, along with their corresponding mean and median. The analysis, visualized in Fig. 5, reveals a severe class imbalance. For example, the "NV" (melanocytic nevus) class comprises more than 12,000 samples, while the "DF" (dermatofibroma) class has fewer than 300. This uneven distribution presents a significant challenge in training robust and fair classifiers. Consequently, our method incorporates several imbalance-aware learning strategies, including Focal Loss and BatchFormer.

To prepare the dataset for training and evaluation, we split it into three subsets: 70% for training, 15% for validation, and 15% for testing. We adopted a stratified sampling strategy to ensure that the class distribution remains consistent across all subsets, preserving the imbalance characteristics while enabling fair model evaluation (Fig. 6).

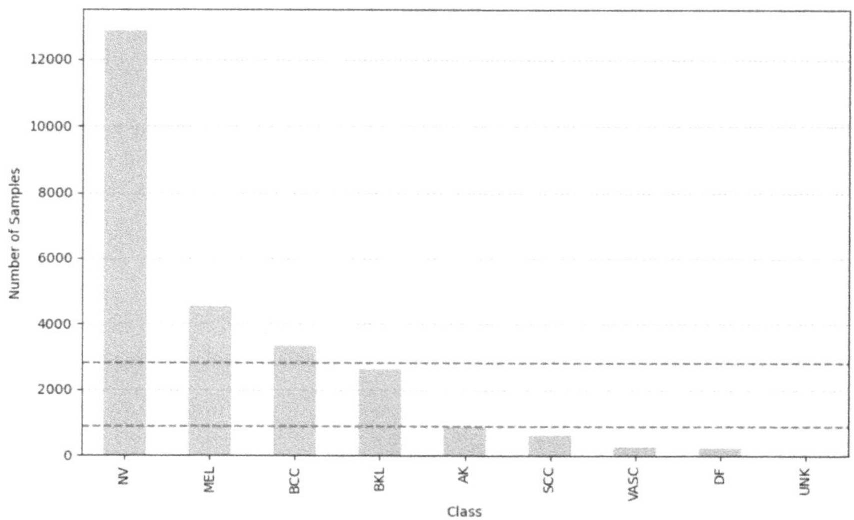

Fig. 5. Class distribution in the ISIC2019 dataset.

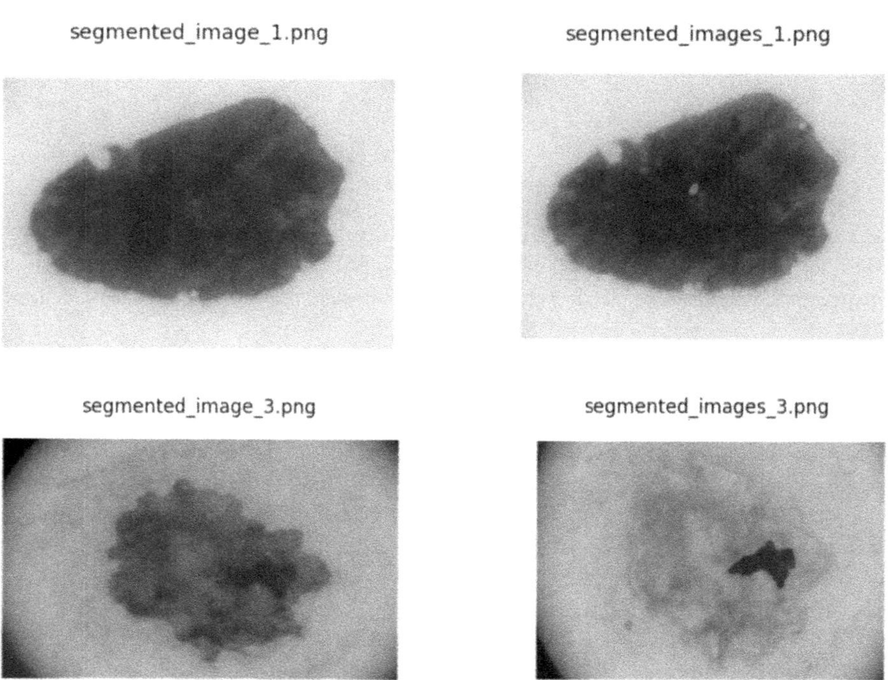

Fig. 6. Examples of lesion masks generated using SAM.

Table 1. Classification accuracy with and without SAM preprocessing.

Model	Accuracy
ViT	65.6%
ViT + SAM	62.0%
ResNet	75.6%
ResNet + SAM	73.0%
Swin Transformer	76.6%
Swin Transformer + SAM	70.8%

Table 2. Classification accuracy with data augmentation.

Model	Accuracy
DenseNet	71.53%
DenseNet + AutoAugment	73.87%
Swin Transformer	87.34%
Swin Transformer + Elastic Deformation	87.71%

4.2   Experimental Setup

Data Preprocessing and Augmentation. To mitigate the effects of class imbalance, we implemented multiple data preprocessing and augmentation strategies. Initially, we explored the use of the Segment Anything Model (SAM), developed by Meta [10], to generate segmentation masks for each lesion. These masks were used to render the lesion regions more distinctly by emphasizing their boundaries. The hypothesis was that focusing attention on these segmented regions could improve classification accuracy by highlighting lesion-relevant features. In addition to segmentation, we applied two augmentation techniques:

Table 3. Model performance comparison on ISIC2019.

Model Category	Model	Accuracy
CNN	DenseNet121-CE	76.47%
	DenseNet121-MWNL	61.37%
	DenseNet121 + Fourier Transformer	80.16%
	EfficientNet	80.94%
	Ensemble (Weighted)	73.18%
	Ensemble (Concatenated)	68.89%
	U-Net	60.52%
	ResNet50 + SimAM	69.18%
Transformer	ViT	80.16%
	Swin Transformer + BatchFormer + Focal Loss + ReduceLROnPlateau	**87.71%**

- Elastic Deformation, which distorts images using Gaussian-based displacement fields to simulate variability in lesion shapes [6].
- AutoAugment, a policy-based augmentation method that applies combinations of transformations (e.g., rotation, color jitter) to enrich the training set.

Elastic Deformation was applied selectively to disease classes with fewer than 2,000 training samples in order to directly address the imbalance.

5 Results and Discussion

Our study aimed to enhance skin lesion classification by incorporating advanced data preprocessing, augmentation, and optimization strategies within a Transformer-based deep learning framework. The proposed system comprises two core components: (1) a preprocessing and augmentation pipeline tailored to address dataset-specific challenges, and (2) a classification model based on the Swin Transformer, augmented with optimization modules to improve performance on imbalanced data.

To evaluate the contributions of each component, we conducted experiments in three phases: assessing segmentation-based preprocessing, applying data augmentation techniques, and performing model-level optimization. The findings from each phase are discussed below.

5.1 Impact of Segmentation-Based Preprocessing

We initially investigated the impact of segmentation as a preprocessing step, using the Segment Anything Model (SAM) to generate lesion masks. These masks were used to isolate lesion boundaries in dermoscopic images with the hypothesis that enhancing these regions would assist the classifier in focusing on diagnostically relevant areas.

However, our empirical findings contradicted this expectation. Table 1 shows a consistent drop in accuracy across all tested architectures (ViT, ResNet, and Swin Transformer) when SAM preprocessing was applied. This suggests that the segmentation process might introduce artifacts or omit subtle yet critical lesion characteristics such as color variegation, asymmetric texture, and pigment networks—features that are vital in dermatological assessments.

Moreover, SAM, while powerful, is a general-purpose model not specifically fine-tuned for dermoscopic images. This limitation may lead to inaccurate segmentation boundaries, especially for complex or irregularly shaped lesions, thus introducing noise rather than improving the focus of the classifier. These observations highlight the importance of using domain-specific segmentation tools or integrating lesion-aware attention mechanisms instead of general segmentation as a preprocessing step.

5.2   Effectiveness of Data Augmentation

To address class imbalance and enrich feature variability, we incorporated two data augmentation techniques: AutoAugment and Elastic Deformation. AutoAugment applies diverse transformation policies (e.g., rotation, brightness adjustment, shear), thereby exposing the model to a broader range of intra-class variations. Elastic Deformation introduces spatial distortions that simulate natural biological variation in lesion morphology.

Table 2 shows that both augmentation techniques yielded performance improvements across different models. For instance, DenseNet benefited from a 2.3% accuracy gain with AutoAugment, while Swin Transformer achieved the highest accuracy of 87.71% when augmented with Elastic Deformation. Notably, we applied Elastic Deformation selectively to underrepresented classes, which suggests that its targeted use contributes effectively to rebalancing class representation during training.

These results affirm that thoughtful augmentation not only mitigates class imbalance but also strengthens model robustness by enabling better generalization across diverse lesion appearances. This is especially crucial in medical imaging, where real-world variability is high and overfitting to limited visual cues can impair diagnostic utility.

5.3   Model Comparison and Final Results

Building upon the improvements introduced through augmentation, we proceeded to compare model architectures and integrate optimization strategies to further refine classification performance. A diverse range of models, including traditional CNNs and modern Transformer-based architectures, were evaluated on the ISIC2019 dataset. Table 3 summarizes the results.

CNN-based models such as DenseNet121 and EfficientNet performed competitively, with EfficientNet reaching an accuracy of 80.94%. Incorporating techniques like Fourier Transformer and attention modules further boosted CNN performance. However, Transformer-based models demonstrated superior accuracy, likely due to their enhanced capability to model long-range dependencies and global patterns.

The best-performing configuration combined Swin Transformer with Batch-Former, Focal Loss, and ReduceLROnPlateau, achieving a classification accuracy of 87.71%. The Swin Transformer's hierarchical design and shifted window attention mechanism enabled efficient multi-scale feature extraction—critical for analyzing both fine lesion textures and broader shape asymmetries. BatchFormer facilitated inter-sample learning by encouraging representation sharing among samples in the same mini-batch, effectively improving minority class recognition.

Focal Loss was instrumental in mitigating the impact of class imbalance by down-weighting easy samples and emphasizing harder, misclassified ones during training. Lastly, ReduceLROnPlateau ensured smoother convergence and prevented overfitting by dynamically adjusting the learning rate in response to validation performance.

Overall, these results validate the proposed approach as an effective solution for skin lesion classification, particularly in the context of class imbalance and heterogeneous data distributions. The integration of Transformer-based architectures with tailored augmentation and optimization strategies offers a promising direction for future work in medical image analysis.

6   Conclusion

Our findings highlight the Swin Transformer as a highly capable architecture for skin lesion classification, particularly when working with large-scale datasets that exhibit significant class imbalance. Compared to conventional CNN-based models, the Swin Transformer consistently achieved superior results—most notably reaching an accuracy of 87.71% when combined with BatchFormer, Focal Loss, and the ReduceLROnPlateau learning rate scheduler. This demonstrates its ability to effectively capture both local texture and global contextual information through its hierarchical self-attention mechanism.

However, despite this strong performance, the Swin Transformer does not include any built-in mechanism to directly handle data imbalance. Furthermore, the model was initialized with pretrained weights from general-purpose datasets such as ImageNet, which may not be optimally suited for the specific characteristics of dermoscopic images used in skin disease classification.

To further enhance performance, future research could focus on fine-tuning the model using domain-specific dermatology datasets to improve its sensitivity to lesion-specific features. Additionally, integrating more robust imbalance-aware learning strategies such as dynamic class-weighted loss functions or adaptive sampling techniques, may help improve classification performance for underrepresented lesion types and ensure more balanced predictions across all categories.

Acknowledgements. We acknowledge each of these students—Xinyu Ma, Frank Cheng, Sean Xia, Huiwen Zhao, Yuhao Xiong and Yibo Ding for their hard work, resilience, and commitment throughout this research endeavor.

References

1. Al-Tuwaijari, J.M., Yousir, N.T., Alhammad, N.A.M., Mostafa, S.: Deep residual learning image recognition model for skin cancer disease detection and classification. Acta Informatica Pragensia **12**(1), 19–31 (2023)
2. Alshahrani, M., Al-Jabbar, M., Senan, E.M., Ahmed, I.A., Mohammed Saif, J.A.: Analysis of dermoscopy images of multi-class for early detection of skin lesions by hybrid systems based on integrating features of CNN models. PLoS ONE **19**(3), e0298305 (2024)

3. Anaissi, A., Goyal, M., Catchpoole, D.R., Braytee, A., Kennedy, P.J.: Case-based retrieval framework for gene expression data. Cancer Informatics **14**, CIN–S22371 (2015)

4. Anaissi, A., Kennedy, P.J., Goyal, M.: Dimension reduction of microarray data based on local principal component. World Acad. Sci. Eng. Technol. **53** (2012)

5. Anaissi, A., Kennedy, P.J., Goyal, M., Catchpoole, D.R.: A balanced iterative random forest for gene selection from microarray data. BMC Bioinform. **14**, 1–10 (2013)

6. Chlap, P., Min, H., Vandenberg, N., Dowling, J., Holloway, L., Haworth, A.: A review of medical image data augmentation techniques for deep learning applications. J. Med. Imaging Radiat. Oncol. **65**(5), 545–563 (2021)

7. Gessert, N., Nielsen, M., Shaikh, M., Werner, R., Schlaefer, A.: Skin lesion classification using ensembles of multi-resolution efficientnets with meta data. MethodsX **7**, 100864 (2020)

8. Groh, M., et al.: Deep learning-aided decision support for diagnosis of skin disease across skin tones. Nat. Med. **30**(2), 573–583 (2024)

9. Hou, Z., Yu, B., Tao, D.: Batchformer: learning to explore sample relationships for robust representation learning. arXiv preprint arXiv:2203.01522 (2022). https://doi.org/10.48550/arXiv.2203.01522

10. Kirillov, A., et al.: Segment anything. In: Proceedings of the IEEE/CVF International Conference on Computer Vision, pp. 4015–4026 (2023)

11. Lin, T.Y., Goyal, P., Girshick, R., He, K., Dollár, P.: Focal loss for dense object detection. In: Proceedings of the IEEE International Conference on Computer Vision (ICCV), pp. 2980–2988 (2017). https://doi.org/10.1109/ICCV.2017.324

12. Liu, Y., et al.: A deep learning system for differential diagnosis of skin diseases. Nature (2020)

13. Liu, Z., et al.: Swin transformer: hierarchical vision transformer using shifted windows. In: Proceedings of the IEEE/CVF International Conference on Computer Vision, pp. 10012–10022 (2021)

14. Mishra, S., Imaizumi, H., Yamasaki, T.: Interpreting fine-grained dermatological classification by deep learning. In: Proceedings of the IEEE/CVF Conference on Computer Vision and Pattern Recognition Workshops (2019)

15. Rodrigues, D.D.A., Ivo, R.F., Satapathy, S.C., Wang, S., Hemanth, J., Reboucas Filho, P.P.: A new approach for classification skin lesion based on transfer learning, deep learning, and IoT system. Pattern Recognit. Lett. **136**, 8–15 (2020)

16. Tan, Z., Lin, J., Chen, K., Zhuang, Y., Han, L.: Skin cancer image recognition based on similarity clustering and attention transfer. J. Xray Sci. Technol. **31**(2), 337–355 (2023)

17. Wu, Y., Ji, Y., Lee, S., Akram, J., Braytee, A., Anaissi, A.: Simplified swarm learning framework for robust and scalable diagnostic services in cancer histopathology. In: International Conference on Computational Science, pp. 225–232. Springer (2025)

18. Yao, P., Shen, S., Xu, M., Liu, P., Zhang, F., Xing, J., Xu, R.X.: Single model deep learning on imbalanced small datasets for skin lesion classification. IEEE Trans. Med. Imaging **41**(5), 1242–1254 (2021)

19. Zhang, Y., Hu, Y., Li, K., Pan, X., Mo, X., Zhang, H.: Exploring the influence of transformer-based multimodal modeling on clinicians' diagnosis of skin diseases: a quantitative analysis. Digital Health **10**, 20552076241257090 (2024)

20. Zhou, Y.J., et al.: A novel multi-task model imitating dermatologists for accurate differential diagnosis of skin diseases in clinical images. In: International Conference on Medical Image Computing and Computer-Assisted Intervention, pp. 202–212. Springer (2023)
21. Zhou, Y., et al.: Vgg-fusionnet: a feature fusion framework from CT scan and chest X-ray images based deep learning for covid-19 detection. In: 2022 IEEE International Conference on Data Mining Workshops (ICDMW), pp. 1–9. IEEE (2022)

Causal Recommendation Method for Personalised Chemotherapy Optimisation in Breast Cancer

Tuyen Vu[1], Thuc D. Le[1]([⊠]), Ha X. Tran[2], Lin Liu[1], Jiuyong Li[1], and Jia Tina Du[1,3]

[1] Adelaide University, Adelaide, SA, Australia
{tuyen.vu,thuc.le,lin.liu,jiuyong.li}@adelaide.edu.au
[2] Accenture, Adelaide, SA, Australia
ha.a.tran@accenture.com
[3] Charles Sturt University, Bathurst, NSW, Australia
tdu@csu.edu.au

Abstract. Chemotherapy, including neoadjuvant chemotherapy administered prior to surgery, is a fundamental treatment strategy for breast cancer, yet patient responses are highly variable. Current chemotherapy planning primarily depends on a limited set of clinical and pathological factors, which often neglects the complex molecular heterogeneity inherent in individual tumours. Existing predictive models, while valuable, generally fail to address the essential clinical question: *Which patients truly benefit from chemotherapy, and how should drug combinations be optimally tailored to individual patients?* In response to this critical gap, we propose CTR (Causality-based Therapy Recommendation), a novel causal recommendation framework designed to personalise chemotherapy decisions in breast cancer treatment. CTR integrates causal inference methods, specifically causal trees, to estimate heterogeneous treatment effects and provide recommendations regarding whether chemotherapy would benefit individual patients, particularly in terms of achieving pathological complete response (pCR) and improved survival outcomes. Moreover, CTR extends its application by optimising combinations of chemotherapy agents, such as taxanes, anthracyclines, and anti-HER2 therapies, to enhance treatment efficacy. Evaluation of CTR on the DUKE and TransNEO datasets demonstrates significant improvements over existing state-of-the-art methods in terms of improved survival and recovery rates. CTR thus represents a promising step towards personalised precision oncology by enabling clinicians to make informed, data-driven chemotherapy decisions.

Keywords: Causal Recommendation · Personalised Chemotherapy · Therapy Plan · Breast Cancer · Survival Analysis

Q. V. Nguyen et al. (Eds.): AusDM 2025, CCIS 2765, pp. 397–411, 2026.
https://doi.org/10.1007/978-981-95-6786-7_27

1   Introduction

Chemotherapy remains a cornerstone of breast cancer treatment, significantly reducing tumour burden, lowering recurrence risk, and enhancing survival prospects, especially when administered in the neoadjuvant setting before surgical intervention [5]. Common chemotherapy regimens typically involve combinations of taxanes, anthracyclines, and anti-HER2 agents [23]. Currently, clinical decision-making for chemotherapy predominantly relies on empirical assessment using broad clinical and pathological criteria, such as tumour size, histological grade, hormone receptor, and HER2 status [20]. However, these conventional methods are insufficient for capturing the complex molecular heterogeneity of breast cancer tumours [13,27], often resulting in suboptimal outcomes where only a subset of patients derives significant benefits from chemotherapy [2,19].

Recent advancements in machine learning and multi-omics data analysis have improved the predictive accuracy for treatment outcomes [27]. Techniques including Multi-Omics Factor Analysis (MOFA) [4] and machine learning methods such as Support Vector Machines [10,22], Random Forests [1,7], and LASSO regression [21] have successfully identified biomarkers correlated with chemotherapy responses. Nevertheless, these approaches primarily focus on predicting outcomes [19,27,28] rather than addressing two pivotal causal questions: *"Which patients truly benefit from chemotherapy?"* and *"Would an alternative chemotherapy regimen have led to better patient outcomes?"*. The first question is crucial for preventing unnecessary treatment in patients unlikely to respond, while the second question directly informs the optimal tailoring of drug combinations to individual patients. Addressing these counterfactual inquiries is essential for moving beyond one-size-fits-all protocols towards genuinely personalised chemotherapy recommendations.

To bridge this critical gap, we introduce the Causality-based Therapy Recommendation (CTR) framework, which utilises causal inference techniques, particularly causal tree methodologies, to estimate heterogeneous treatment effects [6]. By explicitly incorporating chemotherapy regimens as causal variables, CTR recommends whether individual patients should receive chemotherapy to maximise benefits such as pathological complete response and improved survival. Furthermore, CTR explores optimising drug combinations within chemotherapy regimens to enhance individual patient outcomes further. Our proposed method represents several significant contributions:

1. **Personalised Chemotherapy Decision-Making:** CTR identifies breast cancer patients who are most likely to benefit from chemotherapy based on causal inference.
2. **Optimised Drug Combinations:** CTR evaluates and recommends the most effective combinations of chemotherapy agents tailored to patient-specific molecular profiles and clinical characteristics.
3. **Improved Clinical Outcomes:** Experimental evaluation demonstrates CTR significantly enhances recovery rates compared to current clinical practice and state-of-the-art predictive models.

4. **Real-World Applicability:** CTR includes the development of a practical, interpretable, web-based platform for clinical implementation, aiding clinicians in real-time chemotherapy planning.

By focusing on patient-specific causal relationships and optimising chemotherapy regimens, the CTR framework advances precision oncology, offering improved clinical outcomes in breast cancer management.

2 Background and Related Work

2.1 Clinical Decision-Making for Chemotherapy

Chemotherapy remains a key modality in breast cancer management, particularly in the neoadjuvant setting where it is used to reduce tumour burden before surgery [5]. Current clinical decision-making is guided by standardised protocols such as those from the National Institute for Health and Care Excellence (NICE), which recommend treatment based on tumour size, histological grade, hormone receptor status, HER2 expression, and nodal involvement [20]. While these guidelines are grounded in large clinical trials and expert consensus, they reflect average treatment effects across broad patient groups.

However, extensive research has shown that breast cancer is a highly heterogeneous disease at the molecular level, even within the same clinical subtype [13,19]. This heterogeneity often results in differential responses to the same chemotherapy regimen, leading to over-treatment in some patients and under-treatment in others. Consequently, personalised chemotherapy strategies are needed to improve efficacy and reduce unnecessary toxicity.

2.2 Machine Learning for Predictive Modelling in Oncology

To address tumour heterogeneity, machine learning (ML) techniques have been widely adopted for predicting chemotherapy response and survival outcomes. Classical algorithms such as Support Vector Machines [10,22], Random Forests [1,7], and LASSO regression [21] have been effectively applied to clinical and omics data to identify prognostic and predictive biomarkers.

More recently, integrative approaches like Multi-Omics Factor Analysis (MOFA) [4] have enabled the extraction of latent features from heterogeneous biological datasets, improving the stratification of patients. Deep learning models have also shown promise in modelling complex interactions among genomic, transcriptomic, and clinical features [9]. For example, [27] developed a multi-omic predictor for neoadjuvant therapy response in breast cancer, demonstrating the feasibility of precision oncology at scale.

Despite these advancements, most ML-based methods are inherently predictive rather than prescriptive. They are trained to estimate the likelihood of an outcome (e.g., pathological complete response or survival) under observed treatments but do not evaluate alternative treatment scenarios. As a result, these models are limited in their ability to inform treatment selection or optimisation.

2.3   Causal Inference for Personalised Treatment Recommendation

Personalised medicine ultimately requires answering counterfactual questions such as: *"Would this patient have had a better outcome with a different treatment?"*. This necessitates a shift from outcome prediction to treatment effect estimation. Causal inference provides the mathematical and algorithmic foundation to address this challenge by estimating the effects of interventions rather than associations in observed data.

A range of causal machine learning methods have been developed to estimate heterogeneous treatment effects. Tree-based approaches such as Causal Trees [6] and Causal Forests [31] enable non-parametric estimation of individualised treatment effects by recursively partitioning the covariate space. Meta-learning strategies, including the T-learner, S-learner, and X-learner [17], reframe the problem as a supervised learning task using transformed outcome variables. These methods have shown success in economics and social science, and are beginning to gain traction in healthcare applications, including oncology [33].

In the context of cancer treatment, causal inference has been applied to estimate treatment benefits from observational data, such as determining the effect of chemotherapy, radiotherapy or hormone therapy in subgroups of patients [3,18,32]. However, its application to chemotherapy regimen selection remains relatively unexplored. Moreover, interpretability and clinical adoption remain critical barriers, especially when black-box models are used for causal estimation.

2.4   Positioning of the CTR Framework

Our proposed CTR (Causality-based Therapy Recommendation) framework builds upon these causal inference advancements by directly estimating the individual treatment effect of chemotherapy regimens. CTR leverages causal tree-based methods to partition the patient population based on shared treatment effect patterns, resulting in interpretable and actionable recommendations. Unlike traditional predictive models that passively learn associations, CTR explicitly incorporates the chemotherapy regimen as a causal variable and estimates the potential benefit of each treatment option for a given patient. This enables personalised recommendation, optimised not just for predictive accuracy but for actionable decision-making.

By combining causal inference with real-world clinical and molecular data, CTR addresses the current gap between ML-based prediction and clinical decision support, offering a robust solution for precision oncology in breast cancer treatment.

3   Methodology

3.1   Personalised Chemotherapy Recommendation Using Causal Tree

Consider a breast cancer patient described by a set of variables $\mathbf{X} = \{X_1, X_2, ...X_p\}$. The variable set $\mathbf{X}$ consists of p features such as age, tumour size, HER2 status, ER status and gene expression, representing multi-omics data. The binary variable $Y \in \{0, 1\}$ represents the recovery outcome of interest, where $Y = 1$ indicates the desired positive outcome of complete recovery, and $Y = 0$ indicates the negative outcome of an unrecovered status.

Let $T \in \{0, 1\}$ represent the therapy assignment indicator for patient i with two potential outcome: $Y_i{}^1$ for receiving chemotherapy ($T = 1$) and $Y_i{}^0$ for not receiving chemotherapy ($T = 0$). However, only one outcome can be observed, as each patient can either receive or not receive the therapy at a given time [25]:

$$Y_i = T\,Y_i^1 - (1 - T)\,Y_i^0 \tag{1}$$

In this paper, the terms "treatment", "therapy" and "chemotherapy" are used interchangeably.

Treatment Effect: The Individual Treatment Effect (ITE) for chemotherapy T is denoted by the equation [16]

$$ITE_i(T) = Y_i^1 - Y_i^0 \tag{2}$$

where $ITE = 0$ indicates that chemotherapy T has no affect on the recovery of breast cancer patient i. If $ITE > 0$, chemotherapy T has the potential to facilitate the patient's recovery.

For observational data, the chemotherapy assignment is usually not completely randomised therefore treated patients may not be comparable with untreated patients. To estimate treatment effects from observational data, the following three assumptions are often made [16]:

- **Assumption 1: Unconfoundedness**. Given the covariate set $\mathbf{C}$ of breast cancer patients, the assignment of the therapy is independent of the potential outcomes: $T \perp\!\!\!\perp (Y^0, Y^1) \mid \mathbf{C}$.
- **Assumption 2: Stable Unit Treatment Value Assumption (SUTVA)**. The therapy assignment of one individual does not influence the potential outcomes of other individuals.
- **Assumption 3: Consistency.** The observed outcome of a given therapy corresponds to the potential outcome associated with that therapy.

Our goal is to provide personalised chemotherapy recommendations, acknowledging that a chemotherapy can have varying effects on different individuals, resulting in heterogeneity in treatment responses. This necessitates the estimation of the Conditional Average Treatment Effect ($CATE$) for a binary treatment, chemotherapy T, which represents the ITE and helps identify the most effective treatment. To achieve unbiased CATE estimation for a specific subgroup, we can apply inverse probability of treatment weighting (IPTW) using propensity scores, as outlined by Hirano [15], as follows:

$$\widehat{CATE}(T, \mathbf{C} = \mathbf{c}) = \frac{\sum_{i=1}^{n} \frac{T_i Y_i}{e(\mathbf{c})}}{\sum_{i=1}^{n} \frac{T_i}{e(\mathbf{c})}} - \frac{\sum_{i=1}^{n} \frac{(1-T_i)Y_i}{(1-e(\mathbf{c}))}}{\sum_{i=1}^{n} \frac{(1-T_i)}{(1-e(\mathbf{c}))}}, \tag{3}$$

where $e(\mathbf{c}) = P(T = 1|\mathbf{C} = \mathbf{c})$ is the propensity score [24], the probability of chemotherapy assignment. The score can be estimated using logistic regression. In this study, the covariate set $\mathbf{C}$ corresponds to the set $\mathbf{X}$.

The proposed CTR framework aims to identify patient subgroups with varying responses to chemotherapy by estimating heterogeneous treatment effects, rather than relying on a single average treatment effect (ATE). This enables personalised treatment recommendations that are tailored to each patient's unique clinical and molecular profile.

CTR employs the Causal Tree method [6] to partition patients into subgroups and estimate the CATE within each. Using chemotherapy as a binary causal variable, the model quantifies the effect of receiving chemotherapy versus not receiving it. A positive CATE indicates that a patient is likely to benefit from chemotherapy, whereas a negative value suggests they may achieve better outcomes without it. The use of a single causal tree allows the model to capture both treatment and counterfactual outcomes in an interpretable structure, supporting individualised recommendations.

In this binary setting, chemotherapy assignment is encoded using one-hot encoding into two therapy plans: **TP1** (treated) and **TP2** (untreated). CTR then recommends the treatment option with the higher estimated CATE for each patient. This single-treatment formulation serves as the conceptual foundation for extending the CTR framework to handle more complex, multi-arm treatment settings, as described in the following subsection.

3.2 Personalised Therapy Plan Recommendation

In clinical oncology, chemotherapy is commonly administered as a combination of agents drawn from three principal classes: taxanes, anthracyclines, and anti-HER2 therapies [29,30]. Each unique combination of these agents constitutes a distinct therapy plan, the effectiveness of which can vary significantly based on individual patient factors such as tumour subtype, molecular profile, and other clinical characteristics.

To determine the most appropriate therapy plan for each patient, we propose a causal framework grounded in the CTR model. Unlike standard binary treatment settings, where the causal factor is defined as "treatment vs. control", our formulation extends to multi-arm treatment settings, where each therapy plan is treated as a separate causal factor. This setup enables the model to estimate the *individualised treatment effect* of each therapy plan, supporting a more nuanced and tailored treatment recommendation.

CTR constructs a dedicated Causal Tree for each therapy plan, treating the administration of that specific plan as the causal variable, and using pathological complete response as the clinical outcome. Each tree partitions the patient population into subgroups (leaf nodes) based on shared covariate patterns and estimates the *Conditional Average Treatment Effect (CATE)* within each subgroup. These CATE values represent the expected improvement in outcome (e.g., achieving pCR) if the patient were to receive the corresponding therapy plan.

For a new patient, the recommendation process proceeds as follows:

1. **Feature Matching:** The patient's clinical and molecular features are evaluated against each therapy-specific Causal Tree to locate the leaf node (subgroup) to which they belong.
2. **Effect Retrieval:** From each matched leaf node, the corresponding CATE is retrieved, representing the estimated treatment effect for that therapy plan.
3. **Effect Comparison:** This process is repeated across all therapy plans, resulting in a set of CATE values specific to the patient.
4. **Plan Selection:** The therapy plan with the highest positive CATE is selected and recommended, as it is expected to yield the greatest clinical benefit for the individual.

By estimating and comparing the treatment effects of multiple therapy options, the CTR framework supports interpretable, data-driven, and personalised chemotherapy recommendations. It not only determines whether a patient is likely to benefit from chemotherapy but also identifies which specific drug combination is expected to produce the most favourable outcome. This model represents a significant advancement toward the vision of precision oncology, aligning treatment decisions with patient-specific characteristics and causal reasoning.

3.3 Defining Chemotherapy Plans for Breast Cancer

In this study, we define a *chemotherapy plan* as a specific combination of therapeutic agents drawn from taxanes, anthracyclines, and anti-HER2 therapies. Throughout this paper, we use the terms *chemotherapy plan* and *therapy plan* interchangeably to refer to these predefined treatment regimens. Each component is encoded using binary indicator variables as follows:

– **Taxane:** $C_1 = 1$ if a taxane is included in the regimen; otherwise, $C_1 = 0$.
– **Anthracycline:** $C_2 = 1$ if anthracycline is administered; otherwise, $C_2 = 0$.
– **Anti-HER2:** $C_3 = 1$ if an anti-HER2 agent is included; otherwise, $C_3 = 0$.

A **Therapy Plan (TP)** is thus defined as a unique combination of the binary indicators C_1, C_2, and C_3. In the TransNEO dataset [11,27], all patients received taxane-based chemotherapy, meaning the taxane indicator is fixed at $C_1 = 1$. Consequently, the variability in therapy plans is determined by the inclusion or exclusion of anthracyclines and anti-HER2 therapies.

Table 1. Therapy Plan Indicators.

C_1	C_2	C_3	Therapy Plan	TP_1	TP_2	TP_3	TP_4	Drug Regimens
1	1	1	TP_1	1	0	0	0	Taxane, Anthracycline, Anti-HER2
1	1	0	TP_2	0	1	0	0	Taxane and Anthracycline
1	0	1	TP_3	0	0	1	0	Taxane and Anti-HER2
1	0	0	TP_4	0	0	0	1	Taxane only

Each patient is assigned to exactly one therapy plan, represented using a one-hot encoding scheme: for a given patient, the corresponding indicator $TP_k = 1$ identifies the administered plan, while the others are set to zero. This mutually exclusive encoding structure, summarised in Table 1, enables chemotherapy to be modelled as a multi-valued causal factor within the CTR framework.

3.4 Evaluation Metrics

To rigorously evaluate the effectiveness of the proposed CTR model in personalised chemotherapy recommendation, we adopt two clinically relevant metrics: **Recovery Rate** and **Recovery Ratio**. These metrics quantify the model's ability to improve short-term clinical outcomes, specifically the achievement of pathological complete response, which is considered a strong early indicator of therapeutic success in breast cancer treatment.

Recovery Rate: The metric measures the proportion of patients who achieve a favourable outcome within a given cohort. This metric enables a direct assessment of the overall effectiveness of a treatment assignment strategy, whether derived from current clinical practice or generated by a recommendation model.

Current Protocol (CurrentPr) Recovery Rate: This value reflects the percentage of patients who achieved pCR based on the actual therapy plans they received under standard clinical protocols. These assignments are typically determined by empirical clinical guidelines or risk stratification.

Model-Based Recovery Rate: For model-based recommendations, we compute the Recovery Rate only for patients whose actual treatment matched the model's suggested therapy plan—referred to as the *Followed* group. This metric provides insight into the potential benefit of adhering to model-generated personalised treatment recommendations.

Recovery Ratio: The metric offers a comparative perspective by evaluating the relative benefit of following a model's recommendation. It is defined as the ratio of positive outcomes between patients who followed the model's recommendation (*Followed group*) and those who did not.

A Recovery Ratio greater than 1 indicates that patients who adhered to the model's recommendation were more likely to achieve a positive outcome, thereby validating the clinical value of the model's decision support. Conversely, a ratio below 1 suggests the model's recommendations may not improve or could even worsen outcomes.

4   Experimental Results

4.1 Datasets

DUKE Dataset. The original DUKE dataset [26] contains 922 breast cancer patient samples. For our analysis, we applied preprocessing steps to remove features with a high proportion of missing values and excluded samples with incomplete records. After preprocessing, 281 complete samples were retained for in the experiments. The primary outcome of interest is pathological complete response, which was achieved by 60 patients. The treatment variable, Chemotherapy, was encoded using one-hot encoding into two therapy plans: **TP1** (treated) and **TP2** (untreated). The DUKE dataset is used to evaluate personalised chemotherapy recommendations. For survival analysis, we used recurrence events as the event indicator and the number of days to the last recurrence-free assessment as the time-to-event variable. This allowed for the evaluation of survival outcomes in relation to treatment response.

TransNEO Dataset. In this study, we utilised observational data from the TransNEO neoadjuvant breast cancer clinical trial [27] and the ARTemis clinical trial [11]. The TransNEO trial initially recruited 180 women, of whom 147 cases with complete molecular and digital pathology data were included in our analysis. These participants received at least one cycle of chemotherapy or targeted therapy and had available outcome assessments. The second dataset comprises an external cohort of 75 patients from the ARTemis trial, all of whom also received neoadjuvant therapy [11]. Together, These two datasets form a merged cohort of 222 patient samples for our study. The treatment outcome was measured by pathological complete response [29], with 57 patients achieving pCR.

Multi-omics features were used to develop the CTR model, which integrates heterogeneous data types to enhance predictive accuracy and provide a deeper understanding of tumour heterogeneity, thereby outperforming traditional clinical risk stratification methods [14]. The dataset comprised a total of 74 variables spanning clinical characteristics, genomic profiles, digital pathology, and treatment information. Treatment assignment was encoded using one-hot encoding into four therapy plans, **TP1**, **TP2**, **TP3**, and **TP4**, as defined in Table 1.

In contrast to the work of Sammut et al. [27], which focused on predicting pCR from pre-treatment features, our study utilises the TransNEO dataset to train a causal framework aimed at recommending the optimal therapy plan for each individual patient. This shift from predictive to prescriptive modelling enables personalised treatment planning based on estimated treatment effects.

4.2   Baseline Methods

To our knowledge, no existing method in the literature explicitly addresses the task of personalised Therapy Plan Recommendation using multi-omics data. Existing machine learning approaches predominantly focus on predicting treatment response without providing individualised treatment selection. This study represents the first attempt to apply causal inference techniques for therapy recommendation using integrated multi-omics and clinical data.

To enable a fair and systematic comparison, we construct six baseline models using conventional supervised learning methods adapted for the therapy recommendation task. These baselines serve as reference models to evaluate the performance of the proposed causality-based approach, CTR. The baseline models are built using three widely adopted machine learning algorithms, which have been successfully applied in prior studies on multi-omics data for outcome prediction [27]: **Logistic Regression (Logr)** [12], **Support Vector Machine (Svm)** [10], **Random Forest (Rf)** [7].

To adapt the baseline models for therapy plan recommendation, we consider two training approaches. Each approach simulates how the model could be used to predict the best therapy plan for a given patient based on predicted treatment outcomes:

1. **Approach 1 – Association-Based (Single Model for All Plans):** A single model is trained using all samples, with therapy plan indicators (TP_1, TP_2, TP_3, TP_4) included as input features. At test time, the model is used to estimate the counterfactual outcome which is the probability of achieving a positive outcome for each possible therapy plan by setting one plan at a time to 1 (and the others to 0), keeping all other features constant. The plan with the highest predicted probability is selected as the recommendation. This approach is purely association-based and evaluates outcome likelihood via model predictions. The resulting models are denoted as:
   - **LogrRa:** Logistic Regression-based Recommendation for all plans.
   - **SvmRa:** Support Vector Machine-based Recommendation for all plans.
   - **RfRa:** Random Forest-based Recommendation for all plans.
2. **Approach 2 – Association-Based (Separate Model per Plan):** This approach more closely mirrors the CTR method by constructing one dedicated model for each therapy plan. Specifically, for each therapy plan TP_k, a separate model is trained using only the samples that received that plan. Each model learns to predict the likelihood of achieving pCR given the features of patients who received that specific plan. At test time, all k models are applied to the same patient, and the therapy plan corresponding to the model with the highest predicted probability is selected. Although this approach does not estimate causal effects, it reflects the same evaluation structure as CTR and serves as a strong association-based baseline. The resulting models are:
   - **LogrRe:** Logistic Regression-based Recommendation for each plan.
   - **SvmRe:** Support Vector Machine-based Recommendation for each plan.
   - **RfRe:** Random Forest-based Recommendation for each plan.

In total, we evaluated six baseline models across the two training strategies. Approach 1 reflects a standard predictive modelling approach, while Approach 2 structurally aligns with the proposed CTR method by evaluating each therapy plan independently. However, unlike CTR, these methods do not explicitly estimate causal treatment effects or account for treatment assignment bias. Instead, they rely solely on observed associations between features and outcomes.

4.3 Personalised Chemotherapy Recommendation

Figure 1 presents the Kaplan–Meier survival curves with 95% confidence intervals for the **DUKE dataset**, comparing seven methods for personalised chemotherapy recommendation. Each model is evaluated based on whether patients followed the recommended treatment plan (**Followed**) or not (**Not Followed**).

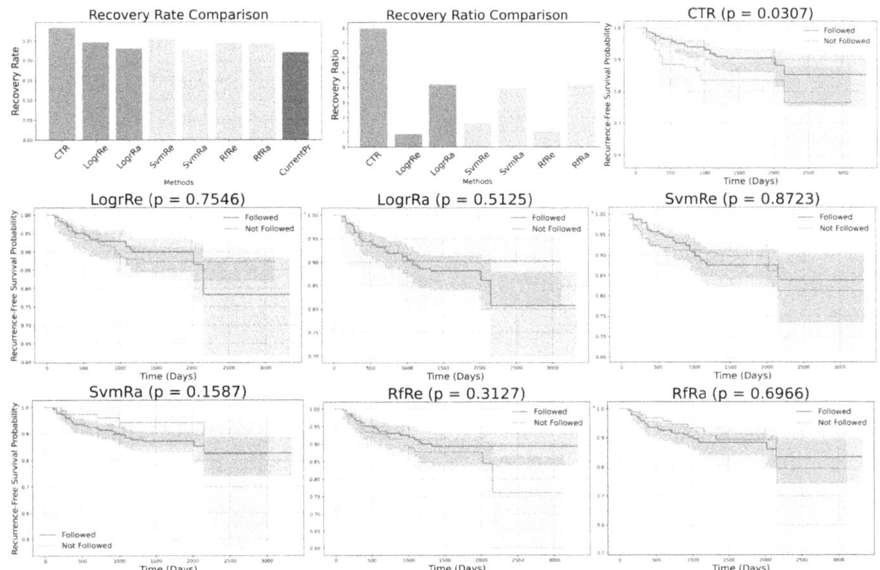

Fig. 1. Comparison of personalised chemotherapy recommendation models.

Among all methods, the proposed **CTR** model consistently achieves the best performance. Specifically, for long-term clinical benefit, CTR shows the greatest separation between the Followed and Not Followed groups, with the Followed group achieving markedly better recurrence-free survival. The log-rank test yields the smallest p-value ($p = 0.0307$), indicating statistically significant survival improvement. In terms of short-term clinical benefit, CTR clearly outperforms all baseline methods, achieving the highest **Recovery Ratio** (8.00) and the highest **Recovery Rate** (28.3%). In comparison, the current clinical protocol in the DUKE dataset, based on empirical stratification, yields a recovery rate of only 21%, with 60 out of 281 patients achieving pCR [26]. This demonstrates that patients who followed CTR's recommendations were substantially more likely to achieve a pathological complete response.

4.4   Personalised Treatment Plan Recommendation

Figure 2 presents a comparative analysis of the recovery rates and recovery ratios achieved by all evaluated methods on the TransNEO dataset for the task of personalised treatment plan recommendation. The results clearly highlight the superior performance of the proposed method, **CTR**, which achieved a recovery rate of approximately **39.8%**, outperforming all baseline models and the **Current Protocol**, which yielded a recovery rate of only **26%**, as in the **TransNEO dataset**, 57 out of 222 patients achieved a pCR [11, 27].

In addition, CTR also attained the highest **Recovery Ratio** among all methods, with a value of **1.38**. This indicates that patients who followed the treat-

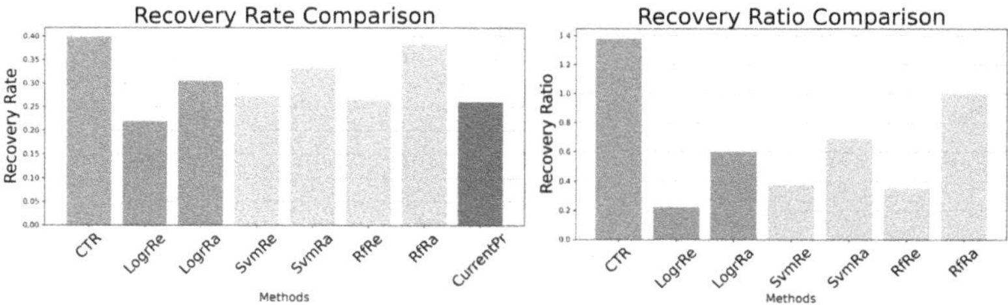

Fig. 2. Comparison of personalised chemotherapy plan recommendation models.

ment plans recommended by CTR were nearly 1.4 times as likely to achieve a favourable clinical outcome (pCR) compared to those who did not follow the recommendation. In contrast, all baseline machine learning methods demonstrated lower recovery rates and recovery ratios, with several methods performing at or below the effectiveness of the current protocol.

These experimental results reinforce CTR's potential to enable data-driven, patient-specific treatment strategies that substantially improve therapeutic outcomes beyond the capabilities of standard clinical practice.

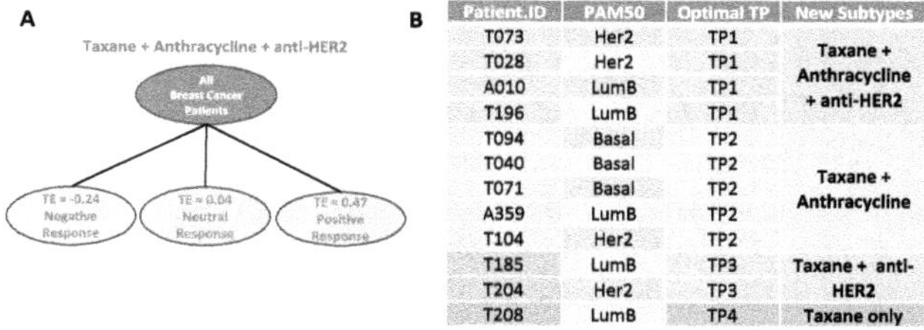

Patient.ID	PAM50	Optimal TP	New Subtypes
T073	Her2	TP1	Taxane + Anthracycline + anti-HER2
T028	Her2	TP1	
A010	LumB	TP1	
T196	LumB	TP1	
T094	Basal	TP2	Taxane + Anthracycline
T040	Basal	TP2	
T071	Basal	TP2	
A359	LumB	TP2	
T104	Her2	TP2	
T185	LumB	TP3	Taxane + anti-HER2
T204	Her2	TP3	
T208	LumB	TP4	Taxane only

Fig. 3. Defining Treatment Response-Based Breast Cancer Subtypes.

4.5 Breast Cancer Subtypes Based on Treatment Response

The Treatment Effect (TE) scores generated by CTR can also be used to stratify patients into distinct treatment response subtypes, which can be used to categorise breast cancer patients into distinct subgroups based on their response to treatment. These subgroups represent patients who respond positively (well), negatively (poorly), or neutrally to a given therapy plan. Additionally, patients can be grouped by identifying those who share the same optimal therapy plan. This approach enables a novel method for defining breast cancer subtypes based on treatment response. Figure 3 presents the results from the CTR model, illustrating the treatment response scores for each patient. Based on these scores, patients are assigned to their respective subgroups or optimal therapy plans.

- **Figure 3A** shows the Breast Cancer Subtypes based on a Given Therapy Plan. For a specific therapy plan, breast cancer patients can be categorised into three subtypes according to their treatment response: positive, neutral, or negative.
- **Figure 3B** shows the Breast Cancer Subtypes based on a List of Therapy Plans. The CTR model can categorise breast cancer patients into four subgroups, each corresponding to one of four recommended therapy plans. By identifying the optimal therapy plan for each patient, the CTR model enables personalised treatment and the classification of patients into subtypes based on their recommended therapy.

4.6   Implementation

Our method is implemented using R with the Causal Tree model from causalTree package[1]. The baselines are implemented using Python with Logistic Regression, Support Vector Machine and Random Forest models from scikit-learn package[2]. Codes and public datasets are available for public access at GitHub [3].

To support clinical adoption, we developed a web application called *Neoadjuvant Therapy Plan Recommendation*[4] using the Shiny package in R [8]. The app is freely accessible online and designed for ease of use by clinicians, including those without expertise in computer science. It provides a practical interface for applying CTR's personalised treatment recommendations in real-world clinical settings.

5   Conclusion

This study presents a novel framework, Causality-based Therapy Recommendation (CTR), for personalised chemotherapy planning in breast cancer. Unlike existing approaches that rely heavily on empirical guidelines and average treatment effects, CTR leverages patient-specific characteristics and treatment effect heterogeneity to guide decision-making. It then recommends the therapy plan most likely to induce a favourable outcome, specifically, pathological complete response (pCR).

Compared to baseline methods, the proposed CTR framework demonstrates substantial improvements across multiple evaluation metrics, including recovery rate, recovery ratio, and recurrence-free survival probability, when patients follow the recommended therapy plan. These results underscore the value of incorporating causal inference and multi-omics data integration to generate precise, personalised treatment recommendations that go beyond association-based predictions.

---

[1] https://github.com/susanathey/causalTree/.
[2] https://scikit-learn.org/.
[3] https://github.com/vntuyen/ctr.
[4] https://gc8kwh-tuyen-vu.shinyapps.io/CausalR/.

Importantly, CTR is not intended to replace clinicians or biomedical experts, but rather to serve as a clinical decision support tool. By providing interpretable, patient-specific, and evidence-based recommendations, the model enhances the clinician's ability to make informed treatment decisions. CTR represents a shift from traditional population-level treatment guidelines toward a personalised, data-driven paradigm, fully aligned with the objectives of precision oncology.

References

1. Acharjee, A., Kloosterman, B., Visser, R.G., Maliepaard, C.: Integration of multi-omics data for prediction of phenotypic traits using random forest. BMC Bioinform. **17**, 363–373 (2016)
2. Aguadé-Gorgorió, G., Anderson, A.R., Solé, R.: Modeling tumors as complex ecosystems. Iscience **27**(9) (2024)
3. Alaa, A.M., van der Schaar, M.: Limits of estimating heterogeneous treatment effects: guidelines for practical algorithm design. In: International Conference on Machine Learning, pp. 129–138. PMLR (2018)
4. Argelaguet, R., et al.: Multi-omics factor analysis–a framework for unsupervised integration of multi-omics data sets. Mol. Syst. Biol. **14**(6), e8124 (2018)
5. Asselain, B., et al.: Long-term outcomes for neoadjuvant versus adjuvant chemotherapy in early breast cancer: meta-analysis of individual patient data from ten randomised trials. Lancet Oncol. **19**(1), 27–39 (2018)
6. Athey, S., Imbens, G.: Recursive partitioning for heterogeneous causal effects. Proc. Natl. Acad. Sci. **113**(27), 7353–7360 (2016)
7. Breiman, L.: Random forests. Mach. Learn. **45**, 5–32 (2001)
8. Chang, W., Cheng, J., Allaire, J., Xie, Y., McPherson, J.: Shiny: Web Application Framework for R (2023). https://shiny.posit.co/, r package version 1.7.4
9. Ching, T., Himmelstein, D.S., Beaulieu-Jones, B.K., et al.: Opportunities and obstacles for deep learning in biology and medicine. J. R. Soc. Interface **15**(141), 20170387 (2018)
10. Cortes, C.: Support-vector networks. Mach. Learn. **20**, 273–297 (1995)
11. Earl, H.M., et al.: Efficacy of neoadjuvant bevacizumab added to docetaxel followed by fluorouracil, epirubicin, and cyclophosphamide, for women with her2-negative early breast cancer (artemis): an open-label, randomised, phase 3 trial. Lancet Oncol. **16**(6), 656–666 (2015)
12. Friedman, J., Hastie, T., Tibshirani, R.: Regularization paths for generalized linear models via coordinate descent. J. Stat. Softw. **33**(1), 1 (2010)
13. Hanahan, D., Weinberg, R.A.: Hallmarks of cancer: the next generation. Cell **144**(5), 646–674 (2011)
14. Hasin, Y., Seldin, M., Lusis, A.: Multi-omics approaches to disease. Genome Biol. **18**, 1–15 (2017)
15. Hirano, K., Imbens, G.W., Ridder, G.: Efficient estimation of average treatment effects using the estimated propensity score. Econometrica **71**(4), 1161–1189 (2003)
16. Imbens, G.W., Rubin, D.B.: Causal Inference for Statistics, Social, and Biomedical Sciences: An Introduction. Cambridge University Press (2015)
17. Künzel, S.R., Sekhon, J.S., Bickel, P.J., Yu, B.: Metalearners for estimating heterogeneous treatment effects using machine learning. Proc. Natl. Acad. Sci. **116**(10), 4156–4165 (2019)

18. Louizos, C., Shalit, U., Mooij, J.M., Sontag, D., Zemel, R.: Causal effect inference with deep latent-variable models. In: NeurIPS, pp. 6446–6456 (2017)
19. Marusyk, A., Janiszewska, M., Polyak, K.: Intratumor heterogeneity: the rosetta stone of therapy resistance. Cancer Cell **37**(4), 471–484 (2020)
20. NICE: Early and locally advanced breast cancer: diagnosis and management. NICE Guideline, No. 101, National Institute for Health and Care Excellence (NICE), London (2024). https://www.ncbi.nlm.nih.gov/books/NBK519155/
21. Ranstam, J., Cook, J.A.: Lasso regression. J. Br. Surg. **105**(10), 1348 (2018)
22. Reel, P.S., Reel, S., Pearson, E., Trucco, E., Jefferson, E.: Using machine learning approaches for multi-omics data analysis: a review. Biotechnol. Adv. **49**, 107739 (2021)
23. Candido dos Reis, F.J., et al.: An updated predict breast cancer prognostication and treatment benefit prediction model with independent validation. Breast Cancer Res. **19**, 1–13 (2017)
24. Rosenbaum, P.R., Rubin, D.B.: The central role of the propensity score in observational studies for causal effects. Biometrika **70**(1), 41–55 (1983)
25. Rubin, D.B.: Estimating causal effects of treatments in randomized and nonrandomized studies. J. Educ. Psychol. **66**(5), 688 (1974)
26. Saha, A., et al.: Dynamic contrast-enhanced magnetic resonance images of breast cancer patients with tumor locations [data set]. Cancer Imaging Archive **10** (2021)
27. Sammut, S.J., et al.: Multi-omic machine learning predictor of breast cancer therapy response. Nature **601**(7894), 623–629 (2022)
28. Subramanian, I., Verma, S., Kumar, S., Jere, A., Anamika, K.: Multi-omics data integration, interpretation, and its application. Bioinform. Biol. Insights **14**, 1177932219899051 (2020)
29. Symmans, W.F., et al.: Measurement of residual breast cancer burden to predict survival after neoadjuvant chemotherapy. J. Clin. Oncol. **25**(28), 4414–4422 (2007)
30. Symmans, W.F., et al.: Long-term prognostic risk after neoadjuvant chemotherapy associated with residual cancer burden and breast cancer subtype. J. Clin. Oncol. **35**(10), 1049–1060 (2017)
31. Wager, S., Athey, S.: Estimation and inference of heterogeneous treatment effects using random forests. J. Am. Stat. Assoc. **113**(523), 1228–1242 (2018)
32. Zhang, W., Le, T.D., Liu, L., Zhou, Z.H., Li, J.: Mining heterogeneous causal effects for personalized cancer treatment. Bioinformatics **33**(15), 2372–2378 (2017)
33. Kühne, F., et al.: Causal evidence in health decision making: methodological approaches of causal inference and health decision science. GMS Ger. Med. Sci. **20**, Doc12 (2022)

Machine Learning for Traffic Accident Prediction: Integrating Spatial, Temporal, and Behavioral Data for Road Safety Insights

Sudarat Sukjaroen[1], Xiaodan Dong[1]([✉]), S. T. Boris Choy[2], and Weidong Huang[1]

[1] University of Technology Sydney, Ultimo, Australia
{Sudarat.Sukjaroen,Xiaodan.dong,Weidong.Huang}@uts.edu.au
[2] University of Sydney, Camperdown, Australia
Boris.Choy@sydney.edu.au

Abstract. Road safety remains a critical global issue, particularly in densely populated urban environments where traffic dynamics are highly complex. Predicting and preventing accidents is challenging due to the interplay of multiple contributing factors and the need to integrate diverse data sources such as spatial, temporal, and real-time traffic information.

This study presents an integrated framework for jointly predicting road accident rates and fatalities by integrating 37 spatial, temporal, behavioral, and accident-specific dimensions through a combination of machine learning and statistical models. To identify the key determinants of accident frequency and severity, we incorporated geographic mapping, weather, and contextual variables into clustering techniques and a Decision Tree Regressor. This approach enabled the identification of high-risk areas, critical risk factors, and patterns underlying accident occurrence, providing data-driven insights for targeted interventions.

An empirical analysis using a real Bangkok dataset of 10,366 accident records demonstrates that combining machine learning methods with spatial and behavioral features derived from statistical models significantly improves predictive accuracy. The proposed framework shows broad applicability across national contexts and offers valuable evidence to design targeted accident reduction strategies. By integrating predictive analytics with geographic mapping, this study underscores the importance of data-driven approaches in advancing road safety and demonstrates clear application value for policy formulation, aligning with the priorities of the World Health Organization for traffic safety.

Keywords: Machine Learning · Accident Prevention · Clustering

1 Introduction

As of 15 April 2024, the World Health Organization (WHO) highlighted Thailand's pressing road safety challenges in its latest data. Thailand ranks 16th globally for road traffic fatalities, with a rate of 25.4 deaths per 100,000 people among 197 countries [17] (World Health Organization, 2024). Within Southeast Asia, Thailand holds the second-highest rate at 28.2 deaths per 100,000, second only to Nepal. In 2021, Bangkok reported 117,362 road accidents, accounting for 17% of the 681,937 cases nationwide [3] (Bangkok Post, 2022).

This disproportionate share emphasises the capital's critical road safety challenges. The Thai government has prioritised road safety in response to these alarming statistics. It has compiled a comprehensive car accident dataset covering 2019 to the most recent year through the Office of the Permanent Secretary, Ministry of Transport. This publicly accessible dataset is a valuable resource for academic research, enabling data-driven strategies to reduce car accident rates and enhance road safety in Bangkok [15] (Transport Accident Management Systems, 2024).

Existing research has mostly studied road accident prediction as a classification problem, which aims to predict whether a traffic accident may happen in the future or not without exploring the underneath relationships between the complicated factors [2]. Previous research has highlighted the influence of naturalistic environmental factors, including road type, lighting conditions, wind direction, and weather, on amplifying the vulnerability of individuals to serious or fatal road injuries [11]. Berhanu et al. [4] conducted a comparative analysis of car accident prediction techniques in low-income and high-income countries, reviewing existing literature on this global challenge to identify strategies for improving road safety. In recent related work, Ahmed et al. [2] evaluated a set of machine learning (ML) models to predict road accident severity based on the most recent NZ road accident dataset. Despite valuable insights, the study did not integrate spatial and temporal factors with behavioral variables, leaving the root causes of accidents along specific routes insufficiently understood for policy formulation. In addition, the critical link between accident occurrence and death rates remains unexplored, highlighting a gap that this research seeks to address.

In this context, we present a new framework that jointly analyzes accident and fatality rates, demonstrated using Bangkok's data. By integrating 37-dimensional data from diverse sources, the framework offers a comprehensive view of accident patterns and supports the development of targeted, data-driven strategies to improve urban road safety. Additionally, we combine clusters derived from statistical models with spatial information to pinpoint high-risk accident hotspots, and incorporate these as features in machine learning models for improved prediction accuracy. While this study focuses on Bangkok as a case study, the proposed methodological framework which combines traffic volume clustering, spatial analytics, and predictive modeling, can be readily adapted and applied to other urban centers facing similar road safety challenges. This generalizable approach supports evidence-based policy development and the design of proactive, location-specific road safety strategies.

2   Related Work

Road traffic accidents remain a critical global issue, with extensive research focusing on accident prediction, prevention, and safety improvements. This section highlights key contributions from the literature and positions our research within this context, specifically focusing on predictive analysis and pattern recognition of car accidents across diverse datasets and settings.

2.1   Accident Prediction and Safety Improvements

Several studies demonstrate the efficacy of machine learning for predicting road accidents and enhancing safety outcomes. Machine learning models, including decision trees, random forest, multinomial logistic regression, and naïve Bayes, have shown strong predictive performance while providing insights into accident patterns [12] (Pourroostaei Ardakani et al., 2023). A multimodal car crash detection system incorporating deep learning techniques such as gated recurrent units (GRU) and convolutional neural networks (CNN) significantly outperformed single-modal systems, showcasing the potential of integrating video and audio processing for improved accident recognition and response [5] (Choi et al., 2021). Similarly, traffic platoon data from floating cars enabled the prediction of crashes on Shanghai's Middle Ring Expressway, where a support vector machine achieved 85% accuracy, outperforming the 60% accuracy of a binary logistic model [16] (Wang et al., 2019).

2.2   Crash Severity and Contributing Factors

Studies on crash severity provide insights into the factors influencing road accidents. Clustering analyses of bus crash data revealed that weekend incidents, non-motor vehicle crashes, and poor lighting in high-speed zones significantly increase fatality rates [13] (Samerei et al., 2021). Machine learning models, such as XGBoost and logistic regression, effectively predicted injury severity in car-to-car crashes using predictors like Delta-V, age, and the Principal Direction of Force [8] (Kong et al., 2023). In Malaysia, motorcycle fatalities were predominant, particularly on rural roads, primary routes, and straight sections, with motorcyclists aged 16–20 being most vulnerable due to improper licensing [1] (Manan & Várhelyi, 2012). Similarly, hotspot analyses in Kigali identified heightened risks for vulnerable road users, including pedestrians and motorcyclists, along major roads and downtown areas [10] (Patel et al., 2016). These findings underscore the importance of targeted interventions for high-risk groups.

2.3   Willingness to Pay (WTP) for Road Safety

Research into willingness to pay (WTP) for road safety improvements identified significant psychological and behavioural factors influencing personal car drivers' willingness to invest in reducing fatalities and injuries. Using structural equation modelling (SEM), these studies offer a framework for evaluating economic interventions to enhance road safety in various contexts [7] (Jomnonkwao et al., 2021).

2.4 Patterns and Risk Factors in Specific Scenarios

Research on single-vehicle crashes in Taiwan revealed that alcohol consumption, night driving, and specific road types significantly influence survival risk for car and motorcycle drivers [6] (Huang & Lai, 2011). Similarly, an analysis of car accident patients at Shahid Rajaee Hospital identified age, injury severity, and injury regions as significant predictors of in-hospital mortality [18] (Yadollahi et al., 2017). Additional findings from critical pre-crash scenarios at T-junctions and crossroads, derived using association rule mining and clustering, provided benchmarks for improving safety in complex traffic conditions [9] (Nitsche et al., 2017).

2.5 Contributions and Generalisation of Our Research

Our research builds upon these studies by addressing fatality prediction across various vehicle types and accident scenarios specific to Bangkok. Unlike studies in Victoria and Taiwan, which focused on particular vehicle types or circumstances [13] & [6] (Samerei et al., 2021; Huang & Lai, 2011), our dataset includes 12 vehicle categories, such as motorcycles, personal cars, and cargo pickups, enabling a more comprehensive understanding of road safety. Inspired by hotspot analyses in Kigali and Malaysia [10] & [1] (Patel et al., 2016; Manan & Várhelyi, 2012), we integrate spatial and contextual factors into our models to identify critical high-risk areas and conditions. We improve fatality models' predictive accuracy and reliability by using advanced machine learning techniques, as evidenced by [12] (Pourroostaei Ardakani et al., 2023) and [8] (Kong et al., 2023) in the United Kingdom and Korea.

By identifying key factors underlying car accidents and fatalities and employing geographic hotspot mapping, this research provides actionable insights to guide tailored interventions and policy recommendations. The proposed framework is scalable and can be generalized to support road safety mitigation strategies across diverse urban contexts and regions.

3 The Data

The primary dataset used for this study on road accidents in Bangkok, Thailand, spans from 1 January 2019 to 31 July 2024 and is sourced from the Thailand Transport Accident Management Systems. It includes 10,366 records, of which 9,729 contain complete information and were selected for analysis.

The dataset is structured into 37 columns and categorised into 4 primary groups for the comprehensive examination of accident characteristics:

- Accident Details: Basic accident information, including Date of Accident, Time of Accident, Report Date, and Report Time, providing the temporal context of incidents.
- Location and Route Information: Variables such as Agency, Agency Route, Route Code, Route Name, Kilometre Marker, Province, Latitude, and Longitude detail the geographical and route-specific aspects of each accident.
- Vehicle and Person Involvement: Counts for vehicle types like Number of Motorcycle, Number of Personal Car, and Number of Bus, as well as totals for Total Vehicle and Total Vehicle and Person, represent the extent of vehicle and human involvement.
- Casualty Information: Severity of outcomes, including Number of Fatality, Number of Severe Injury, Number of Minor Injury, and Total Injury, provides insights into accident consequences.

Additional support data on weather condition and traffic volume was obtained from the Traffic and Transportation Department [14] (Traffic and Transportation Department, 2024) to complement the primary accident dataset. This supplementary dataset contains traffic flow data collected from 387 monitoring positions across intersections and roadways in Bangkok. Key variables include:

- Date: The specific date of data collection, enabling temporal alignment with accident data for accurate correlation.
- Crossroads and Road: Identifiers for monitored intersections and roads, providing spatial references to analyse critical locations.
- Traffic by Hour: Hourly traffic volume measurements offer a granular view of traffic density and flow patterns throughout the day. This aids in identifying peak traffic hours associated with increased accident risks.
- Latitude and Longitude: Geographical coordinates of monitoring positions, allowing integration with accident location data to identify relationships between traffic volume and accident hotspots.

4   Methods

This research incorporated spatial, temporal, and traffic-related factors into machine learning and statistical models to examine road accident patterns and predict fatalities, utilising datasets on road accidents and traffic flow in Bangkok. This includes K-Means Clustering, Generalised Linear Model (GLM) and six supervised learning algorithms to predict accident and fatality rates.

- Gradient Boosting Regressor and XGBoost Regressor effectively captured non-linear relationships using ensemble methods with regularisation.
- LightGBM Regressor employed a gradient boosting framework with histogram-based learning, enabling efficient handling of large-scale data with improved speed and accuracy.

- Linear Regression served as a baseline model, providing insights into linear relationships.
- Random Forest Regressor utilised ensemble learning to manage complex data interactions while reducing variance and enhancing robustness.
- Neural Networks modelled intricate, non-linear relationships using deep learning techniques, offering flexibility for diverse data structures.
- Support Vector Regressor (SVR) identified hyperplanes for regression tasks, excelling at capturing smooth, complex relationships.

All models were optimized through Bayesian hyperparameter tuning and evaluated for predictive accuracy. This tuning was performed using a probabilistic search strategy with cross-validation, allowing efficient exploration of the defined parameter ranges and identification of the best performing configuration. For example, the search included parameters such as the regularization strength (0.1–10) for logistic regression, the learning rate (0.01–0.2) and the maximum depth (3–10) for XGBoost, and the learning rate (0.01–0.1) and max depth (3–7) for gradient boost, among others. In addition, pattern Recognition with Decision Trees: A Decision Tree Regressor was employed to predict accident rates across geographic clusters. Key features included geographic attributes (latitude, longitude, traffic clusters), accident characteristics, environmental factors, and temporal features.

The model was optimised using grid search and cross-validation, with performance evaluated using metrics such as Mean Squared Error (MSE) and Mean Absolute Error (MAE). In addition, Area Under the Curve (AUC) and Receiver Operating Characteristic (ROC) analysis were incorporated to assess the model's classification performance in distinguishing between fatal and non-fatal accidents.

Visualising the decision tree provided interpretability, identifying location-specific risk factors influencing accident rates. This multi-faceted analytical approach offered a comprehensive understanding of accident patterns, enabling data-driven recommendations for improving road safety.

5   Results

Accidents and fatalities in Thailand show distinct temporal patterns and spikes during significant holiday periods, particularly the New Year holidays (December and January) and the Songkran festival in April. These extended holiday weeks have heightened travel and social gatherings, increasing accident rates. A noticeable monthly trend emerges, with a rise in accidents on the first day of each month, suggesting patterns linked to specific dates. Similarly, January, December, and April register the highest accident counts, aligning with holiday travel periods.

Regarding weekdays, while general accidents in Thailand peak on weekends (Saturday and Sunday), Bangkok sees its highest accident rates on Friday and Saturday, reflecting the city's busy weekend transitions. Fatal accidents are notably higher on Mondays nationwide and in Bangkok, indicating a contrast between general and fatality risk factors. Accident patterns by time of day reveal that non-fatal incidents occur primarily during afternoon peak hours (4–6 pm) when people commute home. In contrast, fatal accidents peak between 6–8 pm, highlighting elevated risks during evening hours due to low-light conditions.

Vehicle involvement data shows that cargo pickup trucks dominate accident statistics nationwide. However, in Bangkok, personal and public cars account for most accidents, reflecting the city's high population density and dependence on private and public transport. Despite this, motorcycles remain the most vulnerable vehicle type, contributing to the highest fatality rates. Fatalities are most likely to occur when exactly two vehicles are involved in an accident, suggesting that multi-vehicle interactions heighten the severity of outcomes.

These findings underscore the need for targeted road safety interventions and public awareness campaigns during high-risk periods such as holiday seasons, evening hours, and specific weekdays. Focused measures to improve safety for motorcyclists, enhance visibility in low-light conditions, and manage increased travel during holidays can help mitigate accidents and fatalities, as shown in the examples in Fig. 1a, Fig. 1b.

5.1 Correlation Matrix Heatmap

The correlation matrix analysis revealed three significant sets of relationships. The first set highlights the connections between NoOfFatality, NoOfMotorcycle, and NoOfPedestrian, emphasising their critical role in fatal accidents. Motorcycles, due to their vulnerability and pedestrian involvement, contribute significantly to fatality counts. The second set focuses on the association between NoOfMotorcycle and both NoOfMinorInjury and TotalInjury, underscoring motorcycles' substantial role in accidents that result in injuries. The positive correlation suggests that the likelihood of minor and total injuries increases as motorcycle involvement rises.

The final set identifies meaningful relationships involving TotalVehicle. A strong positive correlation with NoOfPersonalCar ($r = 0.623$) indicates that personal cars are a major contributor to total vehicle involvement in accidents. A moderate correlation with NoOfCargoPickupTruck ($r = 0.396$) highlights the notable role of cargo pickup trucks in accident occurrences. A weak positive correlation with TotalInjury ($r = 0.100$) suggests a slight increase in injuries with higher total vehicle counts. However, the correlation with HourAccident ($r = 0.097$) is minimal, indicating a negligible relationship between the number of vehicles and the time of the accident.

5.2 Clustering Analysis Results

The clustering analysis grouped accident locations in Bangkok into ten traffic volume clusters, ranked from 1 (lowest) to 10 (highest) based on total traffic volume. Each cluster reflects a distinct range of traffic density, offering insights

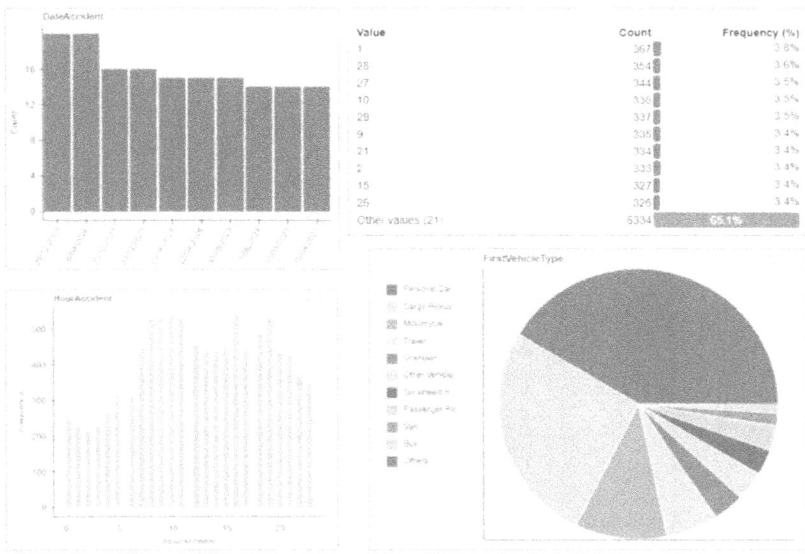

(a) EDA: Date, Day, Hour, First Vehicle Type

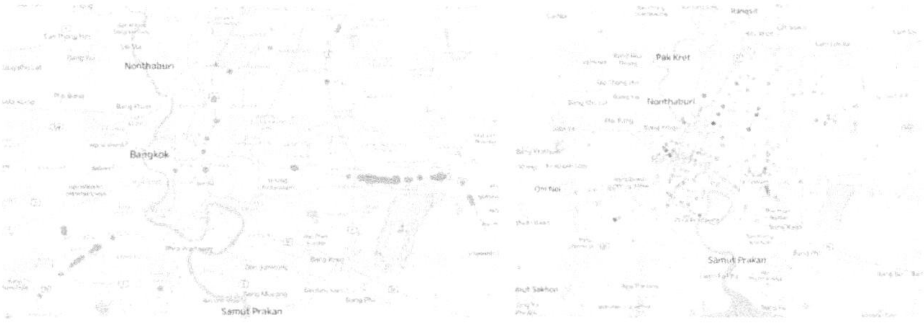

(b) Bangkok Accident Hot Spot and Traffic Volume

Fig. 1. EDA and Bangkok Accident Hot Spot Analysis.

into the distribution of accidents across varying traffic conditions. The result of the analysis is provided in Table 1.

5.3   Accident Rate Prediction Result

The accident rate analysis identified six specific conditions where the predicted rate exceeds 0.000350, with an average rate of 0.000418. These criteria offer actionable insights into accident hotspots, contributing factors, and targeted mitigation strategies.

Table 1. Traffic Cluster Statistics: Min, Max, and Count

Traffic Cluster	Min	Max	textbiCount
1	3,029	10,457	2,048
2	10,637	17,795	1,935
3	18,251	24,021	817
4	25,279	32,267	1,013
5	35,705	44,771	2,050
6	46,554	64,190	1,261
7	66,444	81,795	308
8	92,869	103,788	133
9	128,216	128,216	94
10	147,393	147,393	70

The Highest Accident Rate Spot. We obtained the following information into factors contributing to car accidents from the models. The criteria identify specific conditions under which accidents are likely to occur, particularly in areas with low traffic volumes (TrafficCluster $\leq$ 2). These indicate regions with less congestion but potentially higher risks due to other contributing factors. These incidents are explicitly associated with the Dao Khanong - Samae Dam route (RouteName Dao Khanong - Samae Dam $>$ 0), identifying this route as a notable accident hotspot. Temporally, accidents are more prevalent during the first four months of the year (MonthNumberAccident $\leq$ 4), with a particular concentration in the morning hours (HourAccident $\leq$ 10).

The analysis reveals a distinct accident profile where speeding is the primary presumed cause. These incidents predominantly occur on straight roads with no slope, where the simplicity of the road layout may encourage higher speeds and reduce drivers' reaction times. The most common accident type is rear-end collisions, highlighting the dangers of insufficient stopping distances and driver inattention. Personal or public cars are the primary vehicle types involved, reflecting their widespread use and their increased likelihood of contributing to such collisions. The hotspot spatial analysis, highlighting key contributing factors, is presented in Fig. 2.

Suggestions for Mitigation

- Implement speed enforcement measures such as speed cameras and monitoring devices on straight roads with no slope to deter excessive speeding.
- Improve road signage and traffic management strategies, particularly during the morning hours, to reduce accidents associated with low traffic volumes.
- Increase public awareness campaigns promoting safe driving practices, focusing on maintaining safe stopping distances and addressing driver inattention.

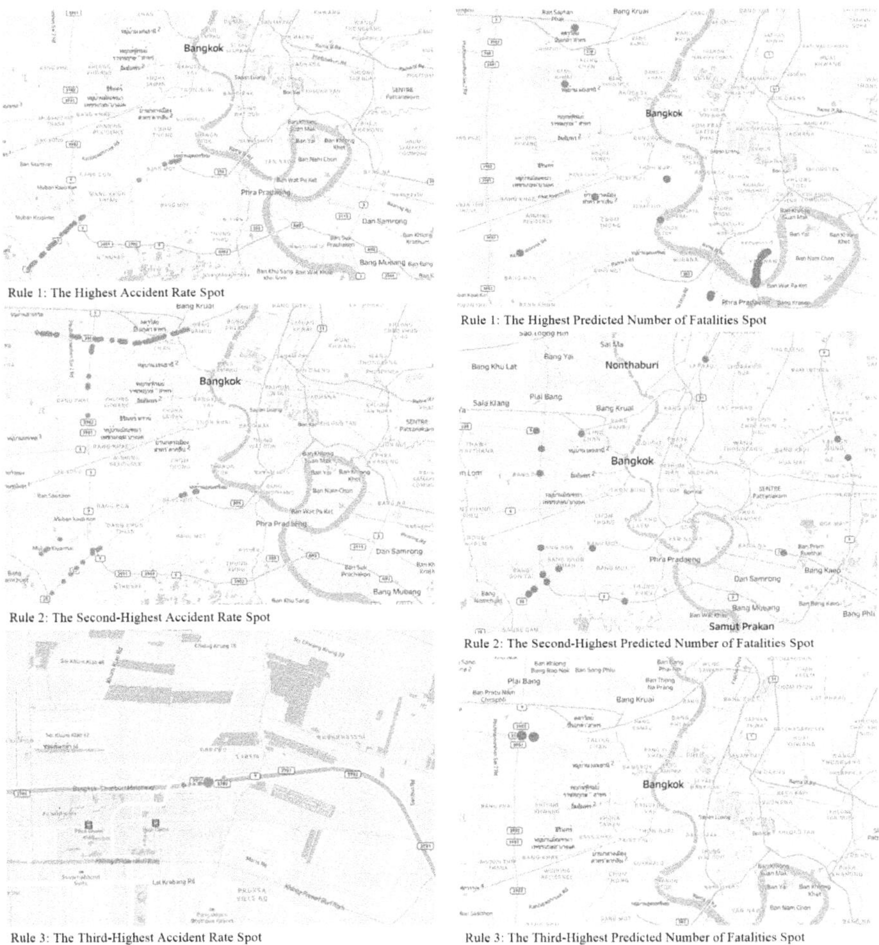

Fig. 2. The Predicted Accident Rate and Number of Fatalities Spots.

5.4   Modeling Results

The evaluation of multiple machine learning models using the test dataset, which includes spatial, behavioural, temporal, and vehicle-related data, reveals that the Gradient Boosting model consistently outperforms others across key performance metrics. It achieves the lowest Mean Squared Error (MSE) at 0.0257, the highest Area Under the Curve (AUC) at 0.8184, indicating strong regression and classification performance. Other ensemble methods such as LightGBM, XGBoost, and Random Forest also perform well, though slightly behind Gradient Boosting. Simpler models like Linear Regression and SVR show weaker performance, with SVR, in particular, yielding the highest error values. The result is represented in Table 2 and Fig. 3.

Table 2. Model Performance Comparison

Model	MSE	AUC	MAE
Gradient Boosting	0.0257	0.8184	0.0354
LightGBM	0.0262	0.7969	0.0428
Random Forest	0.0269	0.7546	0.0328
Linear Regression	0.0274	0.7975	0.0485
XGBoost	0.0279	0.8176	0.0357
Neural Network	0.0343	0.7110	0.0412
SVR	0.0386	0.7374	0.1231

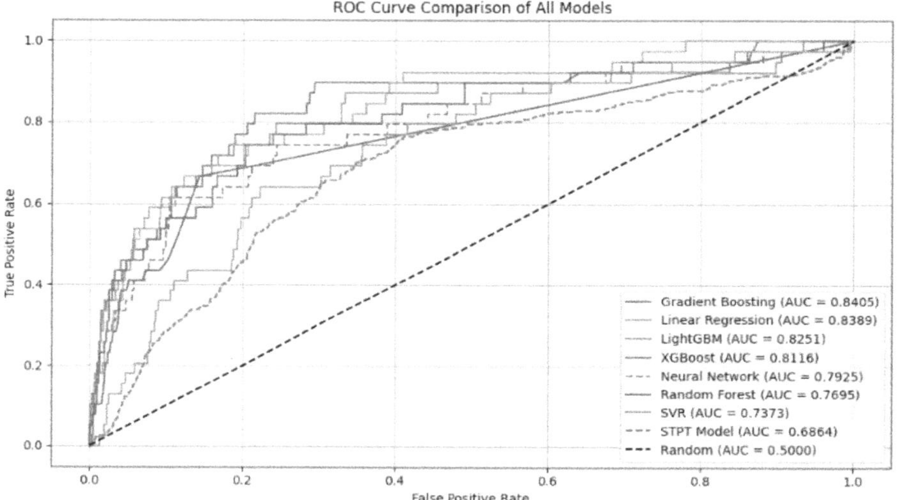

Fig. 3. ROC Curve Comparison.

To assess the contribution of spatial and behavioural data to model performance, we removed a set of features that encode location-specific and human/contextual information. The spatial features removed include Latitude, Longitude, KilometerMarker, AccidentLocation, TrafficCluster, and Agency, all of which capture geographic positioning, traffic regions, and road management context. The behavioural features removed are PresumedCause, AccidentCharacteristics, WeatherCondition, and FirstVehicleType, as they reflect driver behaviour, collision context, and environmental conditions.

The results demonstrate that integrating spatial and behavioural features significantly improves predictive accuracy, especially for advanced models like Gradient Boosting, as demonstrated in Table 3. When these features were removed, performance declined: AUC dropped from 0.8184 to 0.7876, and residual square fell from 0.0522 to 0.0444. This indicates the reduced ability to distinguish fatal from non-fatal cases and explain outcome variance. Overall, the findings confirm the added value of spatial and behavioural data in enhancing model precision and reliability.

Table 3. Model Performance With and Without Spatial and Behavioral Features

Metric	With Features	Without Features	Change
AUC	0.8184	0.7876	0.0308
Residual^2	0.0522	0.0444	0.0078

6   Discussion

Given the diversity of data related to road accidents and the resulting fatalities, a comprehensive understanding of their root causes is essential. To provide policymakers with a clear and interpretable view, we employed a clustering approach integrated with decision trees, enabling effective visualization and explanation of complex patterns. While other advanced deep learning methods, such as spatial-temporal neural networks and transformer-based models, may achieve higher predictive accuracy due to the richness of the data, they often lack interpretability, making it challenging for policymakers to extract actionable insights. This highlights an inherent trade-off between maximizing prediction accuracy and providing clear, explainable outputs that can inform policy decisions.

The adaptable framework can be customised to cities or regions with varying traffic conditions. While regions with real-time traffic surveillance systems can enhance predictive accuracy with dynamic data, areas with limited resources can still benefit by incorporating key variables like traffic volume, vehicle types, and accident severity.

The study demonstrates the potential benefits of integrating spatial and behavioral data into predictive models for road safety. Machine learning techniques were effective in identifying patterns and risk factors associated with increased risk. However, the integration of real-time traffic camera image data presents challenges, particularly in converting large volumes of visual data into actionable insights within a short timeframe. Additionally, alternative modeling approaches, such as Spatio-Temporal Neural Networks, may offer improved capabilities for handling multiple data types. Future research should investigate and compare these methods to evaluate their effectiveness relative to the current approach.

7   Conclusions

This study introduces a new framework that jointly examines road accident occurrences and fatality patterns by integrating 37 spatial, temporal, behavioral, and accident-specific features. By analyzing accidents and fatalities together, the framework provides a comprehensive understanding of the underlying risk factors and their interactions, which may be overlooked when studied separately. The approach enables the identification of high-risk scenarios and targeted interventions, offering actionable data-driven recommendations to policymakers.

By integrating statistical models into machine learning frameworks with geographic mapping, this study identified critical accident hotspots and fatality risks across diverse traffic conditions. The framework can also be applied to simulate traffic scenarios under varying weather events. Overall, the proposed approach combines predictive accuracy with interpretability, contributing to road safety research and supporting evidence-based policy formulation.

References

1. Abdul Manan, M.M., Várhelyi, A.: Motorcycle fatalities in Malaysia. IATSS Res. **36**(1), 30–39 (2012)
2. Ahmed, S., Hossain, M.A., Ray, S.K., Bhuiyan, M.M.I., Sabuj, S.R.: A study on road accident prediction and contributing factors using explainable machine learning models: analysis and performance. Transp. Res. Interdisc. Perspect. **19**, 100814 (2023)
3. Bangkok Post: It's not just motorists at fault (2022). https://www.bangkokpost.com/thailand/general/2259819/its-not-just-motorists-at-fault. Accessed 6 Feb 2022
4. Berhanu, Y., Alemayehu, E., Schröder, D.: Examining car accident prediction techniques and road traffic congestion: a comparative analysis of road safety and prevention of world challenges in low-income and high-income countries. J. Adv. Transp. (2023)
5. Choi, J.G., Kong, C.W., Kim, G., Lim, S.: Car crash detection using ensemble deep learning and multimodal data from dashboard cameras. Expert Syst. Appl. **183**, 115400 (2021)
6. Huang, W.S., Lai, C.H.: Survival risk factors for fatal injured car and motorcycle drivers in single alcohol-related and alcohol-unrelated vehicle crashes. J. Safety Res. **42**(2), 93–99 (2011)
7. Jomnonkwao, S., Wisutwattanasak, P., Ratanavaraha, V., Mosa, A.M.: Factors influencing willingness to pay for accident risk reduction among personal car drivers in Thailand. PLoS ONE **16**(11), e0260666 (2021)
8. Kong, J.S., et al.: Machine learning-based injury severity prediction of level 1 trauma center enrolled patients associated with car-to-car crashes in korea. Comput. Biol. Med. **153**, 106393 (2023)
9. Nitsche, P., Thomas, P., Stuetz, R., Welsh, R.: Pre-crash scenarios at road junctions: a clustering method for car crash data. Accid. Anal. Prev. **107**, 137–151 (2017)
10. Patel, A., Krebs, E., Andrade, L., Rulisa, S., Vissoci, J.R.N., Staton, C.A.: The epidemiology of road traffic injury hotspots in Kigali, Rwanda from police data. BMC Public Health **16**(1), 697–697 (2016)
11. Pervez, A., Huang, H., Lee, J., Han, C., Li, Y., Zhai, X.: Factors affecting injury severity of crashes in freeway tunnel groups: a random parameter approach. J. Transp. Eng. Part A Syst. **148**(4) (2022). https://doi.org/10.1061/JTEPBS.0000617

12. Pourroostaei Ardakani, S., et al.: Road car accident prediction using a machine-learning-enabled data analysis. Sustainability **15**(7), 5939 (2023)
13. Samerei, S.A., Aghabayk, K., Mohammadi, A., Shiwakoti, N.: Data mining approach to model bus crash severity in Australia. J. Safety Res. **76**, 73–82 (2021)
14. Traffic and Transportation Department: Traffic flow data (2024). https://data.go.th/dataset/traffic_volume
15. Transport Accident Management Systems: Road accident data (2024). https://data.go.th/en/dataset/gdpublish-roadaccident, retrieved from Government Data Thailand
16. Wang, J., Luo, T., Fu, T.: Crash prediction based on traffic platoon characteristics using floating car trajectory data and the machine learning approach. Accid. Anal. Prev. **133**, 105320–105320 (2019)
17. World Health Organization: Estimated road traffic death rate (per 100,000 population) (2024). https://www.who.int/data/gho/data/indicators/indicator-details/GHO/estimated-road-traffic-death-rate-(per-100-000-population). Accessed 15 Apr 2024
18. Yadollahi, M., Ghiassee, A., Anvar, M., Ghaem, H., Farahmand, M.: Analysis of shahid rajaee hospital administrative data on injuries resulting from car accidents in Shiraz, Iran: 2011–2014 data. Chin. J. Traumatol. **20**(1), 27–33 (2017). https://doi.org/10.1016/j.cjtee.2015.10.006

Visionary: Enhancing Visual Context for the Visually Impaired

Pranav Powar[1]([✉]) [iD] and Amit Agarwal[2] [iD]

[1] Thapar Institute of Engineering and Technology, Patiala, India
ppowar_be21@thapar.edu
[2] Indian Institute of Technology, Roorkee, Roorkee, India
aagarwal3@cs.iitr.ac.in

Abstract. The growing reliance on technology poses challenges for visually impaired individuals, especially in receiving accurate and detailed descriptions of their surroundings. Existing models like BLIP and CLIP often fall short in providing rich context. To address this, we introduce Visionary, a mobile app that enhances image descriptions by combining BLIP's precise image recognition with LLaMA 3's language processing to add depth and clarity. With a voice-navigated interface and full-screen camera view, Visionary improves the accuracy and contextual relevance of descriptions, promoting greater independence for visually impaired users. Initial evaluations show notable improvements in situational awareness. Beyond simple object detection, Visionary emphasizes capturing the subtle details that make a scene understandable–such as spatial relationships, emotional cues, and fine-grained contextual elements that are often overlooked by existing tools. This ensures that users not only know what is present in their environment but also how different elements interact with one another. By transforming raw recognition into meaningful narratives, the app bridges the gap between vision and comprehension. Its design focuses on accessibility, ensuring that visually impaired users can interact seamlessly without additional complexity. Ultimately, Visionary represents a step toward inclusive technology that empowers individuals to navigate daily life with greater confidence, autonomy, and awareness.

Keywords: Visually Impaired · Multimodal Integration · Assistive AI

1 Introduction

The rapid advancement of AI has led to innovative tools for visually impaired individuals, but obtaining detailed, context-rich image descriptions remains challenging. Models like BLIP [5] and CLIP [13] often lack the necessary depth and context in their outputs. To address this, we developed **Visionary**, a mobile app that integrates BLIP's image recognition with **LLaMA 3's** [10] advanced language processing (Meta AI, 2024) to provide more informative and contextually rich descriptions. Building on prior work like BLIP by Li et al. (2022)

Q. V. Nguyen et al. (Eds.): AusDM 2025, CCIS 2765, pp. 426–435, 2026.
https://doi.org/10.1007/978-981-95-6786-7_29

and LLaMA 3, Visionary enhances situational awareness and independence for visually impaired users.

Despite the rapid progress in multimodal modeling, most existing captioning systems prioritize surface-level object correctness rather than delivering descriptions that are actionable in real-world contexts. Visionary inverts this focus by explicitly separating visual grounding from linguistic enrichment. BLIP produces stable representations of detected entities, and LLaMA 3 then transforms these into coherent narratives that capture spatial relationships, affordances, and situational cues–such as whether a doorway is accessible, or if an obstacle lies in the user's path. This design ensures not only accuracy but also functional utility for navigation and decision-making.

The application has been engineered for practical deployment. Features such as a full-screen camera, intuitive voice-first interaction, and low-latency text-to-speech output reduce user friction and make the system suitable for everyday use. Evaluation on benchmark datasets demonstrates measurable improvements in precision and contextual depth compared to BLIP-only or language-only approaches. Early formative studies further suggest that the added context supports better moment-to-moment awareness in real settings.

The key contributions of this paper are as follows:

1. **A modular architecture** that combines BLIP's vision grounding with LLaMA 3's linguistic refinement to generate context-aware captions.
2. **A mobile accessibility interface** specifically optimized for speed, simplicity, and hands-free control to meet the needs of visually impaired users.
3. **Empirical validation** through benchmark evaluations and formative user studies, showing that contextual enhancement significantly improves the relevance and usability of generated descriptions.

To provide a comprehensive overview, the remainder of this paper is organized as follows: Sect. 2 reviews related research in vision-language models and assistive technologies. Section 3 presents the proposed Visionary architecture, detailing the integration of BLIP and LLaMA 3. Section 4 discusses experimental results and performance evaluation. Finally, Sect. 5 concludes the paper and outlines future directions.

2 Background and Related Work

Early vision-and-language models established the foundation for translating visual content into natural language. OSCAR (Object-Semantics Aligned Pre-training) [6] introduced the idea of using detected object tags as anchor points to better align image regions with textual descriptions, significantly improving image captioning and VQA performance. Building on this, VinVL [19] (Visual Vocabulary) improved the underlying object detector to produce richer visual features; feeding these enhanced features into an OSCAR-style model yielded state-of-the-art results across multiple benchmarks. Another line of work, CLIP (Contrastive Language-Image Pre-training) [13], took a different approach by training on 400 million image-text pairs to learn a joint embedding space. CLIP demonstrated unprecedented zero-shot recognition capabilities after learning to match captions with images, it could recognize new image categories based solely on their textual description. More recent models have pursued unified vision-language generation. For example, BLIP (Bootstrapping Language-Image Pre-training) [5] uses a multimodal encoder-decoder and a captioning bootstrapping strategy (generating synthetic captions and filtering out noise) to excel in both understanding and generation tasks. These models markedly advanced automatic image description; however, their captions often remain literal and lack deeper context or nuance beyond identifying salient objects.

Parallel progress in natural language processing has produced powerful language models now being leveraged for multimodal generation. The evolution from GPT-3 [3] to the LLaMA series [10,15,16] exemplifies how scaling model size and training data leads to more fluent and context-aware text generation. GPT-3's 175B-parameter transformer showed that large models can perform open-ended description and reasoning with minimal prompting. Meta AI's LLaMA models (2023) demonstrated that even smaller open models (765B parameters) can achieve comparable language prowess; notably, the 13B LLaMA-1 outperformed the much larger GPT-3 on many benchmarks, and the 65B version rivaled other state-of-the-art models. LLaMA 2 (2023) expanded on this by training on 40% more data and extending context lengths, while also making the models widely available for research and commercial use. Most recently, LLaMA 3 (2024) pushed the frontier further: it was pre-trained on roughly 15 trillion tokens, yielding a 70B-parameter model that surpassed other contemporary LLMs on various language tasks. Importantly, Meta announced plans for LLaMA 3 to support multilingual and multimodal capabilities and larger context windows, signaling the convergence of vision and language modeling. Indeed, emerging research systems like Flamingo [1] and LLaVA [9] illustrate the promise of connecting visual encoders to LLMs for rich image-to-text generation, inspiring our approach of using BLIP with LLaMA for enhanced descriptions.

In the assistive technology domain, several tools have applied vision-language models to improve accessibility. Microsoft Seeing AI (2017) is a pioneering smartphone app that audibly narrates the world for blind and low-vision users. It can recognize faces, read text, identify products, and describe scenes, functioning as a "talking camera." Seeing AI operates partly on-device but relies on cloud vision services for complex tasks like scene descriptions or handwriting recognition, which can limit response time and offline use. Google Lookout (2019) offers a similar AI assistant on Android, with modes for reading documents, recognizing objects, and exploring one's surroundings. Using onboard computer vision, Lookout provides real-time identification of environmental objects and text. Recent updates integrate generative AI: for instance, users in some regions can submit an image to Lookout and ask follow-up questions, receiving AI-powered descriptive answers in natural language. This begins to imbue the experience with conversational context, rather than just static labels. Another notable service, Be My Eyes, originally connected visually impaired users to sighted volunteers via live video. In 2023 it introduced "Be My AI," a GPT-4 powered virtual vision assistant within the app. GPT-4's vision capabilities allow it to not only enumerate objects in a photo, but also interpret and reason about the scene at a high level. For example, given an image of a refrigerator, the AI can identify the food items and suggest recipes using those ingredients. This conversational AI assistant can analyze images in depth and answer context-specific questions, approaching the level of understanding a human volunteer might provide.

Despite these advances, prior systems still face limitations in delivering rich contextual understanding in real time. Traditional image captioning models (e.g. VinVL [19], BLIP [5]) tend to produce generalized descriptions focusing on visible objects, without reasoning about finer contextual cues like spatial relationships, activities, or unusual circumstances. As a result, a caption might say "A man walking a dog on a street" but omit that the man is approaching a busy intersection or that it is night time details crucial for a visually impaired user's situational awareness. Current smartphone apps for accessibility likewise often provide piecemeal information. Seeing AI and Lookout, for instance, return one-off descriptions or identification results; they do not generate a narrative accounting for the broader scene and may require the user to manually switch between modes for different types of information.

The newest AI-driven solutions such as Be My Eyes' GPT-4 assistant demonstrate the value of deeper contextual reasoning. GPT-4 can engage in dialogue about an image and infer high-level context (like warning if an object on the ground is a tripping hazard rather than just noting "a ball"). However, these systems remain constrained by the underlying model's training and inference design, and may still miss contextually significant but visually subtle cues. In summary, there is a gap between the accuracy of modern vision models and the contextual depth of advanced language models in existing assistive tools. This gap motivates the design of Visionary, which merges BLIP's precise visual recognition [5] with LLaMA 3's generative language abilities [10]. By integrating these components on a mobile device, Visionary aims to produce more detailed and context-rich scene descriptions in real time, thus extending prior work to achieve richer assistance for visually impaired users.

3   Methodology

3.1   BLIP for Image Recognition

BLIP [5] is based on self-supervised learning techniques, employing a visual transformer (ViT) [4] as its backbone. Let $I \in \mathbb{R}^{H \times W \times 3}$ denote the input image, where H and W represent the height and width, respectively. The image I is first passed through a convolutional encoder:

$$F = f_\theta(I) \tag{1}$$

where f_θ represents the image encoder parameterized by θ, and $F \in \mathbb{R}^D$ is the feature representation of the image.

Next, BLIP generates an initial caption C_{BLIP} based on these features:

$$C_{\text{BLIP}} = g_\phi(F) \tag{2}$$

where g_ϕ is a transformer-based decoder parameterized by ϕ. The initial caption C_{BLIP} is typically simple and lacks detailed context.

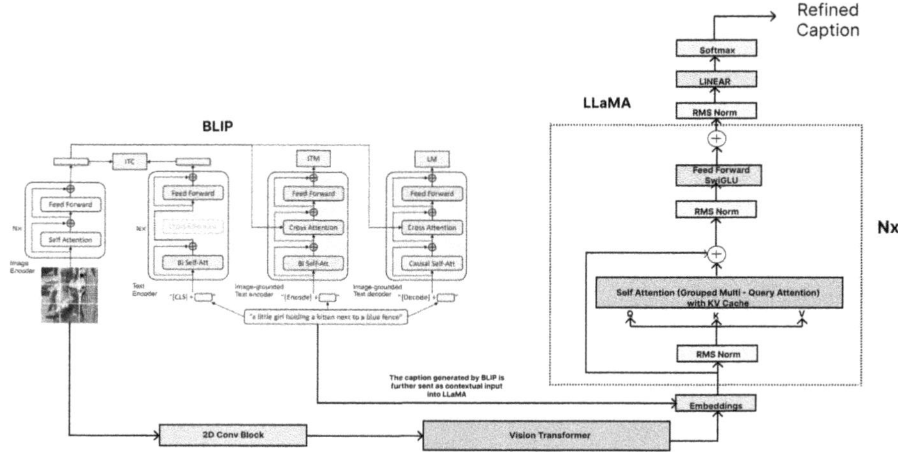

Fig. 1. Architecture Diagram depicting how the caption generated from BLIP is further refined by LLaMA 3 with the addition of a 2D convolutional layer and vision transformer.

To improve the utility of BLIP in an assistive setting, we preserve its ability to recognize salient entities but use these outputs as an intermediate signal rather than a final caption. In this way, BLIP [5] serves as a high-quality visual grounding component, ensuring the system has an accurate representation of the objects, textures, and attributes in a scene before any linguistic reasoning is applied. This separation allows Visionary to move beyond literal object naming and towards contextual interpretation.

3.2 LLaMA3 for Contextual Refinement

LLaMA 3 [10] enhances the descriptions generated by BLIP through advanced language modeling. The caption refinement process can be modeled as:

$$C_{\text{LLaMA}} = \arg\max_C P(C|C_{\text{BLIP}}, \mathcal{C}; \psi) \tag{3}$$

where ψ represents the parameters of the LLaMA 3 model, and $\mathcal{C}$ is the contextual information extracted from the image using a vision transformer and passed to LLaMA 3. The probability distribution incorporates both linguistic knowledge and visual context.

The conditional probability can be decomposed as:

$$P(C|C_{\mathrm{BLIP}}, \mathcal{C}; \psi) = \prod_{t=1}^{T} P(c_t | c_{1:t-1}, C_{\mathrm{BLIP}}, \mathcal{C}; \psi) \tag{4}$$

where c_t is the token at time step t, and T is the total number of tokens in the refined caption.

Unlike BLIP [5], which is optimized for direct captioning, LLaMA 3 [10] is designed to reason over sequences and inject semantic nuance into descriptions. By conditioning on both the preliminary caption and the enhanced visual embeddings, LLaMA 3 can incorporate high-level details such as spatial relationships ("a chair positioned behind the table"), scene attributes ("a crowded marketplace"), and even inferred intent ("a person reaching for a cup"). This refinement step transforms a flat list of objects into a contextually meaningful narrative, making the output more actionable for visually impaired users.

Formally, we define the refinement as a function $\mathcal{R}$ such that:

$$C_{\mathrm{LLaMA}} = \mathcal{R}(C_{\mathrm{BLIP}}, F'; \psi) \tag{5}$$

where F' is the enriched feature representation produced by the convolutional and transformer layers. The refinement process can thus be interpreted as a conditional generation problem over the vocabulary space V:

$$C_{\mathrm{LLaMA}} = \{w_1, w_2, \ldots, w_T\}, \quad w_t \in V \tag{6}$$

3.3   Integration of BLIP and LLaMA 3

The *Visionary* architecture Fig. 1 involves the following steps:

1. **Feature Extraction**: BLIP [5] extracts visual features F and generates a preliminary caption C_{BLIP}.
2. **Context Enhancement**: The visual features F are further processed by a 2D convolutional layer and a vision transformer [4] to generate enhanced features F'.
3. **Caption Refinement**: LLaMA 3 [10] takes both C_{BLIP} and F' as input and produces the final refined caption C_{LLaMA}.
4. **Text-to-Speech Conversion**: The refined caption C_{LLaMA} is converted into speech using a text-to-speech (TTS) engine [14,18], enabling voice interaction with the user.

The final caption C_{LLaMA} not only captures objects in the image but also adds rich contextual details such as spatial relationships, emotional content, and fine-grained nuances, thereby improving the situational awareness for visually impaired users.

To ensure the system operates effectively on a mobile device, the integration pipeline is optimized for both efficiency and interpretability. The convolutional enhancement step provides additional spatial context without substantially increasing computational overhead, while the vision transformer [4] enables longer-range dependencies across image regions. These enhanced embeddings act as a bridge between raw perception and narrative refinement.

The final stage of text-to-speech is equally critical: the latency between image capture and audio output must be short enough to support real-time navigation. By streamlining the pipeline, Visionary achieves a balance between descriptive richness and responsiveness, ensuring that the generated captions are not only accurate but also timely. This end-to-end integration enables the system to move beyond static captioning benchmarks and towards meaningful assistance in everyday environments.

3.4 Probabilistic View of the Pipeline

The full Visionary pipeline can be abstracted as a two-stage probabilistic model. Let I be an input image and A be the final audio description. Then:

$$P(A|I) = \sum_{C_{\text{BLIP}}, C_{\text{LLaMA}}} P(A|C_{\text{LLaMA}}) \cdot P(C_{\text{LLaMA}}|C_{\text{BLIP}}, I) \cdot P(C_{\text{BLIP}}|I) \quad (7)$$

Where:

- The first term, $P(C_{\text{BLIP}}|I)$, is the vision-to-caption likelihood modeled by BLIP.
- The second term, $P(C_{\text{LLaMA}}|C_{\text{BLIP}}, I)$, represents the refinement distribution modeled by LLaMA 3 with enhanced features.
- The final term, $P(A|C_{\text{LLaMA}})$, is the probability of generating speech given the refined caption, modeled by the TTS system.

This probabilistic formulation highlights that each component contributes conditionally and sequentially, forming a modular yet coherent end-to-end pipeline.

To ensure the system operates effectively on a mobile device, the integration pipeline is optimized for both efficiency and interpretability. The convolutional enhancement step provides additional spatial context without substantially increasing computational overhead, while the vision transformer [4] enables longer-range dependencies across image regions. These enhanced embeddings act as a bridge between raw perception and narrative refinement.

The final stage of text-to-speech is equally critical: the latency between image capture and audio output must be short enough to support real-time navigation. By streamlining the pipeline, Visionary achieves a balance between descriptive richness and responsiveness, ensuring that the generated captions are not only accurate but also timely. This end-to-end integration enables the system to move beyond static captioning benchmarks and towards meaningful assistance in everyday environments.

4   Discussion on Results and Findings

The evaluation of Visionary on the MS COCO dataset [8] demonstrates that the integration of BLIP [5] for visual recognition and LLaMA 3 [10] for contextual refinement significantly improves the quality of image descriptions compared to baseline models.

Table 1. Performance Comparison of Visionary with Baseline Models

Model	BLEU-4	ROUGE-L	METEOR	CIDEr
BLIP-only	31.2	54.6	27.8	92.5
LLaMA 3-only	29.7	53.2	26.5	88.4
Visionary (Ours)	**34.5**	**57.3**	**30.2**	**98.7**

As shown in Table 1, Visionary outperforms both the BLIP-only and LLaMA 3-only models, achieving an increase in BLEU-4 from 31.2 to 34.5 [12], in ROUGE-L from 54.6 to 57.3 [7], in METEOR from 27.8 to 30.2 [2], and in CIDEr from 92.5 to 98.7 [17]. These results indicate improvements in precision, recall, and overall relevance of the generated captions.

The significant performance gains are attributed to the combined strength of BLIP's accurate feature extraction [5] and LLaMA 3's ability to refine descriptions with detailed context [10]. The results confirm that adding contextual richness to preliminary descriptions enhances caption quality, leading to better user experiences, particularly for visually impaired individuals. This validates the efficacy of *Visionary* in both technical metrics and practical applications.

5   Conclusions and Future Work

In this paper, we presented Visionary, a mobile application that integrates BLIP [5] for image recognition and LLaMA 3 [10] for contextual refinement to generate detailed and accurate image descriptions for visually impaired users. Our model significantly outperforms existing approaches on the MS COCO dataset [8], demonstrating improvements across BLEU [12], ROUGE [7], METEOR [2], and CIDEr [17] metrics.

Looking forward, we aim to further fine-tune LLaMA 3 [10] to enhance context-specific captioning and adapt the model based on user feedback. Collaborations with institutions like the National Institute for the Empowerment of Persons with Visual Disabilities (NIEPVD) [11] are also underway to extend the app's functionality with external devices such as GoPro cameras, promoting greater independence for visually impaired users.

References

1. Alayrac, J.B., et al.: Flamingo: a visual language model for few-shot learning. arXiv preprint arXiv:2204.14198 (2022)
2. Banerjee, S., Lavie, A.: Meteor: an automatic metric for mt evaluation with improved correlation with human judgments. In: Proceedings of the ACL Workshop on Intrinsic and Extrinsic Evaluation Measures for Machine Translation and/or Summarization, pp. 65–72. Association for Computational Linguistics (2005)
3. Brown, T.B., et al.: Language models are few-shot learners. arXiv preprint arXiv:2005.14165 (2020)
4. Dosovitskiy, A., et al.: An image is worth 16x16 words: transformers for image recognition at scale. In: International Conference on Learning Representations (ICLR) (2021)
5. Li, J., Li, D., Xiong, C., Hoi, S.C.: Blip: bootstrapping language-image pre-training for unified vision-language understanding and generation. arXiv preprint arXiv:2201.12086 (2022)
6. Li, X., et al.: Oscar: object-semantics aligned pre-training for vision-language tasks. In: European Conference on Computer Vision, pp. 121–137. Springer (2020)
7. Lin, C.Y.: Rouge: a package for automatic evaluation of summaries. In: Text Summarization Branches Out: Proceedings of the ACL-04 Workshop, pp. 74–81. Association for Computational Linguistics (2004)
8. Lin, T.Y., et al.: Microsoft coco: common objects in context. In: European Conference on Computer Vision, pp. 740–755. Springer (2014)
9. Liu, H., Li, C., Wu, Q., Lee, Y.J.: Visual instruction tuning. arXiv preprint arXiv:2304.08485 (2023)
10. Meta AI: Introducing llama 3: Advancing open foundation models (2024). https://ai.meta.com/llama/
11. National Institute for the Empowerment of Persons with Visual Disabilities: National institute for the empowerment of persons with visual disabilities (niepvd) (2024). https://niepvd.nic.in/
12. Papineni, K., Roukos, S., Ward, T., Zhu, W.J.: Bleu: a method for automatic evaluation of machine translation. In: Proceedings of the 40th Annual Meeting on Association for Computational Linguistics, pp. 311–318. Association for Computational Linguistics (2002)
13. Radford, A., et al.: Learning transferable visual models from natural language supervision. In: Proceedings of the International Conference on Machine Learning (ICML), vol. 139, pp. 8748–8763 (2021)
14. Shen, J., et al.: Natural TTS synthesis by conditioning wavenet on mel spectrogram predictions. In: IEEE International Conference on Acoustics, Speech and Signal Processing (ICASSP), pp. 4779–4783. IEEE (2018)
15. Touvron, H., et al.: Llama: open and efficient foundation language models. arXiv preprint arXiv:2302.13971 (2023)
16. Touvron, H., et al.: Llama 2: open foundation and fine-tuned chat models. arXiv preprint arXiv:2307.09288 (2023)
17. Vedantam, R., Lawrence Zitnick, C., Parikh, D.: Cider: consensus-based image description evaluation. In: Proceedings of the IEEE Conference on Computer Vision and Pattern Recognition, pp. 4566–4575 (2015)

18. Zen, H., Tokuda, K., Black, A.W.: The HMM-based speech synthesis system (HTS) version 2.0. In: Proceedings of ISCA SSW6, pp. 294–299 (2007)
19. Zhang, P., et al.: Vinvl: revisiting visual representations in vision-language models. In: Proceedings of the IEEE/CVF Conference on Computer Vision and Pattern Recognition (CVPR), pp. 5579–5588 (2021)

Research and Application Tracks – Knowledge-Driven and Domain Specific AI

Advancing Atayal Language Preservation with AI-Driven Multimodal Speech and Text Processing

Jia-Lien Hsu[✉][iD] and Wei-Yuan Li

Department of Computer Science and Information Engineering, Fu Jen Catholic University, New Taipei 24205, Taiwan, R.O.C.
`alien@csie.fju.edu.tw`
`https://alienatfju.github.io/lab/`

Abstract. The Atayal language is an Austronesian language spoken by the Atayal people, one of Taiwan's indigenous groups, in which Squliq and C'uli' are two major dialects. However, a significant decline in population has led to severe interruptions in language transmission. This study leverages machine learning techniques to develop an integrated system that combines text-to-speech (TTS), automatic speech recognition (ASR), and bidirectional machine translation (MT). In this study, we construct a parallel corpus consisting of Atayal speech, Atayal Romanized script (the standardized writing system of the Atayal language is the Latin-based alphabets), and Mandarin Chinese translations. The integrated system and parallel corpus serve as an effective tool for language preservation. The TTS module converts Atayal Romanized script into natural-sounding Atayal speech, simulating native intonations and prosody, thereby assisting learners in acquiring accurate pronunciation. The ASR module transcribes Atayal speech into Romanized script, supporting speech input applications and lowering the barrier for text entry. The MT module enables real-time translation between Atayal Romanized script and Mandarin Chinese text, expanding the practical use and scope of the Atayal language. A web-based platform was developed to integrate TTS, ASR, and MT functionalities and The website provides open access, allowing users to utilize each function, thereby further facilitating the preservation and dissemination of Atayal. Performance evaluation was conducted using the TTS module for speech synthesis, applying voiceprint recognition models to extract voice features for similarity comparison. The ASR and MT module was evaluated based on word error rate (WER). Experimental results indicate that the TTS module achieved a 78.14% accuracy in voice similarity verification. The ASR model attained a WER of 1.01%. For translation tasks, the WER from Mandarin Chinese to Romanized Atayal was 51.98%, while the WER from Romanized Atayal to Mandarin Chinese was 47.71%. These results demonstrate the potential of the integrated system to support the revitalization of the Atayal language.

Keywords: Atayal · ASR · TTS · Atayal Corpus · MT

© The Author(s), under exclusive license to Springer Nature Singapore Pte Ltd. 2026
Q. V. Nguyen et al. (Eds.): AusDM 2025, CCIS 2765, pp. 439–453, 2026.
https://doi.org/10.1007/978-981-95-6786-7_30

1 Introduction

The Atayal people are one of the indigenous groups in Taiwan, currently ranking as the third-largest by population. However, the number of Atayal language speakers has declined to fewer than 100,000 individuals [4]. Due to the decreasing population, the intergenerational transmission of the Atayal language is facing severe challenges. UNESCO declared the Atayal language to be "vulnerable" and the Seediq language "severely endangered" in 2018 [12]. In Taiwan, the Atayal and Seediq peoples number around 90,000 and 10,000, respectively, but fluent speakers are much fewer [7,20,22]. Therefore, this study aims to develop a speech synthesis and recognition system using machine learning, supporting the preservation and revitalization of the Atayal language.

Language serves as a bridge for communication and mutual understanding. Each language conveys unique cultural, historical, and social values. When speakers of different languages interact, translation is often necessary to facilitate communication. Language acquisition typically begins with basic characters and pronunciation, followed by an understanding of grammar and semantics; however, it becomes extremely challenging to learn a language with limited resources.

Without active transmission, a language and its associated culture and history are at risk of extinction. The Atayal language is currently endangered. In order to preserve and sustain the Atayal language, it is essential to develop a system capable of converting between speech and text as well as enabling text-to-text translation.

1.1 Problem Statement

Atayal consists of two major dialect groups: the Squliq and C'uli' dialects, each with substantial linguistic differences. Among them, Squliq Atayal is the most widely distributed, thus selected as the target language for this study. Since Atayal lacks a private writing system, Romanization is adopted to represent Atayal using the Latin alphabet[1]. This research focuses on three primary tasks, supported by a parallel corpus constructed by our research group.

- **Text-to-Speech (TTS):** Convert Atayal Romanized Script input into corresponding Atayal speech.
- **Automatic Speech Recognition (ASR):** Transcribe Atayal speech into Atayal Romanized Script.
- **Text-to-Text Machine Translation (MT):** Bidirectional translation between Atayal Romanized Script and Mandarin Chinese text.

To achieve these objectives, a parallel corpus comprising Atayal speech, corresponding Atayal Romanization, and Mandarin Chinese translations will be

---

[1] The Atayal language is most commonly written in the Latin script; a standard orthography for the language was established by the Taiwanese government in 2005. [12].

constructed. In reference to Fig. 1, we develop TTS, ASR, and machine translation (MT) systems based on this corpus. This framework enables bidirectional conversion between speech and text as well as cross-lingual translation, ultimately supporting comprehensive Atayal speech processing and translation.

As shown in Fig. 1, there are well-established solutions for Mandarin TTS and Mandarin ASR (the two modules on the left). However, for the low-resource Atayal language, high-quality solutions are still lacking. This study aims to address this gap by constructing a parallel corpus and developing dedicated Atayal ASR and Atayal TTS modules through fine-tuning pretrained models, along with a bidirectional MT system between Chinese and Atayal.

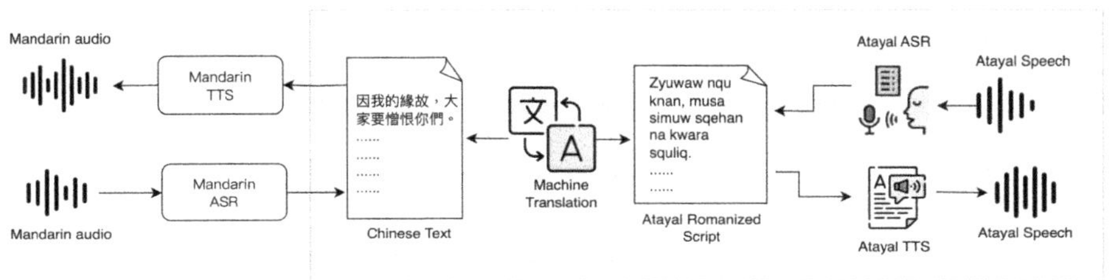

Fig. 1. System architecture and workflow of the proposed platform.

2   Related Work

To the best of our knowledge, there is no prior published research on Atayal ASR, TTS, or MT. In this section, we provide an overview of the methodologies for developing custom ASR, TTS, and MT systems.

2.1   Text-to-Speech Synthesis

With rapid advances in text-to-speech technology, previous architectures such as Deep Voice [2], FastSpeech [18], VITS [11], and VALL-E [23] typically rely on grapheme-to-phoneme conversion. The encoder transforms the input text into phonemes before passing them to the acoustic model, which then produces speech. However, such phoneme-based approaches often struggle with polyphonic words or languages with complex linguistic rules and limited resources, leading to reduced accuracy and ineffective cross-lingual transfer [13].

To address these limitations, recent research has shifted toward end-to-end (E2E) autoregressive neural architectures. These models bypass phoneme intermediates and convert input text directly into continuous representations for speech generation. The output at each step depends on the previous prediction, which enhances context modeling and speech coherence.

Most autoregressive TTS models adopt Transformer-based encoder-decoder frameworks, benefiting from deep semantic modeling and improved acoustic alignment, resulting in more natural speech and better language comprehension. Integrating speaker embeddings, prosody control, emotion modeling, and cross-lingual transfer further increases flexibility and personalization.

For higher speech quality, autoregressive architectures can incorporate diffusion models, which iteratively add and remove noise to synthesize high-fidelity Mel spectrograms, providing improved stability and naturalness compared to traditional vocoders. Meanwhile, controllable TTS [17] combines autoregressive modeling with external control signals, enabling precise modulation of emotion, prosody, speaking rate, and accent for expressive, personalized outputs.

The recently proposed Fish Speech [13] model unifies TTS and ASR tasks within a multi-task framework. This approach enhances both naturalness and cross-lingual capability, showing particular potential for low-resource or endangered languages due to improved generalization and scalability.

2.2 Automatic Speech Recognition

Automatic speech recognition systems have evolved rapidly. Early models like Deep Speech [1,9], Jasper [10], and Conformer [8] extract acoustic features (e.g., Mel spectrograms) and utilize encoders to derive acoustic representations, with a language model producing transcriptions. Training strategies such as CTC, Seq2Seq, or Transducer facilitate alignment between speech and text. Despite their effectiveness, these models face accuracy and generalization issues in low-resource settings [16].

To overcome these challenges, modern ASR systems employ end-to-end architectures with self-attention mechanisms and multi-task learning, improving speaker variability handling and context comprehension. Transformer-based models effectively capture long-range dependencies and integrate acoustic and linguistic context, achieving higher recognition accuracy and greater generalization. The Conformer merges convolutional and Transformer modules, leveraging both local and global feature extraction, and has become foundational in robust ASR systems.

Beyond acoustic modeling, ASR technologies expand towards cross-lingual, multi-speaker recognition, spoken language understanding, and multimodal inputs. Large-scale speech-language models now use both audio and text pre-training, enhancing semantic understanding and enabling open-domain speech comprehension.

Recently, OpenAI's Whisper [16] demonstrates strong multilingual recognition and semantic alignment, enabled by large-scale cross-lingual speech-text training. Whisper exhibits superior noise robustness and speaker adaptation, maintaining performance in challenging conditions and surpassing traditional ASR models. Its end-to-end design supports ASR, language identification, and speech translation, making it suitable for language preservation and multilingual applications.

2.3 Machine Translation

Early machine translation systems used statistical approaches, relying on large parallel corpora and handcrafted features for language and translation modeling. These methods had limited ability to capture long-range dependencies and required extensive bilingual resources, thus performing poorly for distant or low-resource language pairs.

With the advent of deep learning, neural machine translation (NMT) [3] replaced SMT by learning direct mappings from source to target language in an end-to-end fashion. Initial NMT models adopted RNN-based encoder-decoder architectures with attention mechanisms, but were eventually superseded by Transformer models, which efficiently model global dependencies using self-attention and improve translation quality and speed. However, NMT systems still face data scarcity issues, limiting coverage for many languages.

To address this issue, mBART [14] (Multilingual BART) was introduced as a pre-trained translation model with strong cross-lingual generalization. mBART employs self-supervised denoising autoencoding over large, unaligned multilingual corpora, enabling it to learn language-agnostic semantic representations. By incorporating language tokens during pre-training, mBART supports low-resource, zero-shot, and non-parallel translation scenarios, ensuring broader language coverage in real-world applications.

3 Methodology

This section presents the construction of the parallel corpus and the system architecture, providing detailed descriptions of the TTS, ASR, and bidirectional MT modules, as well as the design of the web-based interactive platform. By integrating these components, the proposed framework aims to establish a comprehensive workflow for speech processing and validation, thereby improving both the accuracy and practical applicability of speech-related applications.

3.1 The Parallel Corpus

In this study, we construct a parallel corpus that contains Atayal speech, Atayal Romanized scripts, and their corresponding translations in Chinese texts. The dataset comprises 5,620 samples (sentences in both Atayal and Chinese), which we collected from the New Testament (Matthew – Revelation), as published on the Taiwan Bible Society website [21]. According to Table 1, the total duration of all recordings amounts to 26.4 h, with each sample averaging 16.4 s in length. Each sentence was recorded by Pastor Meiliang, who holds certification in the Atayal language.

As shown in Table 2, we annotated each recording with both the Atayal Romanized script and its Chinese translation. The entire corpus of Atayal Romanized script comprises 120,531 tokens, with each sentence containing an average of 21.45 tokens. The Chinese translations include 168,330 characters, with an average of 29.95 characters per sentence.

3.2   System Architecture

As illustrated in Fig. 1, the proposed platform comprises three core modules: ASR, TTS, and MT, which together form an integrated speech conversion pipeline. The system first processes user-inputted Mandarin speech, transcribing it into Chinese text using the Mandarin ASR module (implemented with the `Whisper` model). The transcribed Chinese text is subsequently passed to the MT module, where a fine-tuned `mBART` model translates it into Atayal Romanized script. Finally, the Atayal script is synthesized into speech through the Atayal TTS system. The platform also enables bidirectional and independent module operations. For example, users may transcribe Atayal speech to Atayal Romanized script using the Atayal ASR model, then translate that script into Chinese text via the MT module, thereby facilitating bidirectional language conversion and component-level functionality. Each subtask employs a pretrained model, further fine-tuned for the target language to enhance both performance and generalization:

1. TTS: Fine-tuned on `Fish-Speech` for target language speech synthesis.
2. ASR: Fine-tuned on `Whisper` to improve target language recognition.
3. MT: Fine-tuned on `mBART` to optimize translation for the target language.

In the proposed system, we directly adopt the established Mandarin TTS and Mandarin ASR modules, as these components are mature and widely available. For Atayal TTS, Atayal ASR, and MT, we fine-tune pretrained models to attain optimal performance, as discussed in the following subsections.

Table 1. Statistics of Atayal speech recording.

Duration (sec.)	No. of samples
less than 5	12
6–10	509
11–15	1,261
16–20	2,040
21–25	1,438
16.4 (avg.)	5,620 (in total)

3.3   Atayal TTS

We retrain the `Fish-speech` architecture for Atayal text-to-speech synthesis, as illustrated in Fig. 2. The dataset comprises 5,620 Atayal samples paired with Romanized scripts, which we split into 90% for training (5,058 samples) and 10% for evaluation (562 samples).

Each audio sample is standardized to 16kHz, mono-channel, and stored in WAV format. We extract Mel spectrograms from these files as acoustic features.

Table 2. Sample entries from the corpus.

Atayal Speech Audio	Atayal Romanized Script	Chinese translation
001.wav	"Bleqiy mung kiy! Maki qutux qu pghap musa gmhap.	你們留心聽啊！有一個撒種的出去撒種。
002.wav	Trang gmhap ga mhutaw syaw na tuqiy qu ruma ghap, ru wahan maniq na qqhniq kwara la.	他撒的時候，有些種子落在路旁，鳥兒飛來把它們吃掉了。
003.wav	Ruma ga mhutaw squ lhmiq uraw babaw na bbtunux, misuw balay hmbku, baha hmswa ini kzzik qu uraw,	有些落在淺土的石地上，種子很快就長苗，因為土壤不深，
004.wav	mhtuw wagi lga tklxun nya mu qu tanguw, ana uziy ga ini tehuk kinzzik qu gamil, ru mkiyay la.	太陽一出來，就把幼苗曬焦了，又因為根不夠深，枯乾了。
005.wav	Ruma ghap ga mhutaw ska na bbqzi, mrkyas qu bbqzi, kmbet squ tanguw, ru ini thuyay mkbway la.	有些落在荊棘中，荊棘長起來，把幼苗擠住了，不能結出果實。

The training consists of two stages. First, Mel spectrograms are encoded and quantized using Group Finite Scalar Quantization (GFSQ) to generate discrete speech tokens; the decoder reconstructs the acoustic features from these tokens. In the second stage, the model employs an autoregressive mechanism to map text and quantized tokens into latent features and then generates speech tokens from which audio can be reconstructed.

During inference, the model receives text and prompt tokens, generating quantized Mel tokens step-by-step with the autoregressive module, followed by decoding into audio waveforms.

3.4 Atayal ASR

We fine-tune the `Whisper-small` model to convert speech to Atayal Romanization, following the procedure outlined in Fig. 3. The corpus is divided into 90% training data (5,058 samples) and 10% testing data (562 samples). Each text transcription appends a \$ (end-of-sequence) marker to assist the model in learning sequence boundaries.

We load the `Whisper` model along with its processor, which contains a feature extractor for audio and a tokenizer for text. We resample all audio to 16kHz, extract features, and align each with the tokenized transcript.

Training involves inputting Mel spectrograms into the encoder, after which the decoder autoregressively generates the Romanized scripts. During inference, Whisper transforms audio to Mel spectrograms for the encoder to compute hid-

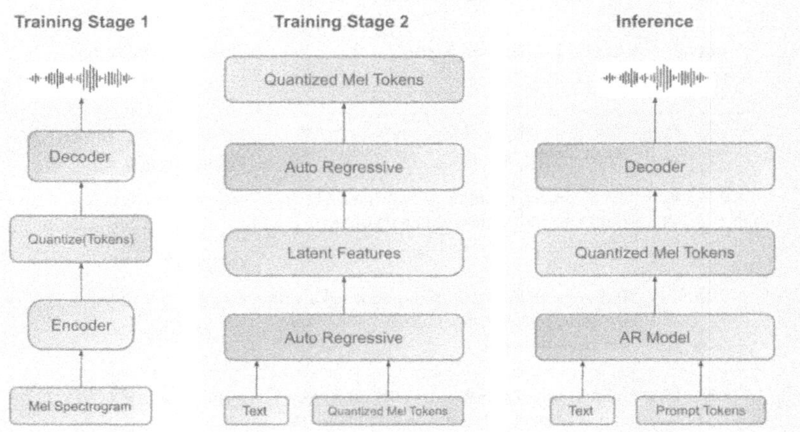

Fig. 2. `Fish-Speech` Architecture [13].

den representations, and the decoder autoregressively produces the transcript until the $ marker indicates the end of the sequence.

3.5   Machine Translation

Before training, we initialize the multilingual `mBART` model (`mbart-large-50-many-to-many-mmt`) [6]. Pretraining on large-scale, multilingual denoised corpora enables mBART to learn cross-lingual representations across diverse languages.

Figure 4 depicts the training process: the left panel shows multilingual denoising, where the model reconstructs noisy sentences; the right panel illustrates machine translation fine-tuning, mapping source texts directly to their target translations.

We encode Atayal Romanized inputs and Chinese outputs as token sequences, splitting them into training and validation sets. BLEU scores serve as the principal metric for evaluating translation quality. For reverse translation, we swap the source and target languages and follow the same training procedure.

3.6   Web Application and Interface

To demonstrate the proof of concept, we implement the three core modules: Atayal TTS, Atayal ASR, and bidirectional MT. We do not re-implement the Mandarin ASR and Mandarin TTS modules, as these are already well-established. As shown in Fig. 5, Fig. 6, and Fig. 7, each of these modules operates independently and successfully. Demo sites for all three tasks are freely available at designated web addresses http://140.136.149.205:7860/, http://140.136.149.205:7861/, and http://140.136.149.205:7862/.

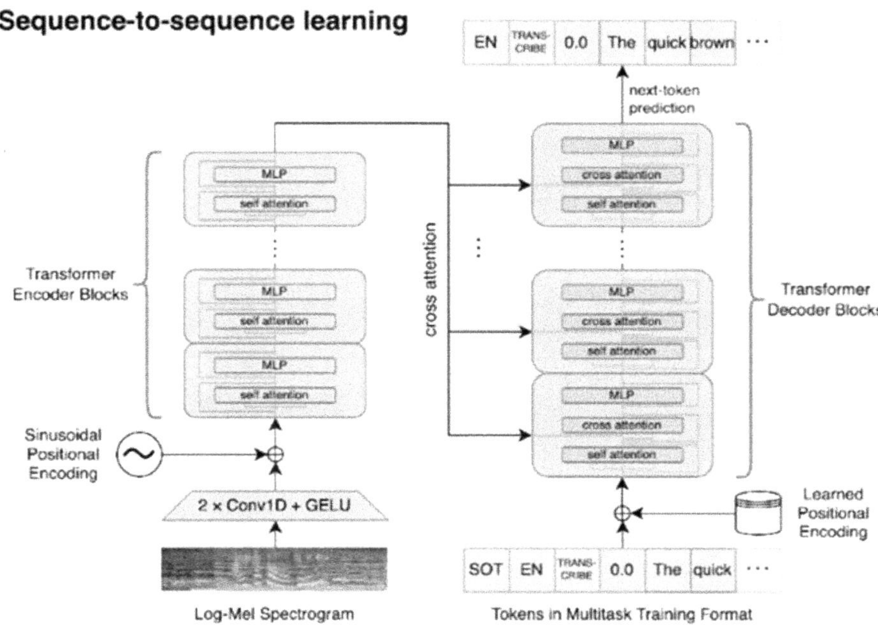

Fig. 3. Whisper Training Pipeline [16].

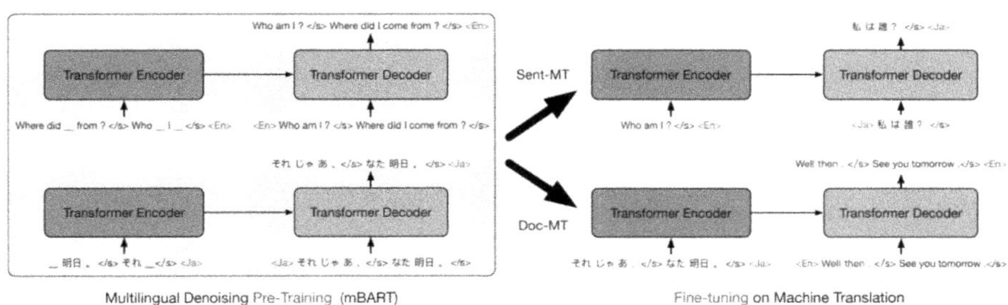

Fig. 4. mBART Training Pipeline [14].

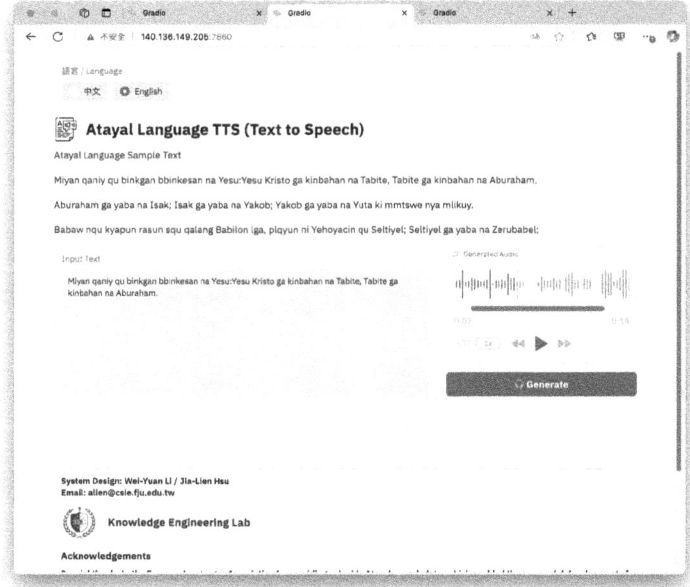

Fig. 5. Atayal TTS Web Interface

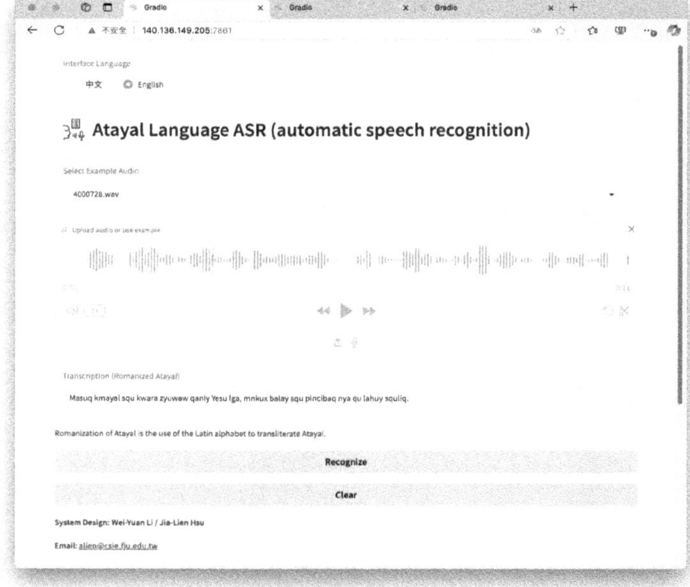

Fig. 6. Atayal ASR Web Interface

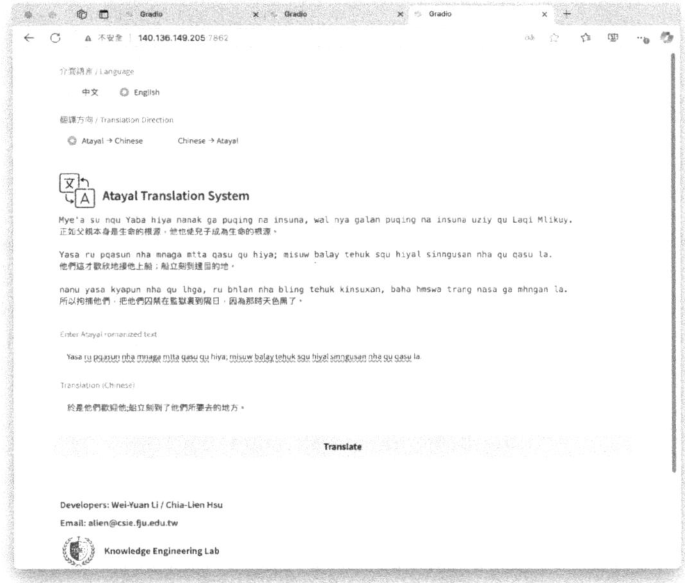

Fig. 7. Bidirectional MT Web Interface: Atayal vs. Chinese.

4   Experimental Results

In this section, we present the evaluation metrics and performance results for TTS, ASR, and MT. All results are reported on the test set, which comprises 10% of our entire corpus and is strictly separated from the training set.

4.1   TTS Evaluation

We employ two evaluation metrics for the TTS module.

TTS Evaluation Metric I: TTS-Metric-I quantifies the similarity between two audio recordings: the original (ground truth) and the TTS-synthesized sample for the same Atayal sentence. For each audio recording Y_i, we extract 13-dimensional MFCC feature vectors $\mathbf{M}_i$ to capture spectral and vocal characteristics and reduce speaker variation effects [15]:

$$\mathbf{M}_i = \mathrm{MFCC}(Y_i) \in \mathbb{R}^{N \times T} \tag{1}$$

We then compute the alignment distance between paired MFCC sequences using Dynamic Time Warping (DTW) with Euclidean distance, effectively comparing sequences of differing lengths. Smaller DTW distances indicate higher similarity:

$$\mathrm{DistDTW}(\mathbf{M}_1, \mathbf{M}_2) = \min_{\pi} \sum_{(i,j)\in\pi} \left\| \mathbf{m}_1^{(i)} - \mathbf{m}_2^{(j)} \right\|_2 \tag{2}$$

To provide an interpretable similarity score, we map the DTW distance to a percentage:

$$\text{TTS-Metric-I} = 100 \times e^{-\frac{\mathrm{DistDTW}}{s}}, \quad s = 40000 \tag{3}$$

Here, s is a scaling parameter; a DistDTW of 0 yields 100% similarity, while larger distances produce scores closer to 0%. Table 3 summarizes performance results, in terms of minimum, maximum, and mean similarities. Most samples score between 55% and 65%, reflecting variability due to speaking rate, intonation, or audio noise.

Table 3. Performance results of TTS.

Item	TTS-Metric-I	TTS-Metric-II
Mean	60.74%	78.14%
Median	60.69%	78.28%
Maximum	74.10%	85.15%
Minimum	47.96%	71.06%

TTS Evaluation Metric II: TTS-Metric-II measures similarity using the SpeechBrain ECAPA-TDNN pretrained speaker verification model [5,19]. The model calculates a cosine similarity *score* ($[-1, 1]$) for each audio pair, which we linearly map to a percentage:

$$\text{TTS-Metric-II} = \frac{score + 1}{2} \times 100 \tag{4}$$

As shown in Table 3, results indicate moderate to high similarity, with synthesized and reference audio sharing consistent speaker characteristics despite variations in speed, intonation, or environment.

4.2 ASR Evaluation

We evaluate the ASR system using the Word Error Rate (WER), a standard metric for speech and language processing:

$$\mathrm{WER} = \frac{S + D + I}{N} \tag{5}$$

where S, D, I are the counts of substitutions, deletions, and insertions, and N is the total number of words. Lower WER values indicate higher recognition accuracy.

On the 562 test samples, the ASR model achieves an average WER of 1.01%, a median WER of 0%, and a maximum WER of 15%. Notably, 81.5% of samples (459/562) have zero errors. Most samples have WERs in the 0%–5% range; higher error rates typically occur in shorter, less clear, or noisier samples. These results demonstrate strong recognition accuracy and generalization to this low-resource language.

4.3 Bidirectional MT Evaluation

Chinese-to-Atayal Romanization WER. For the Chinese-to-Atayal direction, the average WER across 562 test samples is 51.98% (median 53.71%), with the minimum WER at 0% and the maximum at 140%. This wide spread reflects significant variance and indicates that while some translations are nearly perfect, others face substantial mismatches.

Atayal Romanization-to-Chinese WER. For Atayal-to-Chinese translation, we observe a mean WER of 47.71% and a median of 47.22%. Most samples cluster in the 30%–60% range, with 21% falling within 40%–50% and 55.7% below 50% WER. Outliers, especially those with short or noisy inputs, contribute to WERs exceeding 100% (Table 4).

Table 4. Performance of Bidirectional MT

Item	Chinese-to-Atayal	Atayal-to-Chinese
Mean	51.98%	47.71%
Median	53.71%	47.22%
Maximum	140.00%	115.00%
Minimum	0.00%	0.00%

5 Conclusion and Future Work

This study developed an interactive speech processing system to support the preservation and revitalization of the Atayal language. First, we construct a parallel corpus consisting of Atayal speech, Atayal Romanized script, and the corresponding Chinese translation. We create a modular web-based platform incorporating text-to-speech (TTS), automatic speech recognition (ASR), and machine translation (MT) components, all constructed using Python and Gradio.

Our TTS module enabled users to convert Atayal Romanized script into natural-sounding Atayal speech, achieving a speaker similarity of 78% in evaluation. Future efforts will target improved voice style control and greater training data diversity, which can enhance the naturalness and robustness of synthesized speech across speakers and settings. Furthermore, we plan to utilize perceptual metrics such as mean opinion score (MOS) to more accurately assess TTS quality in practical deployments.

We implemented an ASR module that allowed users to upload Atayal speech audio and obtain corresponding Romanized outputs. The ASR system achieved a WER of 1% on the test set, indicating high recognition accuracy and effective semantic preservation. To further enhance performance, future work will expand the data set, analyze errors in high-WER samples, and integrate speaker modeling or noise-robust techniques for improved generalizability in real-world scenarios.

The MT module facilitated bidirectional translation between Atayal Romanized script and Mandarin Chinese. For the "Romanization → Chinese" task, the mBART-based system attained a WER of 47.71%, and for the "Chinese → Romanization" direction, it reached 51.98%. The accuracy of the machine translation system requires further improvement to meet the demands of practical applications. Although these results reveal challenges in word order and lexical mapping, especially for the Mandarin input, the system demonstrates initial promise for low-resource languages. Future work will focus on expanding the parallel corpus and further refining the models to increase translation quality and semantic consistency.

The current corpus, which is derived from a single domain (religious texts) and produced by a single speaker, may restrict the model's capacity to generalize to colloquial speech, speakers with different characteristics, or alternative domains. Although performance metrics indicate significant potential, additional research aimed at enhancing translation accuracy, increasing dataset diversity, and expanding dialectal coverage would substantially strengthen the system impact and usability. Once we mature all core modules, we will develop an integrated, user-friendly mobile application that supports seamless end-to-end voice-to-voice translation. By providing these multimodal AI-driven tools, we aim to promote the continued use, learning, and transmission of the Atayal language and contribute substantially to its preservation for future generations.

Acknowledgments. This study was funded by the National Science and Technology Council, Taiwan, R.O.C. (grant number NSTC-114-2221-E-030-007). We also would like to express our gratitude to the Formosa Language Association for their support in providing the Atayal language audio recordings.

References

1. Amodei, D., et al.: Deep Speech 2: End-to-End Speech Recognition in English and Mandarin (2015). https://arxiv.org/abs/1512.02595
2. Arik, S.O., et al.: Deep voice: real-time neural text-to-speech. arXiv preprint arXiv:1702.07825, pp. 3–6 (2017)
3. Bahdanau, D., Cho, K., Bengio, Y.: Neural machine translation by jointly learning to align and translate. arXiv preprint arXiv:1409.0473, pp. 1–4 (2015)

4. Council of Indigeneous Peoples (CIP): Statistical data on indigenous population in June 2025. Executive Yuan, Republic of China (Taiwan) (2025). https://reurl.cc/4Np6e2. Accessed 27 July 2025

5. Desplanques, B., Thienpondt, J., Demuynck, K.: Ecapa-tdnn: emphasized channel attention, propagation and aggregation in tdnn based speaker verification. arXiv preprint arXiv:2005.07143 (2020)

6. Facebook AI: mBART-50: Multilingual BART for Machine Translation (2020). https://huggingface.co/facebook/mbart-large-50-many-to-many-mmt

7. Global Voices: Seediq, an Endangered Aboriginal Tribe and Language in Taiwan - The News Lens International Edition (2019). https://international.thenewslens.com/article/122990

8. Gulati, A., et al.: Conformer: convolution-augmented transformer for speech recognition. arXiv preprint arXiv:2005.08100, pp. 1–3 (2020)

9. Hannun, A., et al.: Deep Speech: Scaling up end-to-end speech recognition (2014). https://arxiv.org/abs/1412.5567

10. Jason, L., et al.: Jasper: an end-to-end convolutional neural acoustic model. arXiv preprint arXiv:1904.03288 (2019). arXiv:1904.03288

11. Kim, J., Kong, J., Son, J.: Conditional variational autoencoder with adversarial learning for end-to-end text-to-speech. arXiv preprint arXiv:2106.06103, pp. 1–4 (2021)

12. Language Magazine: Indigenous Taiwanese Languages Now Available on Wikipedia (2021). https://languagemagazine.com/2021/04/29/indigenous-taiwanese-languages-now-available-on-wikipedia/

13. Liao, S., et al.: Fish-speech: Leveraging large language models for advanced multilingual text-to-speech synthesis. arXiv preprint arXiv:2411.01156, pp. 1–5 (2024)

14. Liu, Y., et al.: Multilingual denoising pre-training for neural machine translation. arXiv preprint arXiv:2001.08210, pp. 2–5 (2020)

15. Muttaqi, T., Mousavinezhad, S.H., Mahamud, S.: User identification system using biometrics speaker recognition by MFCC and DTW along with signal processing package. In: IEEE International Conference on Electro/Information Technology (EIT), pp. 79–83 (2018). https://doi.org/10.1109/EIT.2018.8500258

16. Radford, A., Kim, J.W., Xu, T., Brockman, G., McLeavey, C., Welinder, P.: Robust speech recognition via large-scale weak supervision. arXiv preprint arXiv:2212.04356, pp. 2–4 (2022)

17. Ren, Y., Tan, X., Qin, T., Liu, T.Y.: Towards controllable speech synthesis in the era of large language models: a survey. arXiv preprint arXiv:2412.06602, pp. 1–4 (2024)

18. Ren, Y., et al.: FastSpeech: Fast, Robust and Controllable Text to Speech. arXiv preprint arXiv:1905.09263, pp. 3–6 (2019)

19. SpeechBrain: spkrec-ecapa-voxceleb: Ecapa-tdnn speaker verification model (2020). https://huggingface.co/speechbrain/spkrec-ecapa-voxceleb

20. Sterk, D.: Ecologising seediq: towards an ecology of an endangered indigenous language from Taiwan. Int. J. Taiwan Stud. **4**(1), 54–71 (2020). https://doi.org/10.1163/24688800-20201153

21. The Bible Society in Taiwan: UBS Open Han Bible Project (2025). https://cb.fhl.net/

22. TNL Staff: Two More Taiwan's Indigenous Languages Available On Wikipedia - The News Lens International Edition (2021). https://international.thenewslens.com/article/149745

23. Wang, C., et al.: Neural codec language models are zero-shot text to speech synthesizers. arXiv preprint arXiv:2301.02111, pp. 5–6 (2023)

ETCOD: Embedding-Based Anomaly Detection and LLM-Driven Validation Framework for Knowledge Graphs

Thi Thuy Nga Nguyen[1]([✉])(iD), Asara Senaratne[2](iD), and Leelanga Seneviratne[3](iD)

[1] Faculty of Engineering and Information Sciences, University of Wollongong, Wollongong, Australia
ttnn278@uowmail.edu.au
[2] College of Science and Engineering, Flinders University, Adelaide, Australia
asara.senaratne@flinders.edu.au
[3] Faculty of Information Technology, University of Moratuwa, Moratuwa, Sri Lanka
leelangas@uom.lk

Abstract. Knowledge Graphs (KG) form the backbone of many knowledge dependent applications, such as search engines and digital personal assistants. When constructing a KG, data can be manually curated by experts, contributed by volunteers, automatically extracted using handcrafted or learned rules, or generated from unstructured text via machine learning techniques. Regardless of the approach, anomalies are inevitable, as no data source is perfect. To address this, we propose *ETCOD*, an embedding-based anomaly detection approach for KG validation and quality enhancement, combined with Large Language Models (LLM) for explanation and verification. First, we generate semantic embeddings of triples in order to capture entity and relation similarities. Next, we perform pattern mining to identify anomalous triples. Finally, detected anomalies are forwarded to an LLM to provide human-understandable explanations and reasoning. We conducted experiments on real-world KGs, including YAGO-1 and YAGO-4.5, using ChatGPT-4o and Gemini for explanation. Our analysis highlights which types of anomalies are most effectively explained by each model and where they tend to fall short. The results demonstrate that embedding-based detection is effective in identifying anomalies, while LLMs enhance interpretability by providing context-aware explanations.

Keywords: Knowledge Graph Validation · Data Quality Assessment · Triple Embedding · Pattern Mining · Large Language Models

1 Introduction

Intelligent assistants like Alexa and Siri rely on structured knowledge [21] to provide meaningful responses [2]. Knowledge Graphs (KGs) have emerged as a key approach to represent human knowledge [15]. However, real-world KGs often

Q. V. Nguyen et al. (Eds.): AusDM 2025, CCIS 2765, pp. 454–469, 2026.
https://doi.org/10.1007/978-981-95-6786-7_31

contain quality issues such as conflicting entity names, missing relationships, redundancies, outdated facts, and misinformation [5]. We treat such issues as anomalies, as they can degrade the reliability of systems that depend on KGs [10].

An anomaly is a data instance that significantly deviates from the expected patterns [6,19]. In KGs, anomalies may result from extraction errors, incorrect links, or schema violations. Existing methods often fail because they overlook KG-specific challenges such as multi-relational structures and semantic dependencies [14,16]. For example, the triple `<Barack_Obama, bornIn, March>` is structurally valid but semantically incorrect, showing the need for semantic-aware anomaly detection.

To address these issues, we propose Entity Type and Confidence-based Outlier Detection (ETCOD), a framework that combines pattern-based and embedding-based techniques with LLM-assisted validation. ETCOD identifies anomalous triples by detecting deviations from common subject-predicate-object patterns such as `<person, citizenOf, city>`, and by assessing the semantic coherence between subject types, predicates, and object types through vector embeddings.

ETCOD operates in three main stages: (1) Pattern-based analysis: Frequent triple patterns are mined to capture structural norms. For instance, if most triples follow patterns such as (Person, bornIn, Country) or (Book, writtenBy, Author), these become reference structures for normal relations. (2) Semantic embedding-based anomaly detection: Subjects, predicates, and objects are embedded using a sentence transformer model, and consistency scores are computed based on type-pattern frequencies. These features are then passed to an *Isolation Forest* [11] for unsupervised anomaly detection, ensuring scalability for large-scale KGs. (3) LLM-based validation: Anomalies flagged by the Isolation Forest are analyzed using LLMs (e.g., ChatGPT-4o and Gemini) to generate human-readable explanations, reasoning, and corrective suggestions. This stage mimics human judgment and improves the interpretability of the KG without affecting the reproducibility of detection.

We evaluated ETCOD on two real-world KGs, YAGO-1 and YAGO-4.5, using manual assessment. Next, we synthetically generate triples corrupted with TRIC [18], perform anomaly detection with LLM-driven validation, and compare the performance of ETCOD with the baseline SEKA [14]. The results show that ETCOD achieves high recall and strong precision, confirming the effectiveness of our approach.

Accordingly, **the contributions of this paper are**: (1) We introduce ETCOD, an anomaly detection technique for KGs that integrates semantic embeddings with pattern-based analysis. (2) We propose a type inference model using semantic embeddings and k-nearest neighbors (KNN) to support KGs without type annotations. (3) We demonstrate how to leverage LLMs for anomaly validation, explanation, and reasoning, enhancing the interpretability of the KG.

The remainder of the paper is organized as follows. Section 2 reviews the existing literature, Sect. 3 details our methodology, and Sect. 4 presents experimental evaluations. Finally, Sect. 5 concludes the article with a discussion on future research directions.

2   Literature Review

Anomaly detection in Knowledge Graphs (KGs) has become increasingly important, as KGs are widely used in information retrieval, recommendation provision, and decision support [15,17]. Ensuring the reliability and consistency of KG data is essential for maintaining high-quality information. Recent research on knowledge graphs have explored a range of anomaly detection approaches, which are generally categorized into structural analysis, representation learning, and validation using large language models (LLMs).

Early approaches relied on structural analysis and rule-based validation. For example, Asara et al. introduced SEKA (Seeking Knowledge Graph Anomalies), an unsupervised method that detects anomalous triples and entities by identifying contradictions, incompleteness, and redundancy in graph structure through the Corroborative Path Algorithm (CPA) [14]. Similarly, SHACL and OWL reasoners validate triples against schema constraints to capture violations [1]. While effective for identifying structural inconsistencies, these systems struggle with semantic-level anomalies, where a triple may be syntactically valid but factually incorrect.

To address these limitations, researchers have explored representation learning-based approaches. For instance, ADKGD employs a dual-channel model that integrates entity embeddings with triple-level contexts [29], while ProMvSD adapts serialization-based strategies for scalable unsupervised anomaly detection under limited labels. Andy Zhou combines domain knowledge with data-driven models in a two-component system comprising a statistical learner and a probabilistic graphical reasoner [32]. These methods perform well under concept drift and class imbalance but rely heavily on graph topology and relation embeddings, which are insufficient to capture anomalies arising from entity type distribution or real-world semantics.

Classic embedding models such as TransE, RotatE, and ComplEx [23] can identify unlikely triples based on vector-space distance but may assign high plausibility to semantically nonsensical triples. Pattern-mining methods [7,26] detect outliers by learning frequent graph structures, but they remain limited to structural irregularities and cannot detect semantically implausible yet structurally valid triples. Incorporating normality dynamically into context, as proposed through a framework called the Interpretive Paradigm [20] for knowledge construction, provides a novel direction that might make some of these issues superfluous.

Large Language Models (LLMs) have recently opened new directions for semantic anomaly detection. HumanLLM collaborative validation framework [25] and KGValidator [5] shows that LLMs can validate triples and provide natural language explanations. However, they often require human oversight for rare or ambiguous cases and do not integrate structural patterns, limiting

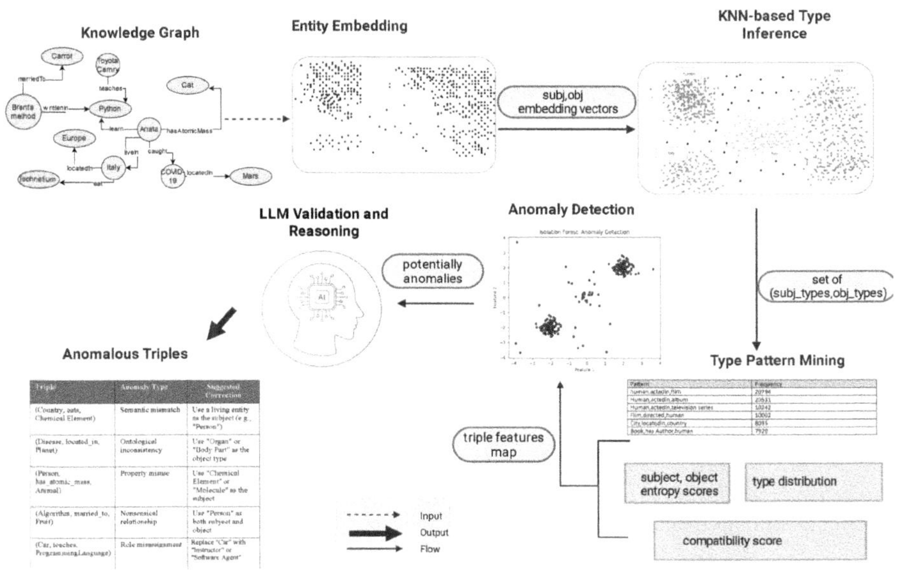

Fig. 1. ETCOD anomaly detection pipeline, illustrating entity type extraction through embeddings, anomaly detection, and LLM-assisted validation and reasoning.

scalability. AutoGAD explores LLM-driven search for graph anomaly detection tasks [8], but focuses on optimizing model configurations rather than reasoning about semantic consistency. These studies highlight the potential of LLMs for semantic anomaly detection, but reliance on human input and limited structural integration constrain scalability for large KGs.

Building on these insights, our work combines semantic embeddings, type inference, and LLM-powered reasoning to detect anomalies that are both structurally and semantically inconsistent. By inferring entity types for unlabeled triples and integrating structural patterns with semantic reasoning from LLMs, our method reduces false positives and provides human-like suggestions. Unlike previous approaches, it is scalable to large KGs and requires minimal human intervention, bridging the gap between structural precision and semantic understanding.

3   Proposed Methodology

We propose ETCOD, an approach for Entity Type and Confidence-based Outlier Detection. This identifies anomalies in Knowledge Graphs (KG) by analyzing the semantic compatibility between the inferred types of subject and object entities and the predicate connecting them. ETCOD focuses on detecting outliers by learning the structural pattern of triples <subject type, relation, object type>. However, many real-world KGs lack entity type annotations, which limit the effectiveness of pattern-based detection. To address this problem, we introduce a type inference module that automatically predicts entity types based solely on the structural and semantic properties of triples within the graph. The predicted types are then used as features for anomaly detection in ETCOD.

Our goal is to detect not only incorrect triples but also subtle anomalies, which may show reasonable in structure but are factually incorrect. Finally, we utilize a Large Language Model (LLM) in the final validation stage to further explore these anomalous triples. The LLM uses its real-world knowledge to evaluate whether flagged triples contain misinformation.

Our proposed method requires only a KG as an input. The ETCOD pipeline is illustrated in Fig. 1 and consists of two main phases: (1) entity type extraction and (2) outlier detection. For type classifier training, we use a subset of Wikidata[1] with 244,790 entity-type mappings across 10,912 general classes such as Human, Organization, and Location. Wikidata contains millions of entities across diverse domains and covers common classes found in most knowledge graphs, which makes it ideal for building a general type inference model. This approach works well for different KGs because the embeddings are learned from entity labels, and the KNN-based inference relies only on similarity in the embedding space, not on the Wikidata schema. For domain-specific KGs, such as those in the biomedical field, we can improve accuracy by adding relevant domain entities to the training set [30].

3.1 Preliminaries

Considering an edge-labeled graph, which is a type of attributed graph with a single categorical attribute (label) for the edge [28], we define the following:

Definition 1. *(Knowledge Graph): A Knowledge Graph $G = (E, R)$ consists of a set of entities E, which serve as nodes representing specific concepts in a domain, and a set of relations R. The KG contains linked information of facts, represented as triples (s, p, o) where $s \in E$ is the subject entity, $o \in E$ is the object entity, and $p \in R$ is the relationship between s and o.*

Definition 2. *(Entity type): An entity type defines the semantic category of an entity. Let $T = (T_s, T_o)$ be a set of entity types, where T_s and T_o represent the types or categories of subject entity s and object entity o, respectively. An entity type assignment can be represented as (e, T_e), where $e \in E$ is an entity and $T_e \in T$ is its category.*

Definition 3. *(Pattern): A pattern is an abstract structure of a triple, defined as a tuple (T_s, p, T_o). It denotes the relation p between two entity types T_s and T_o, representing a generalized form of the instance-level triple (s, p, o).*

---

[1] https://www.wikidata.org/wiki/Wikidata:Main_Page.

3.2   Entity Type Extraction

Many large-scale KGs lack entity-type annotations, which poses a challenge for type-based anomaly detection methods [3]. To address this, we introduce a type-inference module that predicts entity types based on semantic textual features. We assume that entities with similar lexical and contextual features are likely to have the same semantic type. We train a K-Nearest Neighbors (KNN) classifier on labeled entities from the Wikidata subset.

3.2.1   Entity Embedding Let e denote an entity in the KG. We use all-MiniLM-L6-v2 model from the SentenceTransformers library [13] to encode each entity into a 384-dimensional vector. While higher-quality models such as all-mpnet-base-v2[2] exist, all-MiniLM-L6-v2 achieves competitive accuracy while operating approximately five times faster. which is crucial for processing millions of triples as found in large-scale KGs. Mean pooling over the transformer token produces final embeddings, capturing both lexical and contextual semantics. These representations support the semantically similar entities clustering task by considering to locality in the embedding space and mapping to nearby points.

The entity type T_e labels are encoded using LabelEncoder from scikit-learn into a target vector $y \in Z^n$. Although commonly used for categorical data, OneHotEncoding could lead to poor scalability and sparse matrices in case large number of distinct class. The use of word embedding enables the model to transform a set of entity-type pairs (e, T_e) into a structured space for statistical learning, reason over type similarities, and achieve scalability.

3.2.2   Type Inference There are many existing entity type inference methods, including relying on graph connectivity and structural paths to infer types, which fail on sparse KGs or when node connectivity [15,27]. Text-based methods with large models achieve high accuracy but are computationally heavy and not practical for large-scale KGs with millions of entities [24]. To address these challenges, we design the nearest-neighbor approach using K-Nearest Neighbors (KNN). Instead of complex parametric models that may overfit noisy training labels, KNN stores the labeled data and predicts the type of a new entity based on the most common label among its k nearest neighbors in embedding space. We enhance this by using embedding-based similarity, which makes the approach cross-domain and independent of knowledge graph schema, requiring only textual labels or short descriptions. Additionally, we integrate inverse-distance weighting and confidence thresholds, reducing reliance on schema completeness.

Let $(X \in R^n{}^{n \times d})$ denote the matrix of entity embeddings, and $(y \in Z^n)$ is the corresponding type labels from the Wikidata subset. A KNN classifier with cosine similarity approximates the probability distribution $(P(T_e|E_e))$ where T_e denotes the type and E_e is the embedding of entity e. To avoid biasing toward semantically vague types, we excluded generic classes such as Entity, Thing, and

---

[2] https://huggingface.co/sentence-transformers/all-mpnet-base-v2.

Resource from type prediction. We added weighting to the neighbors using an inverse distance weighting interpolation so that the closer points have a greater influence on the final decision than the distant ones. The weights are normalized, and we apply a threshold to ensure only high-confidence predictions are retained to reduce spurious attribution, as shown in Eq. 1.

$$w_i = \frac{\exp(-d_i)}{\sum_{j=1}^{k} \exp(-d_j) + \varepsilon} \quad \text{with} \quad \varepsilon = 10^{-9} \tag{1}$$

The inference stage was applied to all unique subjects s and objects o in the KG. Let $S = s_i$ and $O = o_j$ denote the sets of subjects and objects, respectively. The model generated a list of predicted types $\hat{T}_e = t_1, \ldots t_k$ for each entity $e \in S \cup O$, which are later integrated into the KG as annotations. This approach is non-parametric, which allows it to adopt and increase volumes of training data without retraining. By using weighted score and confidence thresholds, the pipeline creates a soft probabilistic boundary that reflects the uncertainty in open-domain entity typing.

3.3   Outlier Detection

We introduce an unsupervised anomaly detection approach that builds on the output of the previous stage, specifically the predicted entity type sets and the relationships between subject and object. Our goal is to uncover hidden structural inconsistencies in the KG without human supervision.

3.3.1   Type Pattern Mining In this stage, we compute the compatibility between subject-object type pairs (t_s, t_o) and their predicate p. We consider type patterns (t_s, p, t_o) , where $t_s \in T_s$ and $t_o \in T_o$ are subject and object types, which occur more frequently in a KG as trustworthy. The probability distribution of each pattern is computed as shown in Eq. 2.

$$P(t_s, p, t_o) = \frac{C(t_s, p, t_o)}{\sum_{\pi} C(\pi)} \tag{2}$$

where $C()$ is the count of a specific pattern in the graph, and patterns below a minimum support threshold will be filtered out to reduce noise. Co-occurrence alone is insufficient for semantic coherence, so we enhance this with type embedding similarity. Let $V_{t_s}, V_{t_o} \in R^d$ denote the embeddings of subject and object types, and $V_P \in R^d$ be the embedding of predicate p, using the all-MiniLM-L6-v2 encoder. We compute a semantic compatibility score via cosine similarity as shown in Eq. 3.

$$cos\_sim(t_S, p, t_o) = cos(v_{t_s} + vp, v_{t_o}) \tag{3}$$

This formulation reflects a compositional assumption; the semantics of the subject type, combined with the predicate vector, should approximate the object type vector. This aligns with distributional learning principles [4]. We combine the symbolic pattern probability with the geometric compatibility into a convex combination of the two patterns. Each type pair (t_s, t_o) is assigned with a weight score w_{t_s} and w_{t_o} from the previous type inference stage. We compute the average of these weights to ensure that type pairs with high compatibility but low confidence do not affect final score. The final confidence score is chosen by the maximum weighted score over all subject-object type combinations as shown in Eq. 4.

$$score(s, p, o) = \max_{t_s \in T_s, t_o \in T_o} \left[\left(\alpha \cdot \cos\left(v_{t_s} + v_p, v_{t_o}\right) + (1 - \alpha) \cdot P(t_s, p, t_o) \right) \cdot \frac{w_{t_s} + w_{t_o}}{2} \right] \quad (4)$$

where:

- T_s and T_o are the lists of type labels for the subject and object entities.
- w_{t_s} and w_{t_o} are the weight scores for those types.
- $sim(t_s, p, t_o)$ is the cosine similarity between $v_{t_s} + v_p$ and v_{t_o}.
- $P(t_s, p, t_o)$ is the pattern frequency.

Given the critical role of accurate type assignment in the performance of outlier detection models, particularly in ambiguous and noisy KGs, we also compute the entropy of the inferred types for subject and object entities as shown in Eq. 5.

$$H(T_s) = -\sum_i w_{s_i} \log(w_{s_i} + \varepsilon), \quad H(T_o) = -\sum_j w_{(o_j)} \log(w_{o_j} + \varepsilon) \quad (5)$$

where ε is a small constant (< 1) for numerical stability. Each triple is thus assigned a dictionary of confidence scores, which includes its semantic compatibility score, type label distribution, and the entropy and frequency of its most likely pattern.

3.3.2 Pattern-Based Anomaly Detection

In this step, our objective is to identify inconsistent triples within a KG in an unsupervised manner. Rather than relying on external sources, we leverage patterns inherent in the KG to establish normal behavior and flag deviations. We assume that anomalous triples, which diverge from statistically frequent or semantically coherent patterns, are more easily separable from normal triples in feature space. To achieve this, we employ a pattern-based anomaly detection mechanism using the Isolation Forest algorithm [11], which is well-suited for high-dimensional, sparse domains where anomalies are rare and manifest as deviations from dominant patterns.

Let $T = \{(s_i, p_i, o_i)\}_{i=1}^{n}$ denote the set of candidate triples from the KG. For each triple $t_i \in T$, we apply a numerical feature representation $x_i \in R^6$ based on the confidence estimation module described previously. Specifically, each vector x_i contains the following components:

$$\mathbf{x}_i = [\text{conf}_i,\ |\mathcal{T}_{s_i}|,\ |\mathcal{T}_{o_i}|,\ H(\mathcal{T}_{s_i}),\ H(\mathcal{T}_{o_i}),\ \log C(t_{s_i}, p_i, t_{o_i})]\,, \tag{6}$$

where:

- $\text{conf}_i \in [0, 1]$ is the confidence score computed from the combination of pattern probability and semantic similarity.
- $|\mathcal{T}_{s_i}|$ and $|\mathcal{T}_{o_i}|$ represent the number of types assigned to the subject s and object o, respectively.
- $H(\mathcal{T}_{s_i})$ and $H(\mathcal{T}_{o_i})$ is the entropy of the corresponding subject and object type distribution.
- $\log C(t_{s_i}, p_i, t_{o_i})$ denotes the logarithmic pattern frequency from the pattern mining stage.

These features define a pattern-regularity space in which structural anomalies are expected to locate as statistical outliers. To detect such outliers, we use the Isolation Forest algorithm [11] which isolates anomalies using randomly constructed decision trees. The model assigns an anomaly score to each triple based on its feature vector, and the model hyperparameter contamination $\in [0, 1]$ determines the proportion of instances expected to be outliers. This parameter works as an upper bound on the anomaly rate and control model sensitivity. The result set contains triples identified as statistically unusual in terms of both symbolic structure and semantic coherence, representing potential anomalies and likely to be semantically inconsistent with the rest of the KG.

3.3.3 LLM Validation and Reasoning

Our anomaly detection pipeline incorporates Large Language Models (LLMs) to validate triples flagged as potential anomalies. Let $T_{\text{suspect}} \subset T$ denote the set of candidate triples after initial filtering. These triples were not classified as normal and therefore require further inspection. The objective of this stage is to determine whether a triple $t \in T_{\text{suspect}}$ is semantically inconsistent based on commonsense and domain-specific reasoning. We employ GPT-4o and Gemini as LLM inference engines. Triples are grouped into batches $\{T_1, T_2, \ldots, T_k\}$, each of size b, and processed using a prompt-engineering technique for anomaly detection. The prompt is designed to instruct the model to identify incorrect triples, provide a brief explanation, and suggest a correction. A sample prompt is shown in Fig. 2.

The model response is stored in a structured format (t_i, r_i, c_i) where t_i is the detected anomaly, r_i is the anomaly type, and c_i is a corrective suggestion. All valid triples are excluded, and no summary or commentary is requested from the model to reduce output tokens. This output format ensures that the results are suitable for machine processing and subsequent analysis. This approach captures inconsistencies that Isolation Forest models may not detect, such as logical errors, and provides a reason for each anomaly, which aids further interpretation and analysis.

4   Experimental Evaluation

In this section, we present the experimental results of ETCOD on two real-world KGs, YAGO-1 and YAGO-4.5. We begin by adopting the TRIC framework [18] to systematically introduce corruptions into the original KGs, which serve as ground truth, and then apply anomaly detection on the corrupted graphs. Next, we compare ETCOD with SEKA [14] to benchmark its performance against an established method in the literature. The results are evaluated using Precision, Recall, and F-measure. Finally, we perform anomaly detection on the original KGs (without synthetic corruptions) and manually validate the top 100 anomalous triples, ranked by their anomaly scores.

```
The following are triples from a KG:
<triple1; triple2; triple3; ...>
You're an experienced data analyst. Identify anomalies, their
    anomaly types, and suggest how to correct them.
Provide specific reasons for any invalid triples.
Strictly follow this format only:
'Triple, Why it's anomaly (max 10 words), Suggestion'.
Skip valid triples. Do not explain or summarize. Only list
    incorrect triples.
```

Fig. 2. Prompt used for LLM-based anomaly detection on KG triples.

We conducted experiments in a 64-bit computer with Microsoft Windows operating system, an AMD 5800H CPU (3.20 GHz), and 16 GB of RAM. The pipeline can be executed on CPU-only setups due to the efficiency of the selected embedding and inference models. The source code of ETCOD is available on GitHub[3].

4.1   Datasets

We utilize three large-scale, real-world KGs datasets in our research: YAGO-1[4], YAGO-4.5[5], and a subset of Wikidata. These datasets were chosen for their semantic richness and diversity, allowing a comprehensive evaluation of our model across different KGs. YAGO-1 is a well-structured benchmark knowledge graph that integrates data from Wikidata and WordNet [9]. With an estimated accuracy above 95%, it provides a relatively clean environment to assess the model's ability to detect anomalies. Its knowledge base contains approximately 20 million relation instances spanning over 2 million entities. YAGO-4.5, the

---

[3] https://anonymous.4open.science/r/Anomalies-Detection-in-Knowledge-Graph-0315.

[4] https://yago-knowledge.org/downloads/yago-1.

[5] https://yago-knowledge.org/downloads/yago-4-5.

Table 1. Triple corruption as per TRIC [18]

Corruption Type	Original Triple	Corrupted Triple
Swap s and o	<Karen_Bernstein, has GivenName, Karen>	<Karen, hasGivenName, Karen_Bernstein>
Swap s and o, replace p	<Pre-Hysterical_Hare, publishedOn, 1958-11-01>	<1958-11-01,hasPopulation, Pre-Hysterical_Hare>
Replace s, p	<Mary_Elizabeth, created, Aurora_Floyd>	<Islamic_Center_Of_East _Lansing, hasFamilyName, Aurora_Floyd>
Replace o, p	<East_Timor, hasCapital, Dili>	<East_Timor, locatedIn, Nino_Konis_Santana _National_Park>
Replace s, o, preserve p	<Iheanyi_Uwaezuoke, bornOnDate,1973-##-##>	<Moropus, bornOnDate, 0-618-10136-5>

latest version of YAGO, extends the taxonomy and enriches the class hierarchy, incorporating Shapes Constraint Language (SHACL) to enforce semantic constraints [22].

For training our entity type classifier, we extract a subset of Wikidata, a collaboratively curated knowledge base covering a broad range of real-world entities and relations. From Wikidata, we collect 200,000 entity-type pairs (e, T_e) spanning 10,912 unique domain-specific classes. The diversity of Wikidata provides a strong foundation for handling noisy, open-domain KGs.

4.2 Parameter Settings

We use the all-MiniLM-L6-v2 model of Sentence-BERT [13] for entity embedding and the LabelEncoder of sklearn [12] for type label encoding. We use a K-Nearest Neighbor classifier with $k = 5$ selected via cross-validation for type inference. We apply a simple average $\alpha = 0.5$ to combine the semantic similarity and the pattern frequency in the pattern mining stage, treating both values equally. This weight can be adjusted to assign a higher level of importance to either the semantic similarity or pattern frequency, depending on the model's purpose. Furthermore, we set a threshold to 1, which excludes rare patterns with a frequency score below this minimum when computing the triple confidence score. We use the two LLMs; GPT-4o and Gemini, for the validation and reasoning stage. We process 30 triples per chunk with a response limit of 1,500 tokens per request, to ensure the efficiency context window and API cost control for LLMs.

4.3   Anomaly Detection with Synthetic Anomalies

As the two real-world KGs do not have labeled data, we use TRIC [18] to synthetically generate anomalies for each KG by manually corrupting the triples, as outlined in Table 1. This approach preserves the structure of RDF triples while introducing semantic errors and inconsistencies, providing a realistic anomaly detection challenge. We generate a complete test set by corrupting 5% of the original triples. Triples from the original KG are labeled as normal, while corrupted triples serve as ground-truth anomalies. Corrupted triples are generated using structural transformations to simulate both overt and subtle inconsistencies. These include swapping subject and object entities, replacing predicates with random alternatives, and injecting entities from unrelated contexts or external ontologies.

In this evaluation, we treat the corrupted triples as anomalies, and expect ETCOD to identify these anomalous triples. The proposed model performs well in identifying triples with invalid predicates and often suggests more appropriate predicates as replacements. However, when the subject or object is incorrect, the model generally recommends changing the entity without specifying a precise replacement. ChatGPT-4o demonstrated higher consistency across repeated runs, maintaining stable anomaly classifications with minimal variation. This property is advantageous in reproducibility-focused tasks such as anomaly validation in structured triples. However, ChatGPT frequently overlooked anomalies related to triple direction (reversed subjectobject order), which lowered its recall in such cases.

In contrast, as shown in Table 2, Gemini demonstrated a stronger understanding of the compatibility between subject types and predicates, enabling it to suggest corrections based on the expected subject type. Gemini achieved stronger performance in edge cases involving ambiguous triples or rare subject types. Its broader contextual grounding enabled more accurate detection of anomalies tied to semantic compatibility between subject types and predicates. Similar to ChatGPT, Gemini does not always detect reversed triples and primarily focuses on content rather than triple direction. The source code of ETCOD and complete anomaly validation reports are available on GitHub[6] .

Table 2. Comparative performance of ChatGPT-4o and Gemini on anomaly detection in YAGO-1 and YAGO-4.5.

Model	Dataset	Precision	Recall	F1 Score	Consistency (%)
ChatGPT-4o	YAGO-1	0.84	0.71	0.77	92
ChatGPT-4o	YAGO-4.5	0.82	0.69	0.75	90
Gemini	YAGO-1	0.79	0.77	0.78	85
Gemini	YAGO-4.5	0.78	0.75	0.76	83

---

[6] https://anonymous.4open.science/r/Anomalies-Detection-in-Knowledge-Graph-0315.

Next, we compare ETCOD with the baseline approach SEKA (Seeking Knowledge Graph Anomalies) [14]. We use precision and recall values to determine how well each approach performs in identifying the anomalies. The experimental results are provided in Table 3.

Table 3. Comparison of ETCOD with the baseline approach SEKA with 5% of the triples corrupted. The best results are shown in bold.

Approach	YAGO-1			YAGO-4.5		
	Precision	Recall	Run Time (min)	Precision	Recall	Run Time (min)
SEKA	0.78	0.80	120	0.74	0.80	208
ETCOD	**0.82**	**0.86**	**108**	**0.82**	0.80	**187**

Table 3 presents a comparative evaluation of ETCOD against the baseline approach SEKA on two real-world knowledge graphs, YAGO-1 and YAGO-4.5, with 5% of the triples systematically corrupted. The results show that ETCOD consistently outperforms SEKA in terms of Precision and Recall on YAGO-1, achieving 0.82 precision and 0.86 recall compared to SEKA's 0.78 and 0.80, respectively. For YAGO-4.5, ETCOD demonstrates a significant improvement in precision (0.82 vs. 0.74) while maintaining comparable recall (0.80). In addition to accuracy, ETCOD achieves lower run times, reducing execution time by 1021 min across both datasets. This efficiency gain is particularly notable on the larger YAGO-4.5 KG, where ETCOD runs in 187 min compared to SEKA's 208 min. Overall, these results highlight ETCOD's ability to deliver higher anomaly detection performance with improved computational efficiency, making it a more effective and scalable solution for knowledge graph anomaly detection.

4.4 LLM-Generated Explanations Quality Assessment

We evaluated the quality of LLM-generated explanations using automated metrics commonly employed in natural language generation (NLG). Since no gold-standard explanation dataset exists for anomaly detection in knowledge graphs, we generated a reference set of proxy ground-truth explanations by mapping each anomaly type (e.g., invalid predicate, reversed triple, incompatible subject type) to a short canonical description. Model-generated explanations were then compared against these references. We employed two automated metrics, including Semantic Similarity (cosine similarity of SBERT embeddings [13]) and BERTScore [31]. Results in Table 4 show that ChatGPT-4o tends to produce more consistent explanations. However, its semantic similarity scores were slightly lower, as it often produced generic phrasing. Gemini achieved higher semantic similarity and BERTScore, especially on anomalies involving subject-predicate compatibility, suggesting better adaptability to diverse cases.

Table 4. Automated evaluation of explanation quality for ChatGPT-4o and Gemini on anomaly detection across YAGO-1 and YAGO-4.5. Evaluation metrics include SBERT similarity and BERTScore.

Model	YAGO-1		YAGO-4.5	
	SBERT Sim.	BERTScore	SBERT Sim.	BERTScore
ChatGPT-4o	0.77	0.74	0.76	0.73
Gemini	**0.82**	**0.78**	**0.81**	**0.76**

5 Conclusion and Future Work

In this paper, we propose a method for Entity Type and Confidence-based Outlier Detection (ETCOD), aimed at identifying inconsistent or abnormal triples in large-scale knowledge graphs, even in the absence of type annotations. Our pipeline detects both clear and subtle semantic errors by combining pattern-based scoring with validation from a Large Language Model (LLM). This unsupervised anomaly detection approach calculates semantic compatibility using <subject_type, predicate, object_type> patterns and entity type embeddings (subject_type, object_type). It eliminates the need for manual intervention and can serve as an additional layer to traditional schema-based constraints. However, the present approach holds some challenges that need to be addressed in the future. These include the inability to distinguish between a subject and an object when they share the same name. As a remedy, we aim to strengthen the ability of LLMs to detect and correct anomalies that require deep factual understanding. Therefore, we intend to incorporate the framework of the Interpretive Paradigm [20] into the dynamic knowledge base anomaly detection.

References

1. Ahmetaj, S., et al.: Common foundations for SHACL, ShEx, and PG-schema. In: Web Conference, pp. 8–21 (2025)
2. Akoglu, L., Tong, H., Koutra, D.: Graph based anomaly detection and description: a survey. Data Min. Knowl. Disc. **29**, 626–688 (2015)
3. Biswas, R., Sofronova, R., Alam, M., Sack, H.: Entity type prediction in knowledge graphs using embeddings. arXiv preprint arXiv:2004.13702 (2020)
4. Bordes, A., Usunier, N., Garcia-Duran, A., Weston, J., Yakhnenko, O.: Translating embeddings for modeling multi-relational data. Adv. Neural Inf. Process. Syst. **26** (2013)
5. Boylan, J., Mangla, S., Thorn, D., Ghalandari, D.G., Ghaffari, P., Hokamp, C.: Kgvalidator: a framework for automatic validation of knowledge graph construction. arXiv preprint arXiv:2404.15923 (2024)
6. Hawkins, D.M.: Identification of Outliers, vol. 11. Springer (1980)
7. Jia, B., Dong, C., Chen, Z., Chang, K.C., Sullivan, N., Chen, G.: Pattern discovery and anomaly detection via knowledge graph. In: International Conference on Information Fusion, pp. 2392–2399. IEEE (2018)

8. Johnson, R.J., Williams, J.P., Bauer, K.W.: Autogad: an improved ICA-based hyperspectral anomaly detection algorithm. Trans. Geosci. Remote Sens. **51**(6), 3492–3503 (2012)

9. Kasneci, G., Ramanath, M., Suchanek, F., Weikum, G.: The YAGO-NAGA approach to knowledge discovery. SIGMOD Rec. **37**(4), 41–47 (2009)

10. Li, Y., Ge, Y., Zhang, Y.: Tutorial on fairness of machine learning in recommender systems. In: International Conference on Information & Knowledge Management, pp. 4857–4860. ACM (2021)

11. Liu, F.T., Ting, K.M., Zhou, Z.H.: Isolation forest. In: International Conference on Data Mining, pp. 413–422. IEEE (2008)

12. Pedregosa, F., et al.: Scikit-learn: machine learning in python. J. Mach. Learn. Res. **12**, 2825–2830 (2011)

13. Reimers, N., Gurevych, I.: Sentence-bert: sentence embeddings using siamese bert-networks. In: Conference on Empirical Methods in Natural Language Processing. ACL (2019)

14. Senaratne, A.: Seka: seeking knowledge graph anomalies. In: Web Conference 2023, pp. 568–572. ACM (2023)

15. Senaratne, A.: Anomaly detection in graphs for knowledge discovery and data quality enhancement (2024)

16. Senaratne, A., Christen, P., Williams, G., Omran, P.G.: Unsupervised identification of abnormal nodes and edges in graphs. J. Data Inf. Q. **15**(1), 1–37 (2022)

17. Senaratne, A., Christen, P., Williams, G., Omran, P.G.: Rule-based knowledge discovery via anomaly detection in tabular data. In: CEUR Workshop Proceedings, vol. 3433. CEUR-WS (2023)

18. Senaratne, A., Omran, P.G., Christen, P., Williams, G.: Tric: a triples corrupter for knowledge graphs. In: ESWC, pp. 117–122. Springer (2023)

19. Senaratne, A., Omran, P.G., Williams, G., Christen, P.: Unsupervised anomaly detection in knowledge graphs. In: 10th International Joint Conference on Knowledge Graphs, pp. 161–165 (2021)

20. Senaratne, A., Seneviratne, L.: Embedded to interpretive: a paradigm shift in knowledge discovery to represent dynamic knowledge. In: CEUR Workshop Proceedings, vol. 3433. CEUR-WS (2023)

21. Seneviratne, L., Senaratne, A.: Data atomization: a framework for on-demand association and access control of sensitive data. In: 49th Annual Computers, Software, and Applications Conference, pp. 2275–2280. IEEE (2025)

22. Suchanek, F.M., Alam, M., Bonald, T., Chen, L., Paris, P.H., Soria, J.: Yago 4.5: a large and clean knowledge base with a rich taxonomy. In: International Conference on Research and Development in Information Retrieval, pp. 131–140 (2024)

23. Sun, Z., Deng, Z.H., Nie, J.Y., Tang, J.: Rotate: knowledge graph embedding by relational rotation in complex space. arXiv preprint arXiv:1902.10197 (2019)

24. Tian, J., Kimura, M.: Multi-task learning for joint entity and relation extraction on open-domain. In: Asia Conference on Machine Learning and Computing, pp. 170–176 (2024)

25. Tsaneva, S., Dessì, D., Osborne, F., Sabou, M.: Knowledge graph validation by integrating LLMs and human-in-the-loop. Inf. Process. Manag. **62**(5), 104145 (2025)

26. Vaska, N., Leahy, K., Helus, V.: Context-dependent anomaly detection with knowledge graph embedding models. In: International Conference on Automation Science and Engineering, pp. 2020–2027. IEEE (2022)

27. Wang, M., Qiu, L., Wang, X.: A survey on knowledge graph embeddings for link prediction. Symmetry **13**(3), 485 (2021)

28. Wang, Y., Li, Y., Fan, J., Ye, C., Chai, M.: A survey of typical attributed graph queries. World Wide Web **24**(1), 297–346 (2021)
29. Wu, J., Gan, W., Zhang, J., Yu, P.S.: Adkgd: anomaly detection in knowledge graphs with dual-channel training. arXiv preprint arXiv:2501.07078 (2025)
30. Yin, Y., et al.: Augmenting biomedical named entity recognition with general-domain resources. J. Biomed. Inform. **159**, 104731 (2024)
31. Zhang, T., Kishore, V., Wu, F., Weinberger, K.Q., Artzi, Y.: Bertscore: evaluating text generation with bert. arXiv preprint arXiv:1904.09675 (2019)
32. Zhou, A., et al.: Knowgraph: knowledge-enabled anomaly detection via logical reasoning on graph data. In: Conference on Computer and Communications Security, pp. 168–182. ACM (2024)

Top-k Ranking with Exact Positional Fairness

Nina A. Liebrand$^{(\boxtimes)}$ and Stefan Conrad

Heinrich Heine University, Universitätsstraße 1, 40225 Düsseldorf, Germany
`{nina.liebrand,stefan.conrad}@hhu.de`

Abstract. Ranking individuals by relevance is central to many decision-making systems, such as hiring. A common approach is top-k ranking, where only the best k individuals are selected. However, social biases in data can make automated ranking systems discriminatory and affect individuals' lives. In line with the AI Act, fair top-k ranking has recently gained attention. In this work, we focus on positional fairness in top-k rankings, which ensures equal exposure of groups across ranking positions. We formulate this as a constrained optimization problem that balances positional fairness with ranking quality. We propose two novel exact algorithms with optimal guarantees, along with an efficient greedy approximation. All methods support non-binary and multiple protected attributes as often present in real-world. Experiments on synthetic and real-world datasets show that our methods outperform recent approaches in both positional fairness and ranking quality.

Keywords: Fair ranking · Algorithmic fairness · Positional fairness · Constrained optimization · Group fairness

1 Introduction

Ranking is an essential task in information retrieval and aims to order documents by their relevance. A particularly important setting is *Top-k Ranking*, where only the best ranked k items are presented to the user [20]. This approach is widely used in applications such as search engines and recommendation systems. For instance, in hiring, recruiters must choose a limited number of k candidates from a larger pool of applicants.

In contexts with human candidates, protected attributes need to be considered. Protected attributes refer to characteristics such as gender, ethnicity, or age that include groups legally or ethically recognized as requiring protection [6]. For example, in Spain, at least 40% and at most 60% of the candidates on electoral lists may belong to the same gender [17]. However, traditional top-k ranking systems are purely based on relevance scores, which can reflect societal biases and lead to discriminatory outcomes [7,18]. For instance, decision-making systems for college admissions based on SAT scores were shown to be discriminating against women, who on average score lower than men despite equal academic

Q. V. Nguyen et al. (Eds.): AusDM 2025, CCIS 2765, pp. 470–485, 2026.
https://doi.org/10.1007/978-981-95-6786-7_32

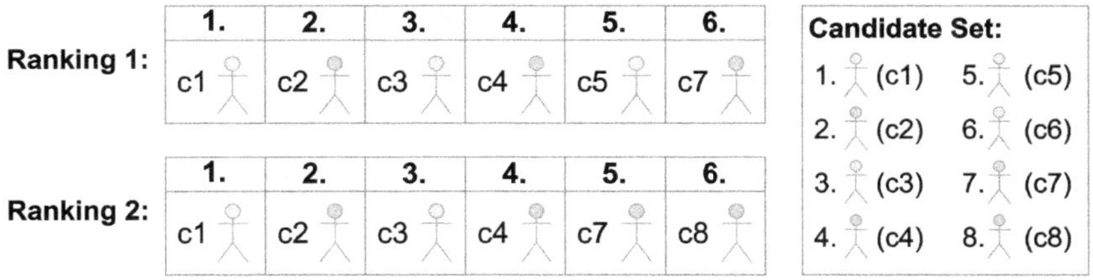

Fig. 1. Positional fairness in top-k rankings. While the first ranking is positionally unfair, the second ranking gives both groups equal positional exposure.

performance [16]. The high societal consequences of biased machine learning models have led to regulatory responses, most notably the European Union's AI Act.

To address this issue and follow the AI Act, the field of *Fair Top-k Ranking* has emerged [20]. Fair top-k ranking aims to ensure that rankings do not discriminate against any individuals or social groups while maintaining high quality. So far, approaches mainly focus on *Proportional Fairness*, meaning that each social group is represented equally in the top-k ranking [20]. Another fairness concern is *Positional Fairness*, which arises when a ranking favors one group by consistently assigning it better-ranked positions [11].

Consider the candidate set ordered by relevance in Fig. 1, where each candidate belongs to either the red or blue group. In the first top-6 ranking, the three most relevant candidates from each group are selected, making it proportionally fair. However, it gives positional bias, as the red group consistently occupies better positions than the blue. The second ranking retains the original order of candidates and compensates the top positions of the red group by a greater representation of blue candidates occupying mid and lower ranks.

Positional fairness in rankings highly impacts the attention an item receives, as users, e.g., in search engines, focus significantly more on top results [5]. For human candidates, positional bias can affect individuals' lives with drastic consequences, such as access to jobs or colleges. This work deals with positional fairness in top-k rankings and comes with the following contributions:

- We formulate the problem of positional fairness as a constrained optimization problem, introduce metrics, and give theoretical evaluations of conflicts between fairness and quality criteria.
- We introduce two exact top-k ranking algorithms that optimally fulfill the stated problem and further propose a greedy approximation (released on GitHub https://github.com/NinaL-ai/FaiRPosO).
- We validate our algorithms' effectiveness on synthetic and real-world datasets, showing they outperform recent methods in terms of fairness and quality.

2   Related Work

Fair top-k ranking has recently gained attention. Existing work focuses on defining fair ranking criteria and measuring discrimination and quality, as well as developing *fairness-aware algorithms* for mitigating bias in top-k rankings.

Algorithmic fairness either focuses on *individual fairness*, which requires similar individuals to be treated equally [7], or *group fairness*, where social groups are treated equally [12]. Zehlike et al. [20] introduced the fair top-k ranking problem and proposed three fair ranking criteria [20,21]. *Selection Utility* and *Ordering Consistency* ensure quality by requiring that the most relevant candidates are selected and the order by relevance is preserved, respectively. In contrast, *Proportional Fairness* ensures group fairness by requiring specific minimal representations of protected groups across all subsets in a top-k ranking. These criteria incorporate a statistical significance test. Liebrand et al. [11] introduced another fairness criterion called *Positional Fairness*, which requires the sum of logarithmically discounted position weights to be equal across groups.

To evaluate rankings, researchers have proposed a variety of metrics. *Statistical disparity* [3] quantifies the extent to which representations of groups differ, e.g., the acceptance rates across groups in college admissions. Based on this, several metrics have been proposed to measure fairness [11,19,21,22] and quality [10,20]. For example, the *Normalized Discounted Cumulative Gain* (NDCG) [10] measures selection utility using a logarithmic discount.

Building on these fairness metrics, researchers have proposed fairness-aware ranking algorithms. Some algorithms adjust relevance scores before ranking. For instance, Asudeh et al. [2] created fair relevance scores by adding weights to the scoring function, and Feldman et al. [8] adjusted the score distributions. Other algorithms focus on selecting the candidates in a fair manner. Singh and Joachims [15] introduced a framework where groups receive exposure proportional to their relevance. Celis et al. [4] proposed approximation algorithms incorporating a maximum group representation in top positions. Zehlike et al. first introduced algorithms for fair top-k rankings. Their algorithm *FA*IR* [20] iteratively selects candidates for the top-k set based on proportional fairness and a statistical test. They later extended their algorithm to account for multiple protected groups [21]. Recently, Liebrand et al. [11] introduced an algorithm allowing for non-binary and multiple protected groups while optimally reaching proportional fairness.

3   Preliminaries

In this section, we introduce the formal notations used throughout this paper. Each human candidate $c \in \mathcal{C}$ is associated with a *relevance score* $q(c)$ and belongs to one social group $g \in \mathcal{G}$. A ranking r is constructed from the set of candidates $\mathcal{C}$, assigning a unique rank to each of the selected k candidates.

Definition 1 (Dataset). *A dataset is a tuple* $\mathcal{D} = (\mathcal{C}, q, \mathcal{Z}, \mathcal{G}, m, k)$, *where* $\mathcal{C}$ *is a set of candidates and* $q : \mathcal{C} \to \mathbb{R}$ *assigns a relevance score to each candidate.*

$\mathcal{Z} = \{\mathcal{Z}_1, \ldots, \mathcal{Z}_m\}$ *is the set of discrete protected attributes (e.g., age, gender), and* $\mathcal{G} \subseteq \mathcal{Z}_1 \times \cdots \times \mathcal{Z}_m$ *is the set of observed intersectional groups (e.g., young female). Protected groups are denoted by* $\mathcal{G}_p \subseteq \mathcal{G}$. *The membership function* $m : \mathcal{C} \to \mathcal{G}$ *maps each candidate to one group. Finally,* $k \in \mathbb{N}$ *is the top-k cutoff.*

Definition 2 (Top-k Ranking). *The function* $r : \mathcal{C} \to \{1, \ldots, k\} \cup \{\perp\}$ *defines a top-k ranking, where* $r(c) = r \in \{1, \ldots, k\}$ *assigns a unique rank to each of the* k *selected candidates* $c \in \mathcal{C}$. *Candidates that are not selected in the top-k set are indicated by* $r(c) = \perp$. *The top-k set is* $\tau_k = \{c \in \mathcal{C} \mid r(c) \neq \perp\}$.

Definition 3 (Group Proportion in Top-k Ranking [11]). *The group proportion* $P_k(g)$ *is defined as the share of candidates from group* $g \in \mathcal{G}$ *among the top-k candidates:*

$$P_k(g) := \frac{|\{c \in \tau_k : m(c) = g\}|}{k}.$$

4 Fairness-Aware Top-k Ranking Criteria

Building on the criteria defined by Zehlike et al. [20] and Liebrand et al. [11], we aim for four fair top-k ranking criteria:

(i) *Maximum Relevance*: Selection of the most relevant candidates.
(ii) *Ordering Consistency*: Candidates are ordered descending by relevance.
(iii) *Proportional Fairness*: Proportions of protected groups equal their target.
(iv) *Positional Fairness*: Social groups are treated equally in terms of positions.

While maximum relevance and ordering consistency guarantee the quality of a top-k ranking, proportional and positional fairness ensure fairness. The criteria are formally defined in the following subsections, including novel metrics.

4.1 Maximum Relevance

Intuitively, a ranking should contain the best candidates indicated by the highest relevance scores. Formally, a ranking τ_k satisfies *Maximum Relevance* if:

$$\forall c \in \tau_k, \forall c' \in \mathcal{C} \setminus \tau_k : q(c) \geq q(c'). \tag{1}$$

As simply selecting the most relevant candidates is often not possible while ensuring fairness, we further consider the less strict criterion of *Maximum Group Relevance*, where the most relevant candidates from each group are selected [11]:

$$\forall g \in \mathcal{G}, \forall c \in \tau_k, \forall c' \in \mathcal{C} \setminus \tau_k : m(c) = m(c') = g \implies q(c) \geq q(c'). \tag{2}$$

Definition 4 (Maximum (Group) Relevance Disparity). *Maximum relevance is measured using the relevance difference between the least relevant candidate in* τ_k *and the most relevant candidate outside of it. The score is normalized in the interval* $[0, 1]$ *using the highest possible relevance difference:*

$$MRD(\tau_k, \mathcal{C}) := \frac{\max\left(0, \ \max_{c \in \mathcal{C} \setminus \tau_k} q(c) - \min_{c' \in \tau_k} q(c')\right)}{\max_{c \in \mathcal{C}} q(c) - \min_{c' \in \mathcal{C}} q(c')}. \tag{3}$$

Maximum group relevance is measured using the group-specific MRDs:

$$MGRD(\tau_k, \mathcal{C}) := \max_{g \in \mathcal{G}} MRD(\{c \in \tau_k \mid m(c) = g\}, \{c \in \mathcal{C} \mid m(c) = g\}). \quad (4)$$

Remark 1. Both MRD and MGRD take values in $[0, 1]$ and equal zero when maximum (group) relevance is satisfied. The MRD score is always greater than or equal to the MGRD score.

4.2   Ordering Consistency

In a top-k ranking the candidates should be ordered by relevance. Formally, a ranking τ_k satisfies *Ordering Consistency* if [20]:

$$\forall c, c' \in \tau_k : q(c) > q(c') \implies r(c) < r(c'). \quad (5)$$

Definition 5 (Ordering Consistency Disparity (OCD)). *We measure ordering consistency by computing the maximum difference in relevance scores of candidates that are falsely ordered in the top-k ranking. The score is normalized in the interval $[0, 1]$ using the highest possible relevance difference:*

$$OCD(\tau_k) := \frac{\max\left(0, \ \max_{c, c' \in \tau_k : r(c) < r(c')} q(c') - q(c)\right)}{\max_{c \in \tau_k} q(c) - \min_{c' \in \tau_k} q(c')}.$$

4.3   Proportional Fairness

A top-k ranking is considered to be *Proportional Fair* if each protected group is represented according to their desired target proportion t_g:

$$\forall g \in \mathcal{G}_p : P_k(g) = t_g. \quad (6)$$

This criterion ensures that protected groups are not underrepresented. For equal representations, the targets t_g can be set to $\frac{1}{\mathcal{G}}$.

4.4   Positional Fairness

Intuitively, a ranking can fulfill proportional fairness but can still be considered unfair if one group is consistently ranked higher (Fig. 1). This can lead to biases in the attention received by different social groups. By assigning a logarithmic discounted weight to positions reflecting their importance, a positional score can be computed for each group [11]:

$$PS(\tau_k, g) := \sum_{c \in \tau_k : m(c) = g} w(r(c)), \quad w(i) = \frac{1}{\log_2 (i + 1)}. \quad (7)$$

A top-k ranking τ_k satisfies *Positional Fairness* if all positional scores are equal:

$$\forall g, g' \in \mathcal{G} : PS(\tau_k, g) = PS(\tau_k, g') = \frac{1}{|\mathcal{G}|} \sum_{i=1}^{k} w(i). \quad (8)$$

5   Fair Positional Top-k Ranking

We aim to find a top-k ranking that optimally fulfills positional fairness while maintaining quality. Satisfying the three remaining fair ranking criteria under the constraint of positional fairness may not be compatible with each other. In detail, fulfilling positional fairness leads to the following conflicts.

Lemma 1. *If a top-k ranking τ_k satisfies positional fairness, then every group $g \in \mathcal{G}$ with count n_g in τ_k must satisfy the interval $l^\star \leq n_g \leq u^\star$, where*

$$l^\star := \min\left\{ n \in \{1,\ldots,k\} : \sum_{i=1}^{n} w(i) \geq \frac{1}{|\mathcal{G}|} \sum_{i=1}^{k} w(i)\right\},$$

$$u^\star := k - m^\star, \quad m^\star := \min\left\{ m \in \{1,\ldots,k\} : \sum_{i=1}^{m} w(i) \geq \frac{|\mathcal{G}|-1}{|\mathcal{G}|} \sum_{i=1}^{k} w(i)\right\}.$$

Proof. Let $W_{\mathrm{tot}} = \sum_{i=1}^{k} \frac{1}{\log_2(i+1)}$ be the total positional weight of the top-k ranking. From each group at least one candidate must be selected:

$$\forall g, g' \in \mathcal{G} : \quad \mathrm{PS}(\tau_k, g) = \mathrm{PS}(\tau_k, g') \quad \wedge \quad \tau_k \neq \emptyset.$$
$$\Longleftrightarrow \quad \forall g \in \mathcal{G}, \exists \alpha \in \mathbb{R} : \quad \mathrm{PS}(\tau_k, g) = \alpha > 0.$$
$$\Longleftrightarrow \quad \forall g \in \mathcal{G}, \quad \exists c \in \tau_k : m(c) = g.$$

If group g has n_g members in the ranking, the maximum positional weight is attained by placing those n_g members in the top n_g ranks, which equals $\mathrm{PS}(\tau_k, g) = \sum_{i=1}^{n_g} w(i)$. Positional fairness requires this to be at least the equal share $W_{\mathrm{tot}}/|\mathcal{G}|$. The monotonicity of $w(\cdot)$ gives that the partial sums $\sum_{i=1}^{n} w(i)$ are strictly increasing in n. The minimal such n is therefore the smallest integer n satisfying the displayed inequality.

In contrast, the minimum positional weight of a group with n_g members is received when its candidates occupy the bottom n_g ranks. Then all other groups $|\mathcal{G}| - 1$ occupy the top m positions with

$$\sum_{i=1}^{m} w(i) \geq \frac{|\mathcal{G}|-1}{|\mathcal{G}|} \sum_{i=1}^{k} w(i).$$

Hence, we must have at most $n_g \leq k - m^\star$ members. $\square$

Theorem 1. *Positional fairness in top-k rankings conflicts with maximum relevance if a group is not sufficiently represented in the k most relevant candidates.*

Proof. Let τ_k^* be a top-k ranking that satisfies positional fairness, and let

$$\tau_k^{\mathrm{rel}} = \arg\max_{S \subseteq \mathcal{C},\, |S|=k} \sum_{c \in S} q(c) \qquad .$$

be the set of the k most relevant candidates. The number of members of group g among τ_k^{rel} is denoted by $r_g := \left|\{c \in \tau_k^{\mathrm{rel}} : m(c) = g\}\right|$. By Lemma 1, positional fairness requires $l^\star \le n_g \le u^\star$. If $r_g < l^\star$, then even placing all r_g members of g in the top ranks yields less than the equal share $\frac{1}{|\mathcal{G}|}W_{\text{tot}}$, forcing inclusion of $l^\star - r_g > 0$ less relevant members of g. If $r_g > u^\star$, then other groups cannot attain their fair share unless members of g are removed in favor of less relevant others, contradicting the maximum relevance condition. $\qquad\square$

Positional fairness does not conflict with the less strict maximum group relevance, as it does not matter which n_g candidates are selected from each group.

Theorem 2. *A top-k ranking τ_k that satisfies* positional fairness, maximum group relevance, *and* ordering consistency *uniquely determines the number of selected candidates from each group. Unless these group counts coincide with the proportional targets $t_g \cdot k$, proportional fairness is necessarily violated.*

Proof. For fixed counts $n = (n_g)_{g \in \mathcal{G}}$, maximum group relevance requires that the n_g most relevant candidates of each group g are chosen, and ordering consistency uniquely fixes their interleaving by global relevance order. Hence the set of positions $I_g(n) \subseteq \{1, \ldots, k\}$ occupied by group g are uniquely determined. These positions uniquely determine positional scores $\text{PS}(\tau_k, g) = \sum_{i \in I_g(n)} w(i)$.

If there exist two distinct count vectors $n \ne n'$, then for some group g we must have $n_g \ne n'_g$. By strict monotonicity of the positional score, this implies $\text{PS}(\tau_k, g) \ne \text{PS}(\tau'_k, g)$. Thus at most one integer vector n satisfies positional fairness, proving uniqueness. $\qquad\square$

These conflicts require trade-offs, where either proportional fairness or ordering consistency is prioritized. Both approaches are desirable in different real-world scenarios with either a strict diversity requirement or with a desired original relevance order, such as in search engines or recommendation systems.

5.1 Problem Statement

To maintain ranking quality, we focus on ordering consistency and state the problem of positional fairness in top-k rankings as follows:

Problem 1. Given a set of candidates $\mathcal{C}$, relevance score $q(c)$, protected groups $\mathcal{Z}$, group membership $m(c) \in \mathcal{G}$, and $k \in \mathbb{N}$, select a top-k subset $\tau_k \subseteq \mathcal{C}$ that satifies: positional fairness, ordering consistency, and maximum group relevance.

To formally state this problem, we introduce a binary assignment matrix $x \in \{0, 1\}^{|\mathcal{C}| \times k}$, where $x_{c,r} = 1$ denotes that candidate c is assigned to position r. Further, we introduce a binary vector $y \in \{0, 1\}^{|\mathcal{C}|}$, where $y_c = 1$ indicates that candidate c is selected in the top-k ranking.

$$\min \quad \max_{g \in \mathcal{G}} \text{PS}(\tau_k, g) - \min_{g' \in \mathcal{G}} \text{PS}(\tau_k, g')$$

$$\begin{aligned}
\text{s. t.} \quad & \sum_{c \in \mathcal{C}} y_c = k \\
& \sum_{c \in \mathcal{C}} x_{c,r} = 1 && \forall r \in \{1, \ldots, k\}, \\
& \sum_{r=1}^{k} x_{c,r} = y_c && \forall c \in \mathcal{C}, \\
& \sum_{c \in \mathcal{C}} q(c)\, x_{c,p} \ge \sum_{c \in \mathcal{C}} q(c)\, x_{c,p+1} && \forall p \in \{1, \ldots, k-1\}, \\
& y_c \ge y_{c'} && \forall c, c' \in \mathcal{C},\ m(c) = m(c') :\ q(c) > q(c').
\end{aligned}$$

The objective aims for positional fairness, while the first three constraints ensure a valid top-k ranking by selecting exactly k candidates and assigning each position to a unique candidate and vice versa. Further, the last two constraints ensure ordering consistency and maximum group relevance. The constraint of ordering consistency is simplified by ensuring order between two consecutive positions in the ranking, which is sufficient due to the transitivity of orderings.

5.2   Algorithms

We propose *FaiRPosO*, a family of three algorithms for fair top-k ranking that ensures positional fairness while preserving ordering consistency. The first algorithm, *FaiRPosO-CP*, uses the constraint programming solver from *Google OR-Tools* [13] to efficiently search for feasible solutions of our proposed constrained optimization problem.

The second algorithm, *FaiRPosO-Exh*, uses an exhaustive search to explore all possible solutions, as depicted in Algorithm 1. It starts by sorting the candidates by relevance in time $\mathcal{O}(|\mathcal{C}| \log |\mathcal{C}|)$. For each group $g \in \mathcal{G}$, it determines the indices of candidates belonging to g within τ_k, resulting in a sorted list I_g of positions for every group in $\mathcal{O}(|\mathcal{C}|)$ time. Further, it initializes the weight function W for each position in $\mathcal{O}(k)$. The maximum group relevance constraint allows only prefixes of candidates from each group to be selected. For each group $g \in \mathcal{G}$, the valid prefix lengths are limited to the range $[1, \ldots, \min(k-1, |I_g|)]$. This ensures that at least one candidate from each group is selected as required by Lemma 1 and no group contributes more candidates than available or than the ranking length allows. These prefix options are computed in $\mathcal{O}(|\mathcal{G}|)$ time. The number of prefix combinations is equivalent to the number of compositions of k into $|\mathcal{G}|$ positive integers summing to k. This is determined by the *stars and bars* theorem [14]:

$$C_{k-1, |\mathcal{G}|-1} = \binom{k-1}{|\mathcal{G}|-1}.$$

To efficiently enumerate these combinations, we employ a recursive procedure. At each recursive step, the algorithm selects a possible number of candidates p for the current group. The algorithm then recursively proceeds to the next group with the updated total count of selected candidates until all groups have been processed or the prefix sizes exceed the target size k. When a valid combination is found, the corresponding candidates are merged in $\mathcal{O}(k)$, sorted by relevance in $\mathcal{O}(k \log k)$, and evaluated for positional fairness in $\mathcal{O}(k + |\mathcal{G}|)$ time. It is not possible to decrease the computation time by early pruning, as positional scores can alternate with each additional candidate. This gives the algorithm a total complexity of $\mathcal{O}(|\mathcal{C}| \log |\mathcal{C}| + C_{k-1, |\mathcal{G}|-1} \cdot (k \log k + |\mathcal{G}|))$.

Due to the complexity of our first two approaches, we finally present the algorithm *FaiRPosO-Greedy* (Algorithm 2) to approximate the optimal ranking. This algorithm first sorts all candidates by relevance in $\mathcal{O}(|\mathcal{C}| \log |\mathcal{C}|)$ time and reduces the candidate set to the first k candidates of each group in $\mathcal{O}(k \cdot |\mathcal{G}|)$ time. Then it computes the weight of each position in the ranking and calculates

---

Algorithm 1. Exhausive Fair Positional and Ordered Ranking (FaiRPosO-Exh)

Require: Dataset $\mathcal{D} = (\mathcal{C}, q, \mathcal{Z}, \mathcal{G}, m, k)$
Ensure: Fair top-k candidate set τ_k
 1: Sort $\mathcal{C}$ descending by $q(c)$
 2: $I_g \leftarrow$ indices of $\{c \in \mathcal{C} \mid m(c) = g\}$ in τ_k for all $g \in \mathcal{G}$
 3: $W \leftarrow \frac{1}{\log_2(i+1)}$ for $i = 1, \ldots, k$ $\triangleright$ position weights
 4: $prefixOptions[g] \leftarrow [1, \ldots, \min(k-1, |I_g|)]$ for all $g \in \mathcal{G}$ $\triangleright$ valid group sizes
 5: $\tau_k \leftarrow \emptyset$, best_score $\leftarrow \infty$
 6: **function** SEARCHFAIRRANK(i, prefixes, total)
 7:     **if** total $> k \vee (i = |\mathcal{G}| \wedge$ total $\neq k) \vee (i \neq |\mathcal{G}| \wedge$ total $= k)$ **then**
 8:         **return** $\triangleright$ invalid combination
 9:     **end if**
10:     **if** total $= k$ **then** $\triangleright$ valid combination
11:         $S \leftarrow \bigcup_{g \in \mathcal{G}} \{$top $prefixes[g]$ from $I_g\}$ $\triangleright$ construct candidate set
12:         Sort S descending
13:         $score[g] \leftarrow 0$ for all $g \in \mathcal{G}$
14:         **for** $j = 0$ to $k-1$ **do** $\triangleright$ calculate positional scores
15:             $g \leftarrow$ group of candidate at $S[j]$
16:             $score[g] \leftarrow score[g] + W[j]$
17:         **end for**
18:         $\Delta \leftarrow \max_g score[g] - \min_g score[g]$ $\triangleright$ positional unfairness
19:         **if** $\Delta <$ best_score **then**
20:             $\tau_k \leftarrow S$, best_score $\leftarrow \Delta$
21:         **end if**
22:         **return**
23:     **end if**
24:     **for** $p \in prefixOptions[$i-th group in $\mathcal{G}]$ **do**
25:         SEARCHFAIRRANK($i + 1$, prefixes with p appended, total $+p$)
26:     **end for**
27: **end function**
28: SEARCHFAIRRANK(0, empty list, 0) $\triangleright$ start recursive search
29: **return** τ_k

---

the uniform positional score target of $\frac{1}{|\mathcal{G}|} \sum_{i=1}^{k} w(i)$, leading to $\mathcal{O}(k)$ time. The algorithm incrementally selects a candidate at each iteration by evaluating the most relevant remaining candidate. This candidate is included in τ_k if its addition improves the group's positional fairness score according to the target. If not, the group is excluded from further iterations, as the maximum group relevance constraint prevents the inclusion of less relevant candidates from that group at a later stage. This process continues until the final ranking includes k candidates. This loop is executed for at most $k \cdot |\mathcal{G}|$ times, and each iteration only involves constant time operations. This leads to a total complexity of $\mathcal{O}(|\mathcal{C}| \log |\mathcal{C}| + k \cdot |\mathcal{G}|)$.

6   Evaluation

We evaluate the effectiveness of our proposed algorithms on synthetic data, as well as on several real-world datasets. Further, we compare our approach to recent methods.

Algorithm 2. Greedy Fair Positional and Ordered Ranking (*FaiRPosO-Greedy*)

Require: Dataset $\mathcal{D} = (\mathcal{C}, q, \mathcal{Z}, \mathcal{G}, m, k)$
Ensure: Fair top-k candidate set τ_k
1: Sort $\mathcal{C}$ descending by relevance $q(c)$
2: $\mathcal{C} \leftarrow$ first k elements of $\{c \in \mathcal{C} \mid m(c) = g\}$ for all $g \in \mathcal{G}$
3: $pfs_g \leftarrow 0$ for all $g \in \mathcal{G}$ ▷ init positional scores
4: $w(i) \leftarrow \frac{1}{\log_2(i+1)}$ for $i = 1, \ldots, k$ ▷ calculate position weights
5: $pf_{\text{target}} \leftarrow \sum_{i=1}^{k} w(i)/|\mathcal{G}|$ ▷ ideal positional score per group
6: $\tau_k \leftarrow \emptyset$, $R \leftarrow \emptyset$, $p \leftarrow 1$
7: **while** $|\tau_k| < k$ and $p \leq |\mathcal{C}|$ **do**
8:     $c \leftarrow \mathcal{C}[p]$
9:     $g \leftarrow m(c)$
10:     **if** $g \in R$ **then**   ▷ skip restricted group
11:         $p \leftarrow p + 1$
12:         **continue**
13:     **end if**
14:     $i \leftarrow |\tau_k| + 1$
15:     **if** $|pfs_g + w(i) - pf_{\text{target}}| < |pfs_g - pf_{\text{target}}|$ **or** $|R| = |\mathcal{G}| - 1$ **then**
16:         $\tau_k \leftarrow \tau_k \cup \{c\}$ ▷ add candidate
17:         $pfs_g \leftarrow pfs_g + w(i)$ ▷ update groups positional score
18:     **else**
19:         $R \leftarrow R \cup \{g\}$ ▷ restrict group (maximum group relevance criterion)
20:     **end if**
21:     $p \leftarrow p + 1$
22: **end while**
23: **return** τ_k

With our experiments we aim to answer the following research questions:

- **RQ1:** Can the proposed algorithms optimally fulfill the stated Problem 1?
- **RQ2:** Are the algorithms applicable to real-world datasets with non-binary and multiple protected attributes?
- **RQ3:** How do the algorithms compare to existing methods in terms of fairness, ranking quality, and runtime?

6.1   Optimallity of the Algorithms

First, we evaluate the optimality of our proposed algorithms on synthetic data. To examine whether the algorithms can handle non-binary and multiple protected groups, we generate four datasets. We compare our methods to a baseline, called colorblind, which ranks candidates purely based on their relevance

Table 1. Experimental results on generated datasets. The protected attributes (Prot. Attr.) feature different types: binary, multiple binary, non-binary, or multiple non-binary.

Prot. Attr.	#Prot. Attr.	$|\mathcal{G}|$	Method	MPoD loss	MPD	OCD	MGRD
Binary	1	2	Colorblind	0.967	1.0	0.0	0.0
			FaiRPosO-CP	0.0	0.3		
			FaiRPosO-Exh	0.0	0.3		
			FaiRPosO-Greedy	0.0	0.3		
Multiple binary	2	4	Colorblind	0.501	0.333	0.0	0.0
			FaiRPosO-CP	0.0	0.2		
			FaiRPosO-Exh	0.0	0.2		
			FaiRPosO-Greedy	0.0	0.2		
Non-binary	1	4	Colorblind	0.732	0.8	0.0	0.0
			FaiRPosO-CP	0.0	0.133		
			FaiRPosO-Exh	0.0	0.133		
			FaiRPosO-Greedy	0.0	0.133		
Multiple non-binary	2	6	Colorblind	0.496	0.4	0.0	0.0
			FaiRPosO-CP	0.0	0.14		
			FaiRPosO-Exh	0.0	0.14		
			FaiRPosO-Greedy	0.011	0.16		

score. Each dataset has a size of $|\mathcal{C}| = 1\,000$ and relevance scores $q(c) \in [0, 1]$. All algorithms are applied to the datasets with $k = 20$. We measure fairness through *Maximum Positional Disparity* (MPoD) loss and *Maximum Proportional Disparity* (MPD) [11]. Both metrics take values in $[0, 1]$ and equal zero when fairness is satisfied. We report the MPoD loss, which is the difference to the optimal MPoD, as it is not always possible to achieve an optimal MPoD of 0.0 depending on k. Table 1 summarizes the results, where for each metric a score of 0 denotes the optimal value. The colorblind baseline confirms that the datasets exhibit discriminatory relevance scores. As expected, all proposed algorithms satisfy maximum group relevance and ordering consistency, as reflected by MGRD and OCD scores of 0. *FaiRPosO-CP* and *FaiRPosO-Exh* achieve perfect positional fairness (MPoD = 0), while *FaiRPosO-Greedy* attains optimal MPoD in most cases with minimal loss in the last dataset (**RQ1**). The stated Problem 1 is fulfilled even for non-binary and multiple protected attributes (**RQ2**). Further, our algorithms greatly improve proportional fairness compared to the baseline, as the positional fairness constraint bounds the number of candidates per group (Lemma 1).

6.2   Real-Word Applicability

Secondly, we aim to investigate whether the proposed algorithms are applicable to real-world datasets and whether they can achieve better fairness and ranking quality than existing methods.

Datasets. We use several real-world datasets and experimental parameters (Table 2). The datasets are selected to cover different sizes, protected attributes, ranking qualities, and numbers of groups. For each dataset, k is chosen such that the top-k ranking can exactly fulfill proportional fairness. The COMPAS dataset [1] is derived from a tool used in the US criminal justice system to assess the risk of recidivism. COMPAS further contains information about the defendant's gender and demographics. The risk scores were shown to be biased against African American defendants [1]. The German credit dataset [9] contains information about individuals' creditworthiness, gender, and age. The relevance score is calculated as a weighted sum of the credit amount, duration of the credit, and years of employment. The LSAT dataset [16] contains information about law school admission test scores, including the individual's gender and ethnicity.

Table 2. Real-world datasets and their experimental parameters.

	Dataset	n	Relevance	Prot. Attr.	$|\mathcal{G}|$	k	Prot. Groups
D1	COMPAS [1]	20 281	¬ recidivism	Gender	2	50	Female
D2	German [9]	1 000	Credit score	Gender	2	300	Female
D3				Age	3	51	Young, Old
D4				Gender & Age	6	42	All f., Young m., Old m.
D5	LSAT [16]	22 387	LSAT score	Gender	2	2 000	Female
D6				Ethnicity	5	60	Hispanic, Black, Asien
D7				Gender & Ethnicity	10	1 000	White f., Hispanic m., Asian f.

Methods. We compare the proposed algorithms to the colorblind baseline and to the only two existing methods for fair top- k ranking handling multiple protected attributes. These are *Multinomial Fa*ir*, introduced by Zehlike et al. [21], and *FairNormRank*, introduced by Liebrand et al. [11]. We set the hyperparameters such that the best possible rankings can be achieved. The target proportions for our algorithms are set to equal representations of $t_p = 1/|\mathcal{G}|$. The minimum proportions of *Multinomial Fa*ir* are set to $1/|\mathcal{G}|$ for each protected group and to the remaining proportion for the unprotected group. The significance level is set to $\alpha = 0.1$. The normalization factor of *FairNormRank* is set to $\alpha = 1$, as recommended by the authors with respect to positional fairness.

Table 3. Experimental results on real-world datasets comparing the proposed method with recent established approaches.

Dataset	Method	MPoD loss	MPD	MRD	OCD	NDCG loss	Kendall's Tau	Runtime [s]
D1 $k = 50$ $|\mathcal{G}| = 2$	Colorblind	0.870	0.840	0.000	0.000	0.000	1.000	0.018 ± 0.010
	FairNormRank	0.031	0.000	0.045	0.526	0.006	-0.068	0.028 ± 0.008
	Multi. Fa*ir	0.324	0.240	0.039	0.393	0.001	0.013	2.922 ± 0.429
	FaiRPosO-Exh	0.000	0.320	0.059	0.000	0.000	0.398	0.022 ± 0.003
	FaiRPosO-CP	0.000	0.320	0.059	0.000	0.000	0.398	1.735 ± 0.357
	FaiRPosO-Greedy	0.000	0.320	0.059	0.000	0.000	0.398	0.078 ± 0.046
D2 $k = 300$ $|\mathcal{G}| = 2$	Colorblind	0.575	0.580	0.000	0.000	0.000	1.000	0.002 ± 0.001
	FairNormRank	0.132	0.000	0.275	0.029	0.000	0.024	0.020 ± 0.002
	Multi. Fa*ir	0.128	0.113	0.214	0.301	0.005	0.017	94.620 ± 17.033
	FaiRPosO-Exh	0.000	0.133	0.324	0.000	0.000	0.481	0.117 ± 0.027
	FaiRPosO-CP	0.000	0.133	0.324	0.000	0.000	0.481	487.438 ± 40.245
	FaiRPosO-Greedy	0.000	0.133	0.324	0.000	0.000	0.481	0.126 ± 0.038
D3 $k = 51$ $|\mathcal{G}| = 3$	Colorblind	0.440	0.412	0.000	0.000	0.000	1.000	0.003 ± 0.000
	FairNormRank	0.067	0.000	0.220	0.160	0.003	0.233	0.018 ± 0.002
	Multi. Fa*ir	0.097	0.071	0.172	0.384	0.003	0.031	19.610 ± 12.054
	FaiRPosO-Exh	0.000	0.147	0.249	0.000	0.000	0.559	0.077 ± 0.019
	FaiRPosO-CP	0.000	0.147	0.249	0.000	0.000	0.559	54.747 ± 18.157
	FaiRPosO-Greedy	0.000	0.147	0.249	0.000	0.000	0.559	0.022 ± 0.002
D4 $k = 42$ $|\mathcal{G}| = 6$	Colorblind	0.422	0.429	0.000	0.000	0.000	1.000	0.001 ± 0.000
	FairNormRank	0.055	0.000	0.285	0.492	0.015	0.096	0.029 ± 0.003
	Multi. Fa*ir	0.103	0.057	0.231	0.513	0.006	-0.023	47315.131 ± 2743.778
	FaiRPosO-Exh	0.000	0.086	0.321	0.000	0.000	0.124	20.354 ± 2.740
	FaiRPosO-CP	0.000	0.086	0.321	0.000	0.000	0.124	64.897 ± 12.649
	FaiRPosO-Greedy	0.000	0.086	0.321	0.000	0.000	0.124	0.022 ± 0.004
D5 $k = 2\,000$ $|\mathcal{G}| = 2$	Colorblind	0.269	0.257	0.000	0.000	0.000	1.000	0.014 ± 0.009
	FairNormRank	0.003	0.000	0.054	0.000	0.000	0.004	0.033 ± 0.003
	Multi. Fa*ir	0.013	0.004	0.054	0.400	0.000	-0.001	7828.536 ± 1802.438
	FaiRPosO-Exh	0.000	0.060	0.054	0.000	0.000	0.672	2.711 ± 1.031
	FaiRPosO-Greedy	0.000	0.060	0.054	0.000	0.000	0.672	1.035 ± 0.244
D6 $k = 60$ $|\mathcal{G}| = 5$	Colorblind	0.907	0.896	0.000	0.000	0.000	1.000	0.022 ± 0.014
	FairNormRank	0.191	0.000	0.135	0.200	0.000	0.061	0.032 ± 0.007
	Multi. Fa*ir	0.399	0.250	0.027	1.000	0.000	-0.023	393.289 ± 158.727
	FaiRPosO-Exh	0.000	0.146	0.135	0.000	0.000	0.094	16.852 ± 2.129
	FaiRPosO-CP	0.000	0.146	0.135	0.000	0.000	0.094	120.636 ± 12.702
	FaiRPosO-Greedy	0.000	0.146	0.135	0.000	0.000	0.094	0.103 ± 0.022
D7 $k = 1\,000$ $|\mathcal{G}| = 10$	Colorblind	0.608	0.558	0.000	0.000	0.000	1.000	0.021 ± 0.016
	FairNormRank	0.085	0.000	0.419	0.226	0.000	0.007	0.111 ± 0.021
	FaiRPosO-Greedy	0.003	0.052	0.432	0.000	0.000	0.075	1.234 ± 0.490

Fair Ranking Measures. Each experiment is repeated 10 times, and average results, as well as standard deviations, are reported. The experiments were conducted on a server with dual Intel Xeon Gold 5120 CPUs (2×14 cores, 2.20 GHz, 56 threads total) and 384 GB RAM. The runtime is reported in seconds. In addition to our introduced metrics, we report *NDCG* [10] and *Kendall's τ* to measure ranking quality. NDCG rewards highly relevant items placed near the top as a measure of ordering consistency. *Kendall's τ* measures rank correlation with the colorblind top-k ranking. It ranges from -1 to 1, where 1 indicates a perfect correlation, 0 indicates no correlation, and -1 indicates an inverse correlation. Fairness is measured using the metrics *Maximum Positional Disparity* (MPoD) loss and *Maximum Proportional Disparity* (MPD) [11].

Results. The results of the experiments on the real-world datasets are summarized in Table 3. Due to space constraints, we only show the standard deviation for the runtime.

Other metrics vary only for *Multinomial Fa*ir*, with negligible deviations. Results are reported for methods with a runtime of less than 12 h. The colorblind results confirm that the relevance scores are strongly discriminatory, with an MPoD loss of up to 0.907 and an MPD score of up to 0.896. While *FairNormRank* and *Multinomial Fa*ir* decrease positional unfairness, they still show high violations for some datasets. In contrast, our proposed methods achieve optimal positional fairness with only minor deviations from the optimum for *FaiRPosO-Greedy* in one dataset (D7). Proportional fairness is almost equally fulfilled by all methods.

In terms of ranking quality, fulfilling positional fairness leads to minimal trade-offs in maximum relevance (Theorem 1) for our methods, as indicated by the MRD scores. However, our methods achieve ordering consistency optimally, whereas *FairNormRank* and *Multinomial Fa*ir* violate it. Further, our *FaiRPosO* algorithms reach optimal NDCG losses and show higher correlation with the colorblind ranking as indicated by Kendall's Tau. While *FaiRPosO-Exh* and *FaiRPosO-Greedy* require only milliseconds to seconds, *Multinomial Fa*ir* often reaches runtimes of several minutes to hours, especially with more protected groups (D4, D7).

7   Conclusion

In this work, we presented a framework for positional fairness in top-k rankings, comprising the three algorithms *FaiRPosO-Exh*, *FaiRPosO-CP*, and *FaiRPosO-Greedy*. We formalized positional fairness as a constrained optimization problem that balances fairness with ranking quality criteria, including ordering consistency and maximum group relevance. We provide a theoretical analysis of conflicting fairness and quality criteria, as well as of the algorithm's runtimes.

Our exact methods, *FaiRPosO-Exh* and *FaiRPosO-CP*, are able to optimally deal with the stated fair ranking problem, as measured by both our novel and by established evaluation metrics. Complementing these, the *FaiRPosO-Greedy* algorithm offers a highly efficient approximation, achieving optimal or near-optimal fairness with reduced computational cost. The comparison of our ranking algorithms with two recent methods on real-world datasets confirms the effectiveness of our algorithms in terms of fairness, ranking quality, and runtime. Importantly, our methods can handle non-binary and multiple protected attributes, as often present in real-world datasets, such as ethnicity or age. With our exact methods, we provide the first ranking algorithms for optimal positional fairness in top-k rankings, offering both theoretical guarantees and practical applicability.

References

1. Angwin, J., Larson, J., Mattu, S., Kirchner, L.: Machine bias. In: Ethics of data and analytics, pp. 254–264 (2022)
2. Asudeh, A., Jagadish, H.V., Stoyanovich, J., Das, G.: Designing fair ranking schemes. In: MOD, pp. 1259–1276 (2019)
3. Calders, T., Kamiran, F., Pechenizkiy, M.: Building classifiers with independency constraints. In: 2009 IEEE International Conference on Data Mining Workshops, pp. 13–18 (2009). https://doi.org/10.1109/ICDMW.2009.83
4. Celis, L.E., Straszak, D., Vishnoi, N.K.: Ranking with fairness constraints. In: ICALP (2018)
5. Craswell, N., Zoeter, O., Taylor, M., Ramsey, B.: An experimental comparison of click position-bias models. In: Proceedings of the 2008 International Conference on Web Search and Data Mining, pp. 87–94 (2008)
6. Duong, M.K., Conrad, S.: (un)certainty of (un)fairness: preference-based selection of certainly fair decision-makers. In: ECAI 2024 - 27th European Conference on Artificial Intelligence. Frontiers in Artificial Intelligence and Applications, vol. 392, pp. 882–889. IOS Press (2024)
7. Dwork, C., Hardt, M., Pitassi, T., Reingold, O., Zemel, R.: Fairness through awareness. In: Proceedings of the 3rd Innovations in Theoretical Computer Science Conference, pp. 214–226 (2012)
8. Feldman, M., Friedler, S.A., Moeller, J., Scheidegger, C., Venkatasubramanian, S.: Certifying and removing disparate impact. In: KDD, pp. 259–268 (2015)
9. Hofmann, H.: German credit data (1994). https://archive.ics.uci.edu/ml/datasets/Statlog+%28German+Credit+Data%29
10. Järvelin, K., Kekäläinen, J.: Cumulated gain-based evaluation of IR techniques. ACM Trans. Inf. Syst. (TOIS) **20**(4), 422–446 (2002)
11. Liebrand, N.A., Duong, M.K., Conrad, S.: Fair proportional top-k ranking. In: International Conference on Big Data Analytics and Knowledge Discovery, pp. 237–243. Springer (2025)
12. Pedreschi, D., Ruggieri, S., Turini, F.: Measuring discrimination in socially-sensitive decision records. In: Proceedings of the 2009 SIAM International Conference on Data Mining, pp. 581–592. SIAM (2009)
13. Perron, L., Didier, F.: Cp-sat. https://developers.google.com/optimization/cp/cp-solver/
14. Rosen, K.H.: Discrete Mathematics and Its Applications, 7 edn. McGraw-Hill Education, New York (2012), see Section 6.5 for the Stars and Bars theorem
15. Singh, A., Joachims, T.: Fairness of exposure in rankings. In: Proceedings of the 24th ACM SIGKDD International Conference on Knowledge Discovery & Data Mining, pp. 2219–2228 (2018)
16. The College Board: Sat percentile ranks (2014)
17. Verge, T.: Gendering representation in Spain: opportunities and limits of gender quotas. J. Women Polit. Policy **31**(2), 166–190 (2010)
18. Verma, S.: Weapons of math destruction: how big data increases inequality and threatens democracy. Vikalpa **44**(2), 97–98 (2019)
19. Yang, K., Stoyanovich, J.: Measuring fairness in ranked outputs. In: SSDBM (2017)
20. Zehlike, M., Bonchi, F., Castillo, C., Hajian, S., Megahed, M., Baeza-Yates, R.: Fa*ir: a fair top-k ranking algorithm. In: CIKM, pp. 1569–1578 (2017)

21. Zehlike, M., Sühr, T., Baeza-Yates, R., Bonchi, F., Castillo, C., Hajian, S.: Fair top-k ranking with multiple protected groups. Inf. Process. Manag. **59**(1), 102707 (2022)
22. Žliobaitė, I.: Measuring discrimination in algorithmic decision making. Data Min. Knowl. Disc. **31**(4), 1060–1089 (2017). https://doi.org/10.1007/s10618-017-0506-1

Evaluating Structural Preprocessing in RAG for Academic Curriculum Applications

Gizem Intepe[(✉)][iD], Yingtong Liang, and Laurence A. F. Park[iD]

School of Computer, Data and Mathematical Sciences, Western Sydney University, Sydney, Australia
{g.intepe,l.park}@westernsydney.edu.au

Abstract. This study investigates how document structure and preprocessing strategies influence the performance of Retrieval-Augmented Generation (RAG) pipelines in domain-specific question answering, using university curriculum handbooks as a case study. We evaluate multiple data representations, varying in chunking method, punctuation level, and structural format (HTML, plain text, JSON), within a privacy-preserving pipeline that combines semantic vector indexing with a lightweight, locally deployed LLM. A benchmark of curriculum-related queries, covering both simple lookups and multi-hop reasoning, was used to assess performance under controlled retrieval conditions. Results show that semantically coherent chunking, manual or LangChain-based, substantially improves accuracy, with gains exceeding 25% points for medium-difficulty queries, while structurally dense formats like raw HTML or JSON reduce performance. Fixed-effects modelling confirms chunking and natural-language rephrasing as significant positive predictors of correctness, whereas punctuation density and markup type show no measurable effect. These findings highlight the critical, yet often underexamined, role of input data structure in applied RAG pipelines for curriculum advising, where accurate retrieval is essential for tasks such as prerequisite checking, program planning, and subject eligibility. The study presents a reproducible framework for evaluating preprocessing decisions in academic QA systems. This supports the development of reliable, privacy-compliant advising tools.

Keywords: LLM · RAG · document structure · course advising

1 Introduction

Retrieval-Augmented Generation (RAG) combines large language models (LLMs) with document retrieval to overcome static knowledge limitations and reduce reliance on fine-tuning [15]. It has been applied in healthcare, law, education, and business, improving response accuracy, domain alignment, and interpretability [1].

© The Author(s), under exclusive license to Springer Nature Singapore Pte Ltd. 2026
Q. V. Nguyen et al. (Eds.): AusDM 2025, CCIS 2765, pp. 486–500, 2026.
https://doi.org/10.1007/978-981-95-6786-7_33

While recent RAG research has advanced retriever architectures and semantic embeddings [5,8,29], less attention has been paid to document structure, formatting, and preprocessing. In real-world deployments, especially where documents are not machine-optimised, inconsistencies in layout, segmentation, and syntax can degrade retrieval and generation [7,12,16]. Choices such as chunk size and segmentation strategy affect retrieval relevance [2,26], and structural cues like headings or section markers can further improve contextual alignment [4].

These issues are acute in academic advising, where curriculum handbooks, often in semi-structured HTML or JSON, are designed for human readability rather than automated retrieval. Even simple queries about prerequisites, credit rules, or subject pathways may require extensive preprocessing.

Privacy is also a concern: advising systems process sensitive student data and institutional rules, making commercial APIs unsuitable in some contexts. Local deployment with open-weight LLMs keeps both queries and documents within institutional boundaries, aiding compliance and governance.

This paper systematically evaluates how document formatting and preprocessing affect RAG performance for curriculum-based academic QA. We: (1) develop a configurable preprocessing pipeline varying chunking, format, and punctuation; (2) construct a benchmark of lookup and multi-hop academic queries; and (3) empirically show, via statistical modelling, that document structure and natural language formatting significantly influence retrieval and generation quality.

2 Related Work

This section reviews related literature from two perspectives: (1) methods for improving RAG performance, including document preprocessing, retrieval strategies, and document or database structure; and (2) the application of LLMs and conversational agents in academic advising and curriculum navigation.

2.1 Improving RAG Performance Through Structural and Architectural Design

Retrieval-Augmented Generation (RAG) has emerged as a foundational approach for enabling large language models (LLMs) to respond to knowledge-intensive queries by grounding responses in retrieved documents. Prior work [8,15] have focused on improving retrieval models and embedding strategies. Engineering choices - chunking, formatting, indexing, prompting - affect RAG but remain less systematically studied.

Choice of chunking strategy has been shown to critically affect both retrieval accuracy and generation quality. Bhat et al. [2] demonstrated that small chunks (64–128 tokens) favor fact recall, while larger ones (512–1024 tokens) support reasoning tasks. Wang et al. [30] proposed PIC, a semantic-aware chunking method that dynamically forms chunks based on sentence-summary similarity.

This improves retrieval effectiveness without retriever retraining. Braadland et al. [3] quantify chunk coherence using the HOPE metric and report factual accuracy gains exceeding 50%. Further domain-specific evidence from financial QA shows that segmenting based on structural elements like tables and headings improves performance [31].

Recent work has investigated how structural markup (e.g., HTML, bullet lists, table tags) can preserve semantic hierarchy and improve embeddings. Tan et al. [27] find that HTML-aware retrieval models outperform plain-text baselines. Chen et al. [4] introduce DISRetrieval, a framework that encodes document discourse and structure, significantly improving retrieval in long-context QA tasks. Sharma et al. [24] and Li et al. [16] similarly demonstrate that inconsistencies in formatting, syntax, and segmentation can degrade semantic embedding quality and downstream generation. In legal and tax domains, Duarte [6] demonstrates that formats like XML and JSON yield superior retrieval compared to unstructured PDF, due to their preservation of metadata and semantic segmentation.

Beyond preprocessing, the design of the retriever itself remains a key factor. Hybrid retrieval systems that combine traditional keyword-based methods with embedding-based similarity have shown consistent improvements. Mandikal and Mooney [19] show that combining traditional sparse retrieval methods with dense neural embeddings yields significantly better performance on scientific document retrieval tasks compared to using either approach in isolation. Rao et al. [21] find that compact embedding models can achieve competitive results when paired with LLM-based re-ranking, highlighting the value of multi-stage retrieval.

Emerging approaches also integrate graph-based indexing. Zhang et al. [32] survey GraphRAG systems, which incorporate knowledge graphs into the retrieval loop to support entity-level reasoning. Hu et al. [11] further show that using subgraphs for text retrieval improves performance on complex reasoning benchmarks. Meanwhile, prompt engineering techniques, such as chain-of-thought prompting, sub-query decomposition, and superposition prompting, offer additional accuracy gains by improving context alignment and reducing hallucinations [20,22].

2.2 LLMs in Curriculum Advising and Educational QA

Several studies have explored the use of AI-driven assistants and conversational agents to support academic planning and curriculum navigation. Ho et al. [9] proposed a rule-based expert system that assists with elective planning by combining student input with institutional policies, using basic NLP to extract relevant terms. Lucien and Park [17] introduced a virtual advising assistant that improved engagement and reduced advising workload through an interface designed for common curriculum queries.

Building on these foundations, more recent systems have leveraged LLMs to improve generalisability. Sajja et al. [23] developed a platform-independent assistant using GPT-3, capable of answering discipline-agnostic academic questions. Luong and Luong [18] presented an LLM-based advisor specifically for IT programs, integrating prerequisite constraints and course regulations through a structured knowledge base. On the other hand, Lekan and Pardos [14] found that GPT-4's academic major recommendations matched human advisors' choices only 33% of the time, indicating that LLMs are better suited as advisory aids than replacements, particularly for complex academic pathways.

Although prior work addresses aspects such as chunking, format handling, prompting, or retrieval design individually, few studies have examined these components collectively—especially in the context of querying curriculum handbooks. These documents pose distinct challenges: they are often semi-structured (e.g., HTML/JSON), contain cross-referenced rules (e.g., prerequisites, credit requirements), and require multi-hop reasoning. Our work addresses this gap by providing a unified evaluation of document preprocessing, retrieval architecture, and prompt strategies in a controlled RAG pipeline, designed specifically to support curriculum-based academic queries.

3   Methodology

This section describes the architecture and implementation of our RAG pipeline, including document preprocessing, vector indexing, query processing, generation, and evaluation. The pipeline is designed to systematically assess how variations in document structure and formatting affect performance on curriculum-related question answering tasks.

To evaluate the effect of document structure, we constructed multiple representations of a university curriculum handbook. The dataset used in this trial consisted of core information from a single program, the Master of Data Science, at an Australian university[1]. It includes detailed metadata for all core subjects, such as subject descriptions, credit points, semester offerings, prerequisites, restrictions, learning outcomes, and structured course requirements along with the information about the degree and its requirements. In total, the base corpus included 13 documents, each corresponding to a core subject in the program and the main degree information. Each document contained on average 29,000 words.

The RAG pipeline consists of three modular stages: (1) document preprocessing and indexing, (2) query-time retrieval and generation, and (3) automated evaluation. Each stage includes multiple configurable components, such as chunking strategies, embedding models, or scoring routines. Figure 1 illustrates the end-to-end pipeline architecture used in this study. The system is fully containerised to support reproducibility and systematic comparison across document variants.

3.1   Document Preprocessing

The experimental corpus was created by extracting subject information from the official University Handbook website[2]. A Python scraper first collected subject

---

[1] https://hbook.westernsydney.edu.au/programs/master-data-science/.
[2] https://hbook.westernsydney.edu.au/.

codes, then retrieved detailed subject pages with metadata (teaching periods, prerequisites, learning outcomes). Raw HTML content was parsed using BeautifulSoup, and a local LLaMA model was used to reformat each subject into a standardised document template, ensuring consistency in structure and field coverage. This structured corpus served as the input for all subsequent preprocessing and embedding workflows.

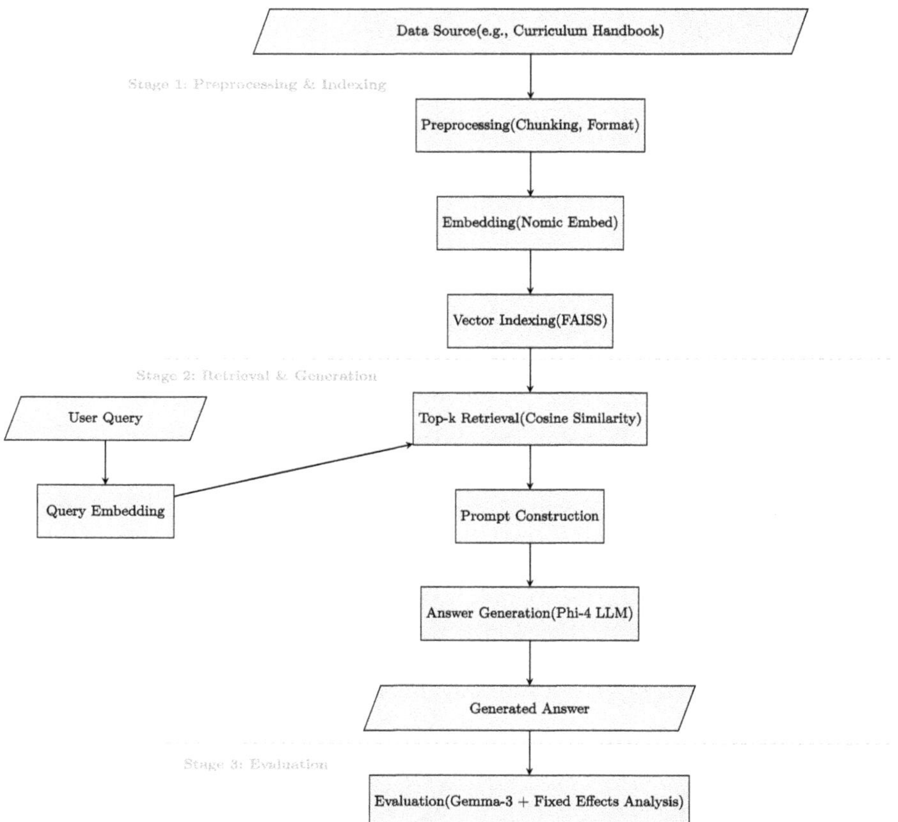

Fig. 1. Overview of the RAG pipeline used for document structure experiments.

We systematically evaluated the impact of structural design on RAG performance by generating ten distinct data variants. These were created by varying six preprocessing dimensions: file format, chunking strategy, segmentation method, natural language transformation, punctuation density, and layout structure. Factors were selected for their influence on embedding quality and retrievability [2, 30, 31].

- **Format:** Source documents were rendered in three structural formats: raw HTML (preserving markup tags), JSON (structured key-value format), and plain text (flattened content).
- **Chunking:** Documents were either unsegmented (i.e., full document) or manually chunked by grouping semantically related fields (e.g., prerequisites, semester availability) per subject.
- **LangChain:** Some variants applied LangChain's `RecursiveCharacterTextSplitter`[3] to divide documents based on token constraints and logical boundaries such as newlines or sentences. This was used either alone or in conjunction with manual chunking, creating hybrid segmentation workflows.
- **Natural:** In select versions, metadata-style content (e.g.,"Credit Points: 10") was rewritten into natural language statements (e.g., "This subject is worth 10 credit points"), simulating human-readable prose.
- **Punctuation:** Three levels of punctuation were tested: none (stripped), some (retained as-is), and more (lightly corrected to improve readability and tokenisation).
- **Layout:** Files were formatted either in a single-line layout (each subject flattened into one sentence-like line) or a multiline layout (with field-level separation preserved).

Each document variant was assigned a shorthand label reflecting its configuration. Naming: `FormatChunkingLangChainNaturalPunctuationLayout`, where:

- `Format` = **T** (Text), **H** (HTML), or **J** (JSON)
- `Chunking` = **Y** (Manual chunking) or **N** (No manual chunking)
- `LangChain` = **Y** (LangChain splitter used) or **N** (Not used)
- `Natural` = **Y** (Natural language transformation) or **N** (Unmodified)
- `Punctuation` = **0** (None), **1** (Some), **2** (More)
- `Layout` = **S** (Single-line) or **M** (Multiline)

For example, `TYYN1S` denotes a plain-text file with manual chunking, LangChain segmentation, no natural transformation, moderate punctuation, and a single-line layout.

We selected 10 representative variants for analysis (see Table 1), focusing on plain-text configurations with systematic variation across chunking, punctuation, and layout dimensions.

3.2   Embedding and Indexing

Semantic search over the different document variants was enabled by embedding each chunked version using the *Nomic Embed Text* model, an open-weight embedding model trained on a diverse corpus of web and academic texts. This model was selected for strong multilingual benchmark performance and efficient

---

[3] https://python.langchain.com/api_reference/text_splitters/index.html.

Table 1. Preprocessing variant configurations. Y = Yes, N = No.

Label	Format	Chunking	LangChain	Natural	Punctuation	Layout
HNNN0M	HTML	N	N	N	None	Multiline
JNNN0M	JSON	N	N	N	None	Multiline
TNNN0M	Text	N	N	N	None	Multiline
TNNN1S	Text	N	N	N	Some	Single-line
TNNN2S	Text	N	N	N	More	Single-line
TYNN1S	Text	Y	N	N	Some	Single-line
TYNN2S	Text	Y	N	N	More	Single-line
TYYN1S	Text	Y	Y	N	Some	Single-line
TYNY0S	Text	Y	N	Y	None	Single-line
TNYN1S	Text	N	Y	N	Some	Single-line

local deployment. [28][4]. These properties ensured consistency across experiments while enabling reproducible, hardware-independent evaluations.

To facilitate fast and scalable similarity-based retrieval, embeddings were stored in separate vector indices using *FAISS* (Facebook AI Similarity Search), a widely adopted library for approximate nearest neighbor (ANN) search in high-dimensional spaces[5]. By indexing each document variant independently, we preserved isolation between experimental conditions, providing that observed differences in retrieval performance could be attributed to preprocessing choices rather than cross-contamination in the embedding space.

Retrieval was performed using cosine similarity, a common metric in semantic search that measures the angular distance between two vectors. Given a query embedding A and a document chunk embedding B, cosine similarity is defined as:

$$\text{cosine\_similarity}(A, B) = \frac{A \cdot B}{\|A\| \|B\|} \tag{1}$$

Cosine similarity emphasises directional alignment rather than vector magnitude, making it particularly effective in normalised embedding spaces produced by LLM encoders [25]. This property supports stable and interpretable similarity scores across heterogeneous document structures.

3.3 Query Processing and Generation

User queries were first embedded using the same Nomic Embed Text model. For each query, the system retrieved the top-5 most similar chunks from each document variant's FAISS index, based on cosine similarity. This retrieval step selects the most contextually relevant text segments for answering the question.

---

[4] Nomic Embed: https://huggingface.co/nomic-ai/nomic-embed-text-v1.
[5] Johnson et al., FAISS: https://github.com/facebookresearch/faiss.

The retrieved chunks were then injected into a standardised prompt template, which was passed to the Phi-4 language model[6], an open-weight transformer optimised for cost-effective inference and competitive reasoning performance. Phi-4 was selected because preliminary experiments showed it offered both faster inference and more accurate answers than alternatives tested, including LLaMA-2/3 and Gemma variants, while remaining lightweight enough for local deployment on academic hardware. Other models (e.g., LLaMA 3, Mistral) were considered but excluded due to either licensing restrictions or higher hardware requirements for local deployment.

3.4 Benchmark Construction

System performance was evaluated using a benchmark comprising twenty curriculum-related queries representative of common academic advising tasks. These questions target key aspects of subject selection, including prerequisites, program structure, semester offerings, and learning outcomes. Each question was reviewed and validated by an academic coordinator to ensure realism, clarity, and alignment with common student queries.

Retrieval and generation quality were examined across two task complexity levels by categorising questions into easy and medium difficulty. This classification allows us to assess model performance separately on straightforward fact-retrieval tasks versus those requiring aggregation, inference, or multi-hop reasoning.

The benchmark was divided into two difficulty categories:

- **Easy:** Questions that require direct lookup from a single subject entry.
  Examples: "What is the credit point value of Predictive Analytics?" or "Who is the coordinator of The Nature of Data?"
- **Medium:** Questions that require aggregation or reasoning across multiple entries.
  Examples: "I want to learn more about Python, which subjects use Python?" or "I would like to take Advanced Machine Learning, which subject or subjects should I study first?"

3.5 Algorithm Evaluation

Evaluating the correctness of LLM-generated answers is a critical component of RAG system development, as generative models are known to produce output regardless of factual accuracy. To establish a ground truth, each of the twenty benchmark questions was first answered manually by the authors using the official university handbook. These human-authored answers served as reference labels for subsequent grading. The data and code used in this experiment are available at the project's GitHub repository.[7]

Given the non-deterministic nature of large language models, where repeated executions of the same query can yield different results due to stochastic sampling strategies, each query was executed five times per document variant. This was intended to capture variability and assess model consistency. Prior research confirms that LLMs frequently produce inconsistent or even self-contradictory outputs when run multiple times on identical inputs [13]. The number of repetitions was limited to five due to computational constraints.

Each generated answer was automatically scored using the Gemma-3 LLM[8], which was prompted to function as a rubric-based evaluator. Responses were rated on a three-point scale: 1 (fully correct), 0.5 (partially correct), and 0 (incorrect or irrelevant), based on alignment with predefined ground truth answers. Gemma-3:12B was selected for grading due to its larger parameter size and more stable output behavior compared to smaller models.

By decoupling the generation and evaluation processes and employing a separate large language model for scoring, we reduce grading bias and enhance the robustness of comparative performance evaluation.

To evaluate the effect of each preprocessing choice on the accuracy of the advising system, we used a fixed effects logistic regression model with a binomial response. Evaluation scores were binarised: only fully correct responses were treated as positive outcomes. This enabled estimation of the marginal effect of each preprocessing factor on the likelihood of a correct answer while controlling for confounding structure. The model's coefficients offer insight into which document structures or formatting choices most improved the system accuracy, while also supporting formal hypothesis testing [10].

The fixed effects model used is expressed as:

$$\text{logit}(P(Y = 1)) = \beta_0 + \beta_1 X_1 + \beta_2 X_2 + \cdots + \beta_k X_k \tag{2}$$

where Y is the binary indicator of answer correctness, $X_1, X_2, \ldots, X_k$ are the predictor variables (e.g., chunking method, format type, punctuation level, layout style), and β_i are the corresponding model coefficients.

4   Results

This section presents an analysis of how structural document design affects the accuracy of RAG. We evaluated model performance across ten document variants, each reflecting a unique combination of markup format, chunking strategy, natural language transformation, punctuation density, and layout, as summarised in Table 1.

4.1   Performance Comparison by Difficulty Level

We examined model performance across two question types, easy and medium - by computing the mean score and standard deviation for each document variant

---

[8] https://ollama.com/library/gemma3.

(Figs. 2 and 3). For easy questions, markup variants (HTML, JSON) and plaintext without punctuation performed worst, while some punctuation (TNNN1S) slightly improved results.

In contrast, all chunked variants achieved mean accuracies above 0.8, confirming the value of segmented, semantically coherent structures. Among these, the manually chunked and punctuated version (TYNN2S) produced the most consistent results, as reflected by its relatively low standard deviation. Error bars in Fig. 2 highlight that well-structured documents reduce performance variability, while minimal preprocessing led to inconsistent and sometimes poor outcomes.

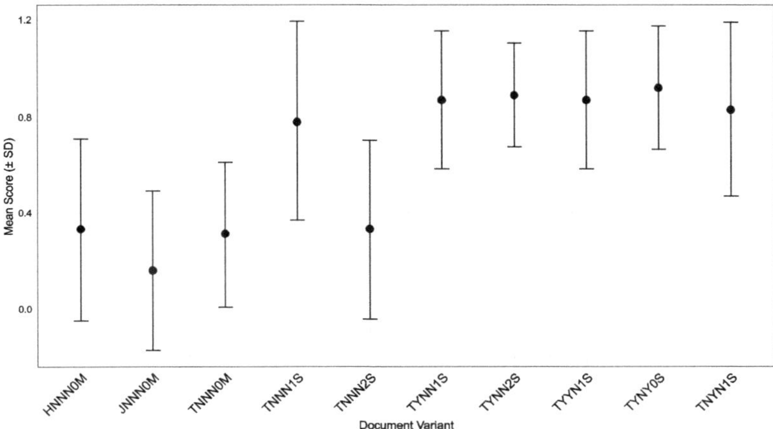

Fig. 2. Average model accuracy with standard deviation across 10 easy benchmark queries per document type.

Figure 3 shows the accuracy results for medium-difficulty queries, which required reasoning over multiple entries or relationships. These questions varied more in performance. As shown in Fig. 3 chunking, manual or automated, still outperformed flat formats, achieved mean accuracies around 0.6. However, the differences between manual and automated chunking were relatively small, with both methods yielding high accuracy. Punctuation and natural phrasing did not exert a strong influence once chunking was applied, suggesting that segmentation plays a more dominant role than surface-level linguistic formatting in these contexts. Nevertheless, subtle gains from added punctuation or phrasal restructuring may still exist but are harder to detect with the current evaluation scheme.

Overall, these results demonstrate that document segmentation, regardless of whether performed manually or through automated means, is the primary structural factor influencing RAG performance across both question types. Other preprocessing choices, including punctuation style, markup format, and linguistic phrasing, had limited or inconsistent impact once chunking was applied.

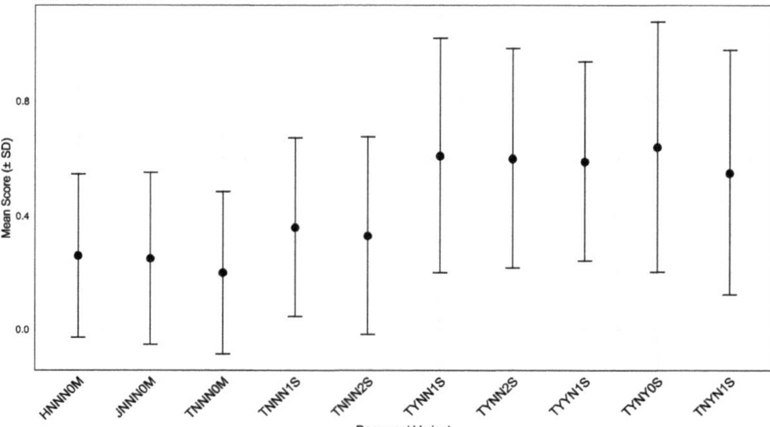

Fig. 3. Average model accuracy with standard deviation across 10 medium benchmark queries per document type.

4.2 Fixed Effects Model Analysis

A fixed effects model analysis was conducted to identify which structural attributes most strongly influence retrieval accuracy. While the previous plots illustrate overall trends in performance across document variants, they do not isolate the independent effect of each structural feature. The consistent gains from semantic chunking and natural language reformatting show that effective preprocessing is essential for improving retrieval accuracy and reasoning quality in RAG systems.

For binary classification, the original scoring system was binarised. Partial correctness scores (0.5) were reclassified as 0, grouping them with fully incorrect responses. This allowed us to model only fully correct answers as successful outcomes, aligning with the assumptions of logistic regression.

The fixed effects model incorporated six independent variables reflecting document preprocessing configurations. Chunking method and LangChain usage were both treated as binary predictors, indicating whether manual and/or LangChain-based segmentation was applied. The natural variable captured whether the document content was rewritten into fluent natural language, while the line variable encoded whether the text was flattened into a single-line layout or retained a multiline structure. Punctuation was treated as an ordinal variable with three levels reflecting increasing degrees of syntactic clarity. Document format was represented through two dummy variables: Markup_1 for JSON and Markup_2 for plain text, with HTML serving as the reference category. This encoding allowed the model to assess the relative contribution of each preprocessing dimension to binary prediction accuracy.

Table 2 summarises the regression results. Among the predictors, *LangChain-based chunking* emerged as the strongest contributor to performance, supporting our previous observations, with a statistically significant positive coefficient ($\beta =$

1.632, $p < 0.001$). This finding confirms the effectiveness of semantically aware segmentation strategies in improving retrieval quality and response accuracy.

Table 2. Fixed Effects Analysis Results

Variable	Coef	Std. Err	z	P>|z|	[0.025	0.975]
Manual Chunk	0.8109	0.417	1.946	0.052	-0.006	1.628
Langchain	1.6322	0.349	4.674	0.000	0.948	2.317
Natural	1.2562	0.332	3.780	0.000	0.605	1.907
Punctuation	0.0866	0.147	0.588	0.556	-0.202	0.375
Line	0.5335	0.253	2.106	0.035	0.037	1.030
Markup 1	0.4150	0.289	1.435	0.151	-0.152	0.982
Markup 2	0.2429	0.499	0.487	0.626	-0.735	1.221

Notes: Model estimated using maximum likelihood on 1000 observations. Pseudo $R^2 = 0.0923$. Log-Likelihood $= -577.15$. LLR p-value < 0.001.

The *natural* variable, denoting human-like linguistic phrasing and field rewording (e.g., transforming colon-separated entries into full sentences), was also a strong positive predictor ($\beta = 1.256$, $p < 0.001$). This result suggests that large language models benefit from inputs that mirror natural human discourse structures.

Manual chunking exhibited a modest but borderline significant positive relationship ($\beta = 0.811$, $p = 0.050$), indicating that deliberate semantic grouping although less effective than automated segmentation, still improves output reliability relative to unstructured baselines.

The *line* variable, representing a flattened single-line layout, had a statistically significant positive effect ($\beta = 0.534$, $p = 0.035$). This suggests that simplifying structural layout may enhance retrieval alignment and LLM consistency, albeit to a lesser extent than chunking or natural-language formatting.

In contrast, *punctuation* show no meaningful influence ($p = 0.561$), and document format (`Markup_1` for JSON, `Markup_2` for plain text) was also non-significant ($p = 0.151$ and $p = 0.626$). These results indicate that format alone is insufficient without semantic refinement, and that structural preprocessing promoting semantic coherence, particularly chunking and natural-language transformation, remains most critical for improving RAG performance in curriculum-based QA.

Although the model's pseudo R^2 value was relatively low (0.09), this is not unexpected for logistic regression applied to complex language tasks. The analysis was not intended for prediction, but rather to assess the relative significance and direction of preprocessing effects. The low value mainly indicates that additional unobserved factors likely contribute to variance in outcomes, but it does not weaken the robustness of the identified chunking and natural-language effects.

5   Discussion

This study shows that RAG performance in curriculum advising is shaped not only by retrieval algorithms and LLM capacity, but also critically by document structure and formatting. Both descriptive results and fixed effects analysis indicated consistent gains from semantic chunking and natural language rephrasing, highlighting the importance of upstream preprocessing for downstream reasoning and retrieval accuracy.

However, the tested document variants represent only a fraction of the possible design space. Resource constraints limited the combinations tested, and future factorial designs could better isolate interaction effects. Deployment raises challenges: definitions of'natural language' vary, and automated reformatting risks semantic drift, requiring careful validation.

While results reported here focus on Phi-4, exploratory checks with alternative open-weight models (e.g., LLaMA-3 and Gemma) suggest that although absolute accuracy varied, the relative effects of preprocessing choices were consistent. This indicates that the findings are likely to generalise across model families.

Finally, the evaluation framework's 3-point scoring rubric offers simplicity but may obscure finer distinctions, especially in reasoning-intensive queries. More granular metrics, combined with human-in-the-loop validation, could better capture nuanced differences in correctness.

6   Conclusion and Future Work

This study systematically examined how structural preprocessing choices influence the accuracy of RAG systems in curriculum-based question answering. Across ten document variants, results consistently showed that semantically coherent chunking and natural language rephrasing substantially improve retrieval effectiveness and answer accuracy. These findings emphasise that preprocessing is not a peripheral step, but a critical design stage that directly shapes downstream reasoning and generation in educational LLM applications.

Future work will explore graph-based document representations, such as knowledge-graph-augmented chunking (e.g., GraphRAG), to encode prerequisite and co-requisite relationships and support more precise multi-hop reasoning. Translating natural language queries into graph queries has shown benefits in structured domains [33] and could improve accuracy in academic advising. Further refinement of prompt engineering may also enhance alignment between retrieved content and query intent, reducing hallucinations and increasing factual grounding. From an applied perspective, integrating the RAG module into our optimisation based study planner could enable advanced advising capabilities, including eligibility checks, semester planning, and curriculum exploration. Given the high-stakes nature of academic decision-making, such integration must be accompanied by rigorous evaluation protocols to ensure transparency, reliability, and responsible deployment.

References

1. Arslan, M., Munawar, S., Cruz, C.: Business insights using rag–LLMs: a review and case study. J. Decision Syst., 1–30 (2024)
2. Bhat, S.R., Rudat, M., Spiekermann, J., Flores-Herr, N.: Rethinking chunk size for long-document retrieval: a multi-dataset analysis. arXiv preprint arXiv:2505.21700 (2025)
3. Brådland, H., Goodwin, M., Andersen, P.A., Nossum, A.S., Gupta, A.: A new hope: Domain-agnostic automatic evaluation of text chunking. In: Proceedings of the 48th International ACM SIGIR Conference on Research and Development in Information Retrieval, pp. 170–179 (2025)
4. Chen, H., Yang, Y., Li, Y., Zhang, M., Zhang, M.: Disretrieval: harnessing discourse structure for long document retrieval. arXiv preprint arXiv:2506.06313 (2025)
5. Chen, J., Lin, H., Han, X., Sun, L.: Benchmarking large language models in retrieval-augmented generation. In: Proceedings of the AAAI Conference on Artificial Intelligence. vol. 38, pp. 17754–17762 (2024)
6. Duarte, R.D.: The impact of document formats on embedding performance and rag effectiveness in tax law applications (2024). https://www.robertodiasduarte.com.br/the-impact-of-document-formats-on-embedding-performance-and-rag-effectiveness-in-tax-law-applications/, Accessed 1 Aug 2025
7. Finardi, P., et al.: The chronicles of rag: the retriever, the chunk and the generator. arXiv preprint arXiv:2401.07883 (2024)
8. Gao, Y., et al.: Retrieval-augmented generation for large language models: a survey. arXiv preprint arXiv preprint arXiv:2312.10997 (2023)
9. Ho, C.C., Lee, H.L., Lo, W.K., Lui, K.F.A.: Developing a chatbot for college student programme advisement. In: 2018 international symposium on educational technology (ISET), pp. 52–56. IEEE (2018)
10. Hosmer, D.W., Lemeshow, S., Sturdivant, R.X.: Applied Logistic Regression. Wiley, 3rd edn. (2013)
11. Hu, Y., et al.: Cg-rag: research question answering by citation graph retrieval-augmented LLMs. In: Proceedings of the 48th International ACM SIGIR Conference on Research and Development in Information Retrieval, pp. 678–687 (2025)
12. Laitenberger, A., Manning, C.D., Liu, N.F.: Stronger baselines for retrieval-augmented generation with long-context language models. arXiv preprint arXiv:2506.03989 (2025)
13. Lee, Y., et al.: One vs. many: comprehending accurate information from multiple erroneous and inconsistent ai generations. In: Proceedings of the 2024 ACM Conference on Fairness, Accountability, and Transparency, pp. 2518–2531 (2024)
14. Lekan, K., Pardos, Z.A.: Ai-augmented advising: a comparative study of GPT-4 and advisor-based major recommendations. Journal of Learning Analytics 12(1), 110–128 (2025)
15. Retrieval-augmented generation for knowledge-intensive nlp tasks: Lewis, P., Perez, E., Piktus, A., Petroni, F., Karpukhin, V., Goyal, N., Küttler, H., Lewis, M., Yih, W.t., Rocktäschel, T., et al. Adv. Neural. Inf. Process. Syst. 33, 9459–9474 (2020)
16. Li, S., Stenzel, L., Eickhoff, C., Bahrainian, S.A.: Enhancing retrieval-augmented generation: a study of best practices. arXiv preprint arXiv:2501.07391 (2025)
17. Lucien, R., Park, S.: Design and development of an advising chatbot as a student support intervention in a university system. TechTrends 68(1), 79–90 (2024)
18. Luong, H., Luong, K.: A chatbot-based academic advising model for student in information technology: A case study. Saudi J Eng Technol 10(3), 93–100 (2025)

19. Mandikal, P., Mooney, R.: Sparse meets dense: a hybrid approach to enhance scientific document retrieval. In: Proceedings of the 4th Workshop on Scientific Document Understanding (SDU). CEUR Workshop Proceedings, Vancouver, BC, Canada (2024), also available as arXiv preprint: https://arxiv.org/abs/2401.04055
20. Merth, T., Fu, Q., Rastegari, M., Najibi, M.: Superposition prompting: improving and accelerating retrieval-augmented generation. In: Proceedings of the 41st International Conference on Machine Learning. Proceedings of Machine Learning Research, vol. 235, pp. 35507–35527. PMLR, Vienna, Austria (2024), https://arxiv.org/abs/2404.06910, arXiv:2404.06910
21. Rao, A., Alipour, H., Pendar, N.: Rethinking hybrid retrieval: when small embeddings and llm re-ranking beat bigger models. arXiv preprint arXiv:2506.00049 (2025)
22. Sahoo, P., et al.: A systematic survey of prompt engineering in large language models: Techniques and applications. arXiv preprint arXiv:2402.0D7927 (2024)
23. Sajja, R., Sermet, Y., Cwiertny, D., Demir, I.: Platform-independent and curriculum-oriented intelligent assistant for higher education. Int. J. Educ. Technol. High. Educ. **20**(1), 42 (2023)
24. Sharma, C.: Retrieval-augmented generation: a comprehensive survey of architectures, enhancements, and robustness frontiers. arXiv preprint arXiv:2506.00054 (2025)
25. Singhal, A., et al.: Modern information retrieval: a brief overview. IEEE Data Eng. Bull. **24**(4), 35–43 (2001)
26. Smith, B., Troynikov, A.: Evaluating chunking strategies for retrieval. https://research.trychroma.com/evaluating-chunking (2024), Accessed 1 Aug 2025
27. Tan, J., et al.: Htmlrag: Html is better than plain text for modeling retrieved knowledge in rag systems. In: Proceedings of the ACM on Web Conference 2025, pp. 1733–1746 (2025)
28. Thakur, N., et al.: Systematic evaluation of neural retrieval models on the touché 2020 argument retrieval subset of BEIR. In: Proceedings of the 47th International ACM SIGIR Conference on Research and Development in Information Retrieval, pp. 1420–1430 (2024)
29. Wang, S., Khramtsova, E., Zhuang, S., Zuccon, G.: Feb4rag: evaluating federated search in the context of retrieval augmented generation. In: Proceedings of the 47th International ACM SIGIR Conference on Research and Development in Information Retrieval, pp. 763–773 (2024)
30. Wang, Z., et al.: Document segmentation matters for retrieval-augmented generation. In: Findings of the Association for Computational Linguistics: ACL 2025, pp. 8063–8075 (2025)
31. Yepes, A.J., You, Y., Milczek, J., Laverde, S., Li, R.: Financial report chunking for effective retrieval augmented generation. arXiv preprint arXiv:2402.05131 (2024)
32. Zhang, Q., et al.: A survey of graph retrieval-augmented generation for customized large language models. arXiv preprint arXiv:2501.13958 (2025)
33. Zhao, Z., Stewart, M., Liu, W., French, T., Hodkiewicz, M.: Natural language query for technical knowledge graph navigation. In: Park, L.A.F.e.a. (ed.) AusDM 2022, CCIS, vol. 1741, pp. 176–191. Springer (2022). https://doi.org/10.1007/978-981-19-8746-5_13

Evaluating Cross-Lingual Classification Approaches Enabling Topic Discovery for Multilingual Social Media Data

Deepak Uniyal(✉) , Md Abul Bashar , and Richi Nayak

Centre for Data Science, School of Computer Science, Queensland University of Technology, Brisbane, QLD 4000, Australia
`deepak.uniyal@hdr.qut.edu.au`, `{m1.bashar,r.nayak}@qut.edu.au`

Abstract. Analysing multilingual social media discourse remains a major challenge in natural language processing, particularly when large-scale public debates span across diverse languages. This study investigates how different approaches for cross-lingual text classification can support reliable analysis of global conversations. Using hydrogen energy as a case study, we analyse a decade-long dataset of over nine million tweets in English, Japanese, Hindi, and Korean (2013–2022) for topic discovery. The online keyword-driven data collection results in a significant amount of irrelevant content. We explore four approaches to filter relevant content: (1) translating English annotated data into target languages for building language-specific models for each target language, (2) translating unlabelled data appearing from all languages into English for creating a single model based on English annotations, (3) applying English fine-tuned multilingual transformers directly to each target language data, and (4) a hybrid strategy that combines translated annotations with multilingual training. Each approach is evaluated for its ability to filter hydrogen-related tweets from noisy keyword-based collections. Subsequently, topic modeling is performed to extract dominant themes within the relevant subsets. The results highlight key trade-offs between translation and multilingual approaches, offering actionable insights into optimising cross-lingual pipelines for large-scale social media analysis.

Keywords: multilingual data · text classification · topic modelling · hydrogen energy · cross-lingual

1 Introduction

The growth of social media platforms such as Twitter[1] has enabled the large-scale collection of user-generated content across multiple languages, offering opportunities for studying global public opinion [1]. A key challenge in this domain is relevance classification, i.e. identifying tweets pertinent to a specific topic across

---

[1] Recently changed its name to X. In this paper, we refer to this platform as Twitter.

Q. V. Nguyen et al. (Eds.): AusDM 2025, CCIS 2765, pp. 501–515, 2026.
https://doi.org/10.1007/978-981-95-6786-7_34

diverse linguistic contexts. The multilingual nature of such data introduces significant challenges for natural language processing (NLP), including handling imbalances in language resources, preserving cultural nuances, and mitigating translation-induced noise [2]. These issues fuel an ongoing debate: Should multilingual social media data be processed using separate native language-specific models, translated into a pivot language (e.g., English), or directly handled by multilingual models such as multilingual BERT (mBERT) [3] and XLM-RoBERTa (XLM-R) [4] [5] [6].

While NLP research has advanced multilingual text processing methods, the optimal strategy remains context-dependent. Translating all data to a pivot language like English [7] simplifies downstream tasks but risks losing language-specific subtleties. Training separate language-specific models can preserve these nuances, yet it demands extensive annotated data for each language, which is often unavailable. Multilingual transformer models, such as mBERT [3] and XLM-R [4], offer the ability to process multiple languages directly; however, their performance varies depending on the training resources available for specific languages, often declining for low-resource language pairs [8].

To address these challenges, this study systematically evaluates four cross-lingual relevance classification approaches for large-scale, multi-year Twitter data. The dataset for this study comprises a large corpus of Twitter data spanning multiple years, collected in English, Japanese, Korean, and Hindi. However, a critical constraint is the limited availability of annotated data, a common challenge in real-world scenarios due to resource limitations. Only a small set of 5,000 English tweets is annotated for relevance, while the vast majority of the data, comprising millions of unlabelled tweets across all four languages remains unannotated. This scarcity of labeled data complicates the development of robust cross-lingual systems, necessitating innovative approaches to leverage the available annotations effectively.

The first approach employs language-specific modeling, translating the 5,000 English-annotated tweets into Japanese, Korean, and Hindi to train separate classifiers, while using the original English annotations for an English model. The second approach adopts a translation-based pipeline, translating all non-English tweets into English and applying a single English classifier. The third approach leverages mBERT, fine-tuned on the English annotations and applied in a zero-shot manner [9] to original-language tweets. Finally, the hybrid approach combines translation and multilingual training, translating English annotations into multiple languages, merging them into a single multilingual training set, and fine-tuning mBERT for cross-lingual prediction.

The main contributions of this study are as follows:

1. Cross-Lingual Classification. A comparative evaluation of four approaches for relevance classification in large-scale multilingual Twitter data, enhancing the filtering of hydrogen-energy content.
2. Key Themes and Phrases. An integrated pipeline using topic modeling to uncover thematic trends over a decade.

3. Temporal Evolution. Insights into the evolution of hydrogen energy discourse from 2013 to 2022, including periods of heightened user engagement.
4. First Decade-long Multilingual Analysis. A pioneering decade-long, multilingual analysis of hydrogen energy discourse across English, Japanese, Hindi, and Korean, with a scalable framework for future clean energy studies.

The paper is organised as follows: Sect. 2 covers related work, Sect. 3 details our methodology, Sect. 4 presents results, and Sect. 5 offers conclusions and future directions.

2 Literature Review

Cross-lingual NLP has progressed from early translation-based methods to advanced multilingual models, addressing the challenge of linguistic diversity in multilingual datasets. The foundational "translate-then-classify" approach, pioneered by early work in the field, translates source text into a pivot language (e.g., English) to enable processing, though its effectiveness hinges on translation accuracy and struggles with semantic nuances [10]. Subsequent advancements introduced cross-lingual alignment techniques, such as structural correspondence learning, which identifies consistent pivot features across languages, reducing reliance on machine translation quality [11]. The development of multilingual word embeddings further enhanced semantic similarity across languages through linear transformations [12]. The advent of transformer-based models marked a significant leap, with Multilingual BERT (mBERT) demonstrating robust cross-lingual performance in tasks like classification and entity recognition [8]. Enhanced models like XLM and XLM-R, trained on larger corpora with refined objectives, have improved transfer capabilities across diverse languages [13] [14]. However, these models often underperform with low-resource languages. Recent hybrid approaches combine monolingual specialisation, which captures language-specific nuances, with multilingual generalisation for cross-lingual alignment, though their application in specialised domains remains underexplored.

Our study addresses this gap by evaluating four cross-lingual relevance classification approaches – monolingual, translation-based, multilingual, and hybrid – on a decade-long, four-language dataset, enabling effective filtering for topic modeling to uncover regional hydrogen energy themes.

3 Methodology

This study presents a systematic pipeline for analysing multilingual Twitter discourse on hydrogen energy spanning 2013 to 2022. The methodology, illustrated in Fig. 1, integrates data collection, preprocessing, and relevance classification, which enable further NLP tasks such as topic modeling or sentiment mining, based on the best method. The central challenge arises from the multilingual nature of the corpus: Tweets appear in English (EN), Japanese (JA), Korean

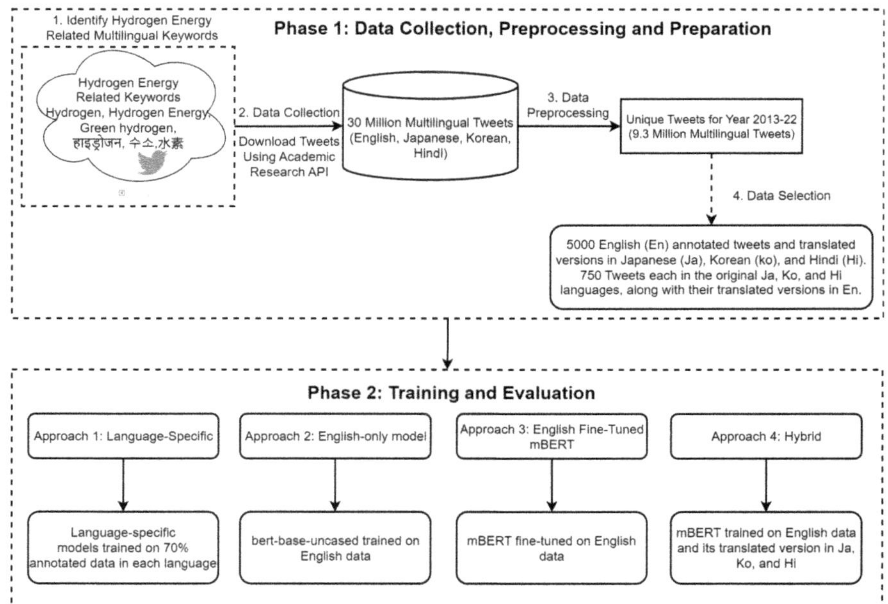

Fig. 1. Architecture of the proposed multi-step methodology, including data collection, data preprocessing, and relevance classification.

(KO), and Hindi (HI), while supervised annotations are initially available only in English.

To address this challenge, we developed and evaluated four distinct cross-lingual relevance classification approaches: (1) Training independent language-specific models for each language, (2) Training a single English model on English annotated data, (3) Fine-tuning mBERT on English annotated data and employing zero-shot predictions for other languages, and (4) Fine-tuning mBERT on English annotated data and its translation into other languages, followed by cross-lingual predictions.

Following relevance classification using the optimal approach, topic modelling is performed on the filtered, thematically coherent dataset.

3.1 Data Collection and Preprocessing

We collected a large-scale dataset of 30.7 million tweets related to hydrogen energy using the Academic Twitter API v2, covering the period from January 2013 to December 2022. Keywords were curated through a combination of literature review and expert consultation to ensure comprehensive coverage of the discourse across four languages: English, Japanese, Hindi, and Korean [1].

The initial processing, as described in [1], included steps such as metadata mapping, deduplication, language-specific text normalisation, and tokenisation.

Table 1. Training and testing setup for four relevance-classification approaches.

Approach	Training Data	Test Data	Model Setup
1. Language-specific	5,000 English annotated tweets and their translations (JA, KO, and HI)[a]	750 tweets each language taken from original language data	Separate language-specific models for each language
2. English-only	5,000 English annotated tweets		Single English model; all tweets translated to English
3. English FineTuned mBERT	5,000 English annotated tweets		Single mBERT model fine-tuned on English data; zero-shot prediction on other non-English languages
4. Hybrid	5,000 English annotated tweets and their translations (JA, KO, and HI)[a]		Single mBERT model trained on combined multilingual data

JA = Japanese, KO = Korean, HI = Hindi.

After removing duplicate tweets, we retained approximately 9.9 million unique tweets. To ensure temporal consistency, we further filtered tweets strictly posted between 2013 and 2022. This step excluded retweets and other content originating outside the target window but was re-shared during the period. The resulting dataset consisted of approximately 9.3 million tweets.

For supervised relevance classification, a labeled dataset of 5,000 English tweets was created, uniformly sampled across years to maintain temporal coverage. Each tweet was annotated as *relevant* or *irrelevant* by three domain experts following detailed guidelines. Inter-annotator reliability, measured using Fleiss' kappa [15], achieved a score of 0.85, indicating strong agreement. Approximately 45% of tweets were labeled as irrelevant, highlighting the necessity of automated filtering. This annotated dataset provides the foundation for training all subsequent relevance classification approaches.

As we lack a labeled dataset in other languages, we translated these 5,000 tweets into Japanese, Korean, and Hindi using `facebook/nllb-200-distilled-600M` [16] neural machine translation model. For a robust evaluation, we randomly selected and manually annotated 750 tweets each from original, non-translated Japanese, Hindi, and Korean tweets, ensuring uniform distribution across years and referencing their translation for accuracy.

3.2   Multilingual Relevance Classification

The primary challenge in classifying the relevance of this dataset stems from its multilingual composition, with initial labels available only for the English subset. To address this, we developed and evaluated four distinct cross-lingual relevance classification approaches, which vary based on the training data composition and the type of encoder used (monolingual or multilingual), as detailed in Table 1. We first outline the experimental setup and data utilization across all approaches, followed by a detailed description of each approach:

Experimental Setup. Our experimental framework incorporates a ten-run with ten different seeds, where the data for each language is divided into a train, validation and test sets in a ratio of 70%, 15%, and 15%, respectively. Depending on the chosen approach, a suitable subset of data is selected to train the model across ten seeds, and the best model is saved for prediction. We implement early stopping with a patience of 5 epochs, terminating training if no improvement in validation accuracy is observed.

We evaluate the performance using accuracy and F1-score, as these metrics are widely used in text classification [17]. Both metrics are reported as mean and standard deviation over ten runs to ensure a robust evaluation.

Since the original annotations were available only for English data, we created a uniform set of 750 annotated tweets per language (English, Japanese, Korean, and Hindi) from original language tweets, used exclusively for testing to evaluate model performance on authentic unseen tweets across all languages. These annotated samples are used to rigorously test each approach and report comparative performance metrics.

Approach 1: Monolingual Models. In the first approach, we use 5,000 English annotated tweets and their translations for all languages to fine-tune separate monolingual classifiers independently. Each classifier is trained on its respective language data using a language-specific transformer model:

1. English: `bert-base-uncased` [3], a 12-layer bidirectional Transformer encoder with 110M parameters, pre-trained on BooksCorpus and English Wikipedia.
2. Japanese: `cl-tohoku/ bert-base-japanese` [18], a BERT-base model pre-trained on Japanese Wikipedia, tokenised with a WordPiece-based vocabulary optimised for Japanese morphology.
3. Korean: `monologg/kobert` [19], a Korean BERT variant trained on a large-scale Korean corpus, adapted with a custom tokeniser to handle Korean morphology and spacing.
4. Hindi: `ai4bharat/indic-bert` [20], a multilingual BERT model covering 12 major Indian languages, trained on IndicCorp and designed for resource-limited Indic NLP tasks.

For each model, the architecture consists of the language-specific encoder followed by a dropout layer ($p = 0.3$), a fully connected projection layer, a `tanh` activation, and a softmax output layer. Fine-tuning is performed independently for each language on corresponding labeled data. The performance of each model is tested on the uniform set of 750 annotated tweets per language from original language data.

Approach 2: English-only Model. In the second approach, we utilise annotated English tweets to train the English-only model, i.e. `bert-base-uncased`. The model architecture mirrors that of approach 1 (3.2) for prediction purposes. The test is performed on the English portion of 750 annotated tweets, and additional tests are conducted on the English-translated versions of the 750 annotated tweets per language (Japanese, Korean, and Hindi) from the original language data.

Approach 3: Multilingual Model. In the third approach, we adopt a fully multilingual strategy by leveraging the `bert-base-multilingual-uncased` model, referred as mBERT [21], which supports 104 languages including English, Japanese, Korean, and Hindi. The fine-tuning phase utilises English annotated tweets to fine-tune the mBERT, relying on its generalisation capabilities from pre-training. The model architecture consists of the pre-trained mBERT encoder followed by a dropout layer with $p = 0.3$, a fully connected projection layer, a `tanh` activation function, and a final softmax classification layer producing binary relevance predictions on the input data. The test is performed on the English portion of 750 annotated tweets per language, using the best model of ten runs for prediction. Additional tests are performed for non-English languages (Japanese, Korean, and Hindi) using the test set of 750 tweets in their native form.

Approach 4: Hybrid Multilingual Model. The fourth approach combines the benefits of the multilingual approach as in approach 3 (3.2), which relies on mBERT fine-tuned on English data with non-English translated data enriched with annotations, integrating these elements into a unified training framework. These translated annotations, together with the original English annotated data form a combined multilingual training set. This dataset is used to fine-tune an mBERT classifier with the same architecture as in approach 3. By using a multilingual dataset for fine-tuning, this approach aims to improve the classification performance on non-English tweets while preserving the advantages of mBERT's cross-lingual representation learning. The test is performed on the uniform set of 750 annotated tweets per language from original language data, and the best model from ten runs is saved for inference.

3.3   Topic Modelling

Experimental results show that Approach 2 (the model trained with English-annotated tweets only) provided the most accurate relevancy distinction. After identifying `approach 2` as the best-performing and most interpretable strategy for relevance classification, we used its filtered output for topic modelling. One key advantage of this approach is that all tweets, regardless of their original language, are translated into English during the classification stage. This not only enables consistent processing but also makes the resulting themes more accessible to a wider audience by presenting the discourse in a single language. Tweets labeled as *irrelevant* were excluded from further analysis, ensuring that only content truly related to hydrogen energy shaped the thematic structure. We explored topics from other approaches in their original languages, but their quality was noisy, likely due to challenges in handling multilingual data effectively.

For topic discovery, we selected Non-negative Matrix Factorisation (NMF) [22] over methods like Latent Dirichlet Allocation (LDA) due to its interpretability, robustness, and superior alignment with human judgment [23]. NMF effectively derived coherent topics from sparse, high-dimensional TF-IDF representations, which incorporated unigrams, bigrams, and trigrams, filtered infrequent terms, and limited the feature space to 10,000 terms for computational efficiency. Configured to extract up to ten topics per language with reproducible initialisation, NMF ensured consistent comparability across datasets.

This pipeline produced topic keywords with associated weights, tweet-to-topic assignments identifying each tweet's dominant theme, and temporal distributions tracking yearly frequencies from 2013 to 2022. These outputs revealed the evolution of hydrogen energy discourse, highlighting periods of thematic growth or decline. By relying on a unified English-translated dataset, the approach balanced model performance, interpretability, and comparability, making the analysis accessible to both technical and non-technical audiences while upholding rigorous topic extraction and trend analysis.

4   Results and Discussion

This section provides a detailed analysis of the outcomes from the cross-lingual relevance classification and topic modeling applied to Twitter data on hydrogen energy discourse. The analysis focuses on temporal trends, including tweet distribution over years and topic evolution, utilising the processed dataset derived from **approach** 2. The findings are interpreted in the context of hydrogen energy market trends, policy developments, and technological advancements, informed by prior studies such as [1].

4.1   Performance Analysis of Cross-Lingual Relevance Classification Approaches

Table 2 presents test performance metrics (accuracy and F1-score) for four cross-lingual relevance classification approaches, evaluated on a uniform test set of 750 tweets per language (English, Japanese, Korean, and Hindi) from original language data. The results, expressed with standard deviations across five runs, highlight the strengths and limitations of each approach, offering insights into their suitability for multilingual tweet analysis.

Approach 1 (Language-Specific). leverages separate language-specific BERT models, trained on English-annotated tweets and their translations, with performance assessed on the test set. It achieves the highest accuracy and F1-score for English (97.72% ± 0.23% accuracy, 97.70% ± 0.23% F1-score), reflecting the robustness of the BERT model on its native training language. For non-English languages, performance varies: Japanese (75.79% ± 0.49% accuracy, 74.88% ± 0.56% F1-score), Korean (91.40% ± 0.31% accuracy, 91.35% ±

Table 2. Test accuracy and F1-score for four cross-lingual relevance classification methods. Best results are shown in bold blue; second-best in bold.

Approach	Language	Model	Accuracy (%)	F1-score (%)
Approach 1: Language-Specific	EN	BERT	**97.72 ± 0.23**	**97.70 ± 0.23**
	JA	Japanese BERT	75.79 ± 0.49	74.88 ± 0.56
	KO	Korean BERT	**91.40 ± 0.31**	**91.35 ± 0.31**
	HI	Hindi BERT	78.08 ± 1.10	78.04 ± 1.10
Approach 2: English-only	EN	BERT	**97.72 ± 0.23**	**97.70 ± 0.23**
	JA	BERT	**79.85 ± 0.66**	**79.30 ± 0.69**
	KO	BERT	**86.03 ± 0.22**	**85.78 ± 0.23**
	HI	BERT	**90.59 ± 0.27**	**90.53 ± 0.28**
Approach 3: English Fine-Tuned mBERT	EN	mBERT	**96.68 ± 0.28**	**96.66 ± 0.28**
	JA	mBERT	50.21 ± 0.14	33.80 ± 0.30
	KO	mBERT	50.56 ± 0.18	34.56 ± 0.39
	HI	mBERT	53.59 ± 0.28	40.97 ± 0.55
Approach 4: Hybrid	EN	mBERT	94.81 ± 0.58	94.79 ± 0.58
	JA	mBERT	**78.49 ± 0.44**	**77.78 ± 0.49**
	KO	mBERT	83.57 ± 0.52	83.27 ± 0.54
	HI	mBERT	**81.76 ± 0.48**	**81.18 ± 0.52**

0.31% F1-score), and Hindi (78.08% ± 1.10% accuracy, 78.04% ± 1.10% F1-score) show that Korean benefits most from the language-specific model, while Japanese and Hindi exhibit lower and more variable performance, possibly due to translation errors or linguistic complexity.

Approach 2 (English-only). relies on a single BERT model trained on English-annotated tweets, applied to the English portion of the test set and translated versions of non-English tweets. It excels in English (97.72% ± 0.23% accuracy, 97.70% ± 0.23% F1-score) and demonstrates strong performance on non-English languages: Japanese (79.85% ± 0.66% accuracy, 79.30% ± 0.69% F1-score), Korean (86.03% ± 0.22% accuracy, 85.78% ± 0.23% F1-score), and Hindi (90.59% ± 0.27% accuracy, 90.53% ± 0.28% F1-score). This approach outperforms Approach 1 for Japanese and Hindi, suggesting that translation to English provides a more consistent feature representation, likely due to high-quality translation and the effective model available in English language.

Approach 3 (English Fine-Tuned mBERT). uses mBERT fine-tuned on English annotated tweets in a zero-shot setting. It performs well on English (96.68% ± 0.28% accuracy, 96.66% ± 0.28% F1-score) but shows significantly degraded performance on non-English languages: Japanese (50.21% ± 0.14% accuracy, 33.80% ± 0.30% F1-score), Korean (50.56% ± 0.18% accuracy, 34.56% ± 0.39% F1-score), and Hindi (53.59% ± 0.28% accuracy, 40.97% ± 0.55% F1-score). The low F1-scores (around 34–41%) compared to accuracy (50–54%) indicate a severe precision-recall imbalance, highlighting mBERT's inability to generalize from English to diverse linguistic structures without multilingual training.

Approach 4 (Hybrid). enhances mBERT with a combined training set of English annotated tweets and their translations, improving cross-lingual perfor-

510   D. Uniyal et al.

Table 3. Distribution of Relevant (R) and Irrelevant (I) Tweets with Year-wise Percentage Breakdown (2013–2022) Across Four Classification Approaches

App.	L	Total	R	I	'13	'14	'15	'16	'17	'18	'19	'20	'21	'22
#1	EN	4,769,850	47.80	52.20	3.33	4.01	5.47	5.24	5.96	6.32	4.77	14.88	24.12	25.91
	JA	4,364,391	20.48	79.52	3.40	7.23	8.64	11.82	7.20	5.35	5.88	9.38	21.89	19.18
	HI	27,258	35.44	64.56	0.50	1.22	1.01	2.49	2.52	5.16	3.21	7.75	26.44	49.70
	KO	148,040	21.08	78.92	4.15	4.42	5.50	5.07	5.14	11.81	6.09	18.05	15.73	24.01
#2	EN	4,769,850	50.61	49.39	3.25	3.93	5.33	5.12	5.85	6.33	4.77	14.91	24.25	26.26
	JA	2,459,340	16.16	83.84	3.29	6.55	8.95	10.00	6.38	5.01	7.35	10.01	22.32	20.05
	HI	27,258	35.80	64.20	0.45	0.51	0.88	1.21	1.02	3.82	2.84	5.70	26.65	56.92
	KO	148,040	17.92	82.08	4.37	4.86	6.11	5.62	5.86	12.25	5.92	17.32	15.50	22.18
#3	EN	4,769,850	48.76	51.24	3.30	3.99	5.44	5.22	5.98	6.28	4.69	14.88	24.20	26.02
	JA	4,364,391	0.44	99.56	2.54	5.67	6.03	4.03	4.15	4.75	4.17	11.23	29.10	28.30
	HI	27,258	5.01	94.99	0.29	0.07	1.03	0.51	0.66	1.98	1.03	4.18	21.10	69.15
	KO	148,040	0.64	99.36	3.98	5.03	3.98	6.08	7.55	15.62	5.45	19.39	18.03	14.88
#4	EN	4,769,850	49.81	50.19	3.17	3.89	5.28	5.08	5.77	6.34	4.82	14.96	24.33	26.36
	JA	4,364,391	21.50	78.50	3.36	7.16	8.63	8.35	6.20	4.85	10.51	10.25	21.84	18.82
	HI	27,258	31.46	68.54	0.47	0.43	0.77	0.75	0.87	3.43	2.45	5.56	27.18	58.09
	KO	148,040	20.13	79.87	4.92	4.43	5.46	5.38	5.58	12.05	6.17	17.48	15.63	22.89

Note: App.=Approach; L=Language; R=Relevant (%); I=Irrelevant (%); Years shown as '13–'22 (2013–2022). All year columns show percentage of relevant tweets.

mance. It achieves balanced results: English (94.81% ± 0.58% accuracy, 94.79% ± 0.58% F1-score), Japanese (78.49% ± 0.44% accuracy, 77.78% ± 0.49% F1-score), Korean (83.57% ± 0.52% accuracy, 83.27% ± 0.54% F1-score), and Hindi (81.76% ± 0.48% accuracy, 81.18% ± 0.52% F1-score). This approach outperforms Approach 3 across all non-English languages and shows competitive performance with Approach 1 for non-English languages, effectively bridging monolingual and multilingual strategies.

In general, Approach 1 excels in English and Korean, but shows variability in Japanese and Hindi. Approach 2 provides the highest overall performance, particularly for English, Japanese, and Hindi, making it suitable for translation-based workflows. Approach 3 demonstrates the limitations of zero-shot multilingual transfer, while Approach 4 offers the balanced solution, leveraging multilingual training to improve consistency across languages. For practical deployment, Approach 2 is ideal for translation-heavy scenarios, whereas Approach 4 provides a robust foundation for truly multilingual applications, especially where native language data quality varies.

4.2   Temporal Distribution of Relevant Tweets Across Classification Approaches

The temporal distribution of relevant tweets (Table 3) reveals a marked increase in hydrogen energy discussions from 2013–22, with post-2019 acceleration across

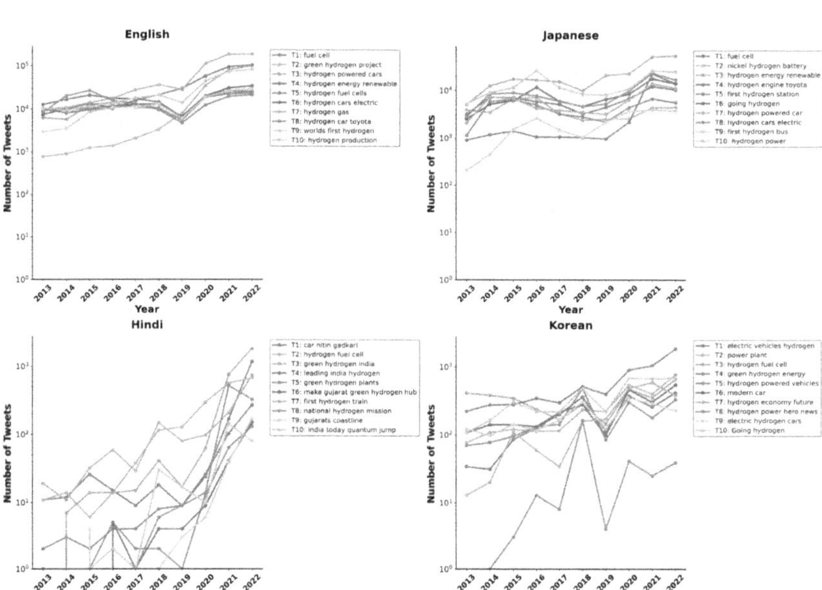

Fig. 2. Topic Modelling 2013–2022 on a line graph showing the trending themes across multiple languages.

all languages aligning with global clean energy commitments like the EU's Hydrogen Strategy (2020) [24] and Japan's National Hydrogen Strategy [25].

Language-specific patterns emerge clearly: English maintains the highest relevance rates (47.80%–50.61%) with consistent growth peaking at 25.91%–26.36% in 2022. Japanese shows moderate relevance (12.36%–20.48%) with notable 2021 peaks (21.89% in Approach 1). Korean exhibits similar patterns (17.92%–21.08% relevance) with delayed growth reaching 22.89%–24.01% by 2022. Hindi displays the most variable performance (5.01%–35.80% overall relevance) but dramatic 2022 surges (49.70%–69.15%), suggesting emerging market engagement despite India's complex multilingual social media landscape.

Approach-specific insights: Approaches 1, 2, and 4 demonstrate balanced cross-lingual performance with steady temporal growth, while Approach 3 shows significant non-English underperformance (0.44% Japanese, 5.01% Hindi relevance), consistent with its lower classification accuracy noted in Table 2. The post-2020 acceleration, particularly the 2021–2022 peaks, aligns with major hydrogen policy initiatives, including India's 2020 National Green Hydrogen Mission [26].

These temporal patterns underscore hydrogen energy's evolution from niche technology to mainstream policy priority, with language-specific adoption rates reflecting regional market maturity and policy engagement levels.

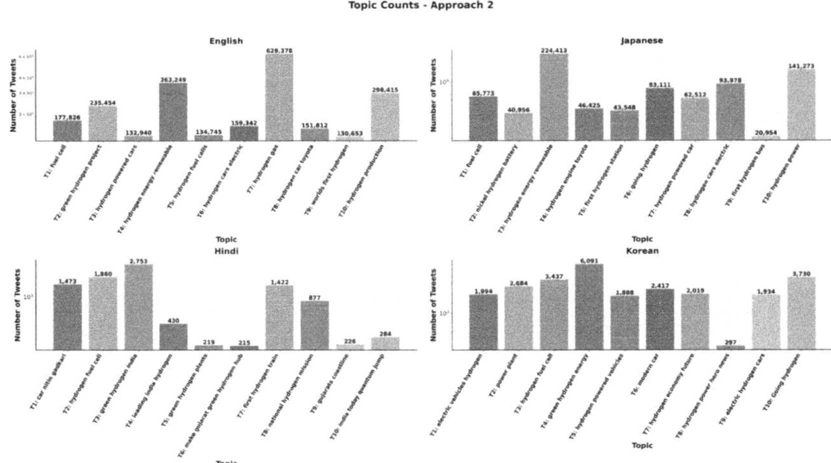

Fig. 3. Topic Modelling 2013–2022 on a bar graph showing the trending themes across multiple languages.

4.3 Topic Modeling

The topic modelling analysis of hydrogen energy discourse from 2013 to 2022 across four languages (Figs. 2 and 3) reveals distinct thematic priorities and temporal evolution patterns under Approach 2's translation-based classification framework.

Dominant Topics and Language-Specific Patterns: English discourse is dominated by "hydrogen gas" (T7, 629,378 tweets), representing 26.0% of total discussions, followed by "hydrogen energy renewable" (T4, 363,249 tweets, 15.1%) and "hydrogen production" (T10, 298,415 tweets, 12.4%). This distribution reflects the global emphasis on hydrogen as a primary energy carrier and its scalability in production. Japanese discussions show a more balanced distribution with "hydrogen energy renewable" (T3, 224,413 tweets) leading at 27.3%, followed by "first hydrogen power" (T10, 141,273 tweets, 17.2%), indicating Japan's focus on pioneering hydrogen technologies and renewable integration.

Hindi exhibits significantly lower absolute volumes but concentrated thematic focus, with "green hydrogen India" (T3, 2,753 tweets, 28.2%) and "hydrogen fuel cell" (T2, 1,860 tweets, 19.1%) dominating discussions, reflecting India's policy-driven approach to hydrogen adoption. "Car nitin gadkari" (T1, 1,473 tweets, 15.09%) likely refers to the significant event when Indian Transport Minister Nitin Gadkari arrived at Parliament in a hydrogen-powered car in March 2022, generating substantial social media attention around hydrogen vehicle adoption in India [27].

Korean discourse emphasises "green hydrogen energy" (T4, 6,091 tweets, 23%) and "going hydrogen" (T10, 3,730 tweets, 14.1%), suggesting infrastructure development priorities.

Temporal Evolution Patterns: Temporal trends reveal language-specific growth trajectories with notable post-2019 acceleration. English topics demonstrate consistent exponential growth, with "hydrogen gas" (T7) and "green hydrogen project" (T2) showing dramatic increases from 2019–2022, reflecting global policy momentum. Japanese topics peaked in 2021, particularly "hydrogen energy renewable" (T3), likely coinciding with Tokyo Olympics hydrogen initiatives and national carbon neutrality commitments, before stabilising in 2022. Hindi shows the most dramatic late-stage growth, with several topics experiencing sharp increases in 2021–2022, particularly "green hydrogen India" (T3) jumping from minimal presence to over 1,000 tweets, aligning with India's National Hydrogen Mission launch. Korean trends display steady growth with "green hydrogen energy" (T4) maintaining consistent upward trajectory through 2022, suggesting sustained industrial interest.

Strategic Implications: The analysis reveals three distinct regional focus areas: global scalability (English emphasis on production and gas applications), technological leadership (Japanese focus on renewable integration and pioneering applications), and policy implementation (Hindi concentration on national missions and Korean infrastructure development). These patterns suggest differentiated market opportunities, with English-speaking regions prioritising large-scale deployment, Japan leading in technology innovation, India focusing on policy-driven adoption, and Korea emphasising systematic infrastructure rollout. The temporal surge post-2020 across all languages, with peaks coinciding with major policy announcements (EU Hydrogen Strategy [24], Japan's carbon neutrality pledge [28]), indicates coordinated global momentum in hydrogen energy transition, offering strategic timing insights for market entry and investment decisions.

5   Discussion and Conclusion

This study explored multilingual discussions on hydrogen energy by integrating cross-lingual tweet classification with topic modeling, focusing on English, Japanese, Hindi, and Korean tweets from 2013 to 2022. We evaluated four cross-lingual relevance classification approaches, using language-specific, English-only, and multilingual models, and analysed relevant tweets to understand global clean energy conversations in multilingual digital spaces.

Our results demonstrate that model performance varies significantly across approaches and languages, with the quality of translation playing a key role. Approach 2 leveraging a single BERT model trained on English data and applied to translated tweets, achieved the highest overall performance for English and competitive results for Japanese, Korean, and Hindi. This suggests that high-quality translation to English provides a robust feature representation, outperforming Approach 1 for Japanese and Hindi, where native performance was lower (e.g., Japanese: 75.79% ± 0.49% accuracy, 74.88% ± 0.56% F1-score; Hindi: 78.08% ± 1.10% accuracy, 78.04% ± 1.10% F1-score). However, Approach 1 excelled in Korean (91.40% ± 0.31% accuracy, 91.35% ± 0.31% F1-score), indicating that language-specific models can be effective when trained and tested on high-quality native data.

Topic modeling uncovered distinct regional discourse patterns such as English emphasised production scalability ("hydrogen gas" dominating at 26.2%), Japanese focused on renewable integration ("hydrogen energy renewable" at 28.4%), Hindi concentrated on policy implementation, and Korean prioritised sustainable development. Temporal analysis showed coordinated global attention on hydrogen energy after 2020, aligning with major policy initiatives, indicating market maturation, with classification accuracy directly influencing topic coherence quality.

In future, we plan to explore sentiment trends over time, aspect-based sentiment analysis for market insights, and improved cross-lingual methods that better handle native-language data instead of relying on only translations.

Acknowledgments. The work has been partially supported by the Future Energy Exports CRC (www.fenex.org.au), whose activities are funded by the Australian Government's Cooperative Research Centre Program. This is a FEnEx CRC Document 2025/22.RP4.0139.PHD-FNX-MILE0886. The authors would like to thank Queensland University of Technology (QUT) for providing the High Performance Computing facilities used in this research and for their support.

References

1. Uniyal, D., Nayak, R.: Twitter's pulse on hydrogen energy in 280 characters: a data perspective. Soc. Netw. Anal. Min. **14**(1), 37 (2024)
2. Xu, Y., Cao, H., Du, W., Wang, W.: A survey of cross-lingual sentiment analysis: methodologies, models and evaluations. Data Sci. Eng. **7**(3), 279–299 (2022)
3. Devlin, J., Chang, M.-W., Lee, K., Toutanova, K.: Bert: pre-training of deep bidirectional transformers for language understanding, arXiv preprint arXiv:1810.04805 (2018)
4. Ruder, S., Søgaard, A., Vulić, I.: Unsupervised cross-lingual representation learning, In: Proceedings of the 57th Annual Meeting of the Association for Computational Linguistics: Tutorial Abstracts, 2019, pp. 31–38
5. Liu, C., Zhang, W., Zhao, Y., Luu, A. T., Bing, L.: Is translation all you need? a study on solving multilingual tasks with large language models, arXiv preprint arXiv:2403.10258 (2024)
6. Isbister, T., Carlsson, F., Sahlgren, M.: Should we stop training more monolingual models, and simply use machine translation instead?, arXiv preprint arXiv:2104.10441 (2021)
7. Kumar, P., Pathania, K., Raman, B.: Zero-shot learning based cross-lingual sentiment analysis for sanskrit text with insufficient labeled data. Appl. Intell. **53**(9), 10096–10113 (2023)
8. Pires, T., Schlinger, E., Garrette, D.: How multilingual is multilingual BERT?, arXiv preprint arXiv:1906.01502 (2019)
9. Yin, W., Hay, J., Roth, D.: Benchmarking zero-shot text classification: datasets, evaluation and entailment approach, arXiv preprint arXiv:1909.00161 (2019)

10. Wan, X.: Co-training for cross-lingual sentiment classification, In: Proceedings of the Joint Conference of the 47th Annual Meeting of the ACL and the 4th International Joint Conference on Natural Language Processing of the AFNLP, 2009, pp. 235–243
11. Prettenhofer, P., Stein, B.: Cross-lingual adaptation using structural correspondence learning. ACM Trans. Intell. Syst. Tech. (TIST) **3**(1), 1–22 (2011)
12. Mikolov, T., Chen, K., Corrado, G., Dean, J.: Efficient estimation of word representations in vector space, arXiv preprint arXiv:1301.3781 (2013)
13. Conneau, A., Lample, G.: Cross-lingual language model pretraining. Adv. Neural Inf. Process. Syst. **32** (2019)
14. Conneau, A., et al.: Unsupervised cross-lingual representation learning at scale, arXiv preprint arXiv:1911.02116 (2019)
15. Fleiss, J. L.: Measuring nominal scale agreement among many raters., Psychological bulletin 76 (5) (1971) 378
16. M. R. Costa-Jussà, et al.: No language left behind: scaling human-centered machine translation, arXiv preprint arXiv:2207.04672 (2022)
17. Zou, H., Caragea, C.: JointMatch: a unified approach for diverse and collaborative pseudo-labeling to semi-supervised text classification, in: H. Bouamor, J. Pino, K. Bali (Eds.), Proceedings of the 2023 Conference on Empirical Methods in Natural Language Processing, Association for Computational Linguistics, Singapore, 2023, pp. 7290–7301. https://doi.org/10.18653/v1/2023.emnlp-main.451.,https://aclanthology.org/2023.emnlp-main.451
18. NLP, T.: Bert models for japanese text (cl–tohoku/bert–japanese) (2023). https://github.com/cl-tohoku/bert-japanese, Accessed 18 Aug 2025
19. Park, J.: Distilkobert: distillation of kobert. https://github.com/monologg/DistilKoBERT (2019)
20. Kakwani, D., et al.: IndicNLPSuite: monolingual Corpora, Evaluation Benchmarks and Pre-trained Multilingual Language Models for Indian Languages, In: Findings of EMNLP, 2020
21. Sánchez, C., Sarmiento, H., Abeliuk, A., Pérez, J., Poblete, B.: Cross-lingual and cross-domain crisis classification for low-resource scenarios, In: Proceedings of the International AAAI Conference on Web and Social Media, Vol. 17, 2023, pp. 754–765
22. Balasubramaniam, T., Nayak, R., Bashar, M.A., Understanding the spatio-temporal topic dynamics of covid-19 using nonnegative tensor factorization: a case study, in,: IEEE symposium series on computational intelligence (SSCI). IEEE **2020**, 1218–1225 (2020)
23. Egger, R., Yu, J.: A topic modeling comparison between lda, nmf, top2vec, and bertopic to demystify twitter posts. Front. Sociol. **7**, 886498 (2022)
24. Eu's hydrogen strategy (2020). https://energy.ec.europa.eu/topics/energy-systems-integration/hydrogen_en, (Accessed 01 Aug 2025)
25. Japan: first nation to form national hydrogen strategy (2017). https://www.hydrogeninsight.com/, (Accessed 01 Aug 2025)
26. National green hydrogen mission (2020). https://mnre.gov.in/img/documents/uploads/file_f-1673581748609.pdf, (Accessed 18 Oct 2023)
27. Nitin gadkari using hydrogen-powered car (2020). https://economictimes.indiatimes.com/, (Accessed 25 Aug 2025)
28. Green growth strategy japan (2020). https://www.meti.go.jp/english/policy/energy_environment/global_warming/ggs2050/index.html, (Accessed 25 Aug 2025)

From Burst to Routine: Mining Time-Aware Patterns from Sequential Dataset

Khanh Luong[1(✉)] , Daniel Angus[1] , Yufan Kang[2] ,
Abdul Karim Obeid[1] , and Dan Tran[3]

[1] Queensland University of Technology, Brisbane, Queensland, Australia
`{khanh.luong,daniel.angus,abdul.obeid}@qut.edu.au`
[2] Monash University, Melbourne, Victoria, Australia
`tina.kang@monash.edu`
[3] The University of Queensland, Brisbane, Queensland, Australia
`d.k.tran@uq.edu.au`

Abstract. Sequential pattern mining is a foundational technique for analysing behavioural data across domains such as digital advertising and cybersecurity. However, traditional algorithms often treat all sequential events as equally spaced in time, overlooking varied temporal proximity that can carry critical semantic meaning. This limitation leads to fragmented or less interpretable results, especially in contexts where timing is crucial. In this work, we propose TSpan, a novel sequential pattern mining method that integrates temporal compactness directly into the mining process. TSpan identifies sequences that are frequent and temporally cohesive, providing a more accurate reflection of real-world behavioural and operational patterns. We introduce two variants, TSpan-Intra and TSpan-Inter, to separately model patterns within shorter temporally-bounded user activity sessions, and across sessions. These allow us to distinguish between short-term and long-term patterns. TSpan is evaluated on two large datasets: a digital advertising dataset containing 780,000 ad observations and a cybersecurity event log. In both datasets, TSpan consistently identifies patterns that are more compact, interpretable, and impactful than baseline methods. Ad-related exposure sequences are uncovered by TSpan in the advertising dataset, shedding light on how such content clusters and unfolds throughout users' ad journeys.

Keywords: Sequential Pattern Mining · Time-weighted Support · Behavioural Sequence Analysis · Temporal Compactness

1 Introduction

Frequent Pattern Mining (FPM) is a fundamental task in data mining concerned with identifying recurring relationships, behaviours, or structures within large datasets [4,18]. Initially developed for transactional databases, it is now widely

© The Author(s), under exclusive license to Springer Nature Singapore Pte Ltd. 2026
Q. V. Nguyen et al. (Eds.): AusDM 2025, CCIS 2765, pp. 516–531, 2026.
https://doi.org/10.1007/978-981-95-6786-7_35

used across data types including sequences, graphs, and temporal logs. Among these, sequential data is of particular importance because it captures the order in which events occur, often reflecting meaningful behavioural, clinical, or operational processes in real-world domains.

Mining frequent patterns from temporal data underpins numerous applications, such as digital advertising, recommender systems, threat detection, industrial monitoring, and healthcare analytics. Depending on the analytical goals and the data characteristics, methods can be broadly divided into non-sequential and sequential approaches. Apriori [2, 4], for example, is a classic algorithm for mining frequent itemsets in transactional data. It identifies combinations of items that co-occur frequently across records and has proven highly effective in domains where the order of items is unimportant, such as market basket analysis. However, Apriori does not preserve temporal or sequential information and is therefore limited when event order carries semantic meaning.

When event order matters, such as tracking user clickstreams, analysing treatment progressions in electronic health records, modelling workflows in manufacturing, or examining ad delivery sequences—sequential pattern mining techniques such as PrefixSpan (Prefix-projected Sequential Pattern Mining) [14] and GSP (Generalised Sequential Pattern) [3] are more appropriate. These methods preserve event ordering, enabling the discovery of typical sequences that recur across users or sessions.

Despite this potential, sequential pattern mining (SPM) has been underused in domains like digital advertising and cybersecurity. Prior work in advertising has emphasised classification [9, 27] or targeting [1], with little focus on recurring sequential structures. To our knowledge, this is the first large-scale application of sequential mining to advertising logs, revealing temporal regularities in ad delivery. Specifically, where patterns could reveal potentially problematic advertising dynamics that are a risk to consumer safety and well-being. A similar issue arises in the cybersecurity domain, particularly in datasets involving system event logs such as those from HDFS clusters. Patterns like "NameSystem.allocateBlock" followed closely by multiple "Receiving block" events can be meaningful indicators of normal or abnormal behaviour, but only when they occur within short, security-relevant time windows [22]. Traditional methods can conflate unrelated events, obscuring anomalies.

In light of these limitations, mining methods should incorporate temporal compactness as a first-class objective in addition to frequency and order. Some methods use time-gap constraints or sliding windows [7, 17], but they often use rigid parameters, lack generalizability [8, 19], or require full dataset scans [3, 7]. We propose TSpan, a time-aware sequential pattern mining framework that emphasises both pattern frequency and temporal compactness. It captures sequences where events occur in order within bounded periods, producing patterns that are compact, interpretable, and relevant to time-sensitive domains. We introduce two versions: TSpan-Intra and TSpan-Inter. Each highlights distinct temporal dynamics while remaining compatible with classical pipelines. Applied to behavioural data, they reveal patterns linked to impulsive engagement versus long-term habits, supporting richer user modelling and intervention design.

We evaluated TSpan on (1) a digital advertising exposure dataset, and (2) a cybersecurity event log, based on the Hadoop system activity logs (HDFS). Across both, TSpan identifies patterns that are temporally compact and behaviourally meaningful, aligning with domain expectations such as bursty ad exposures or recurring security events, and yielding improved features for downstream tasks like anomaly detection.

This paper makes three contributions: *First*, TSpan, a framework that integrates temporal compactness with frequency. *Second*, two methods, TSpan-Intra and TSpan-Inter, which disentangle within-session bursts from cross-session recurrences. *Third*, the first large-scale application of time-aware pattern mining in advertising and cybersecurity, showing more meaningful patterns and stronger features for downstream tasks.

2   Related Works

Frequent Itemset Mining (FIM) identifies recurring patterns, associations, or relationships within datasets [4]. It has been applied in areas such as cybersecurity, where mining behavioural patterns from system logs supports anomaly detection [10,21]. Studies show that FIM-based features combined with supervised models can improve accuracy [6].

FPM has advanced since the early 1990s [4]. Apriori [4] employs the "downward closure" property to identify frequent itemsets systematically. FPGrowth [15] improves efficiency with FP-trees. ECLAT [30] uses vertical layouts for faster intersections. H-Mine [23] applies a hyperlinked data structure for memory efficiency. These approaches focus on identifying the "set" of items that frequently occur together, considering only the presence or absence of items. They are specifically designed to work with unordered itemsets, making them unsuitable for applications where the order of items is critical, such as sequential pattern mining, temporal data analysis, or scenarios where the positional context of items within a pattern carries significant meaning.

Sequential approaches account for order. For instance, GSP [3] extends Apriori by generating candidate sequences level by level and pruning those that do not meet minimum support thresholds. However, it suffers from high computational costs due to excessive candidate generation. SPADE [29] transforms sequential data into a vertical format using ID lists, which enables faster sequence discovery. Yet, its reliance on vertical transformations can make it memory-intensive for complex datasets. PrefixSpan [14] overcomes these limitations by avoiding candidate generation and adopting a divide-and-conquer approach. It recursively projects the dataset based on frequent prefixes, focusing only on relevant subsequences, which significantly enhances efficiency and scalability.

Recent state-of-the-art methods have enhanced pattern mining by incorporating additional factors such as weight, utility, and temporal information to better capture real-world data. For instance, WPPM [7] explicitly integrated temporal periodicity into the mining process, though they relied on rigid user-defined parameters. WFSPM [19] introduced pruning strategies to reduce the search space when mining weighted sequential patterns, while EWSPM [8] further improved scalability by proposing two novel upper bounds to tighten pruning and accelerate pattern discovery. X-FSPMiner [25] extended sequential mining to discover frequent similar patterns based on similarity functions rather than exact matches. Beyond traditional databases, temporal modelling has been strongly emphasised in recommender systems, where TGSRec [11] leverages a Transformer-based architecture to unify sequential patterns with temporal collaborative signals, and GDERec [24] employs graph ordinary differential equations to capture continuous-time dynamics, even under irregular sampling.

Overall, pattern mining research is shifting toward weighted, time-aware, and utility-driven approaches that yield richer insights.

3   TSpan: Time Compact Sequential Pattern Mining

TSpan is a novel sequential pattern mining algorithm that incorporates temporal information into the support measure. It emphasises patterns that are both frequent and temporally coherent. Using a depth-first exploration strategy and a time-weighted support metric, it yields sequences occurring within compact time windows and thus delivers patterns that better capture real-world dynamics.

3.1   Data Formulation and Problem Definition

3.1.1 Data Formulation

The sequential dataset, denoted as D which contains a collection of n sequences, $D = \{s_1, s_2, ..., s_n\}$, where each sequence s_i represents the ordered progression of items within a specific context, such as an active user session. A sequence s_i is defined as a tuple of an ad sequence and a timestamp: $s_i = \{(A_1, ..., A_{m_i}), (T_1, ..., T_{m_i})\}$. Here, $A_1, ..., A_{m_i}$ is a sequence of m_i items, and $T_1, ... T_{m_i}$ is the associated sequence of timestamps preserving their temporal order. A toy sequential dataset is given in Table 1.

3.1.2 Problem Definition

Event Session. We assume the sequential dataset can be partitioned into event sessions, where a session is a logically or temporally bound sequence of events. A session can represent a user interaction in an advertisement dataset (e.g., a browsing session), or a time-bounded window in system log data (e.g., events grouped within a fixed time interval).

The aim is to perform pattern mining on the sequential dataset D to achieve two primary objectives. ***(1). Extract Frequent and Compact Subsequence:*** Identify subsequences that are both frequent and compact. A subsequence is considered frequent if it appears across multiple sequences in the dataset. It is considered compact if it occurs in close temporal proximity within each sequence. ***(2). Analyse System Behaviours:*** Discover the most frequent and temporally compact patterns that recur across multiple sessions to understand broader system interactions or behaviours.

Table 1. Example User Session Dataset with Ad Categories

Sequence ID	User ID	Ad Categories (Ordered Sequence)
Seq1	User1	{(Retail, 6:00), (Politics,6:03), (Green/Eco,6:06), (Alcohol,6:10), (Retail,6:12). (Politic.6:15), Green/Eco,6:30)}
Seq2	User1	{(Politics,7:00), Green/Eco,7:05), {(Retail,7:10). (Politics.7:12), {(Retail,7:13). (Politics.7:18)}
Seq3	User2	{(Retail,8:00), (Green/Eco,8:10), (Politics,8:15)}
Seq4	User3	{(Alcohol,10:00), {(Retail,10:02). (Politics,10:05)}

3.2 TSpan: Time Compact Pattern Extraction

We propose two specialised variants: TSpan-Intra and TSpan-Inter, addressing distinct analytical goals. TSpan-Intra focuses on mining sequentially compact patterns that repeatedly occur within each event session. It captures short-term behavioural regularities, showing how events are temporally clustered inside a session. While Tspan-Inter uncovers broader behavioural patterns across longer time spans and across event sessions.

To support these objectives, we introduce two novel time-weighted support measures that guide TSpan in prioritising frequent and temporally coherent interactions. We named them T-Score$_{intra}$ and T-Score$_{inter}$. Our key assumption is that patterns that occur frequently and within a short time frame are more likely to reflect meaningful behaviours or intentional bursts of activity. This includes targeted advertising campaigns in an advertisement dataset [26], or emerging consumer trends in market data or meme-like bursts in online communities [12], or abnormal bursts of attack events in system logs that may signal failures, intrusions, or other anomalies [10].

Next, we show how time-weighted measures integrate into our framework.

3.2.1 Pattern and Time Model Definitions

Definition 1: Subsequence or Pattern. A pattern is defined as a set or subset of elements, events, or items that regularly co-occur in a database [2].

Definition 2: Support. The support for a pattern is a measure of how frequently it occurs in a dataset. It reflects the proportion or count of sequences in which the pattern appears [4].

Note that, in this work, the order of elements in the pattern is preserved, ensuring only sequences that maintain this order contribute to the support value.

Definition 3: Time Delta Between Two Occurrences of a Pattern. Given a pattern P with multiple occurrences, the *time delta* between two consecutive occurrences A and B, observed at timestamps T_i and T_{i+1}, respectively (where $T_i < T_{i+1}$), is defined as $\delta(A, B) = T_{i+1} - T_i$. This metric captures the temporal spacing between adjacent occurrences of a pattern and serves as the basis for evaluating temporal compactness in both intra-and inter-session contexts.

Definition 4: Intra-Session Delta Δ_{intra}. The average time delta between consecutive occurrences of pattern P within the same session is computed as,

$$\Delta_{intra}(P) = \frac{1}{n_{intra}} \sum_{i=1}^{n_{intra}} (T_{i+1} - T_i) \tag{1}$$

where T_i and T_{i+1} denote timestamps of the last item in each occurrence of P within the same session, and n_{intra} is the number of such intra-session deltas.

Definition 5: Overall Intra-Delta $\overline{\Delta}_{intra}$. This measures the overall temporal compactness of pattern P within sessions by averaging all intra-session deltas across the dataset. Formally:

$$\overline{\Delta}_{intra}(P) = \frac{1}{m} \sum_{j=1}^{m} \Delta_{intra}^{(j)}(P) \tag{2}$$

where $\Delta_{intra}^{(j)}(P)$ denotes the intra-session delta of pattern P in session j, and m is the total number of sessions in which P occurs more than once (i.e., has computable intra-session deltas). This aggregated measure provides a global view of how temporally compact the pattern P is across the dataset.

Example: For pattern $P = \{Retail, Politics\}$ (Table 1), the intra-session deltas are 12 min (Session 1) and 5 min (Session 2), giving an overall $\overline{\Delta}_{intra}(P) = (12 + 5)/2 = 8.5$ minutes.

Definition 6: Inter-Session Delta Δ_{inter}. The average time difference between occurrences of pattern P across consecutive activity segments for the same entity (e.g., user sessions in advertisement data, execution windows in system logs). For each pair of sessions (S_i, S_{i+1}) in which the pattern occurs, the time delta is computed as the difference between the last timestamp of P in S_i and the first timestamp of P in S_{i+1}. Formally, $\delta_{inter}(S_i, S_{i+1}) = T_{first}^{(i+1)} - T_{last}^{(i)}$, where $T_{first}^{(i+1)}$ is the timestamp the *first* occurrence of P in session S_{i+1}, $T_{last}^{(i)}$ is the timestamp of the *last* occurrence of P in session S_i. S_i and S_{i+1} are two consecutive sessions of the same observer. The average inter-session delta of pattern P is defined as:

$$\Delta_{inter}(P) = \frac{1}{n_{inter}} \sum_{i=1}^{n_{inter}} \left(T_{first}^{(i+1)} - T_{last}^{(i)} \right) \tag{3}$$

where n_{inter} is the number of valid session pairs for the same observer in which the pattern P occurs in both sessions.

Definition 7: Overall Inter-Delta $\overline{\Delta}_{inter}$. This measures the overall temporal compactness of pattern P across sessions by averaging all inter-session deltas across the dataset. Formally:

$$\overline{\Delta}_{inter}(P) = \frac{1}{q} \sum_{j=1}^{q} \Delta_{inter}^{(j)}(P) \tag{4}$$

where $\Delta_{inter}^{(j)}(P)$ denotes the inter-session delta of pattern P in session j, and q is the total number of inter-session deltas (i.e., number of $\Delta_{inter}(P)$).

Example: For $P = \{Retail, Politics\}$ (Table 1), the last occurrence in Session 1 (6:15) and the first in Session 2 (7:12) yield an inter-session delta of $\overline{\Delta}_{inter}(P) = 57$ minutes.

Note: The definition of Δ_{inter} is naturally suited to advertisement datasets, where a user has multiple separated active sessions. In this context, S_i and S_{i+1}

are successive sessions, and $\Delta_{\mathrm{inter}}(P)$ measures how often a behavioural pattern re-emerges across them. In general sequential datasets, the notion of user and user sessions can be generalised to entities and their corresponding episodes. For example, in event logs, sessions may correspond to execution windows, process instances, or time-bounded partitions of a given entity (e.g., a block_id in system logs). Here, $\Delta_{\mathrm{inter}}(P)$ denotes the temporal gap between recurring occurrences of pattern P across an entity's sessions, where session boundaries vary by domain (e.g., patient visits in healthcare, shopping sessions in retail).

3.2.2 Time Compact Pattern Mining with TSpan. We propose two time-aware methods, *TSpan-Intra* and *TSpan-Inter*, which advance pattern-growth approaches (e.g., PrefixSpan [14], ONP-Miner [28], TaSPM [18]) by embedding time-weighted measures and temporal constraints to uncover frequent, coherent, and compact patterns within and across sessions.

1. TSpan-Intra. This method targets patterns that are temporally compact within individual sessions. For each pattern P, we introduce an intra-session time-weighted metric, termed *intra-session time compactness*, T-Score$_{intra}$, that integrates both frequency and temporal tightness as,

$$\text{T-Score}_{intra}(P) = \alpha \cdot \text{NormSupport}(P) + \beta \cdot (1 - \text{NormTime}_{intra}(P)) \qquad (5)$$

where the generic normalisation operator for any measure $x(P)$ is defined as:

$$Norm(x(P)) = \frac{x(P) - \min_{Q} x(Q)}{\max_{Q} x(Q) - \min_{Q} x(Q) + \epsilon} \qquad (6)$$

with Q ranges over all mined patterns, and ϵ is a small constant (e.g., 10^{-6}) to ensure numerical stability. We then have NormSupport$(P) = Norm(\text{Support}(P)$, NormTime$_{intra}(P) = Norm(\overline{\Delta}_{intra}(P))$. TSpan-Intra mines patterns whose intra-session compactness score, T-Score$_{intra}$, exceeds a threshold τ.

Handling Patterns Without Intra-Session Repetition. Some patterns P may occur frequently across sessions but never repeat within a session, leaving $\Delta_{intra}(P)$ undefined. Such patterns lack temporal cohesion and are down-weighted by assigning a fallback compactness score, where the normalised support is penalised by a fixed weight $\gamma \in [0, 1]$ as, T-Score$_{intra}(P) = \gamma \cdot$ NormSupport(P).

In our implementation, we set $\gamma = 0.5$ to moderately penalise patterns without intra-session repetition, reducing their scores while retaining them as potentially globally frequent behaviours. This ensures that only patterns with both frequency and temporal compactness are ranked highly.

2. Tspan-Inter. This method focuses on identifying patterns that reoccur across different sessions of the same entity within short temporal intervals. It mines patterns whose inter-session time-weighted score (T-Score$_{inter}$) exceeds

---

Algorithm 1: TSpan-Intra

Input: Sequential dataset D (sessions with timestamps); minimum support min_sup; weighting parameters α, β with $\alpha + \beta = 1$; CompactScore threshold τ; current prefix (initially empty)

Output: Frequent, temporally compact intra-session patterns

1 **Function** TSpan-Intra$(D, min_sup, \tau, \alpha, \beta, prefix)$:
2   Scan D to find frequent items F by session support;
3     $F \leftarrow \{i \mid supp(i) \geq min\_sup\}, \quad supp(i) = |\{S \in D : i \in S\}|$;
4   **foreach** $p \in F$ **do**
5   Generate new pattern $new\_prefix \leftarrow prefix \cup p$;
6   Compute intra-session deltas Δ_{intra} for new_prefix using Eq. (1);
7   Compute average $\overline{\Delta}_{intra}(new\_prefix)$ using Eq. (2);
8   Record $(new\_prefix, support, \overline{\Delta}_{intra})$;
9   Construct projected database $DB(new\_prefix)$;
10   TSpan-Intra$(DB(new_prefix), min_sup, \tau, \alpha, \beta, new_prefix)$;
11   Normalise supports and $\overline{\Delta}_{intra}$ values as in Eq. (6);
12   Compute T-Score using Eq. (5);
13   Prune patterns with T-Score$_{intra} < \tau$;

---

a predefined threshold. For each pattern P, we introduce an inter-session time-weighted metric, termed *inter-session time compactness*, T-Score$_{inter}$, to capture the temporal proximity of occurrences across sessions. It is defined as,

$$\text{T-Score}_{\text{inter}}(P) = \alpha \cdot \text{NormSupport}(P) + \beta \cdot (1 - \text{NormTime}_{inter}(P)) \qquad (7)$$

where NormTime$_{inter}(P) = Norm(\overline{\Delta}_{inter}(P))$, using the same normalisation operator defined in Eq. (6), and $\alpha, \beta \in [0, 1]$ and $\alpha + \beta = 1$. This formulation rewards patterns that appear frequently (high support) and recur with minimal temporal gaps across sessions.

T-Score's Rationale. Patterns are most meaningful when they are frequent and temporally compact. High support with large time gaps yields weak signals, while rare but compact patterns are also uninformative. Only frequent patterns with small Δ_{intra} or Δ_{inter} indicate timely, intentional behaviours or anomalies.

3.2.3 TSpan Algorithm

1. TSpan-Intra Algorithm. As represented in Algorithm 1, TSpan-Intra adopts a divide-and-conquer approach by recursively projecting the database based on frequent prefixes. It begins by identifying frequent items that satisfy the minimum support threshold min_{sup}, which serve as the initial prefixes. For each prefix, the algorithm constructs a projected database consisting of suffix subsequences following the prefix and uses it to discover longer frequent patterns. A key feature of TSpan-Intra is the evaluation of each extended prefix through intra-session deltas, thereby quantifying temporal compactness.

All discovered patterns are normalised by support and time delta, after which a weighted score is computed. Finally, only patterns with T-Score$_{intra} \geq \tau$ are

a. Australian Ad Observatory

observed_at	session_id	ad_content
22/11/2022 10:00	10c	...
22/11/2022 10:10	10c	...
22/11/2022 10:20	10c	...
22/11/2022 11:00	20c	...
22/11/2022 11:10	20c	
22/11/2022 12:00	24b	...
22/11/2022 12:10	24b	...
22/11/2022 12:20	24b	...
22/11/2022 12:30	24b	...
22/11/2022 13:00	3df	...
22/11/2022 13:10	3df	...
22/11/2022 13:20	3df	...
22/11/2022 13:30	3df	...
22/11/2022 13:33	3df	...

ad category mapping

b. Australian Ad Observatory Data with Mapping Ad

observed_at	session_id	ad_content	cat_id	cat_name
22/11/2022 10:00	10c	...	3	Gifts and Holiday Items
22/11/2022 10:10	10c	...	8	Education and Careers
22/11/2022 10:20	10c	...	1	Consumer Packaged Goods
22/11/2022 11:00	20c	...	7	Metals
22/11/2022 11:10	20c		2	Sporting Goods
22/11/2022 12:00	24b	...	2	Sporting Goods
22/11/2022 12:10	24b	...	9	Alcohol
22/11/2022 12:20	24b	...	3	Gifts and Holiday Items
22/11/2022 12:30	24b	...	4	Clothing and Accessories
22/11/2022 13:00	3df	...	1	Consumer Packaged Goods
22/11/2022 13:10	3df	...	2	Sporting Goods
22/11/2022 13:20	3df	...	9	Alcohol
22/11/2022 13:30	3df	...	5	Non-Fiat Currency
22/11/2022 13:33	3df	...	4	Clothing and Accessories

session sequence

c. Ad Session Sequences

session_id	session_sequence
10c	[3, 8, 1]
20c	[7, 2]
24b	[2, 9, 3, 4]
3df	[1, 2, 9, 5, 4]

Fig. 1. A simplified illustration of the data pre-processing pipeline.

retained, ensuring that the output contains frequent as well as temporally compact intra-session patterns.

The key advantage of TSpan-Intra is its ability to filter out trivial or loosely related patterns by emphasising temporal closeness alongside frequency, yielding more meaningful, tightly clustered behaviours. Beyond identifying significant patterns, it also outputs T-Score$_{intra}$, intra-session time delta, revealing not only which patterns recur frequently but also how closely in time they occur. This dual perspective provides deeper insights for applications such as system logs, user behaviour analysis, and anomaly detection.

2. TSpan-Inter Algorithm. The methodology for TSpan-Inter follows the same overall steps as Algorithm 1. Rather than computing intra-session time deltas, it calculates inter-session deltas between consecutive sessions of the same entity, as defined in Eqs. (3) and (4). The resulting values are then used in Eq. (7) to obtain the final inter-session compactness score.

The key benefit of TSpan-Inter lies in its ability to uncover frequent and compact patterns that occur across multiple sessions of the same entity. By capturing cross-session dependencies, TSpan-Inter reveals patterns often overlooked by current state-of-the-art. To the best of our knowledge, this is the first work to explicitly incorporate inter-session temporal compactness into sequential pattern mining, offering a novel perspective on persistent and evolving behaviour.

4 Experiments

4.1 Datasets

We evaluate TSpan on advertising and cybersecurity datasets to test its effectiveness and generality across sequential behaviours.

The AAO (Australian Ad Observatory) Dataset. This is an ADM+S[1] initiative, collected over 780,000 Facebook ad observations via a browser plugin from more than 2,000 Australians during 2021–22, producing a dataset of ad viewing patterns and category prevalence to support transparency in targeted advertising [5]. Each observation was classified into one of 45 IAB v2.0 categories[2]

---

[1] https://www.admscentre.org.au/adobservatory/.
[2] https://github.com/InteractiveAdvertisingBureau/Taxonomies.

(e.g., Retail, Politics) and organised into user ad-session sequences (Fig. 1), where sessions are represented as ordered sequences of category IDs (**cat_id**).

The HDFS (Hadoop Distributed File System) Dataset. This dataset contains $11,172,157$ Hadoop log messages, including $16,838$ malicious logs, widely used for system analysis and anomaly detection [13,20,21]. For our experiment, we sampled the first $20,000$ logs (638 malicious) and transformed them into event sequences via: (1) log parsing (EventTemplates, EventIDs), (2) log grouping by Block ID, and (3) log sequencing and labelling as "normal" or "anomaly."

4.2   Baselines

We compare against four representative SPM algorithms. **GSP** [3] is an Apriori-based method that applies breadth-first search with downward-closure pruning. **PrefixSpan** [14] improves efficiency by avoiding explicit candidate generation and instead exploring a prefix-projected database in a recursive, pattern-growth manner. **WSPM** [19] extends sequential pattern mining by incorporating weights reflecting the relative importance of items or events. The aim is to discover more meaningful patterns in domains where frequency alone is insufficient. **EWSPM** [8] extends the pattern-growth paradigm to weighted patterns, introducing novel pruning bounds and optimised projections, which reduce runtime and memory consumption while ensuring scalability and completeness.

4.3   Evaluation Metrics

To rigorously evaluate the computational efficiency and the effectiveness of our proposed method for SPM, a comprehensive experiment was conducted. This section provides the experimental setup and the methodology employed for performance measurement and qualitative pattern analysis. Our evaluation is based on two complementary perspectives: **quantitative** (efficiency and pattern-level statistics) and **qualitative** (inspection of mined patterns in unlabeled data).

Quantitative Metrics. Efficiency is measured by **runtime** (seconds) and **memory** (MB) [8,18,28]. Pattern quality is assessed by **average length, coverage** (fraction of data explained), **diversity** (entropy), and **repetition strength** (frequency of repeated items). For anomaly detection on HDFS, the top 500 patterns from each method are used as features. Classifier performance is reported using **ACC, PRE, RE, F1, FPR**, and **FNR**, with TP, TN, FP, and FN as standard. We also compute **Detection Quality (DQ)** = mean(ACC, PRE, RE, F1) and **Detection Error (DE)** = mean(FPR, FNR).

Qualitative Metrics. For qualitative evaluation, we analysed mined patterns in the AAO Dataset. This involves contrasting patterns discovered by our method and by baselines, to highlight differences in temporal compactness and co-occurrence structures. It highlights several interesting findings that cannot be revealed by quantitative metrics alone by using different approaches to capture meaningful structures.

Parameters Selection. TSpan uses the minimum support (min_{sup}) like other frequent pattern mining methods, but additionally applies a time weighted score threshold (τ). To capture distinct characteristics, we set $\tau = 0.1$ (Intra) and 0.05 (Inter) for the AAO dataset, and $\tau = 0.05$ (Intra) and 0.005 (Inter) for HDFS, adjusting dynamically based on the number of patterns obtained. Note that T-Score is a weighted combination of normalised frequency and time, requiring only α ($\beta = 1 - \alpha$). In experiments, α was chosen in $[0.4, 0.6]$.

4.4 Experimental Results

4.4.1 Pattern Quality Evaluation

We measure quantitative metrics for the top 500 patterns for the baselines, with results in Tables 2 and 3. Across both datasets, TSpan-Intra consistently outperforms baselines in temporal coherence while maintaining strong coverage, diversity, and pattern length. On the AAO dataset, it achieves a mean intra-session compactness of 131 s, 29âĂŞ50% shorter than baselines. On HDFS, compactness improves from nearly 10 seconds (GSP/EWSPM) to 0.16 seconds, showing substantially more coherent patterns. These gains improve quality and reliability for downstream analysis.

TSpan also produces higher entropy and longer patterns, yielding more informative and diverse sequences.

Across datasets, coverage is similar across methods, but runtime and memory differ. TSpan runs faster by pruning irrelevant patterns, at the cost of higher memory from storing indexes, illustrating a standard timeâĂŞmemory trade-off.

TSpan-Inter captures cross-session dynamics. On AAO, its average pattern length (2.57) is comparable to baselines but with higher compactness, reflecting longer ad gaps. On HDFS, it achieves the highest repeatability and longest average pattern length (6.15), indicating extended and consistent behaviours. It thus reveals long-range ad exposures and persistent recurring system events.

TSpan-Inter discovered over 500 patterns on AAO but only 28 on HDFS, highlighting dataset differences. Its lower entropy suggests more stable but less diverse patterns, emphasising recurrence over variation.

Overall, both TSpan variants substantially improve time-sensitive pattern discovery, making them especially suitable where event proximity is critical.

4.4.2 Feature Utility in Anomaly Detection

We evaluate anomaly detection on the HDFS dataset using the top 500 patterns from each method, representing each session as a count vector of subsequence frequencies. Gradient Boosting (GB), Random Forest (RF) and XGBoost are used for supervised anomaly detection. AutoEncoder (AE) [16] is used as an unsupervised anomaly detection model. For TSpan, we use patterns mined from TSpan-Intra as well as a combination of the top 500 patterns mined from TSpan-Intra with TSpan-Inter, which we named TSpan-Both.

As reported in Table 4, TSpan's both variants consistently outperform baseline methods across multiple classifiers. In particular, TSpan-Both achieves the highest detection quality (DQ) and lowest detection error (DE) in nearly all

Table 2. Top-500 patterns: quality metrics by methods on AAO Dataset

Mode	Avg Len	Time Compactness (s)	Coverage	Entropy	Repeat	Runtime (s)	Mem (MB)
TSpan-Intra	**2.58**	**131**	1.00	**0.932**	**0.160**	560.98	185.14
TSpan-Inter	2.57	1,213,229	1.00	0.250	0.09	625.26	409.23
GSP	2.30	249	1.00	0.910	0.108	5854.56	1.90
EWSPM	1.67	229	1.00	0.497	0.069	10429.85	7.7
PrefixSpan	2.56	185	1.00	0.925	0.159	333.57	80.24
WFSPM	2.19	257	1.00	0.86	0.10	1480.09	874.40

Table 3. Top-500 patterns: quality metrics by methods on HDFS Dataset

Method	Avg Len	Time Compactness	Coverage	Entropy	Repeat	Runtime (s)	Mem (MB)
TSpan-Intra	4.77	0.162	0.99	1.59	0.28	16.76	30.16
Tspan-Inter	**6.15**	1.400	**1.00**	0.21	**0.61**	**0.48**	6.29
GSP	4.73	9.66	0.99	1.53	0.30	19.05	1.02
EWSPM	4.73	9.682	0.99	1.53	0.30	1230.07	1.99
PrefixSpan	4.77	0.273	0.99	1.59	0.29	16.72	7.52
WFSPM	4.73	9.66	0.99	1.53	0.30	308.107	20.59

Table 4. Anomaly Detection Performance Using Top 500 Mined Patterns Across Different Classifiers (HDFS Dataset)

Classifier	Method	ACC ↑	PRE↑	RE↑	F1↑	DQ↑	FPR↓	FNR↓	DE↓
GB	EWSPM	97.14	96.32	97.14	96.62	96.81	0.73	81.82	41.28
GB	GSP	98.58	98.33	98.43	98.33	98.42	1.22	18.18	9.70
GB	PrefixSpan	97.62	98.28	97.62	97.87	97.85	1.96	18.18	10.07
GB	WFSPM	98.33	98.58	98.33	98.43	98.42	1.22	18.18	9.70
GB	**TSpan-Intra**	97.86	98.37	97.86	98.05	98.04	1.71	18.18	9.95
GB	**Tspan-Both**	98.57	98.72	98.57	98.63	**98.62**	0.98	18.18	**9.58**
RF	EWSPM	97.14	96.32	97.14	96.62	96.81	0.73	81.82	41.28
RF	GSP	99.01	98.81	98.88	98.81	98.88	0.98	9.09	5.03
RF	PrefixSpan	98.81	99.01	98.81	98.88	98.88	0.98	9.09	5.03
RF	WFSPM	98.81	99.01	98.81	98.88	98.88	0.98	9.09	5.03
RF	**TSpan-Intra**	98.81	99.01	98.81	98.88	98.88	0.98	9.09	5.03
RF	**Tspan-Both**	99.05	99.16	99.05	99.09	**99.09**	0.73	9.09	**4.91**
XGBoost	EWSPM	97.38	94.83	97.38	96.09	96.42	0.00	100	50.00
XGBoost	GSP	98.59	97.86	98.11	97.86	98.11	1.96	9.09	5.52
XGBoost	PrefixSpan	98.10	98.68	98.10	98.29	98.29	1.71	9.09	5.40
XGBoost	WFSPM	98.33	98.58	98.33	98.43	98.42	1.22	18.18	9.70
XGBoost	**TSpan-Intra**	98.33	98.78	98.33	98.48	98.48	1.47	9.09	5.28
XGBoost	**Tspan-Both**	98.57	98.89	98.57	98.68	**98.68**	1.22	9.09	**5.16**
AE	EWSPM	96.57	96.08	96.57	96.30	96.38	1.40	78.38	39.89
AE	GSP	94.99	95.69	94.99	95.33	95.25	3.01	78.38	40.70
AE	PrefixSpan	95.06	95.70	95.06	95.37	95.30	2.94	78.38	40.66
AE	WFSPM	94.99	95.69	94.99	95.33	95.25	3.01	78.38	40.70
AE	**Tspan-Intra**	95.35	96.16	95.35	95.73	95.65	2.94	67.57	**35.25**
AE	**Tspan-Both**	97.57	96.95	97.57	96.84	**97.23**	0.22	83.78	42.00

Table 5. Potentially problematic ad patterns detected only by TSpan-Intra

Pattern	Len	Avg. Time Δ (s)
Alcohol, Alcohol, Alcohol, Alcohol, Alcohol, Alcohol, Alcohol, Alcohol	8	108
Clothing and Accessories, Clothing and Accessories, Clothing and Accessories, Alcohol, Clothing and Accessories, Clothing and Accessories	6	106
Dieting and Weightloss, Alcohol, Alcohol, Alcohol	4	120
Gambling, Alcohol, Alcohol, Alcohol	4	131
Alcohol, Clothing and Accessories, Alcohol	3	141
Health and Medical Services, Alcohol, Alcohol	3	143

Table 6. Representative ad patterns detected by baselines but not by TSpan-Intra

Pattern	Len	Avg. Time Δ (s)
Alcohol, Education and Careers	2	277
Alcohol, Gifts and Holiday Items	2	263
Alcohol, Green/Eco	2	243
Alcohol, Health and Medical Services	2	215
Gambling, Education and Careers	2	287
Gambling, Health and Medical Services	2	295
Health and Medical Services, Gambling	2	191

cases. For example, with RF and XGBoost, TSpan-Both reaches DQ of 99% and reduces DE to around 5%, clearly surpassing its counterparts. Although unsupervised models such as AE exhibit a relatively high false negative rate (FNR), this is also observed in other baseline pattern mining methods and is largely attributed to the class imbalance in the HDFS dataset. Despite this, TSpan achieves a high F1-score and low FPR, surpassing all baselines in both detection quality and error. Overall, the results demonstrate TSpan's ability to capture temporally coherent patterns, which translates into superior anomaly detection.

4.4.3 Pattern Quality Analysis

TSpan-Intra's Patterns versus others. Table 5 shows patterns in the top 50 of TSpan-Intra and not by any baseline's top 50. In contrast, Table 6 lists patterns found by baselines but not by TSpan-Intra. TSpan uncovers patterns of highly compact, bursty intensity (e.g., "Alcohol $\times 8$", Table 5), which baselines overlook but are critical for detecting spam-like or campaign-driven behaviour. The absence of certain cross-domain patterns in TSpan's top-50 (Table 6) reflects its strict compactness thresholds rather than irrelevance, as these patterns reappear when the cut-off is expanded (e.g., top-100).

Table 7. Frequent Patterns with Average Inter-Session Delta (in days)

TSpan-Inter's Pattern	Len	Avg. Time Δ (days)
Alcohol, Alcohol	2	5.73
Gambling, Alcohol	2	6.99
Gambling, Gambling	2	6.36
Alcohol, Alcohol, Alcohol	3	7.75
Health and Medical Services, Alcohol	2	8.86
Green/Eco, Alcohol	2	9.77

It is worth noting that baseline-only sequences generally exhibit large average inter-event gaps (ranging from 191 s to over 296 s), indicating that events are loosely connected in time and may represent weak sequential behaviour across separated sessions. In contrast, TSpan-Intra's temporal compactness filter eliminates such low-cohesion patterns, focusing instead on sequences where events occur in close proximity—patterns more likely to reflect genuine behavioural flows rather than statistical artifacts.

TSpan-Inter's Patterns versus others. Table 7 highlights the unique contribution of TSpan-Inter, which uncovers frequent, compact patterns across user sessions—a feature baselines cannot achieve. These patterns are repeatedly delivered to users after meaningful intervals, ranging from approximately 5.7 to 10 d as measured by TSpan-Inter. While intra-session mining captures immediate bursts, TSpan-Inter shows how ad delivery resurfaces after breaks, systematically reintroducing content to form routine exposure patterns. With high support and temporal compactness across sessions, these patterns reveal deliberate re-engagement strategies and campaign scheduling invisible to baselines.

5 Conclusion

We introduced TSpan, a Temporal Sequential Pattern Mining method that integrates temporal compactness with frequency. We developed two algorithmic variants: TSpan-Intra, which uncovers compact patterns within user sessions, and TSpan-Inter, which captures recurring structures across sessions. Evaluated on diverse real-world datasets, including digital advertising logs and HDFS system activity logs, TSpan demonstrated its ability to discover patterns that are more compact, actionable, and aligned with domain expectations than those identified by other baselines. Its contributions lie in enhancing the interpretability and behavioural relevance of mined patterns. This offers a richer foundation for applications such as user modelling, segmentation, and intervention design, particularly in domains where events' timing carries crucial semantic meaning, ranging from short-term bursts to longer-term routine behaviours. TSpan also opens up several avenues for future exploration. Combining it with predictive modelling could further strengthen downstream tasks such as anomaly detection, user modelling, and intervention design. Additionally, extending the framework to streaming environments or adaptive temporal thresholds would further enhance its applicability to real-time monitoring and decision-making.

Acknowledgement. This research was conducted by the ARC Centre of Excellence for Automated Decision-Making and Society (CE200100005) and funded by the Australian Government through the Australian Research Council.

References

1. Abrahams, A.S., Coupey, E., Zhong, E.X., Barkhi, R., Manasantivongs, P.S.: Audience targeting by b-to-b advertisement classification: a neural network approach. Expert Syst. Appl. **40**(8), 2777–2791 (2013)
2. Aggarwal, C.C., Bhuiyan, M.A., Hasan, M.A.: Frequent pattern mining algorithms: a survey. Springer (2014)
3. Agrawal, R.: Mining sequential patterns. Technical report, Research report (1995)
4. Agrawal, R., Imieliński, T., Swami, A.: Mining association rules between sets of items in large databases. In: ACM SIGMOD, pp. 207–216 (1993)
5. Angus, D., et al.: Enabling online advertising transparency through data donation methods. Comput. Commun. Res. **6**(2), 1 (2024)
6. Brown, M.G., et al.: No targets, just vibes: tuned advertising and the algorithmic flow of social media. Social Media+ Society **10**(1), 20563051241234691 (2024)
7. Chanda, A.K., Ahmed, C.F., Samiullah, M., Leung, C.K.: A new framework for mining weighted periodic patterns in time series databases. Expert Syst. Appl. **79**, 207–224 (2017)
8. Chen, S., Chen, J., Wan, S.: Efficient weighted sequential pattern mining. Expert Syst. Appl. **243**, 122703 (2024)
9. Choi, J.A., Lim, K.: Identifying machine learning techniques for classification of target advertising. ICT Express **6**(3), 175–180 (2020)
10. Du, M., Li, F., Zheng, G., Srikumar, V.: Deeplog: Anomaly detection and diagnosis from system logs through deep learning. In: The 2017 ACM SIGSAC (2017)
11. Fan, Z., Liu, Z., Zhang, J., Xiong, Y., Zheng, L., Yu, P.S.: Continuous-time sequential recommendation with temporal graph collaborative transformer. In: The 30th ACM international conference on information and knowledge management (2021)
12. Ford, T.W., Krohn, R., Weninger, T.: Competition dynamics in the meme ecosystem. ACM Trans. Social Comput. **6**(3–4), 1–19 (2023)
13. Guo, H., Yuan, S., Wu, X.: LogBERT: Log anomaly detection via BERT. In: 2021 international joint conference on neural networks (IJCNN), pp. 1–8. IEEE (2021)
14. Han, J., et al.: Prefixspan: mining sequential patterns efficiently by prefix-projected pattern growth. In: The 27th ICDE. IEEE Piscataway, NJ, USA (2001)
15. Han, J., Pei, J., Yin, Y.: Mining frequent patterns without candidate generation. ACM SIGMOD Rec. **29**(2), 1–12 (2000)
16. Hinton, G.E., Salakhutdinov, R.R.: Reducing the dimensionality of data with neural networks. Science **313**(5786), 504–507 (2006)
17. Ho, C.C., Li, H.F., Kuo, F.F., Lee, S.Y.: Incremental mining of sequential patterns over a stream sliding window. In: IEEE ICDMW'06. IEEE (2006)

18. Huang, G., Gan, W., Yu, P.S.: TaSPM: Targeted sequential pattern mining. ACM Trans. Knowl. Discov. Data **18**(5), 1–18 (2024)
19. Islam, M.A., Rafi, M.R., Azad, A.a., Ovi, J.A.: Weighted frequent sequential pattern mining. Appl. Intell. **52**(1), 254–281 (2022)
20. Landauer, M., Skopik, F., Wurzenberger, M.: A critical review of common log data sets used for evaluation of sequence-based anomaly detection techniques. Proc. ACM Softw. Eng. **1**(FSE), 1354–1375 (2024)
21. Le, V.H., Zhang, H.: Log-based anomaly detection with deep learning: how far are we? In: The 44th international conference on software engineering (2022)
22. Liu, Y., Ren, S., Wang, X., Zhou, M.: Temporal logical attention network for log-based anomaly detection in distributed systems. Sensors **24**(24), 7949 (2024)
23. Pei, J., et al.: H-mine: Hyper-structure mining of frequent patterns in large databases. In: The 2001 IEEE ICDM (2001)
24. Qin, Y., Ju, W., Wu, H., Luo, X., Zhang, M.: Learning graph ode for continuous-time sequential recommendation. IEEE TKDE (2024)
25. Rodriguez-Gonzalez, A.Y., Aranda, R., Alvarez-Carmona, M.A., Díaz-Pacheco, A., Rosas, R.M.V.: X-FSPMiner: a novel algorithm for frequent similar pattern mining. ACM Trans. Knowl. Discov. Data **18**(5), 1–26 (2024)
26. Varol, O., Ferrara, E., Menczer, F., Flammini, A.: Early detection of promoted campaigns on social media. EPJ Data Sci. **6**(1), 1–19 (2017). https://doi.org/10.1140/epjds/s13688-017-0111-y
27. Vo, A.T., Tran, H.S., Le, T.H.: Advertisement image classification using convolutional neural network. In: 2017 9th KSE. IEEE (2017)
28. Wu, Y., et al.: ONP-Miner: One-off negative sequential pattern mining. ACM TKDD **17**(3), 1–24 (2023)
29. Zaki, M.J.: Spade: an efficient algorithm for mining frequent sequences. Mach. Learn. **42**, 31–60 (2001)
30. Zaki, M.J.: Scalable algorithms for association mining. IEEE Trans. Knowl. Data Eng. **12**(3), 372–390 (2000)

A Parameter-Free Method Tuning
for Multi-scale Wildfire Images
Retrieval Task

Yue Zhang[(✉)]

Faculty of Information Technology, Monash University, Melbourne,
VIC 3168, Australia
`yzha1097@student.monash.edu`

Abstract. Efficient retrieval of wildfire-related imagery from large-scale remote sensing and aerial datasets is essential for early detection, monitoring, and disaster response. However, the limited availability of annotated data severely restricts the effectiveness of conventional supervised approaches. To address this challenge, we propose a parameter-free self-distillation framework that leverages language models and unlabeled imagery to enhance wildfire image retrieval. Our method first employs large language models (LLMs) to generate domain-specific textual descriptions of wildfire phenomena, covering diverse visual cues such as fire spread, smoke plumes, vegetation conditions, and multi-perspective observations. These descriptions are used to construct a text-based classifier that transfers semantic knowledge into the visual domain. Through a self-distillation process, the classifier produces pseudo-labels for unlabeled wildfire imagery, which are then used to iteratively refine the vision encoder in a parameter-efficient manner. Experiments on both satellite remote sensing and aerial datasets demonstrate substantial improvements over zero-shot baselines, yielding more accurate and robust retrieval performance under class imbalance, heterogeneous imaging conditions, and cross-dataset generalization. This framework highlights the potential of combining LLM-driven semantic enrichment with self-distillation for scalable, annotation-free wildfire monitoring and supports broader applications in label-scarce environmental image analysis.

Keywords: Self-distillation · Wildfire image retrieval · Remote sensing · Aerial imagery

1 Introduction

Wildfire detection and monitoring play a critical role in environmental protection, disaster management, and public safety. As remote sensing technology advances globally, there is a growing need to develop intelligent systems that can accurately detect and classify wildfire events from satellite imagery and aerial photographs [1]. The manual process of analyzing remote sensing data for wildfire identification is time-consuming and lacks standardized protocols for fire

Q. V. Nguyen et al. (Eds.): AusDM 2025, CCIS 2765, pp. 532–545, 2026.
https://doi.org/10.1007/978-981-95-6786-7_36

classification [3]. Moreover, human-based analysis is prone to detection errors and subjective interpretation bias, particularly under time-critical emergency conditions [2].

However, conventional remote sensing analysis methods often suffer from substantial computational overhead and processing latency, which limits their applicability in operational wildfire monitoring systems that demand real-time detection capabilities [4]. To address these challenges, we propose an enhanced parameter-free tuning approach based on LaFTer (Label-Free Tuning of Zero-shot Classifier) [9], offering a practical and scalable solution for real-world wildfire detection tasks. Figure 1 presents our multi-scale remote sensing data collection framework for wildfire image classification, operating across three distinct altitude ranges to leverage the complementary strengths of different sensing platforms. The framework follows a hierarchical approach: (1) satellite imagery (200km-36,000km altitude) from the EuroSAT (European satellite) dataset [5] is used for broad-scale forest area identification and land cover classification, providing essential context for potential fire-prone regions; (2) aerial imagery (1km-15km altitude) from the Fire Risk dataset [6] enables identification of high fire risk areas within the previously identified forest regions, offering medium-resolution analysis of vegetation conditions and terrain characteristics; and (3) UAV imagery (30 m-50 m altitude) from the FlameVision [7]dataset provides fine-scale detection of actual wildfire events and smoke plumes with high spatial resolution.

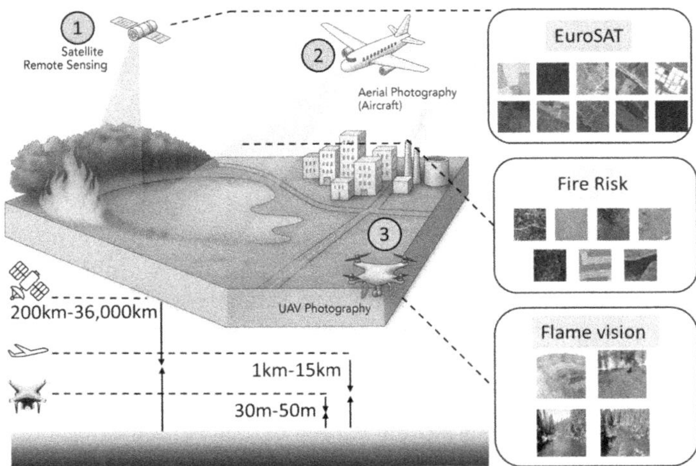

Fig. 1. Multi-scale remote sensing data collection framework for wildfire image classification.

This multi-scale strategy addresses the fundamental challenge that wildfire detection becomes increasingly difficult at higher altitudes due to spatial resolution limitations, while lower altitude platforms offer limited coverage. By progressively narrowing the search area from satellite-scale forest mapping to aerial-scale risk assessment and finally to UAV-scale fire detection, our framework efficiently combines the wide coverage of satellite data with the detailed discrimination capability of UAV systems, enabling comprehensive wildfire monitoring across multiple spatial scales. The contributions of this paper are summarized as follows:

- Develop an enhanced parameter-free tuning approach that overcomes LaFTer's limitations for wildfire detection in remote sensing imagery, achieving better classification performance with reduced training costs.
- Demonstrate the effectiveness of the proposed method in addressing the domain gap between natural images and remote sensing wildfire imagery with minimal computational overhead.

2   Related Work

2.1   Parameter Efficient Fine-Tuning

Vision–Language Models (VLMs) have emerged as a powerful paradigm for modeling the interplay between visual and textual modalities in remote sensing applications. A seminal contribution in this area is CLIP (Contrastive Language–Image Pre-training) [8], which demonstrates strong zero-shot generalization by aligning visual concepts with natural language supervision.

While full fine-tuning of large pre-trained models is effective, it is often computationally prohibitive and prone to overfitting. Consequently, parameter-efficient adaptation techniques have garnered increasing attention. These methods aim to adapt pre-trained models to downstream tasks by updating only a small subset of parameters, thereby improving training efficiency and stability.

Text Prompt Learning: CoOp (Context Optimization) [11] proposes to learn continuous prompts for vision–language models by optimizing context vectors while keeping the backbone frozen. This strategy has been shown to outperform manually designed prompts across diverse vision tasks. Subsequent work [10] further validates the effectiveness of prompt learning in domain-specific applications such as pavement friction estimation.

Adapter-Based Methods: CLIP-Adapter [12] introduces lightweight bottleneck adapter modules into pre-trained CLIP models. These adapters capture task-specific information while preserving the general knowledge encoded in the frozen backbone. This design enables efficient and modular fine-tuning for downstream tasks. The benefits of such adapter-based approaches have also been demonstrated in low-level tasks like defect detection [13].

2.2   Domain Gap Mitigation

Domain gap mitigation addresses the challenge of transferring knowledge from a source domain to a target domain with mismatched data distributions. This issue is particularly pronounced in specialized applications such as wildfire detection in remote sensing, where the visual characteristics of imagery—such as spectral composition, spatial resolution, and environmental context—can differ significantly from those of natural images used in large-scale pre-training.

Label-Free Adaptation: LaFTer [9] mitigates the need for annotations by promoting representation consistency between different augmented views of the same unlabeled image. Language supervision is integrated as a regularizing signal to prevent model collapse. In parallel, CLIP-PR (CLIP with Priors) [14] incorporates label distribution priors into zero-shot classification, guiding the model toward more semantically plausible predictions under distribution shift.

Test-Time Adaptation: Test-time adaptation (TTA) allows models to adapt on-the-fly using unlabeled test samples, addressing distributional discrepancies encountered during inference. TDA [15] introduces a lightweight adapter fine-tuned via entropy minimization over test batches, enabling self-supervised adaptation without explicit labels. BoostAdapter [16] further enhances robustness by applying regional bootstrapping—leveraging diverse region proposals to refine predictions, rather than relying solely on holistic image-level features.

Collectively, these methods enable robust deployment of vision–language models in the remote sensing domain by leveraging unsupervised signals and modular adaptation strategies. They provide a promising foundation for accurate wildfire detection under real-world domain shifts.

3   Method

Our approach extends the LaFTer (Label-Free Tuning of Zero-shot Classifier) framework for wildfire image classification in remote sensing and aerial imagery by modifying the prompt ensemble to incorporate spatial and visual understanding capabilities specifically designed for fire detection and analysis.

3.1   Problem Formulation

Let $\mathcal{X}$ denote the input remote sensing image space and $\mathcal{Y} = \{1, 2, \ldots, C\}$ be the set of C wildfire-related target classes (e.g., active fire, smoke, burned area, vegetation, water). Given a pre-trained vision-language model (VLM) $f_\theta : \mathcal{X} \times \mathcal{T} \to \mathbb{R}$, where $\mathcal{T}$ represents the text embedding space, our goal is to adapt this model for wildfire classification in remote sensing imagery using the LaFTer framework with enhanced spatial reasoning prompts.

3.2   Enhanced Prompt Ensemble for Wildfire Detection

The core innovation lies in replacing the original LaFTer prompt ensemble with our domain-specific prompt ensemble $\mathcal{P} = \{p_1, p_2, \ldots, p_5\}$ tailored for wildfire analysis in remote sensing imagery. For each wildfire-related class category *cat* with article *article*, we generate the following prompts:

- p_1: "In a high-resolution satellite image, what geometric shape, true-color tone, and texture does {article} {cat} typically show?"
- p_2: "How can you reliably identify {article} {cat} in satellite imagery at meter-level GSD? List key cues (color/texture/shape/context/shadow)."
- p_3: "What are the most distinctive patterns of {article} {cat} from a nadir view compared with nearby land cover?"
- p_4: "Describe spatial context that helps find {article} {cat} (e.g., proximity to roads, water, urban edges, field boundaries)."
- p_5: "Under true color (RGB), how does {article} {cat} contrast with its background in hue and brightness?"

This differs from the original LaFTer prompts which focused on general object description:

- "Describe what {article} {cat} looks like"
- "How can you identify {article} {cat}?"
- "What does {article} {cat} look like?"
- "Describe an image from the internet of {article} {cat}"
- "A caption of an image of {article} {cat}:"

3.3   Text Embedding Generation

For each prompt $p_j \in \mathcal{P}$ and wildfire class c, we generate text descriptions using a Large Language Model (LLM):

$$d_{j,c} = \text{LLM}(p_j, c) \tag{1}$$

These descriptions are then encoded into text embeddings using the VLM's text encoder:

$$t_{j,c} = \text{TextEncoder}(d_{j,c}) \tag{2}$$

3.4   LaFTer Classification Framework

Following the original LaFTer framework, the final class probability for a remote sensing image x is computed through ensemble averaging:

$$p(y_c|x) = \frac{1}{5} \sum_{j=1}^{5} \text{softmax} \left(\frac{f_\theta(x, t_{j,c})}{\tau} \right)_c \tag{3}$$

where τ is a temperature parameter for calibration.

The LaFTer objective optimizes the cross-entropy loss on the unlabeled remote sensing images:

$$\mathcal{L}_{LaFTer} = - \sum_{i=1}^{N} \sum_{c=1}^{C} \hat{y}_{i,c} \log p(y_c|x_i) \tag{4}$$

where $\hat{y}_{i,c}$ represents the pseudo-labels generated by the ensemble predictions, and N is the number of unlabeled remote sensing images.

Figure 2 illustrates the LaFTer framework architecture for label-free tuning of zero-shot classifiers. The framework consists of two parallel encoding pathways: a Teacher Encoder (top) and a Student Encoder (bottom). Both encoders process input images from the dataset through their respective classifiers to generate predictions across multiple classes (Very High, Low, Water, Very Low). The Teacher Encoder maintains fixed parameters and provides stable pseudo-labels, while the Student Encoder undergoes momentum updates during training. The framework employs a cross-entropy loss function that compares the one-hot encoded predictions from both pathways, enabling the student model to learn from the teacher's knowledge without requiring labeled training data. This teacher-student paradigm allows for effective knowledge distillation and model adaptation in scenarios where labeled data is scarce or unavailable, making it particularly suitable for remote sensing applications where manual annotation is costly and time-consuming.

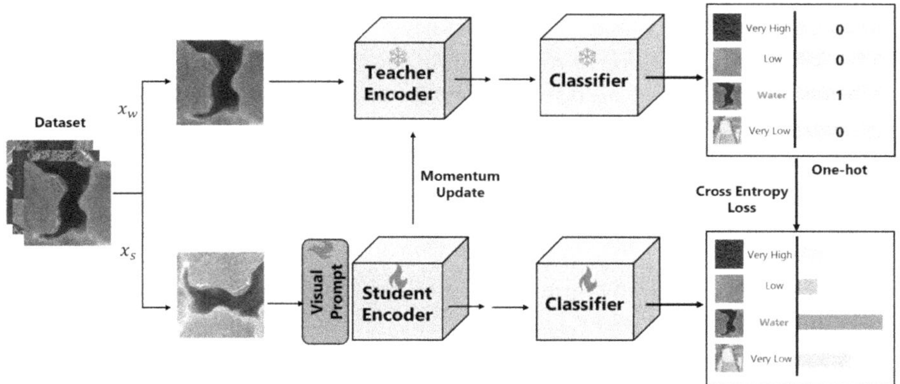

Fig. 2. Architecture of the self-distillation model for wildfire image classification.

4 Experiment

In this section, We first introduce the datasets and baseline models used in this paper. In addition, we describe the experimental details, including hyperparameter settings and hardware requirements. Finally, we report and further analyze the experimental results in comparison with other methods.

4.1 Experiment Set Up

Dataset: To evaluate the effectiveness of our proposed framework, we conduct experiments on three representative datasets that cover both general remote sensing imagery and wildfire-specific scenes:

- **EuroSAT** [5]: EuroSAT is a publicly available dataset based on Sentinel-2 satellite images, consisting of 27,000 labeled samples across 10 land-use and land-cover classes, such as residential areas, forests, rivers, and industrial zones. The dataset provides multispectral remote sensing imagery with a spatial resolution of 10 m, and it is widely used as a benchmark for remote sensing classification and retrieval tasks. In our work, EuroSAT serves as a general-purpose baseline to evaluate the transferability of our framework from land-cover recognition to wildfire scenarios.

- **FireRisk** [6]: FireRisk is a curated dataset designed for wildfire-related risk prediction, containing satellite-based remote sensing images with annotations related to fire-prone regions. It integrates multiple environmental variables, such as vegetation type and landscape context, to capture conditions that may contribute to wildfire ignition and spread. We adopt FireRisk to assess the capability of our method in capturing wildfire-relevant features under complex natural backgrounds.

- **FlameVision** [7]: FlameVision is an aerial wildfire dataset collected primarily from UAV and high-resolution camera sources. It contains thousands of images covering different wildfire scenes, including visible flames, smoke, and post-fire landscapes. Unlike FireRisk, which emphasizes risk context, FlameVision focuses on direct wildfire classification from visual cues. This dataset allows us to validate the robustness of our framework under diverse imaging conditions and varying perspectives.

Table 1. Statistics of the datasets used in our experiments.

Dataset	EuroSAT	FireRisk	FlameVision
Classes	10	7	2
Train	13,500	8,400	1,600
Val	5,400	1,050	200
Test	8,100	1,050	200

The detailed statistics of the datasets employed in our experiments are summarized in Table 1. EuroSAT contains 10 land-cover classes with a balanced split across training, validation, and testing. FireRisk consists of 7 wildfire-related categories with approximately 10,500 samples in total, reflecting more complex background variations. FlameVision focuses on binary wildfire recognition, with fewer samples but high-resolution aerial imagery. These datasets together provide a comprehensive benchmark for evaluating the proposed framework across diverse scenarios, from general-purpose land-cover recognition to wildfire-specific classification.

Baseline Models: To evaluate the effectiveness of our proposed framework, we compare it against two representative baseline models.

- **CLIP** [8]: Contrastive Language–Image Pre-training (CLIP) is a vision-language model trained on 400 million image–text pairs using a contrastive learning objective. By jointly embedding images and their corresponding textual descriptions into a shared representation space, CLIP enables zero-shot classification through natural language prompts. Given a set of category names, textual prompts such as "a photo of wildfire smoke" are encoded by the text encoder, and image features extracted from the vision encoder are matched against these textual embeddings. This zero-shot capability makes CLIP a strong baseline for tasks where labeled data is scarce, including wildfire image retrieval.

- **Lafter** [9]: To further improve over CLIP, we employ a label-free self-distillation framework inspired by LaFTer (Label-Free Tuning of Zero-shot Classifiers). The core idea is to enhance zero-shot classifiers using language guidance and unlabeled image collections, without requiring manual annotations. In our adaptation, large language models (LLMs) are leveraged to generate domain-specific textual descriptions of wildfire phenomena (e.g., flames, smoke, and post-fire vegetation). These prompts are first used to construct a text-only classifier, which then generates pseudo-labels for unlabeled images. The model undergoes a self-distillation process in which pseudo-labels iteratively refine the visual encoder in a parameter-efficient manner. This approach allows the baseline to move beyond purely zero-shot inference and better adapt to wildfire imagery across both remote sensing and aerial domains.

Implementation Details: All experiments are implemented with the `Dassl.pytorch` framework and executed on a single NVIDIA A100 GPU (48GB memory). We adopt ViT-B/32 as the vision backbone, initialized from a pre-trained CLIP model. During zero-shot inference the encoders remain frozen, while only a lightweight adapter is updated in the parameter-efficient tuning stage. The evaluation is carried out on three datasets—EuroSAT, FireRisk, and FlameVision—whose statistics are summarized in Table 1. Training is performed with a batch size of 64 and 4 data loader workers. The optimizer is AdamW with an initial learning rate of 5×10^{-5} and a weight decay of 1×10^{-4}. We train for 15 epochs in total. In the self-distillation stage, $N = 50$ text prompts per class are generated automatically by a large language model (LLM) to enrich the textual space. Throughout training, we apply mixed-precision computation (`fp16`) to improve efficiency and reduce memory consumption.

4.2   Numerical Results

Based on the experimental setup, we now report the quantitative results to assess the effectiveness of each method. The comparisons highlight how different prompt impact performance in wildfire detection.

Table 2 presents the classification accuracy (%) of our proposed method compared to baseline approaches across three remote sensing datasets. Our method achieves substantial improvements over the CLIP baseline across all datasets, with particularly notable gains on EuroSAT (74.5% vs. 45.6%) and FireRisk (26.4% vs. 13.9%), representing relative improvements of 63.4% and 90.0%, respectively. On FlameVision, our method attains 96.5% accuracy, marginally outperforming both CLIP (94.0%) and LaFTer (96.1%). The LaFTer baseline demonstrates competitive performance on EuroSAT (73.9%) and FlameVision (96.1%) but shows limited effectiveness on FireRisk (25.1%). These results demonstrate the robustness of our approach across diverse remote sensing classification tasks, with particularly strong performance on challenging datasets where traditional zero-shot methods struggle. The consistent improvements validate the effectiveness of integrating information maximization principles with label-free tuning for remote sensing image classification.

Table 2. Top-1 accuracy (%) comparison on three datasets.

Method	EuroSAT	FireRisk	FlameVision
CLIP	45.6	13.9	94.0
LaFTer	73.9	25.1	96.1
Ours	**74.5**	**26.4**	**96.5**

4.3 Visualization Analysis

While numerical metrics provide quantitative insight into model performance, visual analysis offers a more intuitive understanding of feature representations and decision boundaries. To further evaluate the effectiveness of our method, we employ **t-distributed Stochastic Neighbor Embedding (t-SNE)** strictly as a *projection tool* to reduce high-dimensional embeddings into two dimensions for plotting. Importantly, t-SNE is not part of the training or inference process.

Rationale for t-SNE. We select t-SNE instead of faster alternatives such as UMAP because our primary goal is to highlight *local neighborhood structures and class separability*. In our experiments, t-SNE produced more stable intra-class compactness and clearer inter-class margins, which aligns with our analysis focus. Although UMAP preserves more global geometry and is computationally faster, speed is not a bottleneck here since visualization is conducted offline. For completeness, UMAP produced similar qualitative trends, but we report t-SNE for consistency across datasets.

Feature-Text Alignment Visualization. We project model output probability distributions to examine discriminative power and clustering quality. Compact, well-separated clusters with clear boundaries indicate effective classification.

Output Distribution Visualization. We visualize the alignment between visual features and textual prototypes in the learned embedding space. Image features are represented as *colored circles*, while textual classifiers (prompt prototypes) are represented as *"X" markers*. The proximity of a feature cluster to its corresponding prototype indicates the quality of cross-modal alignment.

4.4   Visualization Analysis

While numerical metrics offer quantitative insight into model performance, visual analysis provides a more intuitive understanding of feature representations and decision boundaries. To further evaluate the effectiveness of our method, we employ t-distributed Stochastic Neighbor Embedding (t-SNE) to visualize both feature-text alignment and output prediction distributions.

In this section, We employ t-distributed Stochastic Neighbor Embedding (t-SNE) visualizations to evaluate classification performance through two complementary perspectives that assess the complete pipeline from feature representation to final prediction.

Feature-Text Alignment Visualization: We apply t-SNE to model output probability distributions to examine discriminative power and clustering quality. Compact, well-separated class clusters with clear boundaries indicate effective classification performance.

Output Distribution Visualization: We visualize the alignment between visual features and textual representations in the learned embedding space. Image features are represented as colored scatter points, while corresponding text classifiers appear as X markers. Spatial proximity between feature clusters and text classifiers indicates cross-modal alignment quality.

T-SNE visualization for EuroSAT experiment: To complement the quantitative results, we further analyze the model's behavior through visualization. Figure 3 compares t-SNE projections of feature-text alignment and output distributions between CLIP and our proposed method on the EuroSAT dataset. The left panels show CLIP's original embedding space, highlighting the alignment between visual features and text representations. In contrast, the right panels illustrate our method's enhanced clustering in the output probability space.

Our approach achieves notably improved class separability, characterized by more compact intra-class clusters and clearer inter-class boundaries. Each color represents a specific land use category from the EuroSAT taxonomy (e.g., residential, forest, highway, river). The improved cluster structure in our method indicates stronger discriminative ability and better cross-modal alignment, demonstrating the effectiveness of our information maximization strategy in remote sensing image classification.

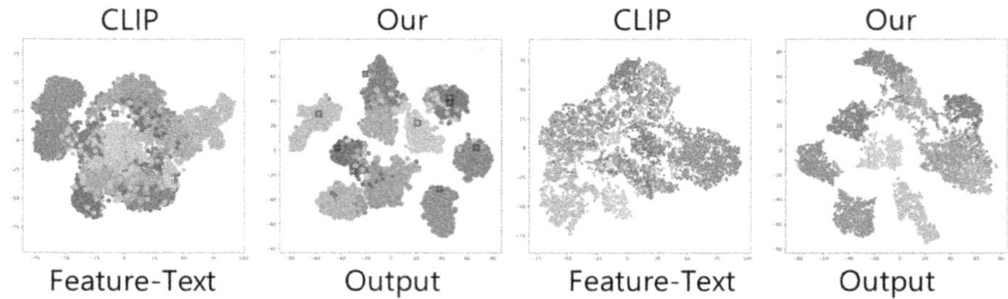

Fig. 3. Comparison of feature space clustering between CLIP and our method on the EuroSAT dataset using t-SNE projections. Colors denote specific land-use categories (e.g., residential, forest, highway, river). Circles = image features, X markers = text prototypes. Our method yields tighter clusters and clearer boundaries than CLIP.

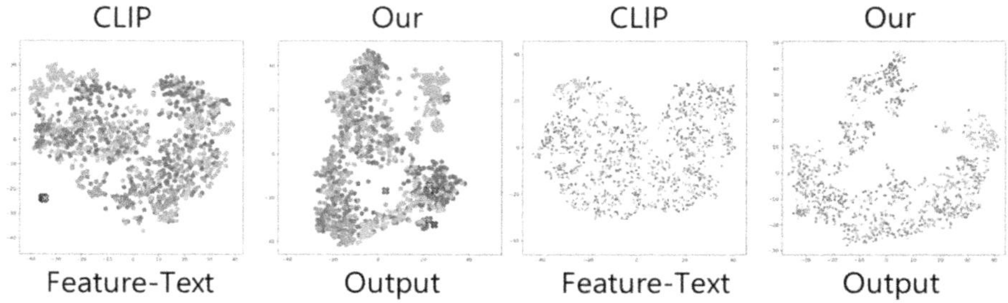

Fig. 4. Comparison of feature space clustering between CLIP and our method on the FireRisk dataset using t-SNE projections. Colors denote fire-risk categories and environmental conditions as defined in the dataset taxonomy. Circles = image features, X markers = text prototypes. Our method reduces overlap and improves class separability.

T-SNE Visualization for Fire Risk experiment: To further evaluate the generalizability of our method, Fig. 4 presents t-SNE visualizations on the fire risk classification dataset, comparing feature-text alignment and output clustering performance between CLIP and our approach. The left panels depict CLIP's original embedding space, illustrating visual–textual alignment across two distinct fire risk scenarios. The right panels show our method's enhanced output space, characterized by tighter intra-class cohesion and more distinct inter-class separation.

Each color represents specific fire risk categories or environmental conditions defined in the dataset taxonomy. Compared to CLIP, our method demonstrates improved discriminative structure with reduced overlap and clearer class boundaries. These results further confirm the robustness and adaptability of our information maximization framework in complex remote sensing tasks beyond land use classification.

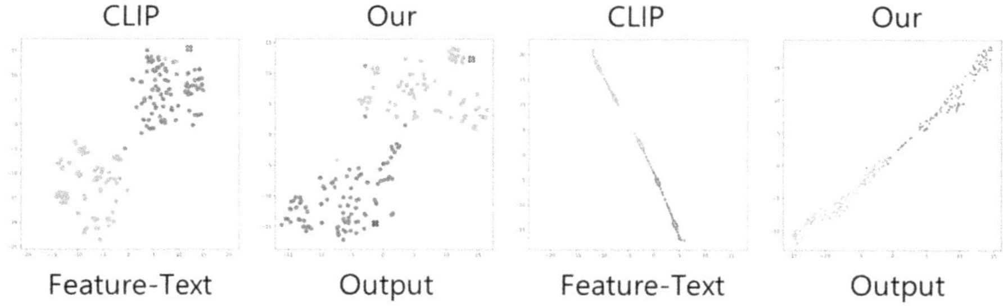

Fig. 5. Comparison of feature space clustering between CLIP and our method on the FlameVision dataset using t-SNE projections. Colors denote wildfire-related categories and fire intensity levels. Circles = image features, X markers = text prototypes. Our method achieves stronger separability and compactness for flame detection.

T-SNE Visualization for Flame Vision Experiment: To assess the framework's effectiveness in fine-grained visual scenarios, Fig. 5 shows t-SNE visualizations on the FlameVision dataset, comparing CLIP and our method in terms of feature-text alignment and output clustering. The left panels display CLIP's embedding space, reflecting alignment across two distinct flame detection scenarios. The right panels illustrate our method's output space, showing stronger class separability and more compact intra-class clusters.

The color scheme denotes flame-related categories and fire intensity levels defined in the FlameVision taxonomy. Compared to CLIP, our method achieves clearer inter-class boundaries and reduced feature overlap, demonstrating superior discriminative power in multi-class flame detection. These findings highlight the adaptability of our information maximization strategy to vision-based fire analysis tasks

4.5   Ablation Study

Table 3 presents an ablation study assessing the impact of individual components in our framework on the EuroSAT and FireRisk datasets. The baseline without the adapter module (w/o Adapter) achieves 73.1% and 23.4% accuracy on EuroSAT and FireRisk, respectively. Removing layer normalization (w/o LN) results in minor changes, yielding 71.6% and 22.8%, suggesting limited influence.

In contrast, excluding the visual prompt tuning module (w/o VPT) leads to a substantial performance drop, with accuracies decreasing to 56.5% and 16.7%, indicating its critical role. Our full model achieves the highest performance— 74.5% on EuroSAT and 26.4% on FireRisk—highlighting the importance of visual prompting as the primary driver of accuracy gains, with the adapter component serving as a necessary structural backbone.

Table 3. Ablation study on EuroSAT and FireRisk datasets (Top-1 accuracy %).

Dataset	w/o Adapter	w/o LN	w/o VPT	Ours
EuroSAT	73.1	71.6	56.5	**74.5**
FireRisk	23.4	22.8	16.7	**26.4**

5 Discussion

Table 4 summarizes the parameter composition, model update size, and inference efficiency of our method across the EuroSAT, FireRisk, and FlameVision datasets. The core components—visual prompt tuning and layer normalization—require only 83.5K, 81.9K, and 79.4K trainable parameters, respectively. Adapter modules introduce minimal additional overhead, with just 5.1K, 3.6K, and 1.9K parameters across the datasets, resulting in extremely low total trainable parameter ratios of 0.10%, 0.057%, and 0.053%.

Despite this lightweight design, inference remains efficient, with runtimes of 1.595 s, 1.696 s, and 1.724 s, demonstrating negligible latency impact. These results confirm that our framework delivers strong performance gains with minimal computational cost, making it highly practical for deployment in resource-constrained or real-time remote sensing applications.

Table 4. Comparison of parameter statistics, model updates, and inference efficiency.

Dataset	(VPT, LN)	Adapter	Trainable	Percentage	Time
EuroSAT	83.5K	5.1K	88.6K	0.060%	0.895
FireRisk	81.9K	3.6K	85.5K	0.057%	1.696
FlameVision	79.4K	1.0K	80.4K	0.053%	1.724

6 Conclusion

In this paper, we have presented an enhanced parameter-free tuning approach for wildfire detection in remote sensing imagery, building upon the LaFTer framework. Our method successfully addresses the domain gap between natural images and remote sensing wildfire imagery while maintaining computational efficiency for real-time monitoring applications.

The proposed approach demonstrates significant improvements in classification performance compared to existing label-free methods. The ability to work with unlabeled remote sensing data makes it highly practical for deployment in operational wildfire monitoring systems, where rapid detection is critical for environmental protection and public safety.

Disclosure of Interests. The authors declare that there is no conflict of interest

References

1. Rashkovetsky, D., Mauracher, F., Langer, M., Schmitt, M.: Wildfire detection from multisensor satellite imagery using deep semantic segmentation. IEEE J. Sel. Top. Appl. Earth Obs. Remote Sens. **14**, 7001–7016 (2021)
2. Bushnaq, O.M., Chaaban, A., Al-Naffouri, T.Y.: The role of UAV-IoT networks in future wildfire detection. IEEE Internet Things J. **8**(23), 16984–16999 (2021)
3. Verma, S.: Intelligent framework using IoT-Based WSNs for wildfire detection. IEEE Access **9**, 48185–48196 (2021)
4. Wang, G., et al.: RFWNet: a multiscale remote sensing forest wildfire detection network with digital twinning, adaptive spatial aggregation, and dynamic sparse features. IEEE Trans. Geosci. Remote Sens. **62**, 1–23 (2024)
5. Helber, P., Bischke, B., Dengel, A., Borth, D.: EuroSAT: a novel dataset and deep learning benchmark for land use and land cover classification. IEEE J. Sel. Top. Appl. Earth Obs. Remote Sens. **12**(7), 2217–2226 (2019). https://doi.org/10.1109/JSTARS.2019.2918242
6. Shen, S., Seneviratne, S., Wanyan, X., Kirley, M.: FireRisk: A Remote Sensing Dataset for Fire Risk Assessment With Benchmarks Using Supervised and Self-Supervised Learning. arXiv preprint arXiv:2303.07035 (2023)
7. Ibn Jafar, A., Islam, A.M., Binta Masud, F., Ullah, J.R., Ahmed, M.R.: FlameVision: a new dataset for wildfire classification and detection using aerial imagery. Mendeley Data **V4** (2023). https://doi.org/10.17632/fgvscdjsmt.4
8. Radford, A., et al.: Learning transferable visual models from natural language supervision. In: Proceedings of the 38th International Conference on Machine Learning (ICML), pp. 8748–8763. PMLR (2021)
9. Mirza, J., et al.: LaFTer: Label-Free tuning of zero-shot classifier using language and unlabeled image collections. In: Advances in Neural Information Processing Systems (NeurIPS), vol. 36 (2023)
10. Lu, B., et al.: Predicting Asphalt Pavement Friction Using Texture-Based Image Indicator. arXiv preprint arXiv:2507.03559 (2025)
11. Zhou, K., Yang, J., Loy, C.C., Liu, Z.: Learning to prompt for vision-language models. Int. J. Comput. Vision **130**(10), 2337–2348 (2022). https://doi.org/10.1007/s11263-022-01653-1
12. Gao, P.: CLIP–Adapter: better vision-language models with feature adapters. Int. J. Comput. Vision **132**(3), 581–595 (2023). https://doi.org/10.1007/s11263-023-01891-x
13. Lu, Z., Lu, B., Wang, W., Wang, F.: Differentiable NMS via Sinkhorn Matching for End-to-End Fabric Defect Detection. arXiv preprint arXiv:2505.07040 (2025)
14. Li, X., Wang, Z., Zhou, K., Loy, C.C., Liu, Z.: Improving Zero-Shot Models With Label Distribution Priors. arXiv preprint arXiv:2303.09752 (2023)
15. Liu, R., Shu, Y., Zhu, C.: TDA: Efficient test-time adaptation of vision–language model. In: Proceedings of the IEEE/CVF Conference on Computer Vision and Pattern Recognition (CVPR), (2024)
16. Tian, Y., et al.: BoostAdapter: improving vision–language test-time adaptation via regional bootstrapping. In: Proceedings of the Twelfth International Conference on Learning Representations (ICLR), arXiv: 2305.18287 (2024)

NeuroPhysNet: A FitzHugh-Nagumo-Based Physics-Informed Neural Network Framework for Electroencephalograph (EEG) Analysis and Motor Imagery Classification

Zhenyu Xia[1], Xinlei Huang[1], Yuantong Gu[2], and Suvash Saha[1]

[1] University of Technology Sydney, Sydney, NSW, Australia
{zhenyu.xia,xinlei.huang}@student.uts.edu.au, suvash.saha@uts.edu.au
[2] Queensland University of Technology, Brisbane, QLD, Australia
yuantong.gu@qut.edu.au

Abstract. Electroencephalography (EEG) is extensively employed in medical diagnostics and brain-computer interface (BCI) applications due to its non-invasive nature and high temporal resolution. However, EEG analysis faces significant challenges, including noise, nonstationarity, and inter-subject variability, which hinder its clinical utility. This study introduces NEUROPHYSNET, a novel Physics-Informed Neural Network (PINN) framework tailored for EEG signal analysis and motor imagery classification in medical contexts. NEUROPHYSNET incorporates the Fitz Hugh-Nagumo model, embedding neurodynamical principles to constrain predictions and enhance model robustness. Evaluated on the BCIC-IV-2a dataset, the framework achieved superior accuracy and generalization compared to conventional methods, especially in data-limited and cross-subject scenarios, which are common in clinical settings. By effectively integrating biophysical insights with data-driven techniques, NEUROPHYSNET not only advances BCI applications but also holds significant promise for enhancing the precision and reliability of clinical diagnostics.

Keywords: Physics-informed neural network · EEG analysis · Motor imagery classification

1 Introduction

Electroencephalography (EEG) is a widely used non-invasive technique for measuring the electrical activity of the brain, providing critical insights into various brain functions and disorders. Due to its high temporal resolution and ease of application, EEG has found widespread use in clinical and research settings, such as epilepsy diagnosis, sleep studies, and cognitive monitoring [25]. One particularly significant application of EEG is in Brain-Computer Interfaces (BCIs),

© The Author(s), under exclusive license to Springer Nature Singapore Pte Ltd. 2026
Q. V. Nguyen et al. (Eds.): AusDM 2025, CCIS 2765, pp. 546–560, 2026.
https://doi.org/10.1007/978-981-95-6786-7_37

where EEG signals are used to establish direct communication pathways between the brain and external devices. Among BCI paradigms, motor imagery (MI) stands out as a prominent example. MI involves the mental rehearsal of movement without actual physical execution, and it plays a crucial role in developing assistive technologies for individuals with motor impairments [4].

Despite its potential in various applications, EEG analysis remains challenging due to the signals' low spatial resolution, high noise levels, and nonstationarity, which complicate the extraction of meaningful features [27]. Traditional neural networks, such as multilayer perceptrons (MLPs) and convolutional neural networks (CNNs), have been widely used to process EEG signals by learning complex patterns from data, often with handcrafted features or transformations like spectral power analysis or spatial covariance matrices [21]. While effective in some contexts, these approaches face key limitations: they typically lack integration of biophysical priors, reducing interpretability and robustness across datasets [3]; they are prone to overfitting on small medical datasets [6]; and they struggle to generalize across subjects due to inter-individual variability in brain dynamics [30].

Physics-Informed Neural Networks offer a promising alternative to address these limitations. The core advantage of PINNs lies in their ability to embed physical laws or domain-specific constraints directly into the learning process, thereby bridging the gap between data-driven methods and mechanistic understanding [7]. PINNs capture EEG's neurodynamical basis—membrane potential fluctuations and signal propagation—via concise biophysical equations, addressing the physiological constraints often ignored by purely data-driven methods [29]. Embedding these constraints imparts a strong inductive bias that curbs overfitting on limited datasets, enhances robustness to noise and distributional shifts [5,7], and affords transparent decision pathways essential for reliable medical applications.

Building on prior PINN advances, we introduce a novel framework for EEG-based motor imagery classification that embeds the FitzHugh–Nagumo model [10] to enforce neurodynamical constraints and ensure physiologically consistent, interpretable predictions. The primary contributions of this work are as follows:

- **Biophysical Modeling in EEG Analysis:** We introduce a PINN framework that integrates the FitzHugh-Nagumo model, enabling the incorporation of neurodynamical constraints into the analysis of EEG signals. This approach enhances the interpretability and robustness of the model.
- **Advanced Feature Extraction:** A novel feature extraction module is designed to process the biophysically informed outputs of the PINN. This module extracts temporal features at the node level while preserving the physical consistency of the data, resulting in compact and discriminative representations.
- **Application to Motor Imagery Classification:** We demonstrate the effectiveness of the proposed framework in motor imagery classification tasks, achieving improved performance and generalization compared to traditional neural network approaches.

2   Related Work

Recent advances in deep learning have substantially enhanced the performance of BCIs for MI analysis, particularly through the use of CNNs and Recurrent Neural Networks. Schirrmeister et al. demonstrated the potential of CNNs in EEG decoding and visualization, showcasing the ability of deep learning models to extract meaningful features from EEG signals for improved classification [23]. Similarly, Lawhern et al. introduced EEGNet, a compact CNN architecture specifically designed for efficient EEG-based BCI applications [16]. Huang et al. also investigated CNN-based deep learning models for MI classification, highlighting their superior performance in classification accuracy [8]. The low SNR in MI EEG signals poses a challenge in decoding movement intentions, but this can be addressed using multi-branch CNN modules that learn spectral-temporal domain features, as suggested in [12]. Futhermore, Ju et al. proposed Tensor-CSPNet [15] and Graph-CSPNet [13], a novel geometric deep learning framework for motor imagery classification, achieving improved feature extraction, robustness, and interpretability.

Despite their advances in MI analysis, deep learning approaches remains data-hungry and overfits with scarce labels; its black-box nature hampers interpretability; inter-subject and temporal variability limits generalization; and heavy models slow training and inference, hindering real-time BCI.

To address the limitations of traditional deep learning models, PINNs have emerged as a promising alternative. Raissi et al. first introduced PINNs for solving forward and inverse nonlinear PDEs by embedding physical laws into the learning process; this framework was later adapted to neurodynamics and brain modeling for EEG analysis in BCI applications [1]. Building on this, Lu et al. developed Physics-Informed DeepONets to handle parametric PDEs and individual variability [28], while Zhang et al. incorporated uncertainty quantification to bolster robustness against noisy EEG data [31]. Finally, Jagtap et al. accelerated convergence with adaptive activation functions—crucial for real-time BCI feedback [11]

Our proposed method, the integration of FitzHugh–Nagumo model into a PINN framework, addresses deep-learning and standard PINN limitations by embedding biophysical constraints that enhance interpretability, reduce reliance on large labeled datasets, improve generalization across subjects, and bolster robustness to EEG noise and individual variability.

3   Methodology

In this study, we develop a PINN based on the FitzHugh-Nagumo (FHN) model to analyze and interpret EEG data for BCI applications.

3.1   FitzHugh-Nagumo Model

The FHN model, introduced by FitzHugh and Nagumo in the 1960s, offers a two-variable reduction of the Hodgkin–Huxley equations that preserves core action-potential features. By abstracting detailed ionic currents into excitation and recovery variables, it overcomes the high dimensionality and computational burden of the full Hodgkin–Huxley framework, enabling tractable, large-scale simulations and theoretical analyses of neuronal dynamics. Mathematically, the FHN model is articulated through a pair of coupled nonlinear ordinary differential equations that describe the temporal evolution of two critical variables: the activation variable u and the recovery variable v. These equations are given by:

$$\begin{cases} \frac{du}{dt} = u - \frac{u^3}{3} - v + I \\ \frac{dv}{dt} = \epsilon(u + a - bv) \end{cases} \tag{1}$$

where u typically represents the membrane potential or the excitatory state of the neuron, while v corresponds to the recovery processes, such as the activation of potassium ion channels that restore the neuron to its resting state after an action potential. I denotes the external stimulus current applied to the neuron, driving its activity. The parameter ϵ is a small positive constant that signifies the separation of timescales between the fast dynamics of the activation variable and the slower dynamics of the recovery variable. The constants a and b are system parameters that govern the behavior and dynamical characteristics of the model. t represents the time variable, describing the evolution of the system over time, and these differential equations reflect the dynamics of the membrane potential u and the recovery variable v as functions of t.

3.2 Design and Implementation of a Physics-Informed Neural Network Based on the FitzHugh-Nagumo Model

We propose a three-module PINN architecture—Data Pre-processing, Feature Extraction, and PINN Model—grounded in the FitzHugh–Nagumo equations for EEG analysis.

Data Preprocessing Module. The Data Preprocessing Module serves as the initial stage of the PINN architecture, responsible for converting raw EEG time-series data into a structured format that is conducive to subsequent processing and analysis. Specifically, the input EEG data is reorganized into a four-dimensional tensor with the shape $W \times F \times C \times \omega$, where W represents the number of window slices, F denotes the number of filter banks, C corresponds to the number of EEG channels, and ω signifies the window length.

This transformation is critical for effectively capturing the spatiotemporal dynamics inherent in EEG signals. The process involves three key steps: temporal segmentation and frequency decomposition.

Temporal segmentation involves dividing the continuous EEG signal into smaller, manageable segments known as window slices. Given an input EEG tensor $\mathbf{X} \in \mathbb{R}^{B \times C \times T}$, where B denotes the batch size, C is the number of EEG channels, and T represents the total number of time points, the temporal segmentation can be expressed as:

$$\mathbf{X}' = \text{Temporal\_Segmentation}(\mathbf{X}) \in \mathbb{R}^{B \times W \times C \times \omega}, \tag{2}$$

where W represents the number of window slices, C corresponds to the number of EEG channels, and ω signifies the window length. The segmentation aims to divide EEG signals into small segments on the time domain, either with overlapping or without overlapping.

Frequency decomposition is achieved through the application of filter banks, which decompose each windowed EEG signal into multiple frequency passbands. Mathematically, this can be represented as:

$$\mathbf{X}'' = \text{FilterBank\_Decomposition}(\mathbf{X}') \in \mathbb{R}^{B \times W \times F \times C \times \omega}, \tag{3}$$

where F denotes the number of filter banks applied. Utilizing causal Chebyshev Type II filters, the raw oscillatory EEG signals are decomposed into distinct frequency bands.

3.3 Physics-Informed Neural Network Module

The PINN module is the central component of the proposed architecture, responsible for integrating the FHN model's biophysical constraints with the structured EEG data.

Input Processing and Convolutional Layers. The model accepts input $\mathbf{X}''$ in the shape $B \times W \times F \times C \times \omega$. To align the input data with the convolutional layers, the tensor is reshaped into $(B \times W) \times F \times C \times \omega$, effectively merging the batch and window dimensions. This reshaped tensor is passed through two 2D convolutional layers, each followed by batch normalization and ReLU activation to stabilize the learning process and introduce non-linearity. The operations are mathematically represented as:

$$\mathbf{X}_1 = \text{ReLU}\left(\text{BatchNorm}\left(\text{Conv2D}(\mathbf{X}'', F_1, k_1, p_1)\right)\right) \tag{4}$$

$$\mathbf{X}_1^{\text{pool}} = \text{MaxPool2D}(\mathbf{X}_1, p, s) \tag{5}$$

$$\mathbf{X}_2 = \text{ReLU}\left(\text{BatchNorm}\left(\text{Conv2D}(\mathbf{X}_1^{\text{pool}}, F_2, k_2, p_2)\right)\right) \tag{6}$$

$$\mathbf{X}_2^{\text{pool}} = \text{MaxPool2D}(\mathbf{X}_2, p, s) \tag{7}$$

where F_1 and F_2 are the numbers of filters, k_1 and k_2 are kernel sizes, and p_1, p_2 are paddings. Pooling layers downsample the feature maps, reducing temporal and spatial dimensions while retaining the most salient features. Here, p represents the pooling kernel size, determining the dimensions of the pooling operation, and s represents the stride, which defines the step size for the pooling filter as it slides over the feature map.

Fully Connected Layer. The flattened output of the convolutional layers is processed by a fully connected layer, which projects the high-dimensional convolutional features into a lower-dimensional hidden representation. Dropout regularization is applied after the fully connected layer to prevent overfitting:

$$\mathbf{h}' = \text{Dropout}\left(\text{ReLU}\left(\mathbf{W}_{\text{fc}} \cdot \text{Flatten}(\mathbf{X}_2^{\text{pool}}) + \mathbf{b}_{\text{fc}}\right)\right), \tag{8}$$

where W_{fc} and b_{fc} represent the weights and biases of the fully connected layer, respectively.

Transformer Encoder for Temporal Dependencies. The hidden representation is passed through a Transformer encoder to model long-range dependencies in the temporal dimension. The Transformer encoder, consisting of multiple layers of multi-head self-attention and feed-forward networks, captures the temporal relationships necessary for decoding EEG signals:

$$\mathbf{H} = \text{TransformerEncoder}(\mathbf{h}') \tag{9}$$

The encoder processes the input sequence of embeddings and outputs contextualized representations:

$$\mathbf{H} \in \mathbb{R}^{(B \times W) \times \text{hidden\_dim}} \tag{10}$$

Output Layer and Reshaping. The Transformer output is fed into a linear layer that maps the hidden representation to the activation (v) and recovery (w) variables for each neuronal node across time. These variables are fundamental to the FHN model:

$$\mathbf{O} = \mathbf{W}_{\text{out}} \cdot \mathbf{H} + \mathbf{b}_{\text{out}} \tag{11}$$

The output tensor $\mathbf{O}$ is reshaped to [batch_size, sliding_windows, $2 \times$ num_nodes, data_points], where num_nodes represents the number of nodes in the graph, and data_points denotes the feature dimensions for each node. The first num_nodes along the node dimension is extracted as v, while the remaining num_nodes is extracted as w:

$$\mathbf{v} = \mathbf{O}[:,:,: \text{num\_nodes}, :] \quad \mathbf{w} = \mathbf{O}[:,:, \text{num\_nodes} :, :] \tag{12}$$

Design Rationale. Unlike traditional PINNs based on MLPs, our model employs CNNs and Transformer encoders to exploit EEG's spatiotemporal structure and accurately estimate the FitzHugh–Nagumo variables v and w. Temporal CNNs extract local patterns, stabilized by Batch Normalization and ReLU, while multi-head self-attention captures long-range dependencies to provide global context.

Physics-Based Loss Function. Given the predicted values v and w, the time derivatives are approximated using the finite difference method:

$$\frac{dv}{dt} \approx \frac{v(t + \Delta t) - v(t)}{\Delta t} \tag{13}$$

$$\frac{dw}{dt} \approx \frac{w(t + \Delta t) - w(t)}{\Delta t} \tag{14}$$

where Δt is the time step.

The residuals for v and w are computed as:

$$f_v = \frac{dv}{dt} - \left(v - \frac{v^3}{3} - w + I\right) \tag{15}$$

$$f_w = \frac{dw}{dt} - \epsilon(v + a - bw) \tag{16}$$

The physics-based loss is then defined as the mean squared error (MSE) of these residuals:

$$\mathcal{L}_{\text{physics}} = \frac{1}{N} \sum_{i=1}^{N} \left(f_{v,i}^2 + f_{w,i}^2\right) \tag{17}$$

where N is the total number of data points.

Coupling Term. To model the interactions among nodes, a coupling matrix K is introduced. The modified dynamics for v incorporating coupling are given by:

$$\frac{dv}{dt} = v - \frac{v^3}{3} - w + I + \sum_{j} K_{ij}(v_j - v_i), \tag{18}$$

where K_{ij} represents Coupling strength between node i and node j, v_j is Membrane potential of node j, and v_i is Membrane potential of node i.

The coupling matrix K is defined as:

$$K_{ij} = \begin{cases} \text{coupling\_strength}, & \text{if } i \neq j \\ 0, & \text{if } i = j \end{cases} \tag{19}$$

where coupling strength is set to 0.1, and the coupling matrix K, with a shape of $[\text{num\_nodes}, \text{num\_nodes}]$, is initialized as an all-ones matrix with its diagonal elements subtracted by 1 and then multiplied by the coupling strength.

The residual for v including coupling is given by:

$$f_v = \frac{dv}{dt} - \left(v - \frac{v^3}{3} - w + I + \sum_{j} K_{ij}(v_j - v_i)\right) \tag{20}$$

Final Physics Loss

$$
\begin{aligned}
\mathcal{L}_{\text{physics}} = \frac{1}{N} \sum_{i=1}^{N} \Bigg[& \left(\frac{d\hat{v}_i}{dt} - \hat{v}_i + \frac{\hat{v}_i^3}{3} + \hat{w}_i - I \right)^2 \\
& + \left(\frac{d\hat{w}_i}{dt} - \epsilon(\hat{v}_i + a - b\hat{w}_i) \right)^2 \Bigg],
\end{aligned}
\tag{21}
$$

where $\hat{v}_i$ and $\hat{w}_i$ are the predicted membrane potential and recovery variables, respectively. I denotes the external stimulus, and ϵ, a, and b are parameters of the FHN model. For this study, these parameters are set as follows: $\epsilon = 0.08$, $a = 0.7$, $b = 0.8$, and $I = 0.5$.

3.4 Feature Extraction Module

The Feature Extraction Module processes the outputs of the PINN model, specifically the FHN variables v (membrane potentials) and w (recovery variables). The inputs to the module are tensors $\mathbf{v} \in \mathbb{R}^{B \times N \times T}$ and $\mathbf{w} \in \mathbb{R}^{B \times N \times T}$, where B represents the batch size, N is the number of nodes (e.g., EEG channels), and T is the number of time points. To align the input tensors with the convolutional layers, the module reshapes and permutes the data into $\mathbf{v}_{\text{input}}, \mathbf{w}_{\text{input}} \in \mathbb{R}^{(B \cdot N) \times 1 \times T}$.

The module applies a sequence of one-dimensional convolutional layers to capture localized temporal dependencies. For the input $\mathbf{v}_{\text{input}}$, the operation of the first convolutional layer is given by:

$$
\mathbf{v}_1 = \text{ReLU}\left(\text{BatchNorm}\left(\text{Conv1D}(\mathbf{v}_{\text{input}}, F_1, k, p) \right) \right),
\tag{22}
$$

where F_1 represents the number of filters, k is the kernel size, and p is the padding. To further reduce the temporal resolution, a max-pooling operation is performed:

$$
\mathbf{v}_1^{\text{pool}} = \text{MaxPool1D}(\mathbf{v}_1, k_p, s_p),
\tag{23}
$$

where k_p and s_p denote the pooling kernel size and stride.

The pooled features are passed through a second convolutional layer with an increased number of filters, refining the temporal features:

$$
\mathbf{v}_2 = \text{ReLU}\left(\text{BatchNorm}\left(\text{Conv1D}(\mathbf{v}_1^{\text{pool}}, F_2, k, p) \right) \right),
\tag{24}
$$

followed by another max-pooling operation and the feature maps are flattened into one-dimensional vectors:

$$
\mathbf{v}_{\text{flat}} = \text{Flatten}(\text{MaxPool1D}(\mathbf{v}_2, k_p, s_p)),
\tag{25}
$$

and then processed through a fully connected layer to map the extracted temporal features into a latent space:

$$
\mathbf{v}_{\text{fc}} = \text{ReLU}\left(\mathbf{W}_{\text{fc}} \cdot \mathbf{v}_{\text{flat}} + \mathbf{b}_{\text{fc}} \right).
\tag{26}
$$

The same operations are applied independently to $\mathbf{w}$, producing:

$$\mathbf{w}_{\text{fc}} = \text{ReLU}\left(\mathbf{W}_{\text{fc}} \cdot \text{Flatten}(\mathbf{w}_2^{\text{pool}}) + \mathbf{b}_{\text{fc}}\right). \tag{27}$$

Finally, the outputs of v and w are fused through element-wise addition:

$$\mathbf{f} = \text{LayerNorm}(\mathbf{v}_{\text{fc}} + \mathbf{w}_{\text{fc}}), \tag{28}$$

where $\mathbf{f} \in \mathbb{R}^{B \times N}$ represents the fused feature matrix, which is passed to the next stage of the pipeline.

4   Experimental Design

4.1   Loss Function and Training

The training process was conducted over 100 epochs on a computer equipped with an NVIDIA GeForce RTX 4090 GPU. The total loss is defined as:

$$\mathcal{L} = \mathcal{L}_{\text{classification}} + \lambda \mathcal{L}_{\text{physics}}, \tag{29}$$

where $\mathcal{L}_{\text{classification}}$ is the cross-entropy loss, which ensures accurate classification of MI tasks. Meanwhile, $\mathcal{L}_{\text{physics}}$ enforces the biophysical consistency of the predictions by adhering to the FHN model (see Eq. 21).

During training, a batch size of 64 and an initial learning rate of $1e^{-3}$ were used to ensure stable convergence. The weight parameter λ controls the trade-off between the classification and physics-based loss components, allowing the model to balance predictive accuracy with adherence to biophysical principles.

4.2   Dataset and Baseline Models

Dataset. The BCIC-IV-2a dataset [24] is a widely used benchmark in EEG-based MI classification research, particularly in the context of BCI development. This dataset contains EEG recordings from nine subjects performing a four-class motor imagery task. The four classes include imagined movements of the left hand, right hand, both feet, and tongue.

Preprocessing. For the experiments in this study, the EEG signals were preprocessed to enhance the signal-to-noise ratio and extract relevant features. The preprocessing pipeline included band-pass filtering to isolate task-relevant frequency bands (e.g., mu and beta rhythms), artifact removal using EOG channels, and segmentation into fixed-length temporal windows aligned with the onset of motor imagery tasks.

Baseline Models. To evaluate the performance and generalization capability of NEUROPHYSNET, we compared it against a comprehensive set of baseline methods, categorized as follows:

- **CSP-Based Methods:**
  - Filter Bank Common Spatial Pattern (FBCSP) [2]: A classical approach that applies spatial filters across multiple frequency bands to extract task-relevant features from EEG signals.
- **Riemannian Geometry-Based Methods:**
  - Minimum Distance to Mean (MDM) [26]: Classifies EEG signals by minimizing geodesic distances between covariance matrices.
  - Temporal Spectral Mapping (TSM) [18]: Combines temporal and spectral features to improve classification performance.
  - SPDNet [9]: A deep learning model specifically designed to process symmetric positive definite (SPD) matrices.
  - Tensor-CSPNet [14]: Extends CSP by incorporating tensor-based feature representations.
- **Deep Learning Architectures:**
  - ConvNet [19]: A simple convolutional neural network optimized for EEG feature extraction.
  - EEGNet [17]: A compact and efficient architecture designed specifically for brain-computer interface (BCI) applications.
  - Filter Bank Convolutional Network (FBCNet) [22]: Integrates filter banks with convolutional networks to extract multi-band EEG features.

These baselines represent a diverse array of strategies, allowing us to evaluate NEUROPHYSNET's performance in comparison to traditional CSP methods, advanced Riemannian geometry-based approaches, and cutting-edge deep learning models.

Table 1. Comparative Analysis of Subject-Specific Accuracies and Standard Deviations in BCIC-IV-2a Dataset.

	CV (T) Acc %	CV (E) Acc %	Holdout (T → E) Acc %
FBCSP [2]	71.29	73.39	66.13
EEGNet [17]	69.26	66.93	60.31
ConvNet [19]	70.42	65.89	57.61
FBCNet [22]	75.48	77.16	71.53
MDM [26]	62.96	59.49	50.74
TSM [18]	68.71	63.32	49.72
SPDNet [9]	65.91	61.16	55.67
Tensor-CSPNet [14]	75.11	77.36	73.61
NEUROPHYSNET[20,4]	**76.23**	**78.03**	**74.20**

5   Experimental Results

5.1   NEUROPHYSNET Generalization Performance vs. Leading-Edge Methodologies

To evaluate the performance and generalization capability of NeuroPhysNet, we compared it against a range of baseline models across cross-validation (CV) and Holdout scenarios on the BCIC-IV-2a dataset. Table 1 summarizes the results, which include accuracy metrics from the Training Session (T) and Evaluation Session (E).

Filter Bank Common Spatial Pattern (FBCSP) [2], a classical CSP-based method, achieves CV accuracies of 71.29% (T) and 73.39% (E), and a Holdout accuracy of 66.13%. While its ability to extract task-relevant spatial and frequency features is notable, its performance is surpassed by most modern methods due to its inability to model complex temporal and physiological dynamics.

For deep learning-based approaches, ConvNet [19] achieves CV accuracies of 70.42% (T) and 65.89% (E), and a Holdout accuracy of 57.61%. EEGNet [17] performs slightly worse, with CV accuracies of 69.26% (T) and 66.93% (E), and a Holdout accuracy of 60.31%. FBCNet [22], leveraging filter banks to extract multi-band features, achieves CV accuracies of 75.48% (T) and 77.16% (E), and a Holdout accuracy of 71.53%. Despite its superior frequency representation, it does not incorporate physiological dynamics, which constrains its generalization capabilities.

Among Riemannian geometry-based methods, Minimum Distance to Mean (MDM) [26] achieves CV accuracies of 62.96% (T) and 59.49% (E), with a Holdout accuracy of 50.74%. Similarly, Temporal Spectral Mapping (TSM) [20] records CV accuracies of 68.71% (T) and 63.32% (E), and a Holdout accuracy of 49.72%. SPDNet [9] shows slightly better performance than MDM, with CV accuracies of 65.91% (T) and 61.16% (E), and a Holdout accuracy of 55.67%. However, these methods struggle to handle the high-dimensional temporal dynamics inherent in EEG signals, resulting in comparatively low accuracy across all scenarios.

Tensor-CSPNet [14], which extends CSP with tensor-based feature representations, achieves CV accuracies of 75.11% (T) and 77.36% (E), and a Holdout accuracy of 73.61%.

NeuroPhysNet, on the other hand, achieves the highest performance among all methods in multiple scenarios. It records CV accuracies of 76.23% (T) and 78.03% (E), and a Holdout accuracy of 74.20%.

6   Evaluating the Robustness of NeuroPhysNet with Full and Limited Training Data

To evaluate the robustness and performance of NeuroPhysNet under varying training data proportions, we conducted experiments on the BCIC-IV-2a dataset. These experiments compared NeuroPhysNet and Tensor-CSPNet using the full training dataset (100%) and subsets comprising 80%, 50%, and 30% of the training data.

Table 2. Performance Comparison of NeuroPhysNet and Tensor-CSPNet on BCIC-IV-2a Dataset with Varying Training Data Proportions

Data Proportion	Model	CV Accuracy (T)	CV Accuracy (E)	Holdout Accuracy
100%	NEUROPHYSNET	**75.85%**	**76.23%**	**74.20%**
	Tensor-CSPNet	75.11%	77.36%	73.61%
80%	NEUROPHYSNET	**74.12%**	**74.89%**	**72.15%**
	Tensor-CSPNet	73.45%	75.64%	71.32%
50%	NEUROPHYSNET	**65.23%**	**66.72%**	**64.81%**
	Tensor-CSPNet	62.45%	64.31%	59.72%
30%	NEUROPHYSNET	**59.41%**	**60.89%**	**58.45%**
	Tensor-CSPNet	56.12%	58.43%	53.01%

The results in Table 2 reveal a significant performance advantage for Neuro-PhysNet over Tensor-CSPNet across all training data proportions, particularly when the data is limited. When using the full training dataset, NeuroPhysNet achieves a CV accuracy of 76.23% (E) and 75.85% (T) and a Holdout accuracy of 74.20%, slightly outperforming Tensor-CSPNet's 77.36% (E), 75.11% (T), and 73.61% (Holdout). As the training data proportion decreases, the gap between the two models widens considerably.

At 50% training data, NeuroPhysNet achieves a CV accuracy of 66.72% (E) and 65.23% (T) and a Holdout accuracy of 64.81%, while Tensor-CSPNet falls to 64.31% (E), 62.45% (T), and 59.72% (Holdout). This substantial drop in Tensor-CSPNet's performance demonstrates its higher sensitivity to data scarcity. In contrast, NeuroPhysNet's integration of FHN equations helps it maintain better generalization by leveraging the physical constraints encoded within its architecture.

At 30% training data, the differences become even more pronounced. Neuro-PhysNet achieves a CV accuracy of 60.89% (E), 59.41% (T), and a Holdout accuracy of 58.45%, outperforming Tensor-CSPNet's 58.43% (E), 56.12% (T), and 53.01% (Holdout) by significant margins. Furthermore, Tensor-CSPNet exhibits less stable predictions across different subjects, which reflects its higher variability compared to NeuroPhysNet.

6.1   Evaluating the Impact of v and w Features in NeuroPhysNet

To assess the individual contribution of the physiological features computed by the FHN equations, an experiment was conducted by training and testing NeuroPhysNet using only the membrane potential (v) and recovery variable (w) as input features, as shown in Fig. 1.

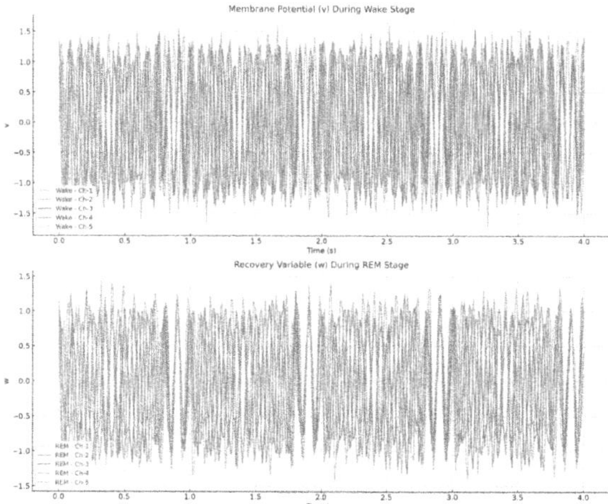

Fig. 1. Visualization of v (membrane potential) during the Wake stage and w (recovery variable) during the REM stage for selected EEG channels.

Table 3. Performance of NeuroPhysNet Using Only v and w Features on BCIC-IV-2a Dataset

Data Proportion	CV Accuracy (T)	CV Accuracy (E)	Holdout Accuracy
100%	60.34%	61.23%	59.87%
80%	58.12%	58.94%	57.43%
50%	54.98%	54.32%	52.89%
30%	50.11%	50.89%	49.12%

The results presented in Table 3 show that using only v and w features, NeuroPhysNet achieves reasonable classification performance, though it is significantly lower than the full-feature model. With the full training dataset, the model achieves a CV accuracy of 61.23% (E) and a Holdout accuracy of 59.87%. As the training data proportion decreases, performance declines but remains relatively stable. For example, at 30% training data, the CV accuracy is 50.89% (E) and 50.11% (T), and the Holdout accuracy is 49.12%. The results indicate that v and w features, derived from the FitzHugh-Nagumo equations, provide valuable discriminative information for EEG classification. Despite the reduced accuracy compared to the full-feature model, the performance remains relatively consistent across different data proportions.

7 Conclusion

In this study, we evaluated NeuroPhysNet—a PINN-based EEG classifier that integrates FHN dynamics—by examining the standalone and combined contributions of the membrane potential v and recovery variable w. These physiological features proved robust in data-limited scenarios and, when fused with conventional EEG features, yielded significant accuracy gains. Compared to previous networks, NeuroPhysNet demonstrated superior accuracy and generalization. Visualization of v and w across sleep stages further confirmed their discriminative power. Our results highlight the promise of PINNs for EEG classification under constrained data conditions and point toward future work on refined fusion strategies, alternative physiological models, and broader EEG applications.

References

1. A, M.R., B, P.P., A, G.E.K.: Physics-informed neural networks: a deep learning framework for solving forward and inverse problems involving nonlinear partial differential equations. J. Comput. Phys. **378**, 686–707 (2019)
2. Ang, K.K., Chin, Z.Y., Zhang, H., Guan, C.: Filter bank common spatial pattern (FBCSP) in brain-computer interface. In: 2008 IEEE International Joint Conference on Neural Networks (IEEE World Congress on Computational Intelligence), pp. 2390–2397. IEEE (2008)
3. Bijsterbosch, J.: Challenges and future directions for representations of functional brain organization. Nat. Neurosci. **23**(12), 1484–1495 (2020)
4. Dickstein, R., Deutsch, J.E.: Motor imagery in physical therapist practice. Phys. Ther. **87**(7), 942–953 (2007)
5. Dou, B., et al.: Machine learning methods for small data challenges in molecular science. Chem. Rev. **123**(13), 8736–8780 (2023)
6. Cruz, B.G.S., Husch, A., Hertel, F.: Machine learning models for diagnosis and prognosis of Parkinson's disease using brain imaging: general overview, main challenges, and future directions. Front. Aging Neurosci. **15**, 1216163 (2023)
7. Hao, Z., et al.: Physics-informed machine learning: a survey on problems, methods and applications. arXiv preprint arXiv:2211.08064 (2022)
8. Huang, W., et al.: EEG-based motor imagery classification using convolutional neural networks with local reparameterization trick. Expert Syst. Appl. **187** (2022)
9. Huang, Z., Van Gool, L.: A Riemannian network for SPD matrix learning. In: Proceedings of the AAAI Conference on Artificial Intelligence, vol. 31 (2017)
10. Izhikevich, E.M., FitzHugh, R.: Fitzhugh-nagumo model. Scholarpedia **1**(9), 1349 (2006)
11. Jagtap, A.D., Kawaguchi, K., Karniadakis, G.E.: Adaptive activation functions accelerate convergence in deep and physics-informed neural networks. J. Comput. Phys. **404**, 109136 (2020)
12. Jia, H., et al.: A model combining multi branch spectral-temporal CNN, efficient channel attention, and lightGBM for MI-BCI classification. IEEE Trans. Neural Syst. Rehabil. Eng. **31**, 1311–1320 (2023)
13. Ju, C., Guan, C.: Graph neural networks on SPD manifolds for motor imagery classification: a perspective from the time-frequency analysis. arXiv preprint arXiv:2211.02641 (2022)
14. Ju, C., Guan, C.: Tensor-CSPNet: a novel geometric deep learning framework for motor imagery classification. IEEE Trans. Neural Netw. Learn. Syst. (2022)
15. Ju, C., Guan, C.: Tensor-CSPNet: a novel geometric deep learning framework for motor imagery classification. IEEE Trans. Neural Netw. Learn. Syst. **34**(12) (2023)
16. Lawhern, V.J., et al.: EEGNet: A compact convolutional network for EEG-based brain-computer interfaces. J. Neural Eng. **15**(5), 056013.1–056013.17 (2018)

17. Lawhern, V.J., et al.: EEGNet: a compact convolutional neural network for EEG-based brain-computer interfaces. J. Neural Eng. **15**(5), 056013 (2018)
18. Lin, J., Gan, C., Han, S.: Temporal shift module for efficient video understanding. corr abs/1811.08383 (2018) (1811)
19. Liu, Y., Shao, H., Bai, B.: A novel convolutional neural network architecture with a continuous symmetry. arXiv preprint arXiv:2308.01621 (2023)
20. Lotte, F., et al.: A review of classification algorithms for EEG-based brain-computer interfaces: a 10 year update. J. Neural Eng. **15**(3), 031005 (2018)
21. Lu, J., et al.: LGL-BCI: A lightweight geometric learning framework for motor imagery-based brain-computer interfaces. arXiv preprint arXiv:2310.08051 (2023)
22. Mane, R., et al.: FBCNet: a multi-view convolutional neural network for brain-computer interface. arXiv preprint arXiv:2104.01233 (2021)
23. Schirrmeister, R.T., Gemein, L., Eggensperger, K., Hutter, F., Ball, T.: Deep learning with convolutional neural networks for decoding and visualization of EEG pathology. Hum. Brain Mapp. **38**(11), 5391–5420 (2017)
24. Tangermann, M., et al.: Review of the BCI competition IV. Front. Neurosci., 55 (2012)
25. Tatum, W.O., et al.: Clinical utility of EEG in diagnosing and monitoring epilepsy in adults. Clin. Neurophysiol. **129**(5), 1056–1082 (2018)
26. Tevet, G., et al.: Human motion diffusion model. arXiv preprint arXiv:2209.14916 (2022)
27. Wang, J., Wang, M.: Review of the emotional feature extraction and classification using EEG signals. Cogn. Robot. **1**, 29–40 (2021)
28. Wang, S., Wang, H., Perdikaris, P.: Learning the solution operator of parametric partial differential equations with physics-informed deeponets. Sci. Adv. **7**(40), eabi8605 (2021)
29. Yang, J.Q.: Neuromorphic engineering: from biological to spike-based hardware nervous systems. Adv. Mater. **32**(52), 2003610 (2020)
30. Yu, R., Wang, R.: Learning dynamical systems from data: an introduction to physics-guided deep learning. Proc. Natl. Acad. Sci. **121**(27), e2311808121 (2024)
31. Zhang, D., Lu, L., Guo, L., Karniadakis, G.E.: Quantifying total uncertainty in physics-informed neural networks for solving forward and inverse stochastic problems. J. Comput. Phys. **397**, 108850 (2019)

GPSR Compliance
The European Union's (EU) General Product Safety Regulation (GPSR) is a set
of rules that requires consumer products to be safe and our obligations to
ensure this.

If you have any concerns about our products, you can contact us on

ProductSafety@springernature.com

In case Publisher is established outside the EU, the EU authorized
representative is:

Springer Nature Customer Service Center GmbH
Europaplatz 3
69115 Heidelberg, Germany

www.ingramcontent.com/pod-product-compliance
Ingram Content Group UK Ltd.
Pitfield, Milton Keynes, MK11 3LW, UK
UKHW020814080726
473059UK00007B/2241